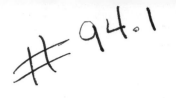

A FIRST COURSE IN ABSTRACT ALGEBRA

A FIRST COURSE IN ABSTRACT ALGEBRA

FIFTH EDITION

John B. Fraleigh
University of Rhode Island

Historical Notes by Victor Katz
University of District of Columbia

ADDISON-WESLEY PUBLISHING COMPANY
Reading, Massachusetts • Menlo Park, California • New York • Don Mills,
Ontario • Wokingham, England • Amsterdam • Bonn • Sydney • Singapore
Tokyo • Madrid • San Juan • Milan • Paris

Sponsoring Editor: Laurie Rosatone
Production Supervisor: Mona Zeftel
Editorial/Production Services: Barbara Pendergast
Text Design: Mark Ong
Art Coordinator: Bob Forget
Cover Design: Marshall Henrichs
Marketing Manager: Andrew Fisher
Manufacturing Supervisor: Roy Logan

Library of Congress Cataloging-in-Publication Data

Fraleigh. John B.
    A first course in abstract algebra / John B. Fraleigh ; with
historical notes by Victor Katz. — 5th ed.
        p.   cm.
    Includes bibliographical references and index.
    ISBN 0-201-53467-3
    1. Algebra, Abstract.   I. Title.
QA162.F7   1993
512'.02—dc20   93-1997
                                                              CIP

1 2 3 4 5 6 7 8 9 10-MA-9796959493

PREFACE

This is an introduction to abstract algebra. It is anticipated that the students have studied calculus and probably linear algebra. However, these are primarily *mathematical maturity* prerequisites since subject matter from calculus and linear algebra appears only in certain illustrative examples and exercises, not in the definitions and theorems.

As in previous editions of the text, my aim remains to teach students as much about groups, rings, and fields as I can in a first course. For many students, abstract algebra is their first extended exposure to an axiomatic treatment of mathematics. Recognizing this, I have included quite a lot of explanation concerning what we are trying to accomplish, how we are trying to do it, and why we choose these methods. Mastery of this text constitutes a firm foundation for more specialized work in algebra and also provides valuable experience for any further axiomatic study of mathematics.

CHANGES FROM THE FOURTH EDITION
More Applications

Probably influenced by the fact that we devote only one semester to undergraduate abstract algebra at my school, previous editions of the text have not included applications to coding, finite-state machines (automata), graph theory, and isometry groups of the plane with the attendant Escher art works. In a single semester, I barely have time to cover the algebra that I feel is essential. In response to several requests, I have included these applications in this new edition.

I dislike seeing such applications relegated to a final chapter or appendix of a text. If there is time to do one of them, it is surely best to do it as soon as the algebra on which it is based has been presented, rather than delay until the final few weeks of the course. Accordingly, I inserted the applications as soon as feasible. On the other hand, I did not want them to interrupt the flow of the text for those unfortunate people like me who can't find time to do all of them. As a compromise, I placed the new applications after the algebra exercises in the sections where the algebra for the applications was developed.

The table of contents shows where presentation of each application occurs. For example, after a discussion of binary operations and exercises on binary operations in Section 1.1, there is a description of finite-state machines, and the exercise set for that section then continues with exercises on such automata. Further discussion of automata appears at the end of the introduction of permutation groups in Section 2.1. Coding is the only application to which an entire section (Section 2.5) is devoted. The fairly long presentation does not seem to split into parts that could appear naturally at the end of different sections. An occasional exercise involving one of the applications appears in a later section; each such occurrence is labeled.

Examples from Linear Algebra

A number of new examples and exercises are drawn from linear algebra, which many students have studied. Section 0.4 reviews matrix algebra, and some exercises build on this review with a bit of additional explanation. Others that depend more heavily on a student's knowledge of the subject are labeled "Linear Algebra."

Reorganization

The discussion of cyclic groups now precedes groups of permutations. The prompt presentation of generators of a group permits the early presentation of the Cayley digraph application.

Chapter 1 of the previous edition has been split to become Chapters 1 and 2, making the start of the course psychologically less forbidding for both the students and the instructor.

The sections on group action on a set and applications of G-sets to counting have been moved forward from the chapter on advanced group theory to conclude Chapter 3. This is an application that my students have enjoyed, and it is now placed at the end of a chapter that is covered by most instructors.

The theorem that the nonzero elements of a finite field form a cyclic multiplicative group has been moved forward from Section 8.5 to Section 4.6, where it appears as a corollary of Theorem 4.18. Thus instructors who don't have time to cover Chapter 8 can still present this nice result.

The overview section on additional algebraic structures (modules, group rings, and algebras) has been moved from the middle to the end of Chapter 8 so that the natural flow of the chapter is not interrupted.

IMPORTANT FEATURES RETAINED

The style is informal, but explanations and proofs are given in great detail.

I have always felt that extended time spent on preliminaries, especially those of a set-theoretic nature, tends to stifle interest in a course. Accordingly, preliminary material is kept to a minimum, and set-theoretic definitions are made when the need for them arises. This permits a prompt introduction of group theory.

Exercise sets start with computations where feasible, to ease the student's transition from the usual calculus course. A ten-part, true-false exercise appears in most problem sets. These exercises are designed to emphasize definitions, logic, and concepts. They also provide practice in deciding whether a mathematical assertion is true, as opposed to an exercise that just says, "Prove. . . ."

Answers to odd-numbered exercises are given at the end of the text *except for those requesting a proof.* I feel that having a proof readily available renders such an exercise almost worthless.

The historical notes by Victor Katz are not just facts and anecdotes about the contributors to the subject; he actually explains the motivation for its development. This is a difficult task, since the order of topics in the text is roughly the reverse of the order of their development. For example, the motivation for group theory, which is studied first in the text, was the study of zeros of polynomials, which is studied last.

INSTRUCTOR'S SOLUTIONS MANUAL

A manual containing complete solutions, including proofs, for all the exercises is available for the instructor from the publisher.

ACKNOWLEDGMENTS

I wish to express my appreciation to all the reviewers of the manuscript for this edition, including

D. J. Hartfiel, *Texas A & M University*
Johnny A. Johnson, *University of Houston*
Michael M. Kostreva, *Clemson University*
Ronald Solomon, *Ohio State University*

I am very grateful to Victor Katz for providing the excellent historical notes. I also wish to thank Laurie Rosatone and the staff at Addison-Wesley for their help in the preparation of this edition.

Kingston, RI   J.B.F.

Dependence Chart

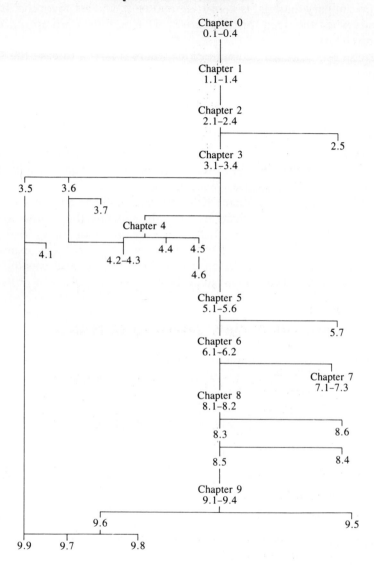

CONTENTS

* Optional applications are shown in italics.
† Not required for the remainder of the text.

‡ This section is required for Sections 3.7 and 4.2 only.

♦

A FEW PRELIMINARIES

♦

It is our experience that spending two or three weeks on background material at the start of an abstract algebra course may destroy interest in the subject. Accordingly, we introduce mathematical tools as they are needed, provided that their presentation can be kept short so that the flow of the text is not interrupted. Since they need longer discussion, we present equivalence relations and partitions of sets in Section 0.2 and proof by mathematical induction in Section 0.3. Section 0.1, which well might be left for students to read on their own, attempts to prepare students for this axiomatic, definition-theorem-proof treatment of algebra. Section 0.4 summarizes some of the algebra of complex numbers and matrices for students who may not be familiar with it.

0.1

Mathematics and Proofs

You have probably never had a laboratory course in mathematics. Mathematics is not considered to be an experimental science, whereas physics, chemistry, and biology are. Research for a chemist can consist of a laboratory experiment designed to validate a conjecture or simply to see what happens. There is little comparable activity in mathematics.

The main business of mathematics is proving theorems.

Just examine any research journal. Few meaningful theorems can be proved by experimentation. In mathematics, experimentation might lead to a *conjecture* which may or may not be correct. If the conjecture is later proved, then it is elevated to the status of a *theorem*. Exercise 10 illustrates experimentation leading to conjecture.

In theory, all of mathematics is an *axiomatic* study, consisting of chains of valid conclusions (theorems) deduced by valid reasoning (proofs) from the axioms of set theory. This is the current view of mathematical logicians who wrestle valiantly with the foundations of mathematics. Probably most research mathematicians would be unable to write down the axioms of set theory or describe all the rules of valid reasoning in a way that would satisfy logicians. In spite of this, logicians would agree that the results in the great majority of mathematical research papers are valid. In case an assertion is not valid, the author of the paper would probably admit to a mistake when the difficulty is pointed out. With perhaps no formal training in mathematical logic, the research mathematician learned as a student the rules of the game and can contribute successfully to the subject.

Keeping in mind the preceding paragraphs, we can try to put one feature of abstract algebra in perspective. Abstract algebra is the most axiomatic study undertaken by the typical mathematics major. It gives a lot of exposure to the rules of the game, mathematics. However, it would be absurd to pretend that this text is a totally axiomatic study. For example, we shall feel free to use familiar properties of the real numbers without any axiomatic verification. An abstract algebra course does represent a big step from the typical freshman-sophomore calculus course toward the modern mathematical method.

Courses in linear algebra vary widely in axiomatic approach. If you used a text that gave axioms for a vector space in the first chapter and then developed the subject from them, your study of linear algebra was similar to the study of abstract algebra in this text. On the other hand, if vector-space axioms did not appear at all or were relegated to an appendix of your linear algebra text, the orientation of the course was probably close to that of the typical first course in calculus.

An axiomatic approach is not used merely to expose students to proofs, although it does serve that function quite well. It is the most efficient way we have found to present algebra. Once a body of theorems has been deduced from axioms, we know that the theorems hold for *every* structure that satisfies the axioms. For example, we will start our study by examining structures called groups, which satisfy three axioms. If we were to prove a theorem in terms of one particular group, perhaps involving addition of real numbers, it might not be clear whether the theorem holds for all groups. We would have to reexamine the proof, doing the same work all over again if we change the group. But if we prove a theorem just in terms of the axioms of a group, without using any other properties, then this single proof allows us to use the theorem freely for any group. This is a virtue of the axiomatic approach, and our study of abstract algebra will illustrate this technique. The adjective *abstract* indicates that algebra is being studied by properties that have been *abstracted* from the subject.

Abstract algebra is often considered an ideal subject for drill in proofs since the lists of axioms used are quite short. However, devising a proof in algebra often amounts to finding just the right method of attack, perhaps considering just the right algebraic expression. If the right method is found, a proof may fall out easily. Otherwise we may struggle a long time without finding a proof. Geometric pictures are usually no help in finding a proof in algebra. For this reason point-set topology might be a better course for training students in proofs, for pictures can often be used in topology as an aid in understanding why a theorem must be true. Emphasis can then be placed on writing a correct proof.

It is not possible for us to give any meaningful outline on how to prove theorems; experience is the best guide. For the remainder of this section we make a few general observations. We will start by pointing out that it is essential to know what we are talking about, that is, to understand *definitions* of the terms we are using.

Definitions

Many students do not realize the great importance of definitions to mathematics. This importance stems partly from the need for mathematicians to communicate with each other about their work. If two people are trying to communicate about some subject, they must have the same understanding of its technical terms.

A very important ingredient of mathematical creativity is the ability to formulate useful definitions, ones that will lead to interesting results. A mathematics student commencing graduate study may find that he or she spends a great deal of time discussing definitions with other graduate students. When I was in graduate school, a physics graduate student once complained to me that at the evening meal the mathematics students always sat together and argued, and that the subject of their argument was always a definition. Graduate students are usually asked to give several definitions on oral examinations. If they cannot explain the meaning of a term, they probably cannot give sensible answers to questions involving that concept.

Every definition is understood to be an *if and only if* type of statement, even though it is customary to suppress the *only if*. Thus we may define an isosceles triangle as follows: "A triangle is **isosceles** if it has two sides of equal length," when we really mean that a triangle is isosceles if and only if it has two sides of equal length.

Do not feel that you have to memorize a definition word for word. The important thing is to *understand* the concept, so that you can define precisely the same concept in your own words. Thus the definition "An **isosceles** triangle is one having two equal sides" is perfectly correct.

Throughout the text, a term that appears in boldface type is being *defined* at that point. Specifically labeled definitions are used for the main algebraic concepts with which we are concerned. Many other terms are defined, using the boldface convention, outside a labeled definition. You will find ideas defined in this fashion in text paragraphs, theorems, and exercises.

Observations on Proofs

Observation 1   If some concept has just been defined and we are asked to prove something concerning the concept, we *must* use the definition as an integral part of the proof.

Immediately after a concept is defined, the definition is the only information one has available regarding the concept.

EXAMPLE 1   An integer n is defined to be **even** if $n = 2m$ for some integer m. It is a theorem that the sum of two even integers is even. The definition of an even integer must be used to prove this theorem. We leave the proof to Exercise 5. ▲

Observation 2   The statement of a theorem consists of two parts: the *hypotheses* and the *conclusion*. If all the hypotheses are needed to prove the theorem, that is, if no hypothesis is redundant, then each hypothesis must be cited somewhere in the proof.

EXAMPLE 2   It is a theorem that the sum of an even integer r and an odd integer s is an odd integer. In proving this theorem, which we leave to Exercise 7, it is essential to use both hypotheses, namely, that r is even and that s is odd. ▲

Observation 3   If even one example can be found for which a statement is not true, then the statement is not a theorem. In fact, the standard way to show that a statement is not a theorem is to provide such a *counterexample*.

EXAMPLE 3   Is the statement "The square of every real number is positive" a theorem? The answer is no, since $0^2 = 0$ and 0 is a real number but is not positive. This is the only counterexample that can be given, but one such example is all that is needed to show that a statement is not a theorem. ▲

Observation 4   Never tacitly assume any hypothesis that is not explicitly stated. Never take for granted any quantifying words or phrases such as *only, for all, for every*, or *for some* that do not actually appear.

EXAMPLE 4   The statement "There are four real numbers whose squares are less than 2" is true (a theorem). As a proof, we need only observe that $(-1)^2$, 0^2, 1^2, and $(\frac{1}{2})^2$ are all less than 2. The statement "There are only four real numbers whose squares are less than 2" is false (not a theorem). We need only observe that $(-1/2)^2 < 2$ also. The word *only* makes all the difference. ▲

Observation 5   A theorem of the form

If hypotheses then conclusion

cannot be proved by giving a specific example where the hypotheses and conclusion all are true. We must show that *for all* examples where the hypotheses are true, the conclusion is true also.

EXAMPLE 5   Consider the statement "If $f(x)$ is continuous, then $f(x)$ is differentiable." Now $f(x) = x^2$ is continuous and is also differentiable, for $f'(x) = 2x$ at every point in the domain of $f(x)$. However, the statement is not a theorem. A counterexample is given by $f(x) = |x|$, which is continuous but not differentiable since $f'(0)$ does not exist. Of course, in classifying this statement as a theorem or not a theorem, we had to know the *definitions* of a continuous function and of a differentiable function. ▲

There are a few types of theorems for which the method of attack for a proof is fairly standard. As our final observation, we mention one type that appears a few times in the text.

Observation 6   Suppose we wish to show that an element having some property exists and is *unique,* that is, that there is one and only one such element. First, show that there is such an element. To show uniqueness, assume that there are two such elements, say r and s, and try to show that r and s must be equal (the same).

EXAMPLE 6   Show that there is a unique real number r such that $rx = r$ for all real numbers x.

Solution   We know that $0x = 0$ for all real numbers x, so that 0 has the property described for the number r. Suppose that a number s also has this property, so that $sx = s$ for all real numbers s. We use the fact that $ab = ba$ for all real numbers a and b and proceed to use an algebraic trick, namely, we consider $0s$. Since both 0 and s have the required property, we see that

$$0s = 0 \qquad \text{and also} \qquad 0s = s0 = s.$$

Thus $0 = s$ since each is equal to $0s$. ▲

The exercises that follow are designed to illustrate the preceding observations further, with special emphasis on the use of the quantifying words and phrases *only, there exists, for all, for every, for each,* and *for some.*

Exercises 0.1

1. The main business of research mathematics is _____ .

2. If followed back to its foundations, mathematics is based on the axioms of _____ theory.

3. Algebra is presented in an axiomatic way for the sake of _____ .

4. To understand what a technical term means, we must know the _____ of the term.

5. Using the definition of an even integer in Example 1, prove that the sum of two even integers is even.

6. Using the definition of an even integer in Example 1, prove that the product of two even integers is an integral multiple of 4.

7. Define an odd integer in a manner similar to the definition of an even integer in Example 1. Prove that the sum of an even integer r and an odd integer s is an odd integer.

8. An example for which a statement does not hold is called a _____ to the statement.

9. Let x be a real variable in the sense explained in calculus. Let $P(x)$ be a property of x. Consider these labeled statements:

   A. $P(x)$ holds for some x.

   B. $P(x)$ holds for an x.

   C. $P(x)$ holds for all x.

   D. $P(x)$ holds for each x.

   E. $P(x)$ holds for one x.

   F. $P(x)$ holds for at least one x.

   G. There exists x such that $P(x)$ holds.

   H. There exists a unique x such that $P(x)$ holds.

   I. $P(x)$ holds for more than one x.

   J. $P(x)$ holds for every x.

   K. $P(x)$ holds for a unique x.

   L. There is no x for which $P(x)$ holds.

   M. There is at least one x for which $P(x)$ holds.

   N. $P(x)$ is false for all x but one.

   O. $P(x)$ is false for all but two values of x.

   Some of these labeled statements are logically equivalent. Arrange the labels A through O into lists such that two statements are logically equivalent if and only if their labels are in the same list.

10. This exercise illustrates experimentation leading to a conjecture. Draw a fairly large circle, and mark one point on the circle. There is just one, undivided region enclosed by the circle; write the number 1 down below the circle.

   a. Place a second point on the circle and draw the chord connecting the two points. Into how many pieces is the region enclosed by the circle divided by this chord? Write down your answer next to the 1 that you wrote below the circle.

   b. Place a third point on the circle and draw the chords connecting it to all the points previously marked on the circle. Into how many parts is the region enclosed by the circle divided by all the chords? Write down your answer next to the last number you entered. Looking at the numbers in your list, conjecture what the next number will be when you place another point on the circle in part (c).

   c. Place a fourth point on the circle and draw the chords connecting it to all the points previously marked on the circle. Into how many parts is the region enclosed by the circle divided by all the chords? Write down your answer next to the last number you entered. Was your conjecture in part (b) correct? Conjecture now how this list of numbers will look if this process of placing another point on the circle, drawing all chords, and counting regions is continued indefinitely. Below the list of four numbers you have generated, write down what you think the first eight numbers in the list would be if you continue the process.

   d. Place a fifth point on the circle in *general position* so that no chord joining it to other points will pass through a point of intersection of other chords. (Such an intersection would reduce the number of parts into which the inside of the circle is divided.) Draw all the chords, count the parts, and write down the number in your tally list. Is this count consistent with your conjecture in part (c)? If not, make a new conjecture. Do you consider your conjecture to be verified? Do you think it is a theorem?

   e. Place a sixth point in general position on the circle, draw all the chords, and answer all the questions in part (d).

In Exercises 11 through 27, decide whether the statement is a theorem. If it is a theorem, prove it. If it is not, give a counterexample.

11. Every even integer that is the square of an integer is an integral multiple of 4.

12. If n is a nonnegative integer, then $(n + 3)^2 > 9$.

13. There exists a unique integer n such that $n^2 + 2 = 3$.

14. There exists one integer n such that $n^2 + 4 = 8$.

15. There exists an integer n such that $n^2 + 5 = 14$.

16. There exists just one integer n such that $n^2 + 5 = 14$.

17. $n^2 > n$ for every integer n.

18. $n^2 > n$ for each negative integer n.

19. $x^2 < x$ for some real number x.

20. $n^2 > n$ for some integer n.

21. $n^2 = n$ for a unique integer n.

22. The square of every odd integer is one more than an integral multiple of 4.

23. If n is one more than an integral multiple of 3, then n^2 is as well.

24. There exists an integer n such that $n^3 < n$.

25. Let n and m be integers such that $n < m$ and $m \neq 0$. Then $(n/m)^2 < 1$.

26. Let n and m be integers such that $n < m$ and $m \neq 0$. Then $(n/m)^2 \leq n/m$.

27. Let n and m be integers such that $n < m$ and $m \neq 0$. Then $(n/m)^3 \leq (n/m)^2$.

0.2

Sets and Equivalence Relations

The Notion of a Set

In Section 0.1, we mentioned the importance of definitions to mathematics. There is a related structural weakness.

It is impossible to define every concept.

Suppose, for example, we define the term set by "A **set** is a well-defined collection of objects." One naturally asks what is meant by a *collection*. We could define it by "A collection is an aggregate of things." What then is an *aggregate*? Now our language is finite, so after some time we will run out of new words to use and have to repeat some words already examined. The definition is then circular and obviously worthless. Mathematicians realize that there must be some undefined or primitive concept with which to start. At the moment they have agreed that *set* shall be such a primitive concept. We shall not define *set*, but shall just hope that when such expressions as "the set of all real numbers" or "the set of all members of the United States Senate" are used, people's various ideas of what is meant are sufficiently similar to make communication feasible.

We summarize briefly some of the things we shall simply assume about sets.

1. A set S is made up of **elements**, and if a is one of these elements, we shall denote this fact by $a \in S$.

2. There is exactly one set with no elements. It is the **empty set** and is denoted by $\varnothing$.

3. We may describe a set either by giving a characterizing property of the elements, such as "the set of all members of the United States Senate," or by listing the elements. The standard way to describe a set by listing elements is to enclose the designations of the elements, separated by commas, in braces, for example, $\{1, 2, 15\}$. If a set is described by a characterizing property $P(x)$ of its elements x, the brace notation

$\{x \mid P(x)\}$ is also often used, and is read "the set of all x such that the statement $P(x)$ about x is true." Thus

$$\{2, 4, 6, 8\} = \{x \mid x \text{ is an even whole positive number} \leq 8\}$$
$$= \{2x \mid x = 1, 2, 3, 4\}.$$

The notation $\{x \mid P(x)\}$ is often called "set-builder notation."

4. A set is **well defined**, meaning that if S is a set and a is some object, then either a is definitely in S, denoted by $a \in S$, or a is definitely not in S, denoted by $a \notin S$. Thus we should never say, "Consider the set S of some positive numbers," for it is not definite whether $2 \in S$ or $2 \notin S$. On the other hand, we can consider the set T of all prime positive integers. Every positive integer is definitely either prime or not prime. Thus $5 \in T$ and $14 \notin T$. It may be hard to actually determine whether an object is in a set. For example, as this book goes to press it is probably unknown whether $2^{(2^{65})} + 1$ is in T. However, $2^{(2^{65})} + 1$ is certainly either prime or not prime.

It is not feasible for this text to push every definition back to the concept of a set. You should be aware that we are building on some very naive definitions, especially at the beginning of the text. The first definition we will meet in Chapter 1 says, "A **binary operation on a set** is a rule ... set." What on earth is a rule?

We now define the notion of a *subset* of a set.

DEFINITION 0.1 (Subset)   A set B is a **subset of a set** A, denoted by $B \subseteq A$ or $A \supseteq B$, if every element of B is in A. The notations $B \subset A$ or $A \supset B$ will be used for $B \subseteq A$ but $B \neq A$.

Note that according to this definition, for any set A, A itself and $\varnothing$ are both subsets of A.

DEFINITION 0.2 (Proper and Improper Subsets)   If A is any set, then A is the **improper subset of** A. Any other subset of A is a **proper subset of** A.

EXAMPLE 1   Let $S = \{1, 2, 3\}$. This set S has a total of eight subsets, namely $\varnothing$, $\{1\}$, $\{2\}$, $\{3\}$, $\{1, 2\}$, $\{1, 3\}$, $\{2, 3\}$, and $\{1, 2, 3\}$. ▲

Throughout this text, much work will be done involving familiar sets of numbers. Let us take care of notation for these sets once and for all:

$\mathbb{Z}$ is the set of all integers (that is, whole numbers: positive, negative, and zero).

$\mathbb{Z}^+$ is the set of all positive integers. (Zero is excluded.)

$\mathbb{Q}$ is the set of all rational numbers (that is, numbers that can be expressed as quotients m/n of integers, where $n \neq 0$).

$\mathbb{Q}^+$ is the set of all positive rational numbers.

$\mathbb{R}$ is the set of all real numbers.

$\mathbb{R}^+$ is the set of all positive real numbers.

$\mathbb{R}^*$ is the set of all nonzero real numbers.

$\mathbb{C}$ is the set of all complex numbers.

$\mathbb{C}^*$ is the set of all nonzero complex numbers.

Partitions and Equivalence Relations

We just described $\mathbb{Q}$ as the set of all numbers that can be expressed as quotients m/n of integers, where $n \neq 0$. It would be incorrect to describe $\mathbb{Q}$ as the set S of all "quotient expressions" m/n for m and n in $\mathbb{Z}$ and $n \neq 0$. For surely $\frac{2}{3}$ and $\frac{4}{6}$ are distinguishable quotient expressions, but we know they represent the *same* rational number. In fact, each element of $\mathbb{Q}$ is represented by an infinite number of different elements of S. When doing arithmetic, we *identify* in our mind elements of S that represent the same rational number in $\mathbb{Q}$.

The illustration in the preceding paragraph is typical of several situations in which we will consider different elements of a set to be arithmetically or algebraically equivalent, so that our set becomes *partitioned* into subsets, each of which we may consider to be a single arithmetic or algebraic entity. If b is an element of such a partitioned set, we usually let $\bar{b}$ represent the subset of all elements being identified with b.

EXAMPLE 2   Let S be the set of quotient expressions m/n for $m, n \in \mathbb{Z}$ and $n \neq 0$, as just described. The subset $\overline{\frac{2}{3}}$ of all elements in S that we identify arithmetically with the number $\frac{2}{3} \in \mathbb{Q}$ is given by

$$\overline{2/3} = \left\{ \frac{2}{3}, \frac{-2}{-3}, \frac{4}{6}, \frac{-4}{-6}, \frac{6}{9}, \frac{-6}{-9}, \cdots \right\}$$

$$= \left\{ \frac{2n}{3n} \,\middle|\, n \in \mathbb{Z} \text{ and } n \neq 0 \right\}. \quad \blacktriangle$$

Let us give a precise definition of a partition of a set.

DEFINITION 0.3 (Partition)   A **partition of a set** S is a decomposition of S into nonempty subsets such that every element of the set is in *one and only one* of the subsets. We call these subsets the **cells** of the partition.

EXAMPLE 3   Let $S = \{1, 2, 3, 4, 5, 6\}$. One partition of S is given by the cells

$$\{1, 6\}, \qquad \{3\}, \qquad \{2, 4, 5\}.$$

The subsets $\{1, 2, 3, 4\}$ and $\{4, 5, 6\}$ do not give a partition of S since 4 is in both subsets. The subsets $\{1, 2, 3\}$ and $\{5, 6\}$ do not give a partition since 4 is in neither subset. ▲

Two sets having no element in common are **disjoint**. Thus the cells in a partition of a set are disjoint.

How do we know whether two quotient expressions m/n and r/s in our partitioned set S in Example 2 are in the same cell, that is, represent the same rational number? One way to decide is to reduce both fractions to lowest terms. This may not be easy to do; for example, 1909/4897 and 1403/3599 represent the same rational number since

$$\frac{1909}{4897} = \frac{23 \cdot 83}{59 \cdot 83} \quad \text{and} \quad \frac{1403}{3599} = \frac{23 \cdot 61}{59 \cdot 61}.$$

However, finding these factorizations, even with a hand calculator, is a somewhat tedious trial-and-error task. But as we know, in fraction arithmetic we have $m/n = r/s$ if and only if $ms = nr$. This gives us a more efficient criterion for our problem, namely

$$(1909)(3599) = (4897)(1403) = 6870491.$$

Let $a \sim b$ denote that a is in the same cell as b for a given partition of a set S containing both a and b. Clearly the following properties are always satisfied:

$a \sim a$. The element a is in the same cell as itself.

If $a \sim b$ then $b \sim a$. If a is in the same cell as b, then b is in the same cell as a.

If $a \sim b$ and $b \sim c$, then $a \sim c$. If a is in the same cell as b and b is in the same cell as c, then a is in the same cell as c.

The theorem that follows is fundamental. It asserts that a relation $\sim$ between elements of a set that satisfies the three properties just described yields a natural partition of the set. Exhibiting a relation with these properties is frequently a convenient way to describe a partition of a set, and it is for this reason that we are discussing this material now.

THEOREM 0.1   Let S be a nonempty set and let $\sim$ be a relation between elements of S that satisfies the following properties for all $a, b, c \in S$.

1. (Reflexive) $a \sim a$.
2. (Symmetric) If $a \sim b$, then $b \sim a$.
3. (Transitive) If $a \sim b$ and $b \sim c$, then $a \sim c$.

Then $\sim$ yields a natural partition of S, where

$$\bar{a} = \{x \in S \mid x \sim a\}$$

is the cell containing a for all $a \in S$. Conversely, each partition of S gives rise to a natural relation $\sim$ satisfying the reflexive, symmetric, and transitive properties if $a \sim b$ is defined to mean that $a \in \bar{b}$.

PROOF   We proved the "converse" part of the theorem before we stated it.

For the direct statement, it only remains to show that the cells defined by $\bar{a} = \{x \in S \mid x \sim a\}$ do constitute a partition of S, that is, that every element of S is in *one and only one* cell. Let $a \in S$. Then $a \in \bar{a}$ by the reflexive condition (1), so a is in *at least one* cell.

Suppose now that a were in a cell $\bar{b}$ also. We need to show that $\bar{a} = \bar{b}$ as sets; this would show that a can't be in more than one cell. There is a standard way to show that two sets are the same:

Show that each set is a subset of the other.

We show that $\bar{a} \subseteq \bar{b}$. Let $x \in \bar{a}$. Then $x \sim a$. But $a \in \bar{b}$, so $a \sim b$. Then, by the transitive condition (3), $x \sim b$, so $x \in \bar{b}$. Thus $\bar{a} \subseteq \bar{b}$. Now we show that $\bar{b} \subseteq \bar{a}$. Let $y \in \bar{b}$. Then $y \sim b$. But $a \in \bar{b}$, so $a \sim b$ and, by symmetry (2), $b \sim a$. Then by transitivity (3), $y \sim a$, so $y \in \bar{a}$. Hence $\bar{b} \subseteq \bar{a}$ also, so $\bar{b} = \bar{a}$ and our proof is complete.   ◆

DEFINITION 0.4 (Equivalence Relation)   A relation $\sim$ on a set S satisfying the reflexive, symmetric, and transitive properties described in Theorem 0.1 is an **equivalence relation on** S. Each cell $\bar{a}$ in the natural partition given by an equivalence relation is an **equivalence class**.

The symbol $\sim$ is usually reserved for an equivalence relation. We will use $\mathscr{R}$ for a relation between elements of a set S which is not necessarily an equivalence relation on S.

The term *natural,* appearing twice in Theorem 0.1, has the following significance. If you start with an equivalence relation, form the partition of equivalence classes, and then consider the relation given by this partition, it is your original equivalence relation. Similarly, starting with a partition, going to the equivalence relation, and then forming the equivalence classes yields the original partition.

EXAMPLE 4   Let us verify directly that

$$m/n \sim r/s \qquad \text{if and only if} \qquad ms = nr$$

is an equivalence relation on the set S of formal quotient expressions we considered earlier.

Reflexive $m/n \sim m/n$ since $mn = nm$.

Symmetric If $m/n \sim r/s$, then $ms = nr$. Consequently, $rn = sm$, so $r/s \sim m/n$.

Transitive If $m/n \sim r/s$ and $r/s \sim u/v$, then $ms = nr$ and $rv = su$. Reordering terms and substituting, we obtain $mvs = vms = vnr = nrv = nsu = nus$. Since $s \neq 0$, we deduce that $mv = nu$, so $m/n \sim u/v$.

Each equivalence class of S is considered to be a rational number.   ▲

Our discussion of the set S of formal quotient expressions, culminating in Example 4, is a special case of work we will do in Section 5.4.

EXAMPLE 5   Let us define a relation $\mathcal{R}$ on the set $\mathbb{Z}$ by $n \mathcal{R} m$ if and only if $nm \geq 0$, and determine whether $\mathcal{R}$ is an equivalence relation.

Reflexive $a \mathcal{R} a$, since $a^2 \geq 0$ for all $a \in \mathbb{Z}$.

Symmetric If $a \mathcal{R} b$, then $ab \geq 0$, so $ba \geq 0$ and $b \mathcal{R} a$.

Transitive If $a \mathcal{R} b$ and $b \mathcal{R} c$, then $ab \geq 0$ and $bc \geq 0$. Thus $ab^2c = acb^2 \geq 0$. If we knew $b^2 > 0$, we could deduce $ac \geq 0$ whence $a \mathcal{R} c$. We have to examine the case $b = 0$ separately. A moment of thought shows $-3 \mathcal{R} 0$ and $0 \mathcal{R} 5$ but we do *not* have $-3 \mathcal{R} 5$, so the relation $\mathcal{R}$ is not transitive, and hence not an equivalence relation. ▲

EXAMPLE 6   Let $S = \{1, 2, 3, 4, 5, 6, 7, 8, 9, 10, 11, 12\}$ and let $n \sim m$ if and only if n and m start with the same letter of the alphabet when written out in English. It is easily seen that $\sim$ is an equivalence relation on S. The subsets in the partition of S arising from this equivalence relation are

$\{1\}$	(start with o)
$\{2, 3, 10, 12\}$	(start with t)
$\{4, 5\}$	(start with f)
$\{6, 7\}$	(start with s)
$\{8, 11\}$	(start with e)
$\{9\}$	(start with n). ▲

For each $n \in \mathbb{Z}^+$ we have a very important equivalence relation on $\mathbb{Z}$, *congruence modulo n*. We proceed to define this relation which will provide us with many examples in future chapters. We leave as an exercise the proof that congruence is an equivalence relation (see Exercise 18).

DEFINITION 0.5 (Congruence Modulo n)   Let h and k be two integers in $\mathbb{Z}$ and let n be any positive integer. We define h **congruent** to k **modulo** n, written $h \equiv k \pmod{n}$, if $h - k$ is evenly divisible by n, so that $h - k = ns$ for some $s \in \mathbb{Z}$. Equivalence classes for congruence modulo n are **residue classes modulo** n.

EXAMPLE 7   We see that $17 \equiv 33 \pmod{8}$ since $17 - 33 = 8(-2)$. ▲

Each residue class modulo $n \in \mathbb{Z}^+$ contains an infinite number of elements.

EXAMPLE 8   Modulo 8, the residue class containing both 17 and 33 is easily seen to be

$$\{\ldots, -47, -39, -31, -23, -15, -7, 1, 9, 17, 25, 33, 41, 49, \ldots\}$$
$$= \{8n + 1 \mid n \in \mathbb{Z}\}.$$

This residue class contains every eighth number, starting with 1. There are actually seven more residue classes in the partition given by congruence modulo 8. Exercise 23 asks us to describe them all. ▲

Exercises 0.2

In Exercises 1 through 4, describe the set by listing its elements.

1. $\{x \in \mathbb{R} \mid x^2 = 3\}$ **2.** $\{m \in \mathbb{Z} \mid m^2 = 3\}$

3. $\{m \in \mathbb{Z} \mid mn = 60 \text{ for some } n \in \mathbb{Z}\}$ **4.** $\{m \in \mathbb{Z} \mid m^2 - m < 115\}$

In Exercises 5 through 10, decide whether the object described is indeed a set (is well defined). Give an alternate description of each set.

5. $\{n \in \mathbb{Z}^+ \mid n \text{ is a large number}\}$

6. $\{n \in \mathbb{Z} \mid n^2 < 0\}$

7. $\{n \in \mathbb{Z} \mid 39 < n^3 < 57\}$

8. $\{x \in \mathbb{Q} \mid \text{the denominator of } x \text{ is greater than } 100\}$

9. $\{x \in \mathbb{Q} \mid x \text{ may be written with denominator greater than } 100\}$

10. $\{x \in \mathbb{Q} \mid x \text{ may be written with denominator less than } 3\}$

In Exercises 11 through 17, determine whether the given relation is an equivalence relation on the set. Describe the partition arising from each equivalence relation.

11. $n \mathrel{\mathscr{R}} m$ in $\mathbb{Z}$ if $nm > 0$ **12.** $x \mathrel{\mathscr{R}} y$ in $\mathbb{R}$ if $x \geq y$

13. $x \mathrel{\mathscr{R}} y$ in $\mathbb{R}$ if $|x| = |y|$ **14.** $x \mathrel{\mathscr{R}} y$ in $\mathbb{R}$ if $|x - y| \leq 3$

15. $n \mathrel{\mathscr{R}} m$ in $\mathbb{Z}^+$ if n and m have the same number of digits in the usual base ten notation

16. $n \mathrel{\mathscr{R}} m$ in $\mathbb{Z}^+$ if n and m have the same final digit in the usual base ten notation

17. $n \mathrel{\mathscr{R}} m$ in $\mathbb{Z}^+$ if $n - m$ is evenly divisible by 2

18. Let n be a particular integer in $\mathbb{Z}^+$. Show that congruence modulo n is an equivalence relation on $\mathbb{Z}$.

In Exercises 19 through 23, describe all residue classes of $\mathbb{Z}$ modulo the given value of n.

19. $n = 1$ **20.** $n = 2$ **21.** $n = 3$ **22.** $n = 4$ **23.** $n = 8$

24. The following is a famous incorrect argument. Find the error. "The reflexive criterion is redundant in the conditions for an equivalence relation, for from $a \sim b$ and $b \sim a$ (symmetry) we deduce $a \sim a$ by transitivity."

In Exercises 25 through 29, find the number of different partitions of a set S having the given number of elements.

25. 1 element   **26.** 2 elements   **27.** 3 elements

28. 4 elements   **29.** 5 elements

0.3

Mathematical Induction

Sometimes we want to prove that a statement about positive integers is true for all positive integers or perhaps for some finite or infinite sequence of consecutive integers. Such proofs are accomplished using *mathematical induction*. The validity of the method rests on this axiom of the positive integers.

Induction Axiom

Let S be a subset of $\mathbb{Z}^+$ satisfying

   1. $1 \in S$, and

   2. if $k \in S$, then $(k + 1) \in S$.

Then $S = \mathbb{Z}^+$.

This axiom leads immediately to the method of mathematical induction.

Mathematical Induction

Let $P(n)$ be a statement concerning the positive integer n. Suppose that

   1. $P(1)$ is true, and

   2. if $P(k)$ is true, then $P(k + 1)$ is true.

Then $P(n)$ is true for all $n \in \mathbb{Z}^+$.

Most of the time, we want to show that $P(n)$ holds for all $n \in \mathbb{Z}^+$. If we wish only to show that it holds for $r, r + 1, r + 2, \ldots, s - 1, s$, then we show that $P(r)$ is true and that $P(k)$ implies $P(k + 1)$ for $r \le k \le s - 1$. Note that r may be any integer, positive, negative, or zero.

EXAMPLE 1   Prove the formula

$$1 + 2 + \cdots + n = \frac{n(n + 1)}{2} \qquad (1)$$

for the sum of the arithmetic progression, using mathematical induction.

Solution   We let $P(n)$ be the statement that Formula (1) is true. For $n = 1$ we obtain

$$\frac{n(n + 1)}{2} = \frac{1(2)}{2} = 1,$$

so $P(1)$ is true.

Suppose $k \geq 1$ and $P(k)$ is true (our *induction hypothesis*), so

$$1 + 2 + \cdots + k = \frac{k(k + 1)}{2}.$$

To show that $P(k + 1)$ is true we compute

$$1 + 2 + \cdots + (k + 1) = (1 + 2 + \cdots + k) + (k + 1)$$

$$= \frac{k(k + 1)}{2} + (k + 1) = \frac{k^2 + k + 2k + 2}{2}$$

$$= \frac{k^2 + 3k + 2}{2} = \frac{(k + 1)(k + 2)}{2}.$$

Thus $P(k + 1)$ holds and Formula (1) is true for all $n \in \mathbb{Z}^+$. ▲

EXAMPLE 2   Show that a set of n elements has exactly 2^n subsets for any nonnegative integer n.

Solution   This time we start the induction with $n = 0$. Let S be a finite set having n elements. We wish to show

$$P(n)\text{: } S \text{ has } 2^n \text{ subsets.} \qquad (2)$$

If $n = 0$, then S is the empty set and has only one subset, namely the empty set itself. Since $2^0 = 1$, we see that $P(0)$ is true.

Suppose $P(k)$ is true. Let S have $k + 1$ elements and let one element of S be c. Let $S - \{c\} = \{x \in S \mid x \neq c\}$. Then $S - \{c\}$ has k elements, and hence 2^k subsets. Now every subset of S either contains c or does not contain c. Those not containing c are subsets of $S - \{c\}$, so there are 2^k of them by the induction hypothesis. Each subset containing c consists of one of the 2^k subsets not containing c, with c adjoined. There are 2^k such subsets also. The total number of subsets of S is then

$$2^k + 2^k = 2^k(2) = 2^{k+1},$$

so $P(k + 1)$ is true. Thus $P(n)$ is true for all nonnegative integers n. ▲

EXAMPLE 3   Let $x \in \mathbb{R}$ with $x > -1$ and $x \neq 0$.   Show   that
$(1 + x)^n > 1 + nx$ for every positive integer $n \geq 2$.

Solution   We let $P(n)$ be the statement

$$(1 + x)^n > 1 + nx. \tag{3}$$

(Note that $P(1)$ is false.) Then $P(2)$ is the statement $(1 + x)^2 > 1 + 2x$. Now
$(1 + x)^2 = 1 + 2x + x^2$, and $x^2 > 0$ since $x \neq 0$. Thus $(1 + x)^2 > 1 + 2x$, so
$P(2)$ is true.
    Suppose $P(k)$ is true, so

$$(1 + x)^k > 1 + kx. \tag{4}$$

Now $1 + x > 0$ since $x > -1$. Multiplying both sides of Eq. (4) by $1 + x$, we
obtain

$$(1 + x)^{k+1} > (1 + kx)(1 + x) = 1 + (k + 1)x + kx^2.$$

Since $kx^2 \geq 0$, we see that $P(k + 1)$ is true. Thus $P(n)$ is true for every
positive integer $n \geq 2$. ▲

    In a frequently used form of induction known as *complete induction*, the
statement

if $P(k)$ is true, then $P(k + 1)$ is true

set off by rules on page 15 is replaced by the statement

If $P(m)$ is true for $1 \leq m \leq k$, then $P(k + 1)$ is true.

Again, we are trying to show that $P(k + 1)$ is true, knowing $P(k)$ is true. But
if we have reached the stage in induction where $P(k)$ has been proved, then
we know $P(m)$ is true for $1 \leq m \leq k$, so the strengthened hypothesis in the
second statement is permissible.
    Our final example involves complete induction. We let P be the set of all
polynomials $f(x)$ with real coefficients. A polynomial $f(x)$ in P is **irreducible**
if there is no factorization $f(x) = g(x)h(x)$ where both $g(x)$ and $h(x)$ are
polynomials in P of degree less than the degree of $f(x)$. We will discuss
irreducible polynomials in much greater depth in Section 5.6. Chapters 7 and
8 deal extensively with polynomials and with solving polynomial equations.

EXAMPLE 4   Let P be the set of all polynomials $f(x)$ with real coefficients.
We shall now show that every polynomial in P of degree $n \in \mathbb{Z}^+$ is either
irreducible itself or can be factored into a product of irreducible polynomials
in P.

Solution   We shall use complete induction. Let $P(n)$ be the statement that is to be proved. Clearly $P(1)$ is true since a polynomial of degree 1 is already irreducible.

Let k be a positive integer. Our induction hypothesis is then: Every polynomial in P of degree less than $k + 1$ is either irreducible or can be factored into irreducible polynomials. Let $f(x)$ be a polynomial of degree $k + 1$. If $f(x)$ is irreducible, we have nothing more to do. Otherwise, we may factor $f(x)$ into polynomials $g(x)$ and $h(x)$ of lower degree than $k + 1$, obtaining $f(x) = g(x)h(x)$. The induction hypothesis indicates that each of $g(x)$ and $h(x)$ can be factored into irreducible polynomials, thus providing such a factorization of $f(x)$. This proves $P(k + 1)$. It follows that $P(n)$ is true for all $n \in Z^+$. ▲

Exercises 0.3

1. Show that $1^2 + 2^2 + 3^2 + \cdots + n^2 = \dfrac{n(n + 1)(2n + 1)}{6}$ for $n \in Z^+$.

2. Show that $1^3 + 2^3 + 3^3 + \cdots + n^3 = \dfrac{n^2(n + 1)^2}{4}$ for $n \in Z^+$.

3. Show that $1 + 3 + 5 + \cdots + (2n - 1) = n^2$ for $n \in Z^+$.

4. Show that $\dfrac{1}{1 \cdot 2} + \dfrac{1}{2 \cdot 3} + \dfrac{1}{3 \cdot 4} + \cdots + \dfrac{1}{n(n + 1)} = \dfrac{n}{n + 1}$ for $n \in Z^+$.

5. Prove by induction that if $a, r \in \mathbb{R}$ and $r \neq 1$, then

$$a + ar + ar^2 + \cdots + ar^n = a(1 - r^{n+1})/(1 - r) \quad \text{for } n \in Z^+.$$

6. Find the flaw in this argument.

We prove that any two integers i and j in Z^+ are equal. Let

$$\max(i, j) = \begin{cases} i & \text{if } i \geq j, \\ j & \text{if } j > i. \end{cases}$$

Let $P(n)$ be the statement

$$P(n): \text{Whenever } \max(i, j) = n, \text{ then } i = j.$$

Note that if $P(n)$ is true for all positive integers n, then any two positive integers i and j are equal. We proceed to prove $P(n)$ for positive integers n by induction.

Clearly $P(1)$ is true since if $i, j \in Z^+$ and $\max(i, j) = 1$ then $i = j = 1$. Assume $P(k)$ is true. Let i and j be such that $\max(i, j) = k + 1$. Then $\max(i - 1, j - 1) = k$ so that $i - 1 = j - 1$ by the induction hypothesis. Therefore $i = j$ and $P(k + 1)$ is true. Consequently $P(n)$ is true for all n.

7. Criticize this argument.

Let us show that every positive integer has some interesting property. Let $P(n)$ be the statement that n has an interesting property. We use complete induction.

Of course $P(1)$ is true, since 1 is the only positive integer that equals its own square, which is surely an interesting property of 1.

Suppose $P(m)$ is true for $1 \leq m \leq k$. If $P(k + 1)$ were not true, then $k + 1$ would be the smallest integer without an interesting property, which would, in itself, be an interesting property of $k + 1$. So $P(k + 1)$ must be true. Thus $P(n)$ is true for all $n \in \mathbb{Z}^+$.

8. We have never been able really to see any flaw in (a). Try your luck with it and then answer (b).

   a. A murderer is sentenced to be executed. He asks the judge not to let him know the day of the execution. The judge says, "I sentence you to be executed at 10 A.M. some day of this coming January, but I promise that you will not be aware you are being executed that day until they come to get you at 8 A.M." The criminal goes to his cell and proceeds to prove he can't be executed in January as follows:

   Let $P(n)$ be the statement that I can't be executed on January $(31 - n)$. I want to prove $P(n)$ for $0 \leq n \leq 30$. Now I can't be executed on January 31, for since that is the last day of the month and I am to be executed that month, I would know that was the day before 8 A.M., contrary to the judge's sentence. Thus $P(0)$ is true. Suppose $P(m)$ is true for $0 \leq m \leq k$ where $k \leq 29$. That is, suppose I can't be executed on January $(31 - k)$ through January 31. Then January $(31 - k - 1)$ must be the last possible day for execution, and I would be aware that was the day before 8 A.M., contrary to the judge's sentence. Thus I can't be executed on January $(31 - (k + 1))$, so $P(k + 1)$ is true. Therefore I can't be executed in January.

   (Of course, the criminal was executed on January 17.)

   b. An instructor teaches a class five days a week, Monday through Friday. She tells her class that she will give one more quiz on one day during the final week of classes, but that the students will not know for sure the quiz will be that day until they come to the classroom. What is the last day of the week she can give the quiz to satisfy these conditions?

0.4

Complex and Matrix Algebra

Complex numbers and matrices provide important examples in this text. For this reason, we now briefly summarize them and their algebraic properties.

Complex Numbers

A real number can be visualized geometrically as a point on a line that we often regard as an x-axis. A complex number can be regarded as a point in

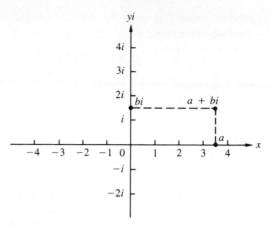

Figure 0.1

the euclidean plane, as shown in Fig. 0.1. Note that we label the vertical axis as the *yi*-axis rather than just the *y*-axis, and label the point one unit above the origin with *i* rather than 1. The point with cartesian coordinates (a, b) is labeled $a + bi$ in Fig. 0.1. The set $\mathbb{C}$ of **complex numbers** is defined by

$$C = \{a + bi \mid a, b \in \mathbb{R}\}.$$

We consider $\mathbb{R}$ to be a subset of the complex numbers by identifying a real number r with the complex number $r + 0i$. For example, we write $3 + 0i$ as 3 and $-\pi + 0i$ as $-\pi$ and $0 + 0i$ as 0. Similarly, we write $0 + 1i$ as i and $0 + si$ as si.

Complex numbers were developed after the development of real numbers. The complex number i was *invented* to provide a solution to the quadratic equation $x^2 = -1$, so we require that

$$i^2 = -1. \tag{1}$$

Unfortunately, i has been called an **imaginary number** and this terminology has led generations of students to view the complex numbers with more skepticism than the real numbers. Actually, *all* numbers, such as 1, 3, π, $-\sqrt{3}$, and i are inventions of our minds. There is no physical entity that *is* the number 1. If there were, it would surely be in a place of honor in some great scientific museum, and past it would file a steady stream of mathematicians, gazing at 1 in wonder and awe. The basic goal of Chapter 8 of this text is to show how we can invent solutions of polynomial equations when the coefficients of the polynomial may not even be real numbers!

Algebra of Complex Numbers

It is customary to use z as a complex variable, much as we use x and y as real variables. Thus we will think of z as $z = x + yi$ when dealing with complex numbers. We might denote two general complex numbers by $z_1 = a + bi$ and $z_2 = c + di$. We now describe how to compute the sum $z_1 + z_2$, the product z_1z_2, and the quotient z_1/z_2 if $z_2 \neq 0$.

For the sum $z_1 + z_2$ to be in $\mathbb{C}$, it must be of the form $r + si$ where r and s are real numbers. We want $r + si$ to be regarded as the sum of $r = r + 0i$ and $si = 0 + si$, and we would like all the formal properties of real arithmetic to hold. This leads to the sum definition

$$z_1 + z_2 = (a + bi) + (c + di) = (a + c) + (b + d)i. \tag{2}$$

Note that this sum is indeed in $\mathbb{C}$ since $a + c$ and $b + d$ are real numbers.

EXAMPLE I   Compute $(2 + 3i) + (4 - 7i)$.

Solution   Of course, we regard $4 - 7i$ as $4 + (-7)i$. We have

$$(2 + 3i) + (4 - 7i) = (2 + 3i) + (4 + (-7)i) = 6 - 4i. \; \blacktriangle$$

Geometrically, we can think of addition of complex numbers as the familiar vector addition in the plane, as illustrated in Fig. 0.2. It is easy to see that the usual properties $z_1 + z_2 = z_2 + z_1$ and $z_1 + (z_2 + z_3) = (z_1 + z_2) + z_3$ hold.

The product $(a + bi)(c + di)$ is defined in the way it must be if we are to enjoy the familiar properties of real arithmetic and require that $i^2 = -1$, in accord with Eq. (1). Namely, we see that we want to have

$$(a + bi)(c + di) = ac + adi + bci + bdi^2$$
$$= ac + adi + bci + bd(-1)$$
$$= (ac - bd) + (ad + bc)i.$$

Consequently, we define multiplication by

$$z_1z_2 = (a + bi)(c + di) = (ac - bd) + (ad + bc)i, \tag{3}$$

which is of the form $r + si$ with $r = ac - bd$ and $s = ad + bc$. It is routine

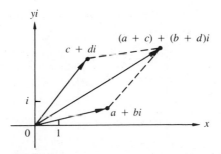

Figure 0.2

to check that the usual properties $z_1 z_2 = z_2 z_1$, $z_1(z_2 z_3) = (z_1 z_2)z_3$ and $z_1(z_2 + z_3) = z_1 z_2 + z_1 z_3$ all hold.

EXAMPLE 2   Compute $(2 - 5i)(8 + 3i)$.

Solution   We don't memorize Eq. (3), but rather we compute the product as we did to motivate that equation. We have

$$(2 - 5i)(8 + 3i) = 16 + 6i - 40i + 15 = 31 - 34i. \ \blacktriangle$$

To establish the geometric meaning of complex multiplication, we first define the **norm** $|a + bi|$ of $a + bi$, also called the **absolute value** (or **modulus**) of $a + bi$, by

$$|a + bi| = \sqrt{a^2 + b^2}. \tag{4}$$

This norm is a nonnegative real number and is the distance from $a + bi$ to the origin in Fig. 0.1. We can now describe a complex number z in the polar-coordinate form

$$z = |z| (\cos \theta + i \sin \theta), \tag{5}$$

where θ is the angle measured counterclockwise from the x-axis to the vector from 0 to z, as shown in Fig. 0.3. Let us set

$$z_1 = |z_1| (\cos \theta_1 + i \sin \theta_1) \qquad \text{and} \qquad z_2 = |z_2| (\cos \theta_2 + i \sin \theta_2)$$

and compute their product in this form. We find that

$$\begin{aligned} z_1 z_2 &= |z_1| |z_2| \left[(\cos \theta_1 \cos \theta_2 - \sin \theta_1 \sin \theta_2) + (\sin \theta_1 \cos \theta_2 + \cos \theta_1 \sin \theta_2)i \right] \\ &= |z_1| |z_2| \left[\cos(\theta_1 + \theta_2) + i \sin(\theta_1 + \theta_2) \right]. \end{aligned} \tag{6}$$

Note that Eq. (6) is in the polar form of Eq. (5) where $|z_1 z_2| = |z_1| |z_2|$ and

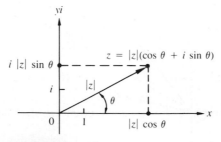

Figure 0.3

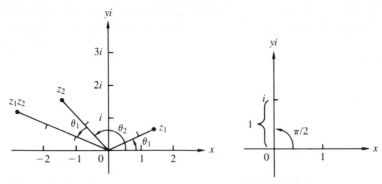

Figure 0.4   **Figure 0.5**

the polar angle θ for $z_1 z_2$ is the sum $\theta = \theta_1 + \theta_2$. Thus, geometrically, we multiply complex numbers by multiplying their norms and adding their polar angles, as shown in Fig. 0.4.

Note that i has polar angle $\pi/2$ and norm 1, as shown in Fig. 0.5. Thus i^2 has polar angle $2(\pi/2) = \pi$ and norm $1 \cdot 1 = 1$, so that $i^2 = -1$.

EXAMPLE 3   Find all solutions in C of the equation $z^2 = i$.

Solution   Writing the equation $z^2 = i$ in polar form and using Formula (6), we obtain

$$|z|^2 (\cos 2\theta + i \sin 2\theta) = 1(0 + i).$$

Thus $|z|^2 = 1$, so $|z| = 1$. The angle θ for z must satisfy $\cos 2\theta = 0$ and $\sin 2\theta = 1$. Consequently, $2\theta = (\pi/2) + n(2\pi)$, so $\theta = (\pi/4) + n\pi$ for an integer n. The values of n yielding values θ where $0 \le \theta < 2\pi$ are 0 and 1, yielding $\theta = \pi/4$ or $\theta = 5\pi/4$. Our solutions are

$$z_1 = 1\left(\cos\frac{\pi}{4} + i \sin\frac{\pi}{4}\right) \quad \text{and} \quad z_2 = 1\left(\cos\frac{5\pi}{4} + i \sin\frac{5\pi}{4}\right)$$

or

$$z_1 = \frac{1}{\sqrt{2}}(1 + i) \quad \text{and} \quad z_2 = \frac{-1}{\sqrt{2}}(1 + i). \ \blacktriangle$$

EXAMPLE 4   Find all solutions of $z^4 = -16$.

Solution   As in Example 3, we write the equation in polar form, obtaining

$$|z|^4 (\cos 4\theta + i \sin 4\theta) = 16(-1 + 0i).$$

Consequently $|z|^4 = 16$, so $|z| = 2$ while $\cos 4\theta = -1$ and $\sin 4\theta = 0$. We find that $4\theta = \pi + n(2\pi)$, so $\theta = (\pi/4) + n(\pi/2)$ for integers n. The different values of θ obtained where $0 \le \theta < 2\pi$ are $\pi/4$, $3\pi/4$, $5\pi/4$, and $7\pi/4$. Thus one solution of $z^4 = -16$ is

$$2\left(\cos\frac{\pi}{4} + i\sin\frac{\pi}{4}\right) = 2\left(\frac{1}{\sqrt{2}} + \frac{1}{\sqrt{2}}i\right) = \sqrt{2}\,(1 + i).$$

In a similar way, we find three more solutions,

$$\sqrt{2}\,(-1 + i), \qquad \sqrt{2}\,(-1 - i), \qquad \text{and} \qquad \sqrt{2}\,(1 - i). \ \blacktriangle$$

The last two examples illustrate that we can find solutions of an equation $z^n = a + bi$ by writing the equation in polar form. There will always be n solutions, provided that $a + bi \ne 0$. Exercises 27 through 32 ask you to solve equations of this type.

We now turn to computing z_1/z_2 where $z_2 \ne 0$, that is, to finding a complex number z_3 such that $z_3 z_2 = z_1$. Note that

$$(c + di)(c - di) = c^2 + d^2 = |c + di|^2.$$

The number $c - di$ is called the **conjugate** of the number $c + di$. We usually compute $z_1/z_2 = (a + bi)/(c + di)$ as follows:

$$z_3 = \frac{z_1}{z_2} = \frac{a + bi}{c + di} = \frac{a + bi}{c + di} \cdot \frac{c - di}{c - di} = \frac{(ac + bd) + (bc - ad)i}{c^2 + d^2}$$

$$= \frac{ac + bd}{c^2 + d^2} + \frac{bc - ad}{c^2 + d^2}i. \tag{7}$$

Note that the result is indeed a number of the form $r + si$, where $r = (ac + bd)/(c^2 + d^2)$ and $s = (bc - ad)/(c^2 + d^2)$ are in $\mathbb{R}$. You can verify, if you feel so inclined, that $z_3 z_2 = z_1$. Note that if $z_2 \ne 0$, then $|z_2| = c^2 + d^2 \ne 0$. We ask you to give the geometric interpretation of complex division in Exercise 26.

EXAMPLE 5   Compute $(2 + 3i)/(1 - 5i)$.

Solution   Rather than memorize the final formula in Eq. (7), we simply remember the technique of multiplying numerator and denominator by the conjugate of the denominator. We have

$$\frac{2 + 3i}{1 - 5i} = \frac{2 + 3i}{1 - 5i} \cdot \frac{1 + 5i}{1 + 5i} = \frac{(2 - 15) + (3 + 10)i}{1 + 25}$$

$$= -\frac{13}{26} + \frac{13}{26}i = -\frac{1}{2} + \frac{1}{2}i. \ \blacktriangle$$

Matrix Algebra

Probably you have had a course in linear algebra. We give a brief review of matrix algebra here. Matrices will provide some examples of algebraic structures in subsequent chapters.

A **matrix** is a rectangular array of numbers. For example, the array

$$\begin{bmatrix} 2 & -1 & 4 \\ 3 & 1 & 2 \end{bmatrix} \tag{8}$$

is a matrix having two rows and three columns. A matrix having m rows and n columns is an $m \times n$ matrix, so matrix (8) is a 2×3 matrix. If $m = n$, the matrix is **square**. Entries in a matrix may be any type of number—integer, rational, real, or complex. We let $M_{m \times n}(\mathbb{R})$ be the set of all $m \times n$ matrices with real number entries. If $m = n$, the notation is abbreviated to $M_n(\mathbb{R})$. We can similarly consider $M_n(\mathbb{Z})$, $M_{2 \times 3}(\mathbb{C})$, etc.

Two matrices having the same number m of rows and the same number n of columns can be added in the obvious way: we add entries in corresponding positions.

EXAMPLE 6   In $M_{2 \times 3}(\mathbb{Z})$, we have

$$\begin{bmatrix} 2 & -1 & 4 \\ 3 & 1 & 2 \end{bmatrix} + \begin{bmatrix} 1 & 0 & -3 \\ 2 & -7 & 1 \end{bmatrix} = \begin{bmatrix} 3 & -1 & 1 \\ 5 & -6 & 3 \end{bmatrix}. \ \blacktriangle$$

We will use uppercase letters to denote matrices. If A, B, and C are $m \times n$ matrices, it is easily seen that $A + B = B + A$ and that $A + (B + C) = (A + B) + C$.

Matrix multiplication, AB, is defined only if the number of columns of A is equal to the number of rows of B. That is, if A is an $m \times n$ matrix, then B must be an $n \times s$ matrix for some integer s. We start by defining as follows the product AB where A is a $1 \times n$ matrix and B is an $n \times 1$ matrix:

$$AB = \begin{bmatrix} a_1 & a_2 & \cdots & a_n \end{bmatrix} \begin{bmatrix} b_1 \\ b_2 \\ \vdots \\ b_n \end{bmatrix} = a_1 b_1 + a_2 b_2 + \cdots + a_n b_n. \tag{9}$$

Note that the result is a number. (We shall not distinguish between a number and the 1×1 matrix having that number as its sole entry.) You may recognize this product as the *dot product* of vectors. Matrices having only one *row* or only one *column* are **row vectors** or **column vectors** respectively.

EXAMPLE 7   We find that

$$[3 \quad -7 \quad 2] \begin{bmatrix} 1 \\ 4 \\ 5 \end{bmatrix} = (3)(1) + (-7)(4) + (2)(5) = -15. \; \blacktriangle$$

Let A be an $m \times n$ matrix and let B be an $n \times s$ matrix. Note that the number n of entries in each row of A is the same as the number n of entries in each column of B. The product $C = AB$ is an $m \times s$ matrix. The entry in the ith row and jth column of AB is the product of the ith row of A times the jth column of B as defined by Eq. (9) and illustrated in Example 7.

EXAMPLE 8   Compute

$$AB = \begin{bmatrix} 2 & -1 & 3 \\ 1 & 4 & 6 \end{bmatrix} \begin{bmatrix} 3 & 1 & 2 & 1 \\ 1 & 4 & 1 & -1 \\ -1 & 0 & 2 & 1 \end{bmatrix}.$$

Solution   Note that A is 2×3 and B is 3×4. Thus AB will be 2×4. The entry in its second row and third column is

$$(\text{2nd row } A)(\text{3rd column } B) = [1 \quad 4 \quad 6] \begin{bmatrix} 2 \\ 1 \\ 2 \end{bmatrix} = 2 + 4 + 12 = 18.$$

Computing all eight entries of AB in this fashion, we obtain

$$AB = \begin{bmatrix} 2 & -2 & 9 & 6 \\ 1 & 17 & 18 & 3 \end{bmatrix}. \; \blacktriangle$$

EXAMPLE 9   The expression

$$\begin{bmatrix} 2 & -1 & 3 \\ 1 & 4 & 6 \end{bmatrix} \begin{bmatrix} 2 & 1 \\ 5 & 4 \end{bmatrix}$$

is not defined, since the number of entries in a row of the first matrix is not equal to the number of entries in a column of the second matrix. $\blacktriangle$

For square matrices of the same size, both addition and multiplication are always defined. Exercise 42 asks us to illustrate the following fact.

Matrix multiplication is not commutative.

That is, AB need not equal BA even when both products are defined, as for $A, B \in M_2(\mathbb{Z})$. It can be shown that $A(BC) = (AB)C$ and $A(B + C) = AB + AC$ whenever all these expressions are defined.

We let I_n be the $n \times n$ matrix with entries 1 along the diagonal from the upper-left corner to the lower-right corner, and entries 0 elsewhere. For example,

$$I_3 = \begin{bmatrix} 1 & 0 & 0 \\ 0 & 1 & 0 \\ 0 & 0 & 1 \end{bmatrix}.$$

It is easy to see that if A is any $n \times s$ matrix and B is any $r \times n$ matrix, then $I_n A = A$ and $BI_n = B$. That is, the matrix I_n acts much as the number 1 does for multiplication when multiplication by I_n is defined.

Let A be an $n \times n$ matrix and consider a matrix equation of the form $AX = B$, where A and B are known but X is unknown. If we can find an $n \times n$ matrix A^{-1} such that $A^{-1}A = AA^{-1} = I_n$, then we can conclude that

$$A^{-1}(AX) = A^{-1}B, \quad (A^{-1}A)X = A^{-1}B, \quad I_n X = A^{-1}B, \quad X = A^{-1}B,$$

and we have found the desired matrix X. Such a matrix A^{-1} acts like the reciprocal of a number: $A^{-1}A = I_n$ and $(1/r)r = 1$. This is the reason for the notation A^{-1}.

If A^{-1} exists, the square matrix A is **invertible** and A^{-1} is the **inverse** of A. If A^{-1} does not exist, then A is said to be **singular**. It can be shown that if there exists a matrix A^{-1} such that $A^{-1}A = I_n$, then $AA^{-1} = I_n$ also, and furthermore, there is only one matrix A^{-1} having this property.

EXAMPLE 10   Let

$$A = \begin{bmatrix} 2 & 9 \\ 1 & 4 \end{bmatrix}.$$

We can check that

$$\begin{bmatrix} -4 & 9 \\ 1 & -2 \end{bmatrix}\begin{bmatrix} 2 & 9 \\ 1 & 4 \end{bmatrix} = \begin{bmatrix} 2 & 9 \\ 1 & 4 \end{bmatrix}\begin{bmatrix} -4 & 9 \\ 1 & -2 \end{bmatrix} = \begin{bmatrix} 1 & 0 \\ 0 & 1 \end{bmatrix}.$$

Thus,

$$A^{-1} = \begin{bmatrix} -4 & 9 \\ 1 & -2 \end{bmatrix}. \quad \blacktriangle$$

We leave the problems of determining the existence of A^{-1} and its computation to a course in linear algebra.

Exercises 0.4

In Exercises 1 through 19, compute the given arithmetic expression and give the answer in the form $a + bi$ for $a, b \in \mathbb{R}$.

1. $(2 - 3i) + (4 + 5i)$
2. $i + (5 - 3i)$

3. $(5 + 7i) - (3 - 2i)$
4. $(1 - 3i) - (-4 + 2i)$

5. i^3 **6.** i^4 **7.** i^{23} **8.** $(-i)^{35}$

9. $(4 - i)(5 + 3i)$
10. $(8 + 2i)(3 - i)$

11. $(2 - 3i)(4 + i) + (6 - 5i)$
12. $(1 + i)^3$

13. $(1 - i)^5$ (Use the binomial theorem.)

14. $\dfrac{7 - 5i}{1 + 6i}$ **15.** $\dfrac{i}{1 + i}$ **16.** $\dfrac{1 - i}{i}$ **17.** $\dfrac{i(3 + i)}{2 - 4i}$

18. $\dfrac{3 + 7i}{(1 + i)(2 - 3i)}$ **19.** $\dfrac{(1 - i)(2 + i)}{(1 - 2i)(1 + i)}$

20. Find $|3 - 4i|$.   **21.** Find $|6 + 4i|$.

In Exercises 22 through 25, write the given complex number z in the polar form $|z| (p + qi)$ where $|p + qi| = 1$.

22. $3 - 4i$ **23.** $-1 + i$ **24.** $12 + 5i$ **25.** $-3 + 5i$

26. Let z_1 and z_2 be complex numbers with $z_2 \neq 0$. Describe the geometric meaning of z_1/z_2 in terms of norms and polar angles.

In Exercises 27 through 32, find all complex solutions of the given equation.

27. $z^4 = 1$ **28.** $z^4 = -1$ **29.** $z^3 = -8$ **30.** $z^3 = -27i$

31. $z^6 = 1$ **32.** $z^6 = -64$

In Exercises 33 through 41, compute the given arithmetic matrix expression, if it is defined.

33. $\begin{bmatrix} -2 & 4 \\ 1 & 5 \end{bmatrix} + \begin{bmatrix} 4 & -3 \\ 1 & 2 \end{bmatrix}$

34. $\begin{bmatrix} 1 + i & -2 & 3 - i \\ 4 & i & 2 - i \end{bmatrix} + \begin{bmatrix} 3 & i - 1 & -2 + i \\ 3 - i & 1 + i & 0 \end{bmatrix}$

35. $\begin{bmatrix} i & -1 \\ 4 & 1 \\ 3 & -2i \end{bmatrix} - \begin{bmatrix} 3 - i & 4i \\ 2 & 1 + i \\ 3 & -i \end{bmatrix}$

36. $\begin{bmatrix} 1 & -1 \\ 3 & 1 \end{bmatrix} \begin{bmatrix} 2 & 4 \\ -1 & 3 \end{bmatrix}$ **37.** $\begin{bmatrix} 3 & 1 \\ -4 & 2 \end{bmatrix} \begin{bmatrix} 1 & 5 & -3 \\ 2 & 1 & 6 \end{bmatrix}$

38. $\begin{bmatrix} 4 & -1 \\ 3 & 2 \end{bmatrix} \begin{bmatrix} 1 & 0 \\ -1 & 7 \\ 3 & 1 \end{bmatrix}$ **39.** $\begin{bmatrix} i & 1 \\ -2 & 1 \end{bmatrix} \begin{bmatrix} 3i & 1 \\ 4 & -2i \end{bmatrix}$

40. $\begin{bmatrix} 1 & -1 \\ 1 & 0 \end{bmatrix}^4$ **41.** $\begin{bmatrix} 1 & -i \\ i & 1 \end{bmatrix}^4$

42. Give an example in $M_2(\mathbb{Z})$ showing that matrix multiplication is not commutative.

43. Find $\begin{bmatrix} 0 & 1 \\ -1 & 0 \end{bmatrix}^{-1}$, by experimentation if necessary.

44. Find $\begin{bmatrix} 2 & 0 & 0 \\ 0 & 4 & 0 \\ 0 & 0 & -1 \end{bmatrix}^{-1}$, by experimentation if necessary.

45. Prove that if $A, B \in M_n(\mathbb{C})$ are invertible, then AB and BA are invertible also.

46. Let $z = r(\cos\theta + i\sin\theta)$. Prove, using mathematical induction, that

$$z^n = r^n(\cos n\theta + i\sin n\theta) \qquad \text{for all} \qquad n \in \mathbb{Z}^+.$$

47. (*Representation of complex numbers by* 2×2 *matrices*) There is a subset S of the set $M_2(\mathbb{R})$ of 2×2 matrices that is algebraically identical with the set $\mathbb{C}$ of complex numbers. You may know that multiplication of a column vector $\mathbf{v}$ in the plane on the left by the matrix

$$A = \begin{bmatrix} \cos\theta & -\sin\theta \\ \sin\theta & \cos\theta \end{bmatrix}$$

yields the vector $A\mathbf{v}$ obtained by rotating $\mathbf{v}$ counterclockwise through the angle θ. Our discussion of the geometric meaning of multiplication of complex numbers shows that multiplication of a complex number $v_1 + iv_2$ in the plane on the left by the complex number $\cos\theta + i\sin\theta$ has the same effect. This suggests that the complex number $a + bi$ might be able to be represented algebraically by the matrix

$$\begin{bmatrix} a & -b \\ b & a \end{bmatrix}. \tag{10}$$

We let S be the set of all 2×2 matrices of the form of matrix (10) where a and b are real numbers. Parts (a) and (b) will show that $\mathbb{C}$ and S are algebraically identical. Later in the text, we will call algebraically identical structures like $\mathbb{C}$ and S *isomorphic algebraic structures.*

For each complex number $a + bi$, we denote the corresponding 2×2 matrix (10) by $\phi(a + bi)$, so that ϕ is a function with domain $\mathbb{C}$ and range S. Note that if $a + bi \neq c + di$ then $\phi(a + bi) \neq \phi(c + di)$, so that different complex numbers correspond to different matrices.

a. Show that $\phi((a + bi) + (c + di)) = \phi(a + bi) + \phi(c + di)$. (This shows that addition in $\mathbb{C}$ is mirrored by addition in S.)

b. Show that $\phi((a + bi)(c + di)) = \phi(a + bi)\phi(c + di)$. (This shows that multiplication in $\mathbb{C}$ is mirrored by multiplication in S.)

CHAPTER ONE

◆

GROUPS AND SUBGROUPS

◆

Our study of abstract algebra now begins. As explained in Section 0.1, we will *abstract* from algebra its basic properties, and study the subject in terms of those properties.

Algebra is based on arithmetic, and in Section 1.1, we abstract from arithmetic its core ideas. Section 1.2 abstracts from algebra those properties that enable us to solve a single linear equation in one variable, for example, the equation $7x = 19$. A structure having those properties is called a *group*. A portion of a group that still enjoys all the properties of a group is called a *subgroup*; we discuss subgroups in Section 1.3. Trying to find as small subgroups as possible leads us naturally to the notion of a *cyclic group* and to *generators* for a group, which are the topics of Section 1.4.

Many thousands of pages of research have been written about groups. Our text can barely scratch the surface of the subject. Chapter 2 gives many more examples of groups, while Chapter 3 studies relations between groups. Chapter 4 delves just a bit deeper into the structure of groups, especially those having only a finite number of elements.

1.1

Binary Operations

Suppose that we are visitors to a strange civilization in a strange world and are observing one of the creatures of this world drilling a class of fellow creatures in addition of numbers. Suppose also that we have not been told that the class is learning to add, but were just placed as observers in the room where this was going on. We are asked to give a report on exactly what happens. The teacher makes noises that sound to us approximately like *gloop, poyt*. The class responds with *bimt*. The teacher than gives *ompt, gaft*, and the class responds with *poyt*. What are they doing? We cannot report that they are adding numbers, for we do not even know that the sounds are representing numbers. Of course, we do realize that there is communication going on. All we can say with any certainty is that these creatures know some

31

rule, so that when certain pairs of things are designated in their language, one after another, like *gloop, poyt,* they are able to agree on a response, *bimt.* This same procedure goes on in addition drill in our first grade classes where a teacher may say *four, seven,* and the class responds with *eleven.*

In our attempt to analyze addition and multiplication of numbers, we are thus led to the idea that addition is basically just a rule that people learn, enabling them to associate, with two numbers in a given order, some number as the answer. Multiplication is also such a rule, but a different rule. Note finally that in playing this game with students, teachers have to be a little careful of what two things they give to the class. If a first grade teacher suddenly inserts *ten, sky,* the class will be very confused. The rule is only defined for pairs of things from some specified set.

Definitions and Examples

As mathematicians, let us attempt to collect the core of these basic ideas in a useful definition. As we remarked in Section 0.2 we do not attempt to define a set.

DEFINITION 1.1 (Binary Operation)   A **binary operation** $*$ **on a set** S is a rule that assigns to each ordered pair (a, b) of elements of S some element of S.

The word *ordered* in this definition is very important, for it allows the possibility that the element assigned to the pair (a, b) may be different from the element assigned to the pair (b, a). Also, we were careful not to say that to each ordered pair of elements is assigned *another* or a *third* element, for we wish to permit cases such as occur in addition of numbers where $(0, 2)$ has assigned to it the number 2.

EXAMPLE 1   Our usual addition $+$ is a binary operation on the set $\mathbb{R}$. Our usual multiplication $\cdot$ is a different binary operation on $\mathbb{R}$. In this example, we could replace $\mathbb{R}$ by any of the sets $\mathbb{C}$, $\mathbb{Z}$, $\mathbb{R}^+$, or $\mathbb{Z}^+$. ▲

Note that we require a binary operation on a set S to be defined for *every* ordered pair (a, b) of elements from S.

EXAMPLE 2   Let $M(\mathbb{R})$ be the set of all matrices with real entries. The usual matrix addition $+$ is *not* a binary operation on this set since $A + B$ is not defined for an ordered pair (A, B) of matrices having different numbers of rows or of columns. ▲

A binary operation on S must assign to each ordered pair (a, b) an element *that is again in* S. This requirement that the element be again in S is known as the **closure condition**; we require that S be **closed** under a binary operation on S.

EXAMPLE 3   Our usual addition $+$ is *not* a binary operation on the set $\mathbb{R}^*$ of nonzero real numbers because $2 + (-2)$ is not in the set $\mathbb{R}^*$; that is, $\mathbb{R}^*$ is not closed under $+$. ▲

EXAMPLE 4   The usual matrix multiplication $\cdot$ is a binary operation on the set $M_4(\mathbb{C})$ of all 4×4 matrices with complex entries, for $\cdot$ is defined for every ordered pair (A, B) of 4×4 matrices, and the product is again a 4×4 matrix. ▲

EXAMPLE 5   Let F be the set of all real-valued functions f having as domain the set $\mathbb{R}$ of real numbers. We are familiar from calculus with the binary operations $+$, $-$, $\cdot$, and $\circ$ on F. Namely, for each ordered pair (f, g) of functions in F, we define for each $x \in \mathbb{R}$

$$(f + g)(x) = f(x) + g(x) \qquad \text{function addition,}$$

$$(f - g)(x) = f(x) - g(x) \qquad \text{function subtraction,}$$

$$(f \cdot g)(x) = f(x)g(x) \qquad \text{function multiplication,}$$

and

$$(f \circ g)(x) = f(g(x)) \qquad \text{function composition.}$$

All four of these functions are again real valued with domain $\mathbb{R}$, so F is closed under all four operations $+$, $-$, $\cdot$, and $\circ$. ▲

The binary operations described in Examples 1 through 5 are very familiar to you. In this text, we want to *abstract* basic structural concepts from our familiar algebra. To emphasize this concept of *abstraction* from the familiar, we should illustrate these structural concepts with unfamiliar examples as well. Consequently, we now invent some examples of binary operations that are not familiar additions, subtractions, or the like. For the first two sections of this chapter, we will denote such an operation by $*$ and let $a * b$ be the element assigned to (a, b) by $*$. If we have several different binary operations under simultaneous discussion, we shall use subscripts or superscripts on the $*$ to distinguish them. The most important method of describing a particular binary operation $*$ on a given set is to characterize the element $a * b$ assigned to each pair (a, b) by some property defined in terms of a and b.

EXAMPLE 6   On $\mathbb{Z}^+$, we define a binary operation $*$ by $a * b$ equals the smaller of a and b or the common value if $a = b$. Thus $2 * 11 = 2$; $15 * 10 = 10$; and $3 * 3 = 3$. ▲

EXAMPLE 7   On $\mathbb{Z}^+$, we define a binary operation $*'$ by $a *' b = a$. Thus $2 *' 3 = 2$, $25 *' 10 = 25$, and $5 *' 5 = 5$. ▲

EXAMPLE 8   On $\mathbb{Z}^+$,   we   define   a   binary   operation $*''$ by $a *'' b = (a * b) + 2$, where $*$ is defined in Example 6. Thus $4 *'' 7 = 6$; $25 *'' 9 = 11$; and $6 *'' 6 = 8$. ▲

It may seem that these examples are of no importance, but consider for a moment. Suppose we go into a store to buy a large, delicious chocolate bar. Suppose we see two identical bars side by side, the wrapper of one stamped $1.67 and the wrapper of the other stamped $1.79. Of course we pick up the one stamped $1.67. Our knowledge of which one we want depends on the fact that sometime we learned the binary operation $*$ of Example 6. It is a *very important operation*. Likewise the binary operation $*'$ of Example 7 is defined using our ability to distinguish order. Think what a problem we would have if we tried to put on our shoes first, and then our socks! Thus we should not be hasty about dismissing some binary operation as being of little significance. Of course, our usual operations of addition and multiplication of numbers have a practical importance well known to us.

Examples 6 and 7 were chosen to demonstrate that a binary operation may or may not depend on the order of the given pair. Thus in Example 6, $a * b = b * a$ for all $a, b \in \mathbb{Z}^+$, and in Example 7 this is not the case, for $5 *' 7 = 5$ but $7 *' 5 = 7$.

DEFINITION 1.2 (Commutative Operation)   A binary operation on a set S is **commutative** if (and only if) $a * b = b * a$ for all $a, b \in S$.

As was pointed out in Section 0.1, it is customary in mathematics to omit the words *and only if* from a definition. Definitions are always understood to be if and only if statements. *Theorems are not always if and only if statements, and no such convention is ever used for theorems.*

Now suppose we wish to consider an expression of the form $a * b * c$. A binary operation $*$ enables us to combine only two elements, and here we have three. The obvious attempts to combine the three elements are to form either $(a * b) * c$ or $a * (b * c)$. With $*$ defined as in Example 6, $(2 * 5) * 9$ is computed by $2 * 5 = 2$ and then $2 * 9 = 2$. Likewise $2 * (5 * 9)$ is computed by $5 * 9 = 5$ and then $2 * 5 = 2$. Hence $(2 * 5) * 9 = 2 * (5 * 9)$, and it is not hard to see that for this $*$,

$$(a * b) * c = a * (b * c),$$

so there is no ambiguity in writing $a * b * c$. But for $*''$ of Example 8

$$(2 *'' 5) *'' 9 = 4 *'' 9 = 6,$$

while

$$2 *'' (5 *'' 9) = 2 *'' 7 = 4.$$

Thus $(a *'' b) *'' c$ need not equal $a *'' (b *'' c)$, and an expression $a *'' b *'' c$ may be ambiguous.

DEFINITION 1.3 (Associative Operation)   A binary operation on set S is **associative** if $(a * b) * c = a * (b * c)$ for all $a, b, c \in S$.

It can be shown that if $*$ is associative, then longer expressions such as $a * b * c * d$ are not ambiguous. Parentheses may be inserted in any fashion for purposes of computation; the final results of two such computations will be the same.

Although we do not define *functions* formally until Section 2.1, we wish to build here on your background with them in calculus. Some of the most important binary operations we will consider are defined by function composition. It is important to know that function composition is always associative whenever it is defined. That is,

---

Associativity of Function Composition

$f \circ (g \circ h) = (f \circ g) \circ h$ whenever this composition is defined.

To see this, we need only note that if this composition is defined, then for each x in the domain of h, we have

$$(f \circ (g \circ h))(x) = f((g \circ h)(x)) = f(g(h(x)))$$
$$= (f \circ g)(h(x)) = ((f \circ g) \circ h)(x).$$

The functions, like f, may even have values $f(x)$ that are in a different set than the domain where the values of x occur. For example, linear algebra shows that if A is an $m \times n$ matrix, then A provides a linear transformation (function) T with domain $\mathbb{R}^n$ and values in $\mathbb{R}^m$ defined by $T(\mathbf{v}) = A\mathbf{v}$ for each column vector $\mathbf{v} \in \mathbb{R}^n$. The linear transformation provided by the product AB of two matrices can be shown to be the function composition of the two transformations. Since function composition is associative whenever it is defined, it follows immediately that matrix multiplication is associative, so that $A(BC) = (AB)C$ whenever the product of the matrices A, B, and C is defined. Proving the associativity of matrix multiplication by the definition given in Section 0.4 is a fairly painful exercise in summation notation. As an application to our work in this text, we should remember that *every binary operation defined as function composition is associative*.

*	a	b	c
a	b	c	b
b	a	c	b
c	c	b	a

Table 1.1

Tables

For a finite set, a binary operation on the set can also be defined by means of a table. The next example shows how this will be done in this text.

EXAMPLE 9   Table 1.1 defines the binary operation $*$ on $S = \{a, b, c\}$ by the following rule:

(*i*th *entry on the left*) $*$ (*j*th *entry on the top*)

= (*entry in the* ith *row and* jth *column of the table body*).

Thus $a * b = c$ and $b * a = a$, so $*$ is not commutative. ▲

We can easily see that *a binary operation defined by a table is commutative if and only if the entries in the table are symmetric with respect to the diagonal that starts at the upper left corner of the table and terminates at the lower right corner.* We always assume that the elements of the set are listed across the top of a table in the same order as they are listed at the left.

EXAMPLE 10   Complete Table 1.2 so that $*$ is a commutative binary operation on the set $S = \{a, b, c, d\}$.

Solution   From Table 1.2, we see that $b * a = d$. For $*$ to be commutative, we must have $a * b = d$ also. Thus we place d in the appropriate square defining $a * b$, which is located symmetrically across the diagonal in Table 1.3

*	a	b	c	d
a	b			
b	d	a		
c	a	c	d	
d	a	b	b	c

Table 1.2

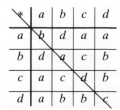

*	a	b	c	d
a	b	d	a	a
b	d	a	c	b
c	a	c	d	b
d	a	b	b	c

Table 1.3

from the square defining $b * a$. We obtain the rest of Table 1.3 in this fashion to give our solution. ▲

Some Words of Warning

Classroom experience shows the chaos that may result if a student is given a set and asked to define some binary operation on it. Remember that in an attempt to define a binary operation $*$ on a set S we must be sure that

1. *exactly one element is assigned to each possible ordered pair of elements of S,*
2. *for each ordered pair of elements of S, the element assigned to it is again in S.*

Regarding Condition 1, a student will often give a rule that assigns an element of S to "most" ordered pairs, but for a few pairs the rule determines no element. In this event, $*$ has **not been defined** on S. It may also happen that for some pairs, the rule could assign any of several elements of S, that is, there is ambiguity. In any case of ambiguity, $*$ is **not well defined**. If Condition 2 is violated, then S is **not closed under** $*$.

Here are several illustrations of attempts to define binary operations on sets. Some of them are worthless. The symbol $*$ is used for the attempted operation in all these examples.

EXAMPLE 11   On $\mathbb{Q}$, "define" $*$ by $a * b = a/b$. Here $*$ is *not defined* on $\mathbb{Q}$, for no rational number is assigned by this rule to the pair $(2, 0)$. ▲

EXAMPLE 12   On $\mathbb{Q}^+$, define $*$ by $a * b = a/b$. Here both Conditions 1 and 2 are satisfied, and $*$ is a binary operation on $\mathbb{Q}^+$. ▲

EXAMPLE 13   On $\mathbb{Z}^+$, "define" $*$ by $a * b = a/b$. Here Condition 2 is violated, for $1 * 3$ is not in $\mathbb{Z}^+$. Thus $*$ is not a binary operation on $\mathbb{Z}^+$, since $\mathbb{Z}^+$ is *not closed under* $*$.

EXAMPLE 14   Let F be the set of all real-valued functions with domain $\mathbb{R}$ as in Example 5. Suppose we "define" $*$ to give the usual quotient of f by g, that is, $f * g = h$, where $h(x) = f(x)/g(x)$. Here Condition 2 is violated, for the functions in F were to be defined for *all* real numbers, and for some $g \in F$, $g(x)$ will be zero for some values of x in $\mathbb{R}$ and $h(x)$ would not be defined at those numbers in $\mathbb{R}$. For example, if $f(x) = \cos x$ and $g(x) = x^2$, then $h(0)$ is undefined, so $h \notin F$. ▲

EXAMPLE 15   Let F be as in Example 14 and "define" $*$ by $f * g = h$, where h is the function greater than both f and g. This "definition" is completely worthless. In the first place, we have not defined what it means for one function to be greater than another. Even if we had, any sensible definition would result in there being many functions greater than both f and g, and $*$ would still be *not well defined*. ▲

EXAMPLE 16   Let S be a set consisting of twenty people, no two of whom are of the same height. Define $*$ by $a * b = c$, where c is the tallest person among the twenty in S. This is a perfectly good binary operation on the set, although not a particularly interesting one. ▲

EXAMPLE 17   Let S be as in Example 16 and "define" $*$ by $a * b = c$, where c is the shortest person in S who is taller than both a and b. This $*$ is *not defined*, since if either a or b is the tallest person in the set, $a * b$ is not determined. ▲

Exercises 1.1

Computations

Exercises 1 through 4 concern the binary operation $*$ defined on $S = \{a, b, c, d, e\}$ by means of Table 1.4.

1. Compute $b * d$, $c * c$, and $[(a * c) * e] * a$.

$*$	a	b	c	d	e
a	a	b	c	b	d
b	b	c	a	e	c
c	c	a	b	b	a
d	b	e	b	e	d
e	d	b	a	d	c

Table 1.4

*	a	b	c	d
a	a	b	c	
b	b	d		c
c	c	a	d	b
d	d			a

Table 1.5

2. Compute $(a * b) * c$ and $a * (b * c)$. Can you say on the basis of this computation whether $*$ is associative?

3. Compute $(b * d) * c$ and $b * (d * c)$. Can you say on the basis of this computation whether $*$ is associative?

4. Is $*$ commutative? Why?

5. Complete Table 1.5 so as to define a commutative binary operation $*$ on $S = \{a, b, c, d\}$.

6. Table 1.6 can be completed to define an associative binary operation $*$ on $S = \{a, b, c, d\}$. Assume this is possible and compute the missing entries.

In Exercises 7 through 11, determine whether the binary operator $*$ defined is commutative and whether $*$ is associative.

7. $*$ defined on $\mathbb{Z}$ by $a * b = a - b$

8. $*$ defined on $\mathbb{Q}$ by $a * b = ab + 1$

9. $*$ defined on $\mathbb{Q}$ by $a * b = ab/2$

10. $*$ defined on $\mathbb{Z}^+$ by $a * b = 2^{ab}$

11. $*$ defined on $\mathbb{Z}^+$ by $a * b = a^b$

12. Let S be a set having exactly one element. How many different binary operations

*	a	b	c	d
a	a	b	c	d
b	b	a	c	d
c	c	d	c	d
d				

Table 1.6

can be defined on S? Answer the question if S has exactly 2 elements; exactly 3 elements; exactly n elements.

13. How many different commutative binary operations can be defined on a set of 2 elements? on a set of 3 elements? on a set of n elements?

Concepts

In Exercises 14 through 19, determine whether the definition of $*$ does give a binary operation on the set. In the event that $*$ is not a binary operation, state whether Condition 1, Condition 2, or both of these conditions on page 37 are violated.

14. On $\mathbb{Z}^+$, define $*$ by $a * b = a - b$

15. On $\mathbb{Z}^+$, define $*$ by $a * b = a^b$.

16. On $\mathbb{R}$, define $*$ by $a * b = a - b$.

17. On $\mathbb{Z}^+$, define $*$ by $a * b = c$, where c is the smallest integer greater than both a and b.

18. On $\mathbb{Z}^+$, define $*$ by $a * b = c$, where c is at least 5 more than $a + b$.

19. On $\mathbb{Z}^+$, define $*$ by $a * b = c$, where c is the largest integer less than the product of a and b.

20 Mark each of the following true or false.

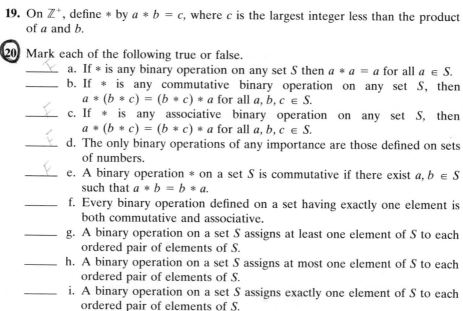

_____ a. If $*$ is any binary operation on any set S then $a * a = a$ for all $a \in S$.
_____ b. If $*$ is any commutative binary operation on any set S, then $a * (b * c) = (b * c) * a$ for all $a, b, c \in S$.
_____ c. If $*$ is any associative binary operation on any set S, then $a * (b * c) = (b * c) * a$ for all $a, b, c \in S$.
_____ d. The only binary operations of any importance are those defined on sets of numbers.
_____ e. A binary operation $*$ on a set S is commutative if there exist $a, b \in S$ such that $a * b = b * a$.
_____ f. Every binary operation defined on a set having exactly one element is both commutative and associative.
_____ g. A binary operation on a set S assigns at least one element of S to each ordered pair of elements of S.
_____ h. A binary operation on a set S assigns at most one element of S to each ordered pair of elements of S.
_____ i. A binary operation on a set S assigns exactly one element of S to each ordered pair of elements of S.
_____ j. A binary operation on a set S may assign more than one element of S to some ordered pair of elements of S.

21. Give a set different from any of those described in the examples of the text and not a set of numbers. Define two different binary operations $*$ and $*'$ on this set. Be sure that your set is _well defined_.

Theory

22. Prove that if $*$ is an associative and commutative binary operation on a set S, then

$$(a * b) * (c * d) = [(d * c) * a] * b$$

for all $a, b, c, d \in S$. Assume the associative law only for triples as in the definition, that is, assume only

$$(x * y) * z = x * (y * z)$$

for all $x, y, z \in S$.

In Exercises 23 and 24, either prove the statement or give a counterexample.

23. Every binary operation on a set consisting of a single element is both commutative and associative.

24. Every commutative binary operation on a set having just two elements is associative.

Let F be the set of all real-valued functions having as domain the set $\mathbb{R}$ of all real numbers. Example 5 defined the binary operations $+$, $-$, $\cdot$, and $\circ$ on F. In Exercises 25 through 31, either prove the given statement or give a counterexample.

25. Function addition $+$ on F is associative.

26. Function subtraction $-$ on F is commutative.

27. Function subtraction $-$ on F is associative.

28. Function multiplication $\cdot$ on F is commutative.

29. Function multiplication $\cdot$ on F is associative.

30. Function composition $\circ$ on F is commutative.

31. If $*$ and $*'$ are any two binary operations on a set S, then

$$a * (b *' c) = (a * b) *' (a * c) \qquad \text{for all} \qquad a, b, c \in S.$$

32. Observe that the binary operations $*$ and $*'$ on the set $\{a, b\}$ given by the tables

$*$	a	b
a	a	a
b	a	b

and

$*'$	a	b
a	a	b
b	b	b

provide the *same type of algebraic structure* on $\{a, b\}$, in the sense that if the table for $*'$ is rewritten as

$*'$	b	a
b	b	b
a	b	a

this table for $*'$ looks just like that for $*$, with the roles of a and b interchanged.

a. Try to give a natural definition of a concept of two binary operations $*$ and $*'$ on the same set giving *algebraic structures of the same type*, which generalizes this observation.

b. How many different types of algebraic structures are given by the 16 possible different binary operations on a set of 2 elements?

Finite-State Machines (Automata)

In the URI Mathematics Building, there is a candy and snack vending machine with its wares and their prices temptingly displayed. The user can *input*, by depositing in a slot or pressing a button, the following things:

$$\text{dollar, quarter, dime, nickel, A, B, C, D, E, F,}$$
$$\text{G, H, I, J, 0, 1, 2, 3, 4, 5, 6, 7, 8, 9, }*\text{, \#.}$$

(1)

The machine can *output* a candy bar, a snack, change, and a display telling the amount of money that the user has deposited. If we deposit a total of 55 cents and then press E and press 9, the machine will (usually) output one of the best-selling candy bars. On the other hand, if we deposit 70 cents and press F and press 2, we get a bag of chocolate-chip cookies as output.

The vending machine just described is an example of a *finite-state machine* or *automaton*. Such a machine can be in a finite number $n + 1$ of *internal states* denoted by $s_0, s_1, s_2, \ldots, s_n$, some of which may trigger an output. There is a finite *alphabet* $x_1, x_2, \ldots, x_r$ like the list (1) that can be used to provide input. If the machine is currently in state s_i, input of an x_p may cause it to change to a different state s_j, or may cause the machine to remain in the same state s_i. We denote these possibilities schematically as shown in Fig. 1.1.

A finite sequence of inputs, say x_1 followed by x_3 followed by x_4, is called an **input string** and is denoted by $x_1 x_3 x_4$. Note that we read an input string from left to right. If after inputting $x_1 x_3 x_4$ we input the string $x_5 x_9$, the result is the same as though we input the longer string $x_1 x_3 x_4 x_5 x_9$, which we call the **concatenation of $x_1 x_3 x_4$ followed by** $x_5 x_9$. Thus concatenation is a binary operation on the set X^+ of all finite-length input strings. It is convenient to include the *empty input string*, often denoted by " " in computer languages, in the set X^+ of possible input strings. We will denote the empty string by ϵ. For every string $\mathbf{x}$ in X^+, we define its concatenation with ϵ by $\mathbf{x}\epsilon = \epsilon\mathbf{x} = \mathbf{x}$.

(a)   (b)

Figure 1.1   (a) State changed. (b) State retained.

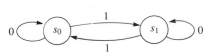

Figure 1.2 A parity-check machine.

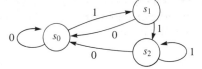

Figure 1.3 Check for a string ending with 11.

Note that concatenation $*$ is not a commutative operation; for example,

$$x_1x_3x_4x_5x_9 = (x_1x_3x_4) * (x_5x_9) \neq (x_5x_9) * (x_1x_3x_4) = x_5x_9x_1x_3x_4.$$

We are asked to show that concatenation is associative in Exercise 34.

A computer is an example of a finite-state machine. We can consider the input alphabet as just consisting of 0 and 1, since the computer regards all data entered as a sequence of 0's and 1's. When storing such a sequence, the computer always appends either 0 or 1 to the end of the sequence, according to whether the number of 1's in the original sequence was even or odd. That is, the computer appends 0 or 1 so that the number of 1's in the augmented sequence is even. For example, it stores 11001111 as 110011110 but stores 10010111 as 100101111. When the computer retrieves this data, it checks that the number of 1's is even. If the number of 1's is odd, the computer may reread that data, and give a PARITY CHECK ERROR message if the number of 1's is still odd. We can diagram such a parity check as a finite-state machine with two states, s_0 denoting an even number of 1's, and s_1 denoting an odd number of 1's. At the start of the check, the number of 1's is zero so the machine is in state s_0 since 0 is an even number. As each successive binary digit 0 or 1 in the sequence is read, the state changes with the input of 0 or 1 as shown in Fig. 1.2. Such diagrams are called *state diagrams*.

Figure 1.3 shows a machine with three states. We can check that with the machine starting in state s_0, the machine will finish in state s_0 if the last two characters of the string input are 00 or 10, will finish in state s_1 if the string ends with 01, and will finish in state s_2 if the string ends with 11.

33. For the vending machine described, we expect the same output for both of these columns of actions.

deposit 25 cents	deposit 25 cents
deposit 5 cents	deposit 5 cents
talk to a friend	deposit 25 cents
deposit 25 cents	press E
press E	talk to a friend
press 9	press 9

Explain how this fact can be regarded as illustrating the associativity of concatenation of input strings.

34. Prove that concatenation of input strings for a finite-state machine is an associative operation.

35. Modify Fig. 1.3 to show a machine with three states, s_0, s_1, s_2, and input alphabet a, b, c such that starting in state s_0, the machine will finish in state s_2 if and only if the input string ends with bb.

36. Give a state diagram for a machine with input alphabet a, b, c that can be used to determine if the input string contains at least two c's.

37. Give a state diagram for a machine with input alphabet a, b, c that can be used to determine if the input string contains exactly three c's.

38. Give a state diagram for a machine with input alphabet 0, 1 that can be used to determine whether the number of 1's in the input string is congruent to 0, 1, or 2 modulo 3.

The change from state s_i to state s_j, or retention in the same state s_i, caused by an alphabet input x_p, can be shown by a table. Such an input table *is shown below for the parity-check machine in Fig.* 1.2.

Present state	0	1 (inputs)
s_0	s_0	s_1
s_1	s_1	s_0

39. Give the input table for the machine in Fig. 1.3.

40. Give the input table for the machine you constructed in Exercise 36.

1.2

Groups

Let us continue the analysis of our past experience with algebra. Once we had mastered the computational problems of addition and multiplication of numbers, we were ready to apply these binary operations to the solution of problems. Often problems lead to equations involving some unknown number x, which is to be determined. The simplest equations are the linear ones of the forms $a + x = b$ for the operation of addition, and $ax = b$ for multiplication. The additive linear equation always has a numerical solution, and so has the multiplicative one, provided $a \neq 0$. Indeed, the need for solutions of additive linear equations such as $5 + x = 2$ is a very good motivation for the negative numbers. Similarly, the need for rational numbers is shown by equations such as $2x = 3$, and the need for the complex number i is shown by the equation $x^2 = -1$.

It is desirable for us to be able to solve linear equations involving our binary operations. This is not possible for every binary operation, however. For example, the equation $a * x = a$ has no solution in $S = \{a, b, c\}$ for the operation $*$ of Example 9 of Section 1.1. Let us abstract from familiar algebra those properties of addition that enable us to solve the equation $5 + x = 2$ in $\mathbb{Z}$. We must not refer to subtraction, for we are concerned with the solution phrased in terms of a single binary operation, in this case addition. The steps in the solution are as follows:

$$5 + x = 2, \qquad \text{given,}$$
$$-5 + (5 + x) = -5 + 2, \qquad \text{adding } -5,$$
$$(-5 + 5) + x = -5 + 2, \qquad \text{associative law,}$$
$$0 + x = -5 + 2, \qquad \text{computing } -5 + 5,$$
$$x = -5 + 2, \qquad \text{property of 0,}$$
$$x = -3, \qquad \text{computing } -5 + 2.$$

Strictly speaking, we have not shown here that -3 is a solution, but rather that it is the only possibility for a solution. To show that -3 is a solution, one merely computes $5 + (-3)$. A similar analysis could be made for the equation $2x = 3$ in the rational numbers with the operation of multiplication:

$$2x = 3, \qquad \text{given,}$$
$$\tfrac{1}{2}(2x) = \tfrac{1}{2}(3), \qquad \text{multiplying by } \tfrac{1}{2},$$
$$(\tfrac{1}{2} \cdot 2)x = \tfrac{1}{2}3, \qquad \text{associative law,}$$
$$1 \cdot x = \tfrac{1}{2}3, \qquad \text{computing } \tfrac{1}{2}2,$$
$$x = \tfrac{1}{2}3, \qquad \text{property of 1,}$$
$$x = \tfrac{3}{2}, \qquad \text{computing } \tfrac{1}{2}3.$$

We can now see what properties a set S and a binary operation $*$ on S would have to have to permit imitation of this procedure for an equation $a * x = b$ for $a, b \in S$. Basic to the procedure is the existence of an element e in S with the property that $e * x = x$ for all $x \in S$. For our additive example, 0 played the role of e, and 1 played the role for our multiplicative example. Then we need an element a' in S that has the property that $a' * a = e$. For our additive example, -5 played the role of a', and $\frac{1}{2}$ played the role for our multiplicative example. Finally we need the associative law. The remainder is just computation. A similar analysis shows that in order to solve the equation $x * a = b$ (remember that $a * x$ need not equal $x * a$) we would like to have an element e in S such that $x * e = x$ for all $x \in S$ and an a' in S such that $a * a' = e$. With all of these properties of $*$ on S, we could be sure of being able to solve linear equations. These are precisely the properties of a *group*.

Definition and Examples

With the preceding as motivation, we now define a *group*, which is one of the simplest but most basic and most widely studied algebraic structures.

DEFINITION 1.4 (Group)   A **group** $\langle G, * \rangle$ is a set G, closed under a binary operation $*$, such that the following axioms are satisfied:

◆   **H I S T O R I C A L   N O T E**   ◆

There are three historical roots of the development of abstract group theory evident in the mathematical literature of the nineteenth century: the theory of algebraic equations, number theory, and geometry. All three of these areas used group-theoretic methods of reasoning, though the methods were considerably more explicit in the first area than in the other two.

One of the central themes of geometry in the nineteenth century was the search for invariants under various types of geometric transformations. Gradually attention became focused on the transformations themselves, which in many cases can be thought of as elements of groups.

In number theory, already in the eighteenth century Leonhard Euler had considered the remainders on division of powers a^n by a fixed prime p. These remainders have "group" properties. Similarly, Carl F. Gauss in his *Disquisitiones Arithmeticae* (1800) dealt extensively with quadratic forms $ax^2 + 2bxy + cy^2$ and in particular showed that equivalence classes of these forms under composition possessed what amounted to group properties.

Finally, the theory of algebraic equations provided the most explicit prefiguring of the group concept. Joseph-Louis Lagrange (1736–1813) in fact initiated the study of permutations of the roots of an equation as a tool for solving it. These permutations, of course, were ultimately considered as elements of a group.

It was Walter von Dyck (1856–1934) and Heinrich Weber (1842–1913) who in 1882 were able independently to combine the three historical roots and give clear definitions of the notion of an abstract group.

$\mathcal{G}_1$: The binary operation $*$ is associative.

$\mathcal{G}_2$: There is an element e in G such that $e * x = x * e = x$ for all $x \in G$. (This element e is an **identity element** for $*$ on G.)†

$\mathcal{G}_3$: For each a in G, there is an element a' in G with the property that $a' * a = a * a' = e$. (The element a' is an **inverse of a with respect to the operation** $*$.)

We should point out right now that we are going to be sloppy in notation. Observe that a group is not just a set G. Rather, a group $\langle G, * \rangle$ is made up of two entities, the set G and the binary operation $*$ on G. There are *two*

---

† Remember that boldface type indicates that a term is being defined. See Section 0.1. Thus an **identity element** for a binary operation $*$ on a set S is any element e satisfying $e * x = x * e = x$ for all $x \in S$.

ingredients involved. Denoting the group by the single set symbol G is logically incorrect. Nevertheless, as you get further into the theory, the logical extensions of the notation $\langle G, * \rangle$ become so unwieldy as to actually make the exposition hard to read. At some point, all authors give up and become sloppy, denoting the group by the single letter G. We choose to recognize this and be sloppy from the start. We emphasize, however, that when you are speaking of a specific group G, you must make it clear what the group operation on G is to be, since a set could conceivably have a variety of binary operations defined on it, all giving different groups. We shall sometimes resort to the notation $\langle G, * \rangle$ for reasons of clarity in our discussions.

DEFINITION 1.5 (Abelian Group)   A group G is **abelian** if its binary operation $*$ is commutative.

Let us give some examples of some sets with binary operations that give groups and also of some that do not give groups.

◆   **H I S T O R I C A L   N O T E**   ◆

Commutative groups are called abelian in honor of the Norwegian mathematician Niels Henrik Abel (1802–1829). Abel was interested in the question of solvability of polynomial equations. In a paper written in 1828 he proved that if all the roots of such an equation can be expressed as rational functions $f, g, \ldots, h$ of one of them, say x, and if for any two of these roots, $f(x)$ and $g(x)$, the relation $f(g(x)) = g(f(x))$ always holds, then the equation is solvable by radicals. Abel showed that each of these functions in fact permutes the roots of the equation; hence these functions are elements of the group of permutations of the roots. It was this property of commutativity in these permutation groups associated with solvable equations that led Camille Jordan in his 1870 treatise on algebra to name such groups abelian; the name since then has been applied to commutative groups in general.

Abel was attracted to mathematics as a teenager and soon surpassed all his teachers in Norway. He finally received a government travel grant to study elsewhere in 1825 and proceeded to Berlin where he befriended August Crelle, the founder of the most influential German mathematical journal. Abel contributed numerous papers to Crelle's Journal during the next several years, including many in the field of elliptic functions, whose theory he created virtually single-handedly. Abel returned to Norway in 1827 with no position and an abundance of debts. He nevertheless continued to write brilliant papers, but died of tuberculosis at the age of 26, two days before Crelle succeeded in finding a university position for him in Berlin.

EXAMPLE I   The Set $\mathbb{Z}^+$ with operation $+$ is *not* a group. There is no identity element for $+$ in $\mathbb{Z}^+$. ▲

EXAMPLE 2   The set of all nonnegative integers (including 0) with operation $+$ is still *not* a group. There is an identity element 0, but no inverse for 2. ▲

EXAMPLE 3   The familiar additive properties of integers and of rational, real, and complex numbers show that $\mathbb{Z}$, $\mathbb{Q}$, $\mathbb{R}$, and $\mathbb{C}$ with the operation $+$ are abelian groups. ▲

EXAMPLE 4   The set $\mathbb{Z}^+$ with operation multiplication is *not* a group. There is an identity 1, but no inverse of 3. ▲

EXAMPLE 5   The familiar multiplicative properties of rational, real, and complex numbers show that the sets $\mathbb{Q}^+$ and $\mathbb{R}^+$ of positive numbers and the sets $\mathbb{Q}^*$, $\mathbb{R}^*$, and $\mathbb{C}^*$ of nonzero numbers with the operation multiplication are abelian groups. ▲

EXAMPLE 6   The set of all real-valued functions with domain $\mathbb{R}$ with operation function addition is a group. This group is abelian. ▲

EXAMPLE 7   (Linear Algebra) Those who have studied vector spaces should note that the axioms for a vector space V pertaining just to vector addition can be summarized by asserting that V with the operation of vector addition is an abelian group. ▲

EXAMPLE 8   The set $M_{m \times n}(\mathbb{R})$ of all $m \times n$ matrices with operation matrix addition is a group. The $m \times n$ matrix with all entries 0 is the identity matrix. This group is abelian. ▲

EXAMPLE 9   The set $M_n(\mathbb{R})$ of all $n \times n$ matrices with operation matrix multiplication is not a group. The $n \times n$ matrix with all entries 0 has no inverse.

EXAMPLE 10   Show that the subset S of $M_n(\mathbb{R})$ consisting of all *invertible* $n \times n$ matrices with operation matrix multiplication is a group.

Solution   We start by showing that S is closed under matrix multiplication. In keeping with the notation given in Section 0.4, we use A^{-1} rather than A'

for the multiplicative inverse of an invertible matrix A. Let A and B be in S, so that both A^{-1} and B^{-1} exist and $AA^{-1} = BB^{-1} = I_n$. Then

$$(AB)(B^{-1}A^{-1}) = A(BB^{-1})A^{-1} = AI_nA^{-1} = I_n,$$

so that AB is invertible and consequently is also in S.

Since matrix multiplication is associative and I_n acts as the identity, and since each element of S has an inverse by definition of S, we see that S is indeed a group. This group is *not* commutative. It is our first example of a *nonabelian group*. ▲

In place of the phrase *with the binary operation of* we often use the word *under*, so that "the group $\mathbb{R}$ with the binary operation of addition" becomes "the group $\mathbb{R}$ under addition."

The group of invertible $n \times n$ matrices described in the preceding example is of fundamental importance in linear algebra. It is the **general linear group of degree** n, and is usually denoted by $GL(n, \mathbb{R})$. Those who have studied linear algebra know that a matrix A in $GL(n, \mathbb{R})$ gives rise to an invertible linear transformation $T:\mathbb{R}^n \to \mathbb{R}^n$, defined by $T(\mathbf{x}) = A\mathbf{x}$, and that conversely, every invertible linear transformation of $\mathbb{R}^n$ into itself is defined in this fashion by some matrix in $GL(n, \mathbb{R})$. Also, matrix multiplication corresponds to composition of linear transformations. Thus all invertible transformations of $\mathbb{R}^n$ into itself form a group under function composition; this group is usually denoted by $GL(\mathbb{R}^n)$.

Thus far our examples of groups all have as binary operation the familiar addition or multiplication of numbers or matrices. We should give at least one example where the operation is less familiar, and check the axioms.

EXAMPLE 11 Let $*$ be defined on $\mathbb{Q}^+$ by $a * b = ab/2$. Then

$$(a * b) * c = \frac{ab}{2} * c = \frac{abc}{4},$$

and likewise

$$a * (b * c) = a * \frac{bc}{2} = \frac{abc}{4}.$$

Thus $*$ is associative. Computation shows that

$$2 * a = a * 2 = a$$

for all $a \equiv \mathbb{Q}^+$, so 2 is an identity element for $*$. Finally,

$$a * \frac{4}{a} = \frac{4}{a} * a = 2,$$

so $a' = 4/a$ is an inverse for a. Hence $\mathbb{Q}^+$ with the operation $*$ is a group. ▲

Elementary Properties of Groups

As we proceed to prove our first theorem about groups, we must use Definition 1.4, which is the only thing we know about groups at the moment. The proof of a second theorem can employ both Definition 1.4 and Theorem 1.1; the proof of a third theorem can use the definition and the first two theorems, etc.

Our first theorem will establish cancellation laws. In real arithmetic we know that $2a = 2b$ implies that $a = b$. We need only divide both sides of the equation $2a = 2b$ by 2, or equivalently, multiply both sides by $1/2$, which is the multiplicative inverse of 2. We parrot this proof to establish cancellation laws for any group. Note that we will also use the associative law.

THEOREM 1.1   If G is a group with binary operation $*$, then the **left and right cancellation laws** hold in G, that is, $a * b = a * c$ implies $b = c$, and $b * a = c * a$ implies $b = c$ for $a, b, c \in G$.

PROOF   Suppose $a * b = a * c$. Then by $\mathcal{G}_3$, there exists a', and

$$a' * (a * b) = a' * (a * c).$$

By the associative law,
$$(a' * a) * b = (a' * a) * c.$$

By the definition of a' in $\mathcal{G}_3$, $a' * a = e$, so

$$e * b = e * c.$$

By the definition of e in $\mathcal{G}_2$,
$$b = c.$$

Similarly, from $b * a = c * a$ one can deduce that $b = c$ upon multiplication on the right by a' and use of the axioms for a group. ◆

Our next proof can make use of Theorem 1.1. We show that a "linear equation" in a group has a *unique* solution. Recall that we chose our group properties to allow us to find solutions of such equations.

THEOREM 1.2   If G is a group with binary operation $*$, and if a and b are any elements of G, then the linear equations $a * x = b$ and $y * a = b$ have unique solutions in G.

PROOF   First we show the existence of *at least* one solution by just computing that $a' * b$ is a solution of $a * x = b$. Note that

$$a * (a' * b) = (a * a') * b, \qquad \text{associative law,}$$
$$= e * b, \qquad \text{definition of } a',$$
$$= b, \qquad \text{property of } e.$$

Thus $x = a' * b$ is a solution of $a * x = b$. In a similar fashion, $y = b * a'$ is a solution of $y * a = b$.

To show uniqueness of y, we use the standard method of assuming that we have two solutions, y_1 and y_2, so that $y_1 * a = b$ and $y_2 * a = b$. Then $y_1 * a = y_2 * a$, and by Theorem 1.1, $y_1 = y_2$. The uniqueness of x follows similarly. ▲

Of course, to prove the uniqueness in the last theorem we could have followed the procedure we used in motivating the definition of a group, showing that if $a * x = b$, then $x = a' * b$. However, we chose to illustrate the standard way to prove an object is unique given in Section 0.1; namely, suppose you have two such objects, and then prove they must be the same. Note that the solutions $x = a' * b$ and $y = b * a'$ need not be the same unless $*$ is commutative.

There is one other result about groups we would like to prove in this section.

THEOREM 1.3   In a group G with binary operation $*$, there is only one identity e such that

$$e * x = x * e = x$$

for all $x \in G$. Likewise for each $a \in G$, there is only one element a' such that

$$a' * a = a * a' = e.$$

In summary, the identity and inverses are unique in a group.

PROOF   To show the identity is unique, we proceed in the standard way (see Section 0.1) and suppose we have two elements e and e_1 in G such that $e * x = x * e = x$ and $e_1 * x = x * e_1 = x$ for all x in G. We want to show that $e_1 = e$. This is easy to do if we spot the right method (see Section 0.1) and consider the product $e * e_1$. We let e and e_1 compete. Now regarding e as identity, $e * e_1 = e_1$. But regarding e_1 as identity, $e * e_1 = e$. Thus

$$e_1 = e * e_1 = e,$$

and the identity of a group is unique.

Turning to the uniqueness of an inverse, suppose that $a \in G$ has inverses a' and a'' so that $a' * a = a * a' = e$ and $a'' * a = a * a'' = e$. Then

$$a * a'' = a * a' = e$$

and, by Theorem 1.1,

$$a'' = a',$$

so the inverse of a in a group is unique. ◆

Note that in a group G, we have

$$(a * b) * (b' * a') = a * (b * b') * a' = (a * e) * a' - a * a' = e.$$

This equation and Theorem 1.3 show that $b' * a'$ is the unique inverse of $a * b'$. That is, $(a * b)' = b' * a'$. We state this as a corollary.

COROLLARY   Let G be a group. For all $a, b \in G$, we have $(a * b)' = b' * a'$. ▲

For your information, we remark that algebraic structures consisting of sets with binary operations for which not all of the group axioms hold have also been studied quite extensively. Of these weaker structures, the **semigroup**, a set with an associative binary operation, has perhaps had the most attention. A **monoid** is a semigroup that has an identity element for the binary operation.

Finally, it is possible to give axioms for a group $\langle G, * \rangle$ that seem at first glance to be weaker, namely:

1. The binary operation $*$ on G, is associative.
2. There exists a **left identity** e in G such that $e * x = x$ for all $x \in G$.
3. For each $a \in G$, there exists a **left inverse** a' in G such that $a' * a = e$.

From this *one-sided definition*, one can prove that the left identity is also a right identity and a left inverse is also a right inverse for the same element. Thus these axioms should not be called weaker, since they result in exactly the same structures being called groups. It is conceivable that it might be easier in some cases to check these *left axioms* than to check our *two-sided axioms*. Of course, by symmetry it is clear that there are also *right axioms* for a group.

Finite Groups and Group Tables

Thus far all our examples have been of infinite groups, that is, groups where the set G has an infinite number of elements. You may wonder whether there can be a group structure on some finite set. The answer is yes, and indeed such structures are very important.

Since a group has to have at least one element, namely the identity, a minimal set that might give rise to a group is a one-element set $\{e\}$. The only possible binary operation $*$ on $\{e\}$ is defined by $e * e = e$. The three group axioms hold. The identity element is always its own inverse in every group.

Let us try to put a group structure on a set of two elements. Since one of the elements must play the role of identity element, we may as well let the set be $\{e, a\}$. Let us attempt to find a table for a binary operation $*$ on $\{e, a\}$ that gives a group structure on $\{e, a\}$. When giving a table for a group operation, we shall always list the elements in the same order across the top as down the left side, with the identity listed first, as in the following table.

$*$	e	a
e		
a		

Since e is to be the identity, so

$$e * x = x * e = x$$

for all $x \in \{e, a\}$, we are forced to fill in the table as follows, if $*$ is to give a group:

$*$	e	a
e	e	a
a	a	

Also, a must have an inverse a' such that

$$a * a' = a' * a = e.$$

In our case, a' must be either e or a. Since $a' = e$ obviously does not work, we must have $a' = a$, so we have to complete the table as follows:

$*$	e	a
e	e	a
a	a	e

All the group axioms are now satisfied except possibly the associative property. We will show an easy way to verify associativity in Example 12. Checking associativity on a case-by-case basis from a table defining an operation can be a very tedious process.

With these examples as background, we should be able to list some necessary conditions that a table giving a binary operation on a finite set must satisfy for the operation to give a group structure on the set. There must be one element of the set, which we may as well denote by e, that acts as identity. The condition $e * x = x$ means that the row of the table opposite e at the extreme left must contain exactly the elements appearing across the very top of the table in the same order. Similarly, the condition $x * e = x$ means that the column of the table under e at the very top must contain exactly the elements appearing at the extreme left in the same order. The fact that every element a has a right and a left inverse means that in the row opposite a at the extreme left, the element e must appear, and in the column under a at the very top, the e must appear. Thus e must appear in each row and in each column. We can do even better than this, however. By Theorem 1.2, not only the equations $a * x = e$ and $y * a = e$ have unique solutions, but also the equations $a * x = b$ and $y * a = b$. By a similar argument, this means that *each element b of the group must appear once and only once in each row and column of the table.*

Suppose conversely that a table for a binary operation on a finite set is such that there is an element acting as identity and that in each row and each column each element of the set appears exactly once. Then it can be seen that the structure is a group structure if and only if the associative law holds. If a binary operation $*$ is given by a table, the associative law is usually messy to check. If the operation $*$ is defined by some characterizing property of $a * b$, the associative law is often easy to check. Fortunately this second case turns out to be the one usually encountered.

We saw that there was essentially only one group of two elements in the sense that if the elements are denoted by e and a with the identity e

*	e	a
e	e	a
a	a	e

Table 1.7

appearing first, the table must be as shown in Table 1.7. Suppose that a set has three elements. As before, we may as well let the set be $\{e, a, b\}$. For e to be an identity, a binary operation $*$ on this set has to have a table of the form shown in Table 1.8. This leaves four places to be filled in. You can quickly see that Table 1.8 must be completed as shown in Table 1.9 if each row and each column are to contain each element exactly once. Again we will see an easy way to verify associativity in Example 12, so $*$ does give a group structure on $G = \{e, a, b\}$.

Now suppose that G' is any other group of three elements and imagine a table for G' with identity element appearing first. Since our filling out of the table for $G = \{e, a, b\}$ could be done in only one way, we see that if we take the table for G' and rename the identity e, the next element listed a, and the last element b, the resulting table for G' must be the same as the one we had for G. In other words, the *structural* features are the same for the two groups, and one group can be made to look exactly like the other by a renaming of the elements. *Thus any two groups of three elements are structurally the same.* This is our introduction to the concept of *isomorphism*. (See also Exercise 47 in Section 0.4.) The groups G and G' are *isomorphic*. This concept is sometimes a bit sticky. We will make it more precise later, but we will continue to use the term isomorphism informally so that you will develop a feeling for the idea from the context in which it is used.

EXAMPLE 12   It is easy to show that for each $n \in \mathbb{Z}^+$, the n solutions in $\mathbb{C}$ of the equation $x^n = 1$ form a multiplicative group U_n. (See Exercise 27.) Thus

$$U_1 = \{1\}, \qquad U_2 = \{-1, 1\}, \qquad U_3 = \left\{1, -\frac{1}{2} + \frac{\sqrt{3}}{2}i, -\frac{1}{2} - \frac{\sqrt{3}}{2}i\right\},$$

and

$$U_4 = \{1, i, -1, -i\}$$

*	e	a	b
e	e	a	b
a	a		
b	b		

Table 1.8

*	e	a	b
e	e	a	b
a	a	b	e
b	b	e	a

Table 1.9

are abelian groups under complex multiplication. As we showed before, the group U_2 must be structurally the same as (isomorphic to) the group $\{e, a\}$ with Table 1.7. Since we know that the multiplication operation in U_2 is associative, we conclude at once that the operation given by that two-element table is associative. Similarly, the group U_3 is isomorphic to the structure given by Table 1.9, showing that the operation in Table 1.9 is associative. The group U_1 is isomorphic to the group $\{e\}$. Exercise 17 will show that U_4 is one of two possible 4-element group structures. The group U_n is the multiplicative group of nth *roots of unity.* ▲

Exercises 1.2

Computation

In Exercises 1 through 6, determine whether the binary operation $*$ gives a group structure on the given set. If no group results, give the first axiom in the order $\mathcal{G}_1, \mathcal{G}_2, \mathcal{G}_3$ from Definition 1.4 that does not hold.

1. Let $*$ be defined on $\mathbb{Z}$ by $a * b = ab$.

2. Let $*$ be defined on $2\mathbb{Z} = \{2n \mid n \in \mathbb{Z}\}$ by $a * b = a + b$.

3. Let $*$ be defined on $\mathbb{R}^+$ by $a * b = \sqrt{ab}$.

4. Let $*$ be defined on $\mathbb{Q}$ by $a * b = ab$.

5. Let $*$ be defined on the set $\mathbb{R}^*$ of nonzero real numbers by $a * b = a/b$.

6. Let $*$ be defined on $\mathbb{C}$ by $a * b = |ab|$.

7. Show that the set of all complex numbers of norm 1 with the operation of multiplication is a group. (Recall that such a number can be expressed in the form $\cos \theta + i \sin \theta$.)

In Exercises 8 through 15, determine whether the given set of matrices under the specified operation, matrix addition or multiplication, is a group. Recall that a **diagonal matrix** is a square matrix whose only nonzero entries lie on the **main diagonal**, from the upper left to the lower right corner. An **upper-triangular matrix** is a square matrix with only zero entries below the main diagonal. Associated with each $n \times n$ matrix A is a number called the determinant of A, denoted by $\det(A)$. If A and B are both $n \times n$ matrices, then $\det(AB) = \det(A) \det(B)$. Also, $\det(I_n) = 1$ and A is invertible if and only if $\det(A) \neq 0$.

8. All $n \times n$ diagonal matrices under matrix addition.

9. All $n \times n$ diagonal matrices under matrix multiplication.

10. All $n \times n$ diagonal matrices with no zero diagonal entry under matrix multiplication.

11. All $n \times n$ diagonal matrices with all diagonal entries 1 or -1 under matrix multiplication.

12. All $n \times n$ upper-triangular matrices under matrix multiplication.

*	e	a	b	c
e	e	a	b	c
a	a	?		
b	b			
c	c			

Table 1.10

13. All $n \times n$ upper-triangular matrices under matrix addition.

14. All $n \times n$ upper-triangular matrices with determinant 1 under matrix multiplication.

15. All $n \times n$ matrices with determinant either 1 or -1 under matrix multiplication.

16. Let S be the set of all real numbers except -1. Define $*$ on S by

$$a * b = a + b + ab.$$

a. Show that $*$ gives a binary operation on S.

b. Show that $\langle S, * \rangle$ is a group.

c. Find the solution of the equation $2 * x * 3 = 7$ in S.

17. This exercise shows that there are two possible different group structures on a set of 4 elements.

   Let the set be $\{e, a, b, c\}$, with e the identity element for the group operation. A group table would then have to start in the manner shown in Table 1.10. The square indicated by the question mark can't be filled in with a. It must be filled in either with the identity e or with an element different from both e and a. In this latter case, it is no loss of generality to assume that this element is b. If this square is filled in with e, the table can then be completed in two ways to give a group. Find these two tables. (You need not check the associative law.) If this square is filled in with b, then the table can only be completed in one way to give a group. Find this table. (Again you need not check the associative law.) Of the three tables you now have, two give the same type of group structure. Determine which two tables these are, and show how the elements in one table would have to be renamed for these two tables to be the same.

a. Are all groups of 4 elements commutative?

b. Which table gives a group isomorphic to the group U_4 of Example 12, so that we know the binary operation defined by the table is associative?

c. Show that the group given by one of the other tables is structurally the same as the group in Exercise 11 for one particular value of n, so that we know that the operation defined by that table is associative also.

18. According to Exercise 12 of Section 1.1, there are 16 possible binary operations on a set of 2 elements. How many of these give a structure of a group? How many of the 19,683 possible binary operations on a set of 3 elements give a group structure?

Concepts

19. Consider our axioms $\mathcal{G}_1$, $\mathcal{G}_2$, and $\mathcal{G}_3$ for a group. We gave them in the order $\mathcal{G}_1 \mathcal{G}_2 \mathcal{G}_3$. Conceivable other orders to state the axioms are $\mathcal{G}_1 \mathcal{G}_3 \mathcal{G}_2$, $\mathcal{G}_2 \mathcal{G}_1 \mathcal{G}_3$, $\mathcal{G}_2 \mathcal{G}_3 \mathcal{G}_1$, $\mathcal{G}_3 \mathcal{G}_1 \mathcal{G}_2$, and $\mathcal{G}_3 \mathcal{G}_2 \mathcal{G}_1$. Of these six possible orders, exactly three are acceptable for a definition. Which orders aren't acceptable, and why? (Remember this. Most instructors ask the student to define a group on at least one test.)

20. The following "definitions" of a group are taken verbatim, including spelling and punctuation, from papers of students who wrote a bit too quickly and carelessly. Criticize them.

    a. A group G is a set of elements together with a binary operation $*$ such that the following conditions are satisfied

    $*$ is associative

    There exists $e \in G$ such that
    $$e * x = x * e = x = \text{identity.}$$
    For every $a \in G$ there exists an a' (inverse) such that
    $$a \cdot a' = a' \cdot a = e$$

    b. A group is a set G such that

    The operation on G is associative.

    there is an identity element (e) in G.

    for every $a \in G$, there is an a' (inverse for each element)

    c. A group is a set with a binary operation such

    the binary operation is defined

    an inverse exists

    an identity element exists

    d. A set G is called a group over the binery operation $*$ such that for all $a, b \in G$

    Binary operation $*$ is associative under addition

    there exist an element $\{e\}$ such that
    $$a * e = e * a = e$$
    Fore every element a there exists an element a' such that
    $$a * a' = a' * a = e$$

21. Give a table for a binary operation on the set $\{e, a, b\}$ of three elements satisfying axioms $\mathcal{G}_2$ and $\mathcal{G}_3$ for a group but not axiom $\mathcal{G}_1$.

22. Mark each of the following true or false.

    _____ a. A group may have more than one identity element.

    _____ b. Any two groups of three elements are isomorphic.

    _____ c. In a group, each linear equation has a solution.

    _____ d. The proper attitude toward a definition is to memorize it so that you can reproduce it word for word as in the text.

    _____ e. Any definition a person gives for a group is correct provided that everything that is a group by that person's definition is also a group by the definition in the text.

    _____ f. Any definition a person gives for a group is correct provided he or she

can show that everything that satisfies the definition satisfies the one in the text and conversely.

_____ g. Every finite group of at most three elements is abelian.

_____ h. An equation of the form $a * x * b = c$ always has a unique solution in a group.

_____ i. The empty set can be considered a group.

_____ j. The text has as yet given no examples of groups that are not abelian.

Theory

23. Show that if G is a finite group with identity e and with an even number of elements, then there is $a \neq e$ in G such that $a * a = e$.

24. Let $\mathbb{R}^*$ be the set of all real numbers except 0. Define $*$ on $\mathbb{R}^*$ by $a * b = |a| b$.

    a. Show that $*$ gives an associative binary operation on $\mathbb{R}^*$.

    b. Show that there is a left identity for $*$ and a right inverse for each element in $\mathbb{R}^*$.

    c. Is $\mathbb{R}^*$ with this binary operation a group?

    d. Explain the significance of this exercise.

25. If $*$ is a binary operation on a set S, an element x of S is an **idempotent for** $*$ if $x * x = x$. Prove that a group has exactly one idempotent element. (You may use any theorems proved so far in the text.)

26. Show that every group G with identity e and such that $x * x = e$ for all $x \in G$ is abelian. [*Hint*: Consider $(a * b) * (a * b)$.]

27. Show that the nth roots of unity in $\mathbb{C}$ do form a multiplicative group, as asserted in Example 12.

28. Let G be an abelian group and let $c^n = c * c * \cdots * c$ for n factors c, where $c \in G$ and $n \in \mathbb{Z}^+$. Give a mathematical induction proof that $(a * b)^n = (a^n) * (b^n)$ for all $a, b \in G$.

29. Let G be a group with a finite number of elements. Show that for any $a \in G$, there exists an $n \in \mathbb{Z}^+$ such that $a^n = e$. See Exercise 28 for the meaning of a^n. [*Hint*: Consider $e, a, a^2, a^3, \ldots, a^m$, where m is the number of elements in G, and use the cancellation laws.]

30. Show that if $(a * b)^2 = a^2 * b^2$ for a and b in a group G, then $a * b = b * a$. See Exercise 28 for the meaning of a^2.

31. Let G be a group and let $a, b \in G$. Show that $(a * b)' = a' * b'$ if and only if $a * b = b * a$.

32. Let G be a group and suppose that $a * b * c = e$ for $a, b, c \in G$. Show that $b * c * a = e$ also.

33. Prove that a set G, together with a binary operation $*$ on G satisfying the left axioms 1, 2, and 3 given on page 52, is a group.

34. Prove that a nonempty set G, together with an associative binary operation $*$ on G such that

$$a * x = b \quad \text{and} \quad y * a = b \quad \text{have solutions in } G \text{ for all } a, b \in G,$$

is a group [*Hint*: Use Exercise 33.]

Automata

35. Let X^+ be the set of all input strings for an automaton. What is the name, mentioned in this section, for the algebraic structure given by string concatenation on X^+ if

a. the empty string ϵ is included in X^+.

b. the empty string ϵ is excluded from X^+?

1.3

Subgroups

Notation and Terminology

It is time to explain some conventional notation and terminology used in group theory. Algebraists as a rule do not use a special symbol $*$ to denote a binary operation different from the usual addition and multiplication. They stick with the conventional additive or multiplicative notation and even call the operation *addition* or *multiplication,* depending on the symbol used. The symbol for addition is of course $+$, and usually multiplication is denoted by juxtaposition without a dot, if no confusion results. Thus in place of the notation $a * b$, we shall be using either $a + b$ to be read "the *sum* of a and b," or ab to be read "the *product* of a and b." There is a sort of unwritten agreement that the symbol $+$ should be used only to designate commutative operations. Algebraists feel very uncomfortable when they see $a + b \neq b + a$. For this reason when developing our group theory in a general situation where the operation may or may not be commutative, we shall always use multiplicative notation.

Algebraists frequently use the symbol 0 to denote an additive identity and the symbol 1 to denote a multiplicative identity, even though they may not be actually denoting the integers 0 and 1. Of course, if they are also talking about numbers at the same time, so that confusion would result, symbols such as e or u are used as identity elements. Thus a table for a group of three elements might be one like Table 1.11 or, since such a group is commutative, the table might look like Table 1.12. In general situations we shall continue to use e to denote the identity element of a group.

	1	a	b
1	1	a	b
a	a	b	1
b	b	1	a

Table 1.11

+	0	a	b
0	0	a	b
a	a	b	0
b	b	0	a

Table 1.12

It is customary to denote the inverse of an element a in a group by a^{-1} in multiplicative notation and by $-a$ in additive notation. From now on we shall be using these notations in place of the symbol a'.

Let us explain one more term that is used so often it merits a special definition.

DEFINITION 1.6 (Order of G)   If G is a finite group, then the **order** $|G|$ of G is the number of elements in G. In general, for any finite set S, $|S|$ is the number of elements in S.

Subsets and Subgroups

You may have noticed that we sometimes have had groups contained within larger groups. For example, the group $\mathbb{Z}$ under addition is contained within the group $\mathbb{Q}$ under addition, which in turn is contained within the group $\mathbb{R}$ under addition. When we view the group $\langle \mathbb{Z}, + \rangle$ as contained in the group $\langle \mathbb{R}, + \rangle$, it is very important to notice that the operation $+$ on integers n and m as elements of $\langle \mathbb{Z}, + \rangle$ produces the same element $n + m$ as would result if you were to think of n and m as elements in $\langle \mathbb{R}, + \rangle$. Thus we should *not* regard the group $\langle \mathbb{Q}^+, \cdot \rangle$ as contained in $\langle \mathbb{R}, + \rangle$, even though $\mathbb{Q}^+$ is contained in $\mathbb{R}$ as a set. In this instance, $2 \cdot 3 = 6$ in $\langle \mathbb{Q}^+, \cdot \rangle$, while $2 + 3 = 5$ in $\langle \mathbb{R}, + \rangle$. We are requiring not only that the set of one group be contained in the set of the other, but also that the group operation on the smaller set assign the same element to each ordered pair from this smaller set as is assigned by the group operation of the larger set. We give a preliminary definition before defining a *subgroup*.

DEFINITION 1.7 (Induced Operation)   Let G be a group and let S be a subset of G. If for every $a, b \in S$ it is true that the product ab computed in G is also in S, then S is **closed** under the group operation of G. The binary operation on S thus defined is the **induced operation on S from** G.

EXAMPLE 1   The subset $\mathbb{Q}$ of $\mathbb{R}$ is closed under the operation $+$ on $\mathbb{R}$ since the sum of two rational numbers is again a rational number. However, the subset $\mathbb{R}^*$ of nonzero real numbers is not closed under $+$ since $2 + (-2) = 0$. ▲

We can now define the subject of this section.

DEFINITION 1.8 (Subgroup)   If a subset H of a group G is closed under the binary operation of G and if H is itself a group, then H is a **subgroup of** G. We shall let $H \le G$ or $G \ge H$ denote that H is a subgroup of G, and $H < G$ or $G > H$ shall mean $H \le G$ but $H \ne G$.

Thus $\langle \mathbb{Z}, + \rangle < \langle \mathbb{R}, + \rangle$ but $\langle \mathbb{Q}^+, \cdot \rangle$ is *not* a subgroup of $\langle \mathbb{R}, + \rangle$, even though as sets, $\mathbb{Q}^+ \subset \mathbb{R}$. Every group G has as subgroups G itself and $\{e\}$, where e is the identity element of G.

DEFINITION 1.9 (Proper and Trivial Subgroups)   If G is a group, then the subgroup consisting of G itself is the **improper subgroup** of G. All other subgroups are **proper subgroups**. The subgroup $\{e\}$ is the **trivial subgroup** of G. All other subgroups are **nontrivial**.

We turn to some illustrations.

EXAMPLE 2   Let $\mathbb{R}^n$ be the additive group of all n-component row vectors with real number entries. The subset consisting of all of these vectors having 0 as entry in the first component is a subgroup of $\mathbb{R}^n$. ▲

EXAMPLE 3 $\mathbb{Q}^+$ under multiplication is a proper subgroup of $\mathbb{R}^+$ under multiplication. ▲

EXAMPLE 4   The nth roots of unity in $\mathbb{C}$ form a subgroup U_n of the group $\mathbb{C}^*$ of nonzero complex numbers under multiplication. See Example 12 and Exercise 27 of Section 1.2. ▲

EXAMPLE 5   There are two different types of group structures of order 4 (see Exercise 17 of Section 1.2). We describe them by their group tables (Tables 1.13 and 1.14). The group V is the **Klein 4-group**, and the notation V comes from the German word *Viergruppe*.

The elements of $\mathbb{Z}_4$ are the possibilities for the remainder when an integer is divided by 4.

The only nontrivial proper subgroup of $\mathbb{Z}_4$ is $\{0, 2\}$. Note that $\{0, 3\}$ is *not* a

$\mathbb{Z}_4$: +	0	1	2	3
0	0	1	2	3
1	1	2	3	0
2	2	3	0	1
3	3	0	1	2

Table 1.13

V:	e	a	b	c
e	e	a	b	c
a	a	e	c	b
b	b	c	e	a
c	c	b	a	e

Table 1.14

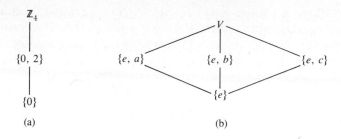

Figure 1.4   (a) Lattice diagram for $\mathbb{Z}_4$. (b) Lattice diagram for V.

subgroup of $\mathbb{Z}_4$, since $\{0, 3\}$ is *not closed* under $+$. For example, $3 + 3 = 2$, and $2 \notin \{0, 3\}$. However, the group V has three nontrivial proper subgroups, $\{e, a\}$, $\{e, b\}$, and $\{e, c\}$. Here $\{e, a, b\}$ is *not* a subgroup, since $\{e, a, b\}$ is not closed under the operation of V. For example, $ab = c$, and $c \notin \{e, a, b\}$. ▲

It is often useful to draw a *lattice diagram* of the subgroups of a group. In such a diagram, a line running downward from a group G to a group H means that H is a subgroup of G. Thus the larger group is placed nearer the top of the diagram. Figure 1.4 contains the lattice diagrams for the groups $\mathbb{Z}_4$ and V of Example 5.

Note that if $H \le G$ and $a \in H$, then by Theorem 1.2, the equation $ax = a$ must have a unique solution, namely the identity element of H. But this equation can also be viewed as one in G, and we see that this unique solution must also be the identity e of G. A similar argument then applied to the equation $ax = e$, viewed in both H and G, shows that the inverse a^{-1} of a in G is also the inverse of a in the subgroup H.

EXAMPLE 6   Let F be the group of all real-valued functions with domain $\mathbb{R}$ under addition. The subset of F consisting of those functions that are continuous is a subgroup of F, for the sum of continuous functions is continuous, the function f where $f(x) = 0$ for all x is continuous and is additive identity, and if f is continuous, then $-f$ is continuous. ▲

It is convenient to have routine steps for determining whether a subset of a group G is a subgroup of G. Example 6 indicates such a routine, and in the next theorem, we demonstrate carefully its validity. While more compact criteria are available, involving only one condition, we prefer this more transparent theorem for a first course.

THEOREM 1.4   A subset H of a group G is a subgroup of G if and only if

1. H is closed under the binary operation of G,
2. the identity e of G is in H.
3. for all $a \in H$ it is true that $a^{-1} \in H$ also.

PROOF   The fact that if $H \le G$ then Conditions 1, 2, and 3 must hold follows at once from the definition of a subgroup and from the remarks preceding Example 6.

Conversely, suppose H is a subset of a group G such that Conditions 1, 2, and 3 hold. By 2 we have at once that $\mathcal{G}_2$ is satisfied. Also $\mathcal{G}_3$ is satisfied by 3. It remains to check the associative axiom, $\mathcal{G}_1$. But surely for all $a, b, c \in H$ it is true that $(ab)c = a(bc)$ in H, for we may actually view this as an equation in G, where the associative law holds. Hence $H \leq G$. ◆

EXAMPLE 7 Let F be as in Example 6. The subset of F consisting of those functions that are differentiable is a subgroup of F, for the sum of differentiable functions is differentiable, the constant function 0 is differentiable, and if f is differentiable, then $-f$ is differentiable. ▲

EXAMPLE 8 Recall from linear algebra that every square matrix A has associated with it a number $\det(A)$ called its determinant, and that A is invertible if and only if $\det(A) \neq 0$. If A and B are square matrices of the same size, then it can be shown that $\det(AB) = \det(A) \cdot \det(B)$. Let G be the multiplicative group of all invertible $n \times n$ matrices with entries in $\mathbb{C}$ and let T be the subset of G consisting of those matrices with determinant 1. The equation $\det(AB) = \det(A) \cdot \det(B)$ shows that T is closed under matrix multiplication. Recall that the identity matrix I_n has determinant 1. From the equation $\det(A) \cdot \det(A^{-1}) = \det(AA^{-1}) = \det(I_n) = 1$, we see that if $\det(A) = 1$, then $\det(A^{-1}) = 1$. Theorem 1.4 then shows that T is a subgroup of G. ▲

Cyclic Subgroups

We remarked in Example 5 that $\{0, 3\}$ is not a subgroup of $\mathbb{Z}_4$. Let us see how big a subgroup H of $\mathbb{Z}_4$ would have to be if it contained 3. It would have to contain the identity 0 and the inverse of 3, which is 1. Also H would have to contain $3 + 3$, which is 2. Hence the only subgroup of $\mathbb{Z}_4$ containing 3 is $\mathbb{Z}_4$ itself.

Let us imitate this reasoning in a general situation. As we remarked before, for a general argument we always use multiplicative notation. Let G be a group and let $a \in G$. A subgroup of G containing a must, by Theorem 1.4, contain aa, which we denote by a^2. Then it must contain a^2a, which we denote by a^3. In general, it must contain a^n, the result of computing products of a and itself for n factors for every positive integer n. (In additive notation we would denote this by na.) These positive integral powers of a do give a set closed under multiplication. It is possible, however, that the inverse of a is not in this set. Of course, a subgroup containing a must also contain a^{-1}, and then $a^{-1}a^{-1}$, which we denote by a^{-2}, and, in general, it must contain a^{-m} for all $m \in \mathbb{Z}^+$. It must contain the identity $e = aa^{-1}$. For obvious symbolic reasons, we agree to let a^0 be e. Summarizing, *a subgroup of G*

containing a must contain all elements a^n *(or na for additive groups) for all* $n \in \mathbb{Z}$. That is, a subgroup containing a must contain $\{a^n \mid n \in \mathbb{Z}\}$. Observe that these powers a^n of a need not be distinct. For example, in the group V of Example 5,

$$a^2 = e, \qquad a^3 = a, \qquad a^4 = e, \qquad a^{-1} = a, \qquad \text{and so on.}$$

It is easy to see that our usual law of exponents, $a^m a^n = a^{m+n}$ for $m, n \in \mathbb{Z}$, holds. For $m, n \in \mathbb{Z}^+$, it is clear. We illustrate another type of case by an example:

$$a^{-2}a^5 = a^{-1}a^{-1}aaaaa = a^{-1}(a^{-1}a)aaaa = a^{-1}eaaaa = a^{-1}(ea)aaa$$

$$= a^{-1}aaaa = (a^{-1}a)aaa = eaaa = (ea)aa = aaa = a^3.$$

We have almost proved the next theorem.

THEOREM 1.5   Let G be a group and let $a \in G$. Then

$$H = \{a^n \mid n \in \mathbb{Z}\}$$

is a subgroup of G and is the smallest subgroup of G that contains a, that is, every subgroup containing a contains H.†

PROOF   We check the three conditions given in Theorem 1.4 for a subset of a group to give a subgroup. Since $a^r a^s = a^{r+s}$ for $r, s \in \mathbb{Z}$, we see that the product in G of two elements of H is again in H. Thus H is closed under the group operation of G. Also $a^0 = e$, so $e \in H$, and for $a^r \in H$, $a^{-r} \in H$ and $a^{-r}a^r = e$. Hence all the conditions are satisfied, and $H \leq G$.

Our arguments prior to the statement of the theorem showed that any subgroup of G containing a must contain H, so H is the smallest subgroup of g containing a. ▲

DEFINITION 1.10 (Cyclic Subgroup $\langle a \rangle$)   The group H of Theorem 1.5 is the **cyclic subgroup of** G **generated by** a, and will be denoted by $\langle a \rangle$.

DEFINITION 1.11 (Generator; Cyclic Group)   An element a of a group G **generates** G and is a **generator for** G if $\langle a \rangle = G$. A group G is **cyclic** if there is some element a in G that generates G.

---

† We may find occasion to distinguish between the terms *minimal* and *smallest* as applied to subsets of a set S that have some property. A subset H of S is minimal with respect to the property if H has the property, and no subset $K \subset H$, $K \neq H$, has the property. If H has the property and $H \subseteq K$ for every subset K with the property, then H is the smallest subset with the property. These may be many minimal subsets, but there can only be one smallest subset. To illustrate, $\{e, a\}$, $\{e, b\}$, and $\{e, c\}$ are all minimal nontrivial subgroups of the group V. (See Fig. 1.4.) However, V contains no smallest nontrivial subgroup.

EXAMPLE 9 Let $\mathbb{Z}_4$ and V be the groups of Example 5. Then $\mathbb{Z}_4$ is cyclic and both 1 and 3 are generators, that is,

$$\langle 1 \rangle = \langle 3 \rangle = \mathbb{Z}_4.$$

However, V is *not* cyclic, for $\langle a \rangle$, $\langle b \rangle$, and $\langle c \rangle$ are proper subgroups of two elements. Of course $\langle e \rangle$ is the trivial subgroup of one element. ▲

EXAMPLE 10 The group $\mathbb{Z}$ under addition is a cyclic group. Both 1 and -1 are generators for the group. ▲

EXAMPLE 11 Consider the group $\mathbb{Z}$ under addition. Let us find $\langle 3 \rangle$. Here the notation is additive, and $\langle 3 \rangle$ must contain

$$3, \qquad 3 + 3 = 6, \qquad 3 + 3 + 3 = 9, \qquad \text{and so on,}$$

$$0, \qquad -3, \qquad -3 + -3 = -6, \qquad -3 + -3 + -3 = -9, \qquad \text{and so on.}$$

In other words, the cyclic subgroup generated by 3 consists of all multiples of 3, positive, negative, and zero. We denote this subgroup by $3\mathbb{Z}$ as well as $\langle 3 \rangle$. In a similar way, we shall let $n\mathbb{Z}$ be the cyclic subgroup $\langle n \rangle$ of $\mathbb{Z}$. Note that $6\mathbb{Z} < 3\mathbb{Z}$. ▲

EXAMPLE 12 Let U_n be the multiplicative group of the nth roots of unity in C. These elements of U_n can be represented geometrically by equally spaced points on a circle about the origin, as illustrated in Fig. 1.5. The heavy point represents the number

$$\alpha = \cos\frac{2\pi}{n} + i\sin\frac{2\pi}{n}.$$

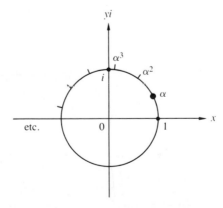

Figure 1.5

The geometric interpretation of multiplication of complex numbers, ex-plained in Section 0.4, shows at once that as α is raised to powers, it works its way counterclockwise around the circle, landing on each of the elements of U_n in turn. Thus U_n under multiplication is a cyclic group, and α is a generator. The group U_n is the cyclic subgroup $\langle \alpha \rangle$ of the group U of all complex numbers z, where $|z| = 1$, under multiplication. ▲

Exercises I.3

Computations

In Exercises 1 through 6, determine whether the given subset of the complex numbers is a subgroup under addition of the group $\mathbb{C}$ of complex numbers under addition.

1. $\mathbb{R}$ **2.** $\mathbb{Q}^+$ **3.** $7\mathbb{Z}$

4. The set $i\mathbb{R}$ of pure imaginary numbers including 0

5. The set $\pi\mathbb{Q}$ of rational multiples of π **6.** The set $\{\pi^n \mid n \in \mathbb{Z}\}$

In Exercises 7 through 12, determine whether the given set of invertible $n \times n$ matrices with real number entries is a subgroup of $GL(n, \mathbb{R})$.

7. The $n \times n$ matrices with determinant 2

8. The diagonal $n \times n$ matrices with no zeros on the diagonal

9. The upper-triangular $n \times n$ matrices with no zeros on the diagonal

10. The $n \times n$ matrices with determinant -1

11. The $n \times n$ matrices with determinant -1 or 1

12. The set of all $n \times n$ matrices A such that $(A^T)A = I_n$ [These matrices are called **orthogonal**. Recall that A^T, the *transpose* of A, is the matrix whose jth column is the jth row of A, and that the transpose operation has the property $(AB)^T = (B^T)(A^T)$.]

Let F be the set of all real-valued functions with domain $\mathbb{R}$ and let $\tilde{F}$ be the subset of F consisting of those functions that have a nonzero value at every point in $\mathbb{R}$. In Exercises 13 through 18, determine whether the given subset of F with the induced operation is (a) a subgroup of the group F under addition, (b) a subgroup of the group $\tilde{F}$ under multiplication.

13. The subset $\tilde{F}$

14. The subset of all $f \in F$ such that $f(1) = 0$

15. The subset of all $f \in \tilde{F}$ such that $f(1) = 1$

16. The subset of all $f \in \tilde{F}$ such that $f(0) = 1$

17. The subset of all $f \in \tilde{F}$ such that $f(0) = -1$

18. The subset of all constant functions in F.

19. Nine groups are given below. Give a *complete* list of all subgroup relations, of the form $G_i \leq G_j$, that exist between these given groups $G_1, G_2, \ldots, G_9$.

$G_1 = \mathbb{Z}$ under addition

$G_2 = 12\mathbb{Z}$ under addition

$G_3 = \mathbb{Q}^+$ under multiplication

$G_4 = \mathbb{R}$ under addition

$G_5 = \mathbb{R}^+$ under multiplication

$G_6 = \{\pi^n \mid n \in \mathbb{Z}\}$ under multiplication

$G_7 = 3\mathbb{Z}$ under addition

$G_8 =$ the set of all integral multiples of 6 under addition

$G_9 = \{6^n \mid n \in \mathbb{Z}\}$ under multiplication

20. Write at least 5 elements of each of the following cyclic groups.

a. $25\mathbb{Z}$ under addition

b. $\{(\frac{1}{2})^n \mid n \in \mathbb{Z}\}$ under multiplication

c. $\{\pi^n \mid n \in \mathbb{Z}\}$ under multiplication

In Exercises 21 through 24, describe all the elements in the cyclic subgroup of $GL(2, \mathbb{R})$ generated by the given 2×2 matrix.

21. $\begin{bmatrix} 0 & -1 \\ -1 & 0 \end{bmatrix}$ **22.** $\begin{bmatrix} 1 & 1 \\ 0 & 1 \end{bmatrix}$ **23.** $\begin{bmatrix} 3 & 0 \\ 0 & 2 \end{bmatrix}$ **24.** $\begin{bmatrix} 0 & -2 \\ -2 & 0 \end{bmatrix}$

25. Which of the following groups are cyclic? For each cyclic group, list all the generators of the group.

$G_1 = \langle \mathbb{Z}, + \rangle$ $\quad$ $G_2 = \langle \mathbb{Q}, + \rangle$ $\quad$ $G_3 = \langle \mathbb{Q}^+, \cdot \rangle$ $\quad$ $G_4 = \langle 6\mathbb{Z}, + \rangle$

$G_5 = \{6^n \mid n \in \mathbb{Z}\}$ under multiplication

$G_6 = \{a + b\sqrt{2} \mid a, b \in \mathbb{Z}\}$ under addition

In Exercises 26 through 34, find the order of the cyclic subgroup of the given group generated by the indicated element.

26. The subgroup of $\mathbb{Z}_4$ generated by 3 (see Table 1.13)

27. The subgroup of V generated by c (see Table 1.14)

28. The subgroup of U_6 generated by $\cos \dfrac{2\pi}{3} + i \sin \dfrac{2\pi}{3}$ (see Example 12)

29. The subgroup of U_5 generated by $\cos \dfrac{4\pi}{5} + i \sin \dfrac{4\pi}{5}$ (see Example 12)

30. The subgroup of U_8 generated by $\cos \dfrac{3\pi}{2} + i \sin \dfrac{3\pi}{2}$ (see Example 12)

31. The subgroup of U_8 generated by $\cos \dfrac{5\pi}{8} + i \sin \dfrac{5\pi}{8}$ (see Example 12)

32. The subgroup of the multiplicative group G of invertible 4×4 matrices generated by

$$\begin{bmatrix} 0 & 0 & 1 & 0 \\ 0 & 0 & 0 & 1 \\ 1 & 0 & 0 & 0 \\ 0 & 1 & 0 & 0 \end{bmatrix}$$

33. The subgroup of the multiplicative group G of invertible 4×4 matrices generated by

$$\begin{bmatrix} 0 & 0 & 0 & 1 \\ 0 & 0 & 1 & 0 \\ 1 & 0 & 0 & 0 \\ 0 & 1 & 0 & 0 \end{bmatrix}$$

34. The subgroup of the multiplicative group G of invertible 4×4 matrices generated by

$$\begin{bmatrix} 0 & 1 & 0 & 0 \\ 0 & 0 & 0 & 1 \\ 0 & 0 & 1 & 0 \\ 1 & 0 & 0 & 0 \end{bmatrix}$$

35. Study the structure of the table for the group $\mathbb{Z}_4$ of Example 5.

a. By analogy, complete Table 1.15 to give a cyclic group $\mathbb{Z}_6$ of 6 elements. (You need not prove the associative law.)

b. Compute the subgroups $\langle 0 \rangle$, $\langle 1 \rangle$, $\langle 2 \rangle$, $\langle 3 \rangle$, $\langle 4 \rangle$, and $\langle 5 \rangle$ of the group $\mathbb{Z}_6$ given in part (a).

c. Which elements are generators for the group $\mathbb{Z}_6$ of part (a)?

d. Give the lattice diagram for the part (b) subgroups of $\mathbb{Z}_6$. (We will see later that these are all the subgroups of $\mathbb{Z}_6$.)

$\mathbb{Z}_6$: +	0	1	2	3	4	5
0	0	1	2	3	4	5
1	1	2	3	4	5	0
2	2					
3	3					
4	4					
5	5					

Table 1.15

Concepts

36. Mark each of the following true or false.
_____ a. The associative law holds in every group.
_____ b. There may be a group in which the cancellation law fails.
_____ c. Every group is a subgroup of itself.
_____ d. Every group has exactly two improper subgroups.
_____ e. In every cyclic group, every element is a generator.
_____ f. This text has still given no example of a group that is not abelian.
_____ g. Every set of numbers that is a group under addition is also a group under multiplication.
_____ h. A subgroup may be defined as a subset of a group.
_____ i. $\mathbb{Z}_4$ is a cyclic group.
_____ j. Every subset of every group is a subgroup under the induced operation.

37. Show by means of an example that it is possible for the quadratic equation $x^2 = e$ to have more than two solutions in some group G with identity e.

Theory

38. Show that if H and K are subgroups of an abelian group G, then
$$\{hk \mid h \in H \text{ and } k \in K\}$$
is a subgroup of G.

39. Find the flaw in the following argument: "Condition 2 of Theorem 1.4 is redundant, since it can be derived from 1 and 3, for let $a \in H$. Then $a^{-1} \in H$ by 3, and by 1, $aa^{-1} = e$ is an element of H, proving 2."

40. Show that a nonempty subset H of a group G is a subgroup of G if and only if $ab^{-1} \in H$ for all $a, b \in H$. (This is one of the *more compact criteria* referred to prior to Theorem 1.4.)

41. Prove that a cyclic group with only one generator can have at most 2 elements.

42. Prove that if G is an abelian group with identity e, then all elements x of G satisfying the equation $x^2 = e$ form a subgroup H of G.

43. Repeat Exercise 42 for the general situation of the set H of all solutions x of the equation $x^n = e$ for a fixed integer $n \geq 1$ in an abelian group G with identity e.

44. Show that if $a \in G$, where G is a finite group with identity e, then there exists $n \in \mathbb{Z}^+$ such that $a^n = e$.

45. Let a nonempty finite subset H of a group G be closed under the binary operation of G. Show that H is a subgroup of G.

46. Let G be a group and let a be a fixed element of G. Show that
$$H_a = \{x \in G \mid xa = ax\}$$
is a subgroup of G.

47. Generalizing Exercise 46, let S be any subset of a group G.

a. Show that $H_S = \{x \in G \mid xs = sx \text{ for all } s \in S\}$ is a subgroup of G.

b. In reference to part (a), the subgroup H_G is the **center of** G. Show that H_G is an abelian group.

48. Let H be a subgroup of a group G. For $a, b \in G$, let $a \sim b$ if and only if $ab^{-1} \in H$. Show that $\sim$ is an equivalence relation on G.

49. For sets H and K, we define the **intersection** $H \cap K$ by

$$H \cap K = \{x \mid x \in H \text{ and } x \in K\}.$$

Show that if $H \leq G$ and $K \leq G$, then $H \cap K \leq G$.

50. Prove that every cyclic group is abelian.

51. Let G be a group and let $G_n = \{g^n \mid g \in G\}$. Under what hypothesis about G can we show that G_n is a subgroup of G?

52. Show that a group with no proper nontrivial subgroups is cyclic.

1.4
―――

Cyclic Groups and Generators

Recall the following facts and notations from Section 1.3. If G is a group and $a \in G$, then

$$H = \{a^n \mid n \in \mathbb{Z}\}$$

is a subgroup of G (Theorem 1.5). This group is the **cyclic subgroup** $\langle a \rangle$ **of** G **generated by** a. Also, given a group G and an element $a \in G$, if

$$G = \{a^n \mid n \in \mathbb{Z}\},$$

then a is a **generator of** G and the group $G = \langle a \rangle$ is **cyclic**. We introduce one new bit of terminology. Let a be an element of a group G. If the cyclic subgroup $\langle a \rangle$ of G is finite, then the **order of** a is the order $|\langle a \rangle|$ of this cyclic subgroup. Otherwise, we say that a is of **infinite order**. We will see in this section that if $a \in G$ is of finite order m, then m is the smallest positive integer such that $a^m = e$.

The first goal of this section to describe all cyclic groups and all subgroups of cyclic groups. This is not an idle exercise. We will see later that cyclic groups serve as building blocks for all sufficiently small abelian groups, in particular, for all finite abelian groups. Cyclic groups are fundamental to the understanding of abelian groups.

Elementary Properties of Cyclic Groups

We start with a demonstration that cyclic groups are abelian.

THEOREM 1.6   Every cyclic group is abelian.

PROOF   Let G be a cyclic group and let a be a generator of G so that

$$G = \langle a \rangle = \{a^n \mid n \in \mathbb{Z}\}.$$

If g_1 and g_2 are any two elements of G, there exist integers r and s such that $g_1 = a^r$ and $g_2 = a^s$. Then

$$g_1 g_2 = a^r a^s = a^{r+s} = a^{s+r} = a^s a^r = g_2 g_1,$$

so G is abelian.   ◆

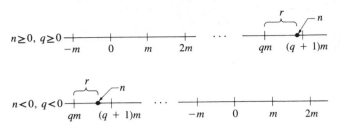

Figure 1.6

We shall continue to use multiplicative notation for our general work on cyclic groups, even though they are abelian.

The *division algorithm* that follows is a seemingly trivial, but very fundamental tool for the study of cyclic groups.

DIVISION ALGORITHM FOR $\mathbb{Z}$ If m is a positive integer and n is any integer, then there exist unique integers q and r such that

$$n = mq + r \quad \text{and} \quad 0 \le r < m.$$

PROOF   We give an intuitive diagrammatic explanation, using Fig. 1.6. On the real x-axis of analytic geometry, mark off the multiples of m and the position of n. Now n falls either on a multiple qm of m and r can be taken as 0, or n falls between two multiples of m. If the latter is the case, let qm be the first multiple of m to the left of n. Then r is as shown in Fig. 1.6. Note that $0 \le r < m$. Uniqueness of q and r follows since if n is not a multiple of m so that we can take $r = 0$, then there is a unique multiple qm of m to the left of n and at distance less than m from n, as illustrated in Fig. 1.6. ◆

In the notation of the division algorithm, we regard q as the **quotient** and r as the nonnegative **remainder** when n is divided by m.

EXAMPLE I   Find the quotient q and remainder r when 38 is divided by 7 according to the division algorithm.

Solution   The positive multiples of 7 are 7, 14, 21, 28, 35, 42, ... Choosing the multiple to leave a nonnegative remainder less than 7, we write

$$38 = 35 + 3 = 7(5) + 3$$

so the quotient is $q = 5$ and the remainder is $r = 3$. ▲

EXAMPLE 2   Find the quotient q and remainder r when -38 is divided by 7 according to the division algorithm.

Solution   The negative multiples of 7 are $-7, -14, -21, -28, -35,$ $-42, \ldots$. Choosing the multiple to leave a nonnegative remainder less than

7, we write

$$-38 = -42 + 4 = 7(-6) + 4$$

so the quotient is $q = -6$ and the remainder is $r = 4$. ▲

We will use the division algorithm to show that a subgroup H of a cyclic group G is also cyclic. Think for a moment what we will have to do to prove this. We will have to use the *definition* of a cyclic group since we have proved little about cyclic groups yet. That is, we will have to use the fact that G has a generating element a. We must then exhibit, in terms of this generator a, some generator $c = a^m$ for H in order to show that H is cyclic. There is really only one natural choice for the power m of a to try. Can you guess what it is before you read the proof of the theorem?

THEOREM 1.7   A subgroup of a cyclic group is cyclic.

PROOF   Let G be a cyclic group generated by a and let H be a subgroup of G. If $H = \{e\}$, then $H = \langle e \rangle$ is cyclic. If $H \neq \{e\}$, then $a^n \in H$ for some $n \in \mathbb{Z}^+$. Let m be the smallest integer in $\mathbb{Z}^+$ such that $a^m \in H$.

We claim that $c = a^m$ generates H; that is,

$$H = \langle a^m \rangle = \langle c \rangle.$$

We must show that every $b \in H$ is a power of c. Since $b \in H$ and $H \leq G$, we have $b = a^n$ for some n. Find q and r such that

$$n = mq + r \qquad \text{for} \quad 0 \leq r < m$$

in accord with the division algorithm. Then

$$a^n = a^{mq+r} = (a^m)^q a^r,$$

so

$$a^r = (a^m)^{-q} a^n.$$

Now since $a^n \in H$, $a^m \in H$, and H is a group, both $(a^m)^{-q}$ and a^n are in H. Thus

$$(a^m)^{-q} a^n \in H; \qquad \text{that is,} \quad a^r \in H.$$

Since m was the smallest positive integer such that $a^m \in H$ and $0 \leq r < m$, we must have $r = 0$. Thus $n = qm$ and

$$b = a^n = (a^m)^q = c^q,$$

so b is a power of c. ◆

As noted in Examples 10 and 11 of Section 1.3, $\mathbb{Z}$ under addition is cyclic and for a positive integer n, the set $n\mathbb{Z}$ of all multiples of n is a subgroup of $\mathbb{Z}$ under addition, the cyclic subgroup generated by n. Theorem 1.7 shows that these cyclic subgroups are the only subgroups of $\mathbb{Z}$ under addition. We state this as a corollary.

COROLLARY The subgroups of $\mathbb{Z}$ under addition are precisely the groups $n\mathbb{Z}$ under addition for $n \in \mathbb{Z}$.

This corollary to Theorem 1.7 gives us an elegant way to define the **greatest common divisor** of two positive integers r and s. Exercise 45 asks us to check that $H = \{nr + ms \mid n, m \in \mathbb{Z}\}$ is a subgroup of the group $\mathbb{Z}$ under addition. Thus H must be cyclic and have a generator d, which we may choose to be positive.

DEFINITION 1.12 Let r and s be two positive integers. The positive generator d of the cyclic group

$$H = \{nr + ms \mid n, m \in \mathbb{Z}\}$$

under addition is the **greatest common divisor** (abbreviated gcd) of r and s.

Note from the definition that d is a divisor of both r and s since both $r = 1r + 0s$ and $s = 0r + 1s$ are in H. Since $d \in H$, we can write

$$d = nr + ms$$

for some integers n and m. We see that every integer dividing both r and s divides the right-hand side of the equation, and hence must be a divisor of d also. Thus d must be the largest number dividing both r and s; this accounts for the name given to d in Definition 1.12.

EXAMPLE 3 Find the gcd of 42 and 72.

Solution The positive divisors of 42 are 1, 2, 3, 6, 7, 14, 21, and 42. The positive divisors of 72 are 1, 2, 3, 4, 6, 8, 9, 12, 18, 24, 36, and 72. The greatest common divisor is 6. Note that $6 = (3)(72) + (-5)(42)$. There is an algorithm for expressing the greatest common divisor d of r and s in the form $d = nr + ms$, but we will not need to make use of it. ▲

Two positive integers are **relatively prime** if their gcd is 1. For example, 12 and 25 are relatively prime. Note that they have no prime factors in common. In our discussion of subgroups of cyclic groups, we will need to know the following:

If r and s are relatively prime and if r divides sm, then r must divide m. **(1)**

Let's prove this. If r and s are relatively prime, then we may write

$$1 = ar + bs \qquad \text{for some} \quad a, b \in \mathbb{Z}.$$

Multiplying by m, we obtain

$$m = arm + bsm,$$

Now r divides both arm and bsm since r divides sm. Thus r is a divisor of the right-hand side of this equation, so r must divide m.

The Structure of Cyclic Groups

Let G be a cyclic group with generator a. We consider two cases.

Case I   *G has an infinite number of elements*; that is, the order of a is infinite. In this case we claim that no two distinct exponents h and k can give equal elements a^h and a^k of G. Suppose $a^h = a^k$ and say $h > k$. Then

$$a^h a^{-k} = a^{h-k} = e,$$

the identity, and $h - k > 0$. Let m be the smallest positive integer such that $a^m = e$ (note the similarity with the construction in the proof of Theorem 1.7). We claim that G would then only have the distinct elements, $e, a, a^2, \ldots, a^{m-1}$. Let $a^n \in G$, and find q and r such that

$$n = mq + r \qquad \text{for} \quad 0 \le r < m$$

by the division algorithm. Then

$$a^n = a^{mq+r} = (a^m)^q a^r = e^q a^r = a^r$$

for $0 \le r < m$. This would mean that G would be finite, contradicting our assumption for Case I. *Thus all powers of a are distinct.*

Suppose that G' is another infinite cyclic group with generator b. Observe that if b^n is renamed a^n, G' can be made to look exactly like G; that is, the groups are isomorphic. [This will be explained precisely in Chapter 3.] *Thus all infinite cyclic groups are just alike except for the names of the elements and operations.* We can take $\mathbb{Z}$ with the operation of addition as a prototype of any infinite cyclic group. We obtain $\mathbb{Z}$ from $G = \{a^i \mid i \in \mathbb{Z}\}$ by renaming a^i by its exponent i.

EXAMPLE 4   It may seem quite odd that $\mathbb{Z}$ and $3\mathbb{Z}$, both being infinite cyclic groups under addition, are structurally identical although $3\mathbb{Z} < \mathbb{Z}$. One might say that $1 \in \mathbb{Z}$ but $1 \notin 3\mathbb{Z}$, so how can they be structurally the same? But names don't matter, and if 1 is renamed 3, 2 is renamed 6, and in general n is renamed $3n$, we have converted $\mathbb{Z}$ into $3\mathbb{Z}$ as an additive group. ▲

Case II   *G has finite order.* In this case, not all positive powers of a generator a of G can be distinct, so for some h and k we must have $a^h = a^k$. Following the argument of Case I, there exists an integer m such that $a^m = e$ and no smaller positive power of a is e. The group G then consists of the distinct elements $e, a, a^2, \ldots, a^{m-1}$, and a is of order m.

Since it is usual to use n for the order of a general finite cyclic group, we shall change notation, setting $m = n$ for what follows.

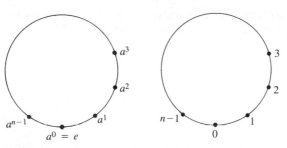

Figure 1.7   **Figure 1.8**

EXAMPLE 5   It is nice to visualize the elements $e = a^0, a^1, a^2, \ldots, a^{n-1}$ of a cyclic group of order n as being distributed evenly on a circle (see Fig. 1.7). The element a^h is located h of these equal units counterclockwise along the circle, measured from the bottom where $e = a^0$ is located. To multiply a^h and a^k diagrammatically, we start from a^h and go k additional units around counterclockwise. To see arithmetically where we end up, find q and r such that

$$h + k = nq + r \quad \text{for} \quad 0 \le r \le n.$$

The nq takes us all the way around the circle q times, and we then wind up at a^r. ▲

Figure 1.8 is essentially the same as Fig. 1.7 but with the points labeled with the exponents on the generator. The operation on these exponents is of course additive, and is known as *addition modulo n*. We give a formal definition.

DEFINITION 1.13   Let n be a fixed positive integer and let h and k be any integers. The remainder r when $h + k$ is divided by n in accord with the division algorithm is the **sum of h and k modulo** n.

EXAMPLE 6   The sum of 23 and 31 modulo 45 is 9, since $23 + 31 = 54 = 45(1) + 9$. ▲

Congruence modulo n was discussed in Section 0.2; we see that if $h + k = r$ in $\mathbb{Z}_n$, then for addition in $\mathbb{Z}$ we have $h + k \equiv r \pmod{n}$.

THEOREM 1.8   The set $\{0, 1, 2, \ldots, n - 1\}$ is a cyclic group $\mathbb{Z}_n$ of elements under addition modulo n.

The proof of Theorem 1.8 gives us practice using the division algorithm and is left to Exercise 50. We have to check $\mathcal{G}_1$, $\mathcal{G}_2$, and $\mathcal{G}_3$. Check them mentally. Think of the diagram in Fig. 1.8 as explained in Example 5.

Thus there is a cyclic group of order n for every positive integer n. Note that if G and G' are two cyclic groups of n elements each and generators a

and b respectively, then if b^r is renamed a^r, G' can be made to look just like G, that is, *any two cyclic groups of the same finite order are isomorphic.* For example, we obtain the group denoted by Fig. 1.8 from that in Fig. 1.7 by renaming each element a^i by its exponent i.

Subgroups of Finite Cyclic Groups

We have completed our description of cyclic groups and turn to their subgroups. The corollary of Theorem 1.7 gives us complete information about subgroups of infinite cyclic groups. Let us give the basic theorem regarding generators of subgroups for the finite cyclic groups.

THEOREM 1.9   Let G be a cyclic group with n elements and generated by a. Let $b \in G$ and let $b = a^s$. Then b generates a cyclic subgroup H of G containing n/d elements, where d is the greatest common divisor (abbreviated gcd) of n and s.

PROOF   That b generates a cyclic subgroup H of G is known from Theorem 1.5. We need show only that H has n/d elements. Following the argument of Case I, we see that H has as many elements as the smallest positive power m of b that gives the identity. Now $b = a^s$, and $b^m = e$ if and only if $(a^s)^m = e$, or if and only if n divides ms. What is the smallest positive integer m such that n divides ms? Let d be the gcd of n and s. Then there exists integers u and v such that

$$d = un + vs.$$

Since d divides both n and s, we may write

$$1 = u(n/d) + v(s/d)$$

where both n/d and s/d are integers. This last equation shows that n/d and s/d are relatively prime, for any integer dividing both of them must also divide 1. We wish to find the smallest positive m such that

$$\frac{ms}{n} = \frac{m(s/d)}{(n/d)} \text{ is an integer.}$$

From Property (1), we conclude that n/d must divide m, so the smallest such m is n/d. Thus the order of H is n/d. ◆

EXAMPLE 7   Consider $\mathbb{Z}_{12}$, with the generator $a = 1$. Since the greatest common divisor (gcd) of 3 and 12 is 3, $3 = 3 \cdot 1$ generates a subgroup of $\frac{12}{3} = 4$ elements, namely

$$\langle 3 \rangle = \{0, 3, 6, 9\}.$$

Since the gcd of 8 and 12 is 4, 8 generates a subgroup of $\frac{12}{4} = 3$ elements, namely

$$\langle 8 \rangle = \{0, 4, 8\}.$$

Since the gcd of 12 and 5 is $1, 5$ generates a subgroup of $\frac{12}{1} = 12$ elements, that is, 5 is a generator of the whole group $\mathbb{Z}_{12}$. ▲

The following corollary follows immediately from the theorem.

COROLLARY If a is a generator of a finite cyclic group G of order n, then the other generators of G are the elements of the form a^r, where r is relatively prime to n.

EXAMPLE 8 Let us find all subgroups of $\mathbb{Z}_{18}$ and give their lattice diagram. All subgroups are cyclic. By the corollary of Theorem 1.9, the elements 1, 5, 7, 11, 13, and 17 are all generators of $\mathbb{Z}_{18}$. Starting with 2,

$$\langle 2 \rangle = \{0, 2, 4, 6, 8, 10, 12, 14, 16\}$$

is of order 9 and has as generators elements of the form $h2$, where h is relatively prime to 9, namely $h = 1, 2, 4, 5, 7$, and 8, so $h2 = 2, 4, 8, 10, 14$, and 16. The element 6 of $\langle 2 \rangle$ generates $\{0, 6, 12\rangle$ and 12 also is a generator of this subgroup.

We have thus far found all subgroups generated by 0, 1, 2, 4, 5, 6, 7, 8, 10, 11, 12, 13, 14, and 16. This leaves just 3, 9, and 15 to consider.

$$\langle 3 \rangle = \{0, 3, 6, 9, 12, 15\},$$

and 15 also generates this group of order 6, since $15 = 5 \cdot 3$, and the gcd of 5 and 6 is 1. Finally,

$$\langle 9 \rangle = \{0, 9\}.$$

The lattice diagram for these subgroups of $\mathbb{Z}_{18}$ is given in Fig. 1.9.

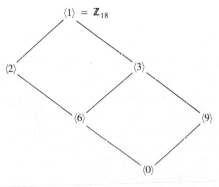

Figure 1.9   Lattice diagram for $\mathbb{Z}_{18}$.

This example is straightforward; we are afraid we wrote it out in such detail that it may look complicated. The exercises give some practice along these lines. ▲

Generators of a Group

Let G be a group, and let $a \in G$. We have described the cyclic subgroup $\langle a \rangle$ of G, which is the smallest subgroup of G that contains the element a. Suppose we want to find as small a subgroup as possible that contains both a and b for another element b in G. By Theorem 1.5, we see that any subgroup containing a and b must contain a^n and b^m for all $m, n \in \mathbb{Z}$, and consequently must contain all finite products of such powers of a and b. For example, such an expression might be $a^2b^4a^{-3}b^2a^5$. Note that we cannot "simplify" this expression by writing first all powers of a, followed by the powers of b, since G may not be abelian. However, products of such expressions are again expressions of the same type. Furthermore, $e = a^0$ and the inverse of such an expression is again of the same type. For example, the inverse of $a^2b^4a^{-3}b^2a^5$ is $a^{-5}b^{-2}a^3b^{-4}a^{-2}$. By Theorem 1.4, this shows that all such products of integral powers of a and b form a subgroup of G, which surely must be the smallest subgroup containing both a and b. We call a and b **generators** of this subgroup. If this subgroup should be all of G, then we say that $\{a, b\}$ *generates* G. Of course, there is nothing sacred about taking just two elements $a, b \in G$. We could have made similar arguments for three, four, or any number of elements of G, as long as we take only finite products of their integral powers.

EXAMPLE 9   The Klein 4-group $V = \{e, a, b, c\}$ of Example 5 in Section 1.3 is generated by $\{a, b\}$ since $ab = c$. It is also generated by $\{a, c\}$, $\{b, c\}$, and $\{a, b, c\}$. If a group G is generated by a subset H, then every subset of G containing H also generates G. ▲

EXAMPLE 10   The group $\mathbb{Z}_6$ is generated by $\{1\}$ and by $\{5\}$. It is also generated by $\{2, 3\}$ since $2 + 3 = 5$, so that any subgroup containing 2 and 3 must contain 5 and must therefore be $\mathbb{Z}_6$. It is also generated by $\{3, 4\}$, $\{2, 3, 4\}$, $\{1, 3\}$, and $\{3, 5\}$, but it is not generated by $\{2, 4\}$ since $\langle 2 \rangle = \{0, 2, 4\}$ contains 2 and 4. ▲

We have given an intuitive explanation of the subgroup of a group G generated by a subset of G. What follows is a detailed exposition of the same idea approached in another way, namely via intersections of subgroups. After we get an intuitive grasp of a concept, it is nice to try to write it up as carefully as possible. We give a set-theoretic definition and generalize a theorem that was in Exercise 49 of Section 1.3.

DEFINITION 1.14 (Intersection) Let $\{S_i \mid i \in I\}$ be a collection of sets. Here I may be any set of indices. The **intersection** $\bigcap_{i \in I} S_i$ **of the sets** S_i is the set of all elements that are in all the sets S_i; that is,

$$\bigcap_{i \in I} S_i = \{x \mid x \in S_i \text{ for all } i \in I\}.$$

If I is finite, $I = \{1, 2, \ldots, n\}$, we may denote $\bigcap_{i \in I} S_i$ by

$$S_1 \cap S_2 \cap \cdots \cap S_n.$$

THEOREM 1.10 The intersection of subgroups H_i of a group G for $i \in I$ is again a subgroup of G.

PROOF Let us show closure. Let $a \in \bigcap_{i \in I} H_i$ and $b \in \bigcap_{i \in I} H_i$, so that $a \in H_i$ for all $i \in I$ and $b \in H_i$ for all $i \in I$. Then $ab \in H_i$ for all $i \in I$, since H_i is a group. Thus $ab \in \bigcap_{i \in I} H_i$.

Since H_i is a subgroup for all $i \in I$, we have $e \in H_i$ for all $i \in I$, and hence $e \in \bigcap_{i \in I} H_i$.

Finally, for $a \in \bigcap_{i \in I} H_i$, we have $a \in H_i$ for all $i \in I$, so $a^{-1} \in H_i$ for all $i \in I$, which implies that $a^{-1} \in \bigcap_{i \in I} H_i$. ◆

Let G be a group and let $a_i \in G$ for $i \in I$. There is at least one subgroup of G containing all the elements a_i for $i \in I$, namely G is itself. Theorem 1.10 assures us that if we take the intersection of all subgroups of G containing all a_i for $i \in I$, we will obtain a subgroup H of G. This subgroup H is the smallest subgroup of G containing all the a_i for $i \in I$.

DEFINITION 1.15 (Generators) Let G be a group and let $a_i \in G$ for $i \in I$. The smallest subgroup of G containing $\{a_i \mid i \in I\}$ is the **subgroup generated by** $\{a_i \mid i \in I\}$. If this subgroup is all of G, then $\{a_i \mid i \in I\}$ **generates** G and the a_i are **generators of** G. If there is a finite set $\{a_i \mid i \in I\}$ that generates G, then G is **finitely generated**.

Note that this definition is consistent with our previous definition of a generator for a cyclic group. Our next theorem gives the structural insight into the subgroup of G generated by $\{a_i \mid i \in I\}$ that we discussed for two generators before Example 9.

THEOREM 1.11 If G is a group and $a_i \in G$ for $i \in I$, then the subgroup H of G generated by $\{a_i \mid i \in I\}$ has as elements precisely those elements of G that are finite products of integral powers of the a_i, where powers of a fixed a_i may occur several times in the product.

PROOF Let K denote the set of all finite products of integral powers of the a_i. Then $K \subseteq H$. We need only observe that K is a subgroup and then, since H is the smallest subgroup containing a_i for $i \in I$, we will be done. Observe that a product of elements in K is again in K. Since $(a_i)^0 = e$, we have $e \in K$. For every element k in K, if we form from the product giving k a new product with the order of the a_i reversed and the opposite sign on all exponents, we have k^{-1}, which is thus in K. For example,

$$[(a_i)^3(a_2)^2(a_1)^{-7}]^{-1} = (a_1)^7(a_2)^{-2}(a_1)^{-3},$$

which is again in K. ◆

Exercises 1.4

Computations

In Exercises 1 through 4, find the quotient and remainder, according to the division algorithm, when n is divided by m.

1. $n = 42, m = 9$ **2.** $n = -42, m = 9$

3. $n = -50, m = 8$ **4.** $n = 50, m = 8$

In Exercises 5 through 7, find the greatest common divisor of the two integers.

5. 32 and 24   **6.** 48 and 88   **7.** 360 and 420

In Exercises 8 through 11, let $+_n$ denote addition modulo n. Compute the indicated quantity.

8. $13 +_{17} 8$ **9.** $21 +_{30} 19$

10. $26 +_{42} 16$ **11.** $39 +_{54} 17$

In Exercises 12 through 15, find the number of generators of a cyclic group having the given order.

12. 5   **13.** 8   **14.** 12   **15.** 60

In Exercises 16 through 20, find the number of elements in the indicated cyclic group.

16. The cyclic subgroup of $\mathbb{Z}_{30}$ generated by 25

17. The cyclic subgroup of $\mathbb{Z}_{42}$ generated by 30

18. The cyclic subgroup $\langle i \rangle$ of the group C^* of nonzero complex numbers under multiplication

19. The cyclic subgroup of the group C^* of Exercise 18 generated by $(1 + i)/\sqrt{2}$

20. The cyclic subgroup of the group C^* of Exercise 18 generated by $1 + i$

In Exercises 21 through 23, find all subgroups of the given group, and draw the lattice diagram for the subgroups.

21. $\mathbb{Z}_{12}$ **22.** $\mathbb{Z}_{36}$ **23.** $\mathbb{Z}_8$

In Exercises 24 through 28, find all orders of subgroups of the given group.

24. $\mathbb{Z}_6$ **25.** $\mathbb{Z}_8$ **26.** $\mathbb{Z}_{12}$ **27.** $\mathbb{Z}_{20}$ **28.** $\mathbb{Z}_{17}$

In Exercises 29 through 34, list the elements of the subgroup generated by the given subset.

29. The subset $\{2, 3\}$ of $\mathbb{Z}_{12}$ **30.** The subset $\{4, 6\}$ of $\mathbb{Z}_{12}$

31. The subset $\{8, 10\}$ of $\mathbb{Z}_{18}$ **32.** The subset $\{12, 30\}$ of $\mathbb{Z}_{36}$

33. The subset $\{12, 42\}$ of $\mathbb{Z}$ **34.** The subset $\{18, 24, 39\}$ of $\mathbb{Z}$

Concepts

35. Mark each of the following true or false.

    _____ a. Every cyclic group is abelian.

    _____ b. Every abelian group is cyclic.

    _____ c. $\mathbb{Q}$ under addition is a cyclic group.

    _____ d. Every element of every cyclic group generates the group.

    _____ e. There is at least one abelian group of every finite order >0.

    _____ f. Every group of order ≤ 4 is cyclic.

    _____ g. All generators of $\mathbb{Z}_{20}$ are prime numbers.

    _____ h. If G and G' are groups, then $G \cap G'$ is a group.

    _____ i. If H and K are subgroups of a group G, then $H \cap K$ is a group.

    _____ j. Every cyclic group of order >2 has at least two distinct generators.

In Exercises 36 through 40, either give an example of a group with the property described, or explain why no example exists.

36. A finite group that is not cyclic

37. An infinite group that is not cyclic

38. A cyclic group having only one generator

39. An infinite cyclic group having four generators

40. A finite cyclic group having four generators

The generators of the cyclic multiplicative group U_n of all nth roots of unity in $\mathbb{C}$ are the **primitive** nth **roots of unity**. In Exercises 41 through 44, find the primitive nth roots of unity for the given value of n.

41. $n = 4$ **42.** $n = 6$ **43.** $n = 8$ **44.** $n = 12$

Theory

45. Let r and s be positive integers. Show that $\{nr + ms \mid n, m \in \mathbb{Z}\}$ is a subgroup of $\mathbb{Z}$.

46. Let a and b be elements of a group G. Show that if ab has finite order n, then ba also has order n.

47. Let r and s be positive integers.

    a. Define the **least common multiple** of r and s as a generator of a certain cyclic group.

    b. Under what condition is the least common multiple of r and s their product, rs?

    c. Generalizing part (b), show that the product of the greatest common divisor and of the least common multiple of r and s is rs.

48. Show that a group that has only a finite number of subgroups must be a finite group.

49. Show by a counterexample that the following "converse" of Theorem 1.7 is not a theorem: "If a group G is such that every proper subgroup is cyclic, then G is cyclic."

50. Let $+_n$ be addition modulo n in $\mathbb{Z}_n = \{1, 2, 3, \ldots, n\}$. Prove that $(\mathbb{Z}_n, +_n)$ is a group. [*Hint:* Associativity is the only nontrivial axiom. Use the division algorithm, and show that both $r +_n (s +_n t)$ and $(r +_n s) +_n t$ are just the remainder of $r + s + t$ when divided by n.]

51. Let G be a group and suppose $a \in G$ generates a cyclic subgroup of order 2 and is the *unique* such element. Show that $ax = xa$ for all $x \in G$. [*Hint:* Consider $(xax^{-1})^2$.]

52. Let p and q be prime numbers. Find the number of generators of the cyclic group $\mathbb{Z}_{pq}$.

53. Let p be a prime number. Find the number of generators of the cyclic group $\mathbb{Z}_{p^r}$, where r is an integer ≥ 1.

54. Show that in a finite cyclic group G of order n, the equation $x^m = e$ has exactly m solutions x in G for each positive integer m that divides n.

55. With reference to Exercise 54, what is the situation if $1 < m < n$ and m does not divide n?

56. Show that $\mathbb{Z}_p$ has no proper subgroups if p is a prime number.

57. Let G be an abelian group and let H and K be finite cyclic subgroups with $|H| = r$ and $|K| = s$.

    a. Show that if r and s are relatively prime, then G contains a cyclic subgroup of order rs.

b. Generalizing part (a), show that G contains a cyclic subgroup of order the least common multiple of r and s.

Cayley Diagraphs

For each generating set S of a finite group G, there is a directed graph representing the group in terms of the generators in S. The term *directed graph* is usually abbreviated as *digraph*. These visual representations of groups were devised by Cayley, and are also referred to as *Cayley diagrams* in the literature.

Intuitively, a **digraph** consists of a finite number of points, called **vertices** of the digraph, and some **arcs** (each with a direction denoted by an arrowhead) joining vertices. In a digraph for a group G using a generating set S we have one vertex, represented by a dot, for each element of G. Each generator in S is denoted by one type of arc. We could use different colors for different arc types in pencil and paper work. Since different colors are not available in our text, we use different style arcs, like solid, dashed, and dotted, to denote different generators. Thus if $S = \{a, b, c\}$ we might denote

$$a \text{ by } \longrightarrow, \qquad b \text{ by } \dashrightarrow, \qquad \text{and} \qquad c \text{ by } \cdots\!\!\blacktriangleright\!\cdots.$$

With this notation, an occurrence of $x\bullet\!\!\longrightarrow\!\!\bullet y$ in a Cayley digraph means that $xa = y$. That is, traveling an arc in the direction of the arrow indicates that multiplication of the group element at the start of the arc *on the right* by the generator corresponding to that type of arc yields the group element at the end of the arc. Of course, since we are in a group, we know immediately that $ya^{-1} = x$. Thus traveling an arc in the direction opposite to the arrow corresponds to multiplication on the right by the inverse of the corresponding generator. If a generator in S is its own inverse, it is customary to denote this by omitting the arrowhead from the arc, rather than using a double arrow. For example, if $b^2 = e$, we might denote b by -------.

EXAMPLE 11 Both of the digraphs shown in Fig. 1.10 represent the group $\mathbb{Z}_6$ with generating set $S = \{1\}$. The length and shape of an arc on the angle between arcs has no significance. ▲

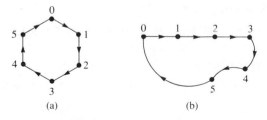

(a) (b)

Figure 1.10 Two digraphs for $\mathbb{Z}_6$ with $S = \{1\}$ using $\longrightarrow_1$.

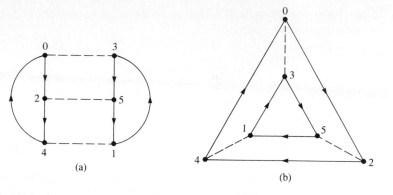

Figure 1.11   Two digraphs for $\mathbb{Z}_6$ with $S = \{2, 3\}$ using ──►── and ─────── 2   3

EXAMPLE 12   Both of the digraphs shown in Fig. 1.11 represent the group $\mathbb{Z}_6$ with generating set $S = \{2, 3\}$. Since 3 is its own inverse, there is no arrowhead on the dashed arcs representing 3. Notice how different these Cayley diagrams look from those in Fig. 1.10 for the same group. The difference is due to the different choice for the set of generators. ▲

Every digraph for a group must satisfy these four properties for the reasons indicated.

Property	**Reason**
1. The digraph is connected, that is, we can get from any vertex g to any vertex h by traveling along consecutive arcs, starting at g and ending at h.	Every equation $gx = h$ has a solution in a group.
2. At most one arc goes from a vertex g to a vertex h.	The solution of $gx = h$ is unique.
3. Each vertex g has exactly one arc of each type starting at g, and one of each type ending at g.	For $g \in G$ and each generator b we can compute gb, and $(gb^{-1})b = g$.
4. If two different sequences of arc types starting from vertex g lead to the same vertex h, then those same sequences of arc types starting from any vertex u will lead to the same vertex v.	If $gq = h$ and $gr = h$, then $uq = ug^{-1}h = ur$.

It can be shown that, conversely, every digraph satisfying these four

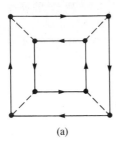

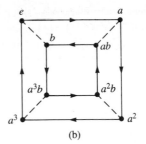

(a)   (b)

Figure 1.12

properties is a Cayley digraph for some group. Due to the symmetry of such a digraph, we can choose labels like a, b, c for the various arc types, name any vertex e to represent the identity, and name each other vertex by a product of arc labels and their inverses that we can travel to attain that vertex starting from the one that we named e. Some finite groups were first constructed (found) using digraphs.

EXAMPLE 13 A digraph satisfying the four properties above is shown in Fig. 1.12(a). To obtain Fig. 1.12(b), we selected the labels

$$\xrightarrow{\hspace{1cm}}_{a} \quad \text{and} \quad \text{------}_{b} \; ,$$

named a vertex e, and then named the other vertices as shown. We have a group $\{e, a, a^2, a^3, b, ab, a^2b, a^3b\}$ of eight elements. Note that the vertex that we named ab could equally well be named ba^{-1}, the vertex that we named a^3 could be named a^{-1}, etc. It is not hard to compute products of elements in this group. To compute $(a^3b)(a^2b)$, we just start at the vertex labeled a^3b and then travel in succession two solid arcs and one dashed arc, arriving at the vertex a, so $(a^3b)(a^2b) = a$. In this fashion, we could write out the table for this 8-element group. ▲

58. How can we tell from a Cayley digraph whether or not the corresponding group is commutative?

59. Referring to the preceding exercise, determine whether the group corresponding to the Cayley digraph in Fig. 1.12(b) is commutative.

60. Is it obvious from a Cayley digraph of a group whether or not the group is cyclic? [*Hint:* Look at Fig. 1.11(b).]

61. The large outside triangle in Fig. 1.11(b) exhibits the cyclic subgroup $\{0, 2, 4\}$ of $\mathbb{Z}_6$. Does the smaller inside triangle similarly exhibit a cyclic subgroup of $\mathbb{Z}_6$? Why?

62. The generating set $S = \{1, 2\}$ for $\mathbb{Z}_6$ contains more generators than necessary, since 1 is a generator for the group. Nevertheless, we can draw a Cayley digraph for $\mathbb{Z}_6$ with this generating set S. Draw such a Cayley digraph.

63. Draw a Cayley digraph for $\mathbb{Z}_8$ taking as generating set $S = \{2, 5\}$.

64. A **relation** on a set S of generators of a group G is an equation that equates some product of generators and their inverses to the identity e of G. For example, if $S = \{a, b\}$ and G is commutative so that $ab = ba$, then one relation is $aba^{-1}b^{-1} = e$. If moreover b is its own inverse, then another relation is $b^2 = e$.

  a. Explain how we can find some relations on S from a Cayley digraph of G.

  b. Find three relations on the set $S = \{a, b\}$ of generators for the group described in Example 13.

65. For the group described in Example 13, compute these products.

  a. $(a^2b)a^3$ b. $(ab)(a^3b)$ c. $b(a^2b)$

In Exercises 66 through 68, give the table for the group having the indicated digraph. In each digraph, take e as identity element. List the identity e first in your table, and list the remaining elements alphabetically, so that your answers will be easy to check.

66. The digraph in Fig. 1.13(a)

67. The digraph in Fig. 1.13(b)

68. The digraph in Fig. 1.13(c)

69. Draw digraphs of the two possible structurally different groups of order 4, taking as small a generating set as possible in each case, You need not label vertices.

70. Show that for $n \geq 3$, there exists a nonabelian group with $2n$ elements that is generated by two elements of order 2.

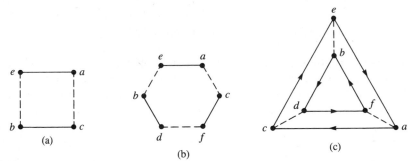

Figure 1.13

CHAPTER TWO

◆

MORE GROUPS AND COSETS

◆

Groups composed of *permutations* play a major role in the study and application of group theory. In fact, we will see in Chapter 3 that every group is structurally the same as some group of permutations. Sections 2.1 and 2.2 discuss permutations and groups whose elements are permutations. Section 2.3 introduces the notion of *cosets* of a subgroup of a group, and shows that the order of a subgroup of a finite group G must be a divisor of the order of G. Section 2.4 shows how new groups can be formed from known groups by taking *direct products,* and explains how all finite abelian groups can be described using direct products. Optional Section 2.5 gives an application of our work to *coding* a message so that errors in its transmission can be detected, and perhaps even corrected.

2.1

Groups of Permutations

We have seen examples of groups of numbers, like the groups $\mathbb{Z}$, $\mathbb{Q}$, and $\mathbb{R}$ under addition. We have also introduced groups of matrices, like the group $GL(2, \mathbb{R})$. Each element A of $GL(2, \mathbb{R})$ yields a transformation of the plane $\mathbb{R}^2$ into itself; namely, if we regard $\mathbf{x}$ as a 2-component column vector, then $A\mathbf{x}$ is also a 2-component column vector. The group $GL(2, \mathbb{R})$ is typical of many of the most useful groups in that its elements *act on things* to transform them. Often, an action produced by a group element can be regarded as a *function*, and the binary operation of the group can be regarded as *function composition.* In this section, we construct some finite groups whose elements, called *permutations,* act on finite sets. These groups will provide us with examples of finite nonabelian groups. We shall show in a later chapter that any finite group is structurally the same as some group of permutations. Unfortunately, this result, which sounds very powerful, does not turn out to be particularly useful to us.

Permutations will be defined as functions. Although most of us have worked with functions already, we shall review them in detail at this time.

$$1 \to 4 \qquad 1 \to 3$$

$$2 \to 2 \qquad 2 \to 2$$

$$3 \to 5 \qquad 3 \to 4$$

$$4 \to 3 \qquad 4 \to 5$$

$$5 \to 1 \qquad 5 \to 3$$

Figure 2.1   Figure 2.2

Functions and Permutations

You may be familiar with the notion of a permutation of a set as a rearrangement of the elements of the set. Thus for the set $\{1, 2, 3, 4, 5\}$, a rearrangement of the elements could be given schematically as in Fig. 2.1, resulting in the new arrangement $\{4, 2, 5, 3, 1\}$. Let us think of this schematic diagram in Fig. 2.1 as a carrying or a *mapping* of each element listed in the left column into a single (not necessarily different) element from the same set listed at the right. Thus 1 is carried into 4, 2 is mapped into 2, and so on. Furthermore, to be a permutation of the set, this mapping must be such that each element appears in the right column once and only once. For example, the diagram in Fig. 2.2 does *not* give a permutation, for 3 appears twice while 1 does not appear at all in the right column. We shall be defining a permutation as such a mapping. Assigning to each element of some set an element of the same or possibly of a different set is the purpose of a *function*.

DEFINITION 2.1 (Function or Mapping) A **function** or **mapping** ϕ **from a set** A **into a set** B is a rule that assigns to each element a of A exactly one element b of B. We say that ϕ **maps** a **into** b, and that ϕ **maps** A **into** B.

The classic notation to denote that ϕ maps a into b is

$$\phi(a) = b.$$

The element b is the **image of** a **under** ϕ. The fact that ϕ maps A into B will be symbolically expressed by

$$\phi : A \to B.$$

It may help you to visualize a function in terms of Fig. 2.3.

If ϕ and ψ are functions with $\phi : A \to B$ and $\psi : B \to C$, then there is a natural function mapping A into C, as illustrated in Fig. 2.4. That is, you can

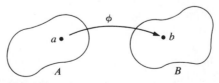

Figure 2.3

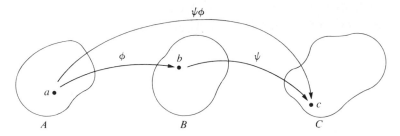

Figure 2.4

get from A to C via B, using the functions ϕ and ψ. This function mapping A into C is the **composite function** consisting of ϕ followed by ψ. We write

$$\psi(\phi(a)) = c.$$

In the context of this section, where we wish to denote a group operation multiplicatively by juxtaposition, it is convenient to denote this composite function by $\psi\phi$. Note that the product $\psi\phi$ for ϕ followed by ψ has to be read from *right to left*. The notation $\psi \circ \phi$ is also used for this composite function.

Returning to permutations, according to our definition, we see that the assignment of Fig. 2.2 is a function from $\{1, 2, 3, 4, 5\}$ into itself. We do not wish to call this a permutation, however. We need to pick out those functions ϕ such that *each* element of the set has *exactly one* element mapped into it. There is again a terminology for a more general situation.

DEFINITION 2.2 (One to One and Onto)   A function from a set A into a set B is **one to one** if each element of B has at most one element of A mapped into it, and is **onto** B if each element of B has at least one element of A mapped into it.†

In terms of Fig. 2.3, a function $\phi: A \rightarrow B$ is one to one if each $b \in B$ has *at most one* arrow coming into it. To say that ϕ is onto B is to say that *every* $b \in B$ has *at least one arrow* coming into it. Since we will often be proving that certain functions are one to one or onto or both, it is worth outlining the technique always used.

1. To show that ϕ is one to one, we show that $\phi(a_1) = \phi(a_2)$ implies $a_1 = a_2$.
2. To show that ϕ is onto B, we show that for each $b \in B$, there exists $a \in A$ such that $\phi(a) = b$.

---

† We should mention another terminology, used by the disciples of N. Bourbaki, in case you encounter it elsewhere. In Bourbaki's terminology, a one-to-one map is an **injection**, an onto map is a **surjection**, and a map that is both one to one and onto is a **bijection**.

Finally, we remark that for $\phi: A \rightarrow B$, the set A is the **domain of** ϕ, the set B is the **codomain of** ϕ, and the set $\phi[A] = \{\phi(a) \,|\, a \in A\}$ is the **image of** A **under** ϕ.

EXAMPLE 1   The function $f: \mathbb{R} \rightarrow \mathbb{R}$ where $f(x) = 2x + 3$ is one to one since different x-values yield different values $f(x)$. It is onto $\mathbb{R}$ since every value y in $\mathbb{R}$ can be obtained by taking f of a suitable x, namely, of $x = (y - 3)/2$. ▲

EXAMPLE 2   The function $f: \mathbb{R} \rightarrow \mathbb{R}$ where $f(x) = x^2$ is not one to one, for $f(2) = f(-2) = 4$. It is not onto since the negative real numbers cannot be obtained by squaring real numbers. ▲

Since for a permutation of a set A we want each element of A to have both at most one element of A and at least one element of A mapped into it, we arrive at the following definition.

DEFINITION 2.3 (Permutation)   A **permutation of a set** A is a function $\phi: A \rightarrow A$ that is both one to one and onto. In other words, a permutation of A is a one-to-one function from A onto A.

One sometimes writes

$$\phi: A \xrightarrow[\text{onto}]{1\text{-}1} B$$

for a one-to-one function ϕ mapping A onto B.

Please take a little time to study and to try to understand these ideas. It will make things easier throughout the course. This is one case in which the names given to the concepts are not as suggestive of their meaning as they might be. (See Exercise 31.)

Permutation Groups

We now show that function composition is a binary operation on the collection of all permutations of a set A. We will call this operation *permutation multiplication*. Let A be a set, and let σ and τ be permutations of A so that σ and τ are both one-to-one functions mapping A onto A. The composite function $\sigma\tau$, defined schematically by

$$A \xrightarrow{\tau} A \xrightarrow{\sigma} A,$$

gives a mapping of A into A. Now $\sigma\tau$ will be a permutation if it is one to one and onto A. Remember that the action of $\sigma\tau$ on A must be read in right-to-left order: first apply τ and then σ. Let us show that $\sigma\tau$ is one to one, following the procedure labeled 1 before Example 1. If

$$(\sigma\tau)(a_1) = (\sigma\tau)(a_2),$$

◆   **H I S T O R I C A L   N O T E**   ◆

One of the earliest recorded studies of permutations occurs in the Sefer
Yetsirah, or Book of Creation, written by an unknown Jewish author
sometime before the eighth century. The author was interested in
counting the various ways in which the letters of the Hebrew alphabet
can be arranged. The question was in some sense a mystical one. It was
believed that the letters had magical powers; therefore suitable arrangements could subjugate the forces of nature. The actual text of the Sefer
Yetsirah is very sparse: "Two letters build two words, three build six
words, four build 24 words, five build 120, six build 720, seven build
5040." Interestingly enough, the idea of counting the arrangements of
the letters of the alphabet also occurs in Islamic mathematics in the
eighth and ninth centuries. By the thirteenth century, in both the
Islamic and Hebrew cultures, the abstract idea of a permutation had
taken root so that both Abu-l-'Abbas ibn al-Banna (1256–1321), a
mathematician from Marrakech in what is now Morocco, and Levi ben
Gerson (1288–1344), a French rabbi, philosopher, and mathematician,
were able to give rigorous proofs that the number of permutations of
any set of n elements is $n!$ as well as prove various results about
counting combinations.

Levi and his predecessors, however, were concerned with permutations as simply arrangements of a given finite set. It was the search for
solutions of polynomial equations that led Lagrange and others in the
late eighteenth century to think of permutations as functions from a
finite set to itself, the set being that of the roots of a given equation.
And it was Augustin-Louis Cauchy (1789–1857) who developed in
detail the basic theorems of permutation theory and who introduced the
standard notation used in this text.

---

then

$$\sigma(\tau(a_1)) = \sigma(\tau(a_2)),$$

and since σ is given to be one to one, we know that $\tau(a_1) = \tau(a_2)$. But then,
since τ is one to one, this gives $a_1 = a_2$. Hence $\sigma\tau$ is one to one. To show that
$\sigma\tau$ is onto A, we follow the procedure labeled 2 before Example 1. Let
$a \in A$. Since σ is onto A, there exists $a' \in A$ such that $\sigma(a') = a$. Since τ is
onto A, there exists $a'' \in A$ such that $\tau(a'') = a'$. Thus

$$a = \sigma(a') = \sigma(\tau(a'')) = (\sigma\tau)(a''),$$

so $\sigma\tau$ is onto A.

EXAMPLE 3   Suppose that

$$A = \{1, 2, 3, 4, 5\}$$

and that σ is the permutation given by Fig. 2.1. We write σ in a more standard notation as

$$\sigma = \begin{pmatrix} 1 & 2 & 3 & 4 & 5 \\ 4 & 2 & 5 & 3 & 1 \end{pmatrix},$$

so that $\sigma(1) = 4$, $\sigma(2) = 2$, and so on. Let

$$\tau = \begin{pmatrix} 1 & 2 & 3 & 4 & 5 \\ 3 & 5 & 4 & 2 & 1 \end{pmatrix}.$$

Then

$$\sigma\tau = \begin{pmatrix} 1 & 2 & 3 & 4 & 5 \\ 4 & 2 & 5 & 3 & 1 \end{pmatrix}\begin{pmatrix} 1 & 2 & 3 & 4 & 5 \\ 3 & 5 & 4 & 2 & 1 \end{pmatrix} = \begin{pmatrix} 1 & 2 & 3 & 4 & 5 \\ 5 & 1 & 3 & 2 & 4 \end{pmatrix}.$$

For example, multiplying in right-to-left order,

$$(\sigma\tau)(1) = \sigma(\tau(1)) = \sigma(3) = 5. \ \blacktriangle$$

We now show that the collection of all permutations of a nonempty set A forms a group under this permutation multiplication.

THEOREM 2.1   Let A be a nonempty set, and let S_A be the collection of all permutations of A. Then S_A is a group under permutation multiplication.

PROOF   We have shown that composition of two permutations of A yields a permutation of A, so S_A is closed under permutation multiplication.

Now permutation multiplication is defined as function composition, and in Section 1.1, we showed that *function composition is associative.* Hence $\mathcal{G}_1$ is satisfied.

The permutation ι such that $\iota(a) = a$, for all $a \in A$ acts as identity. Therefore $\mathcal{G}_2$ is satisfied.

For a permutation σ, define σ^{-1} to be the permutation that reverses the direction of the mapping σ, that is, $\sigma^{-1}(a)$ is to be the element a' of A such that $a = \sigma(a')$. The existence of exactly one such element a' is a consequence of the fact that, as a function, σ is both one to one and onto. (See Exercise 43.) For each $a \in A$ we have

$$\iota(a) = a = \sigma(a') = \sigma(\sigma^{-1}(a)) = (\sigma\sigma^{-1})(a)$$

and also

$$\iota(a') = a' = \sigma^{-1}(a) = \sigma^{-1}(\sigma(a')) = (\sigma^{-1}\sigma)(a'),$$

so that $\sigma^{-1}\sigma$ and $\sigma\sigma^{-1}$ are both the permutation ι. Thus $\mathscr{G}_3$ is satisfied. ◆

Warning: Some texts compute a product $\sigma\mu$ of permutations in left-to-right order, so that $(\sigma\mu)(a) = \mu(\sigma(a))$. Thus the permutation they get for $\sigma\mu$ is the one we would get by computing $\mu\sigma$. Exercise 47 asks us to check in two ways that we still get a group. If you refer to another text on this material, be sure to check its order for permutation multiplication.

There was nothing in our definition of a permutation to require that the set A be finite. However, most of our examples of permutation groups will be concerned with permutations of finite sets. Note that the group S_A is not concerned in any way with what the elements of A really are. Thus if A and B both have the same number of elements, then the group of all permutations of A has the same structure as the group of all permutations of B. One group can be obtained from the other by just renaming elements. This is again the concept of *isomorphic groups* mentioned before and about which more will be said later. Meanwhile, we continue to use the term informally.

DEFINITION 2.4 (Symmetric Group) Let A be the finite set $\{1, 2, \ldots, n\}$. The group of all permutations of A is the **symmetric group on** n **letters**, and is denoted by S_n.

Note that S_n has $n!$ elements, where

$$n! = n(n - 1)(n - 2) \cdots (3)(2)(1).$$

Two Important Examples

EXAMPLE 4   An interesting example for us is the group S_3 of $3! = 6$ elements. Let the set A be $\{1, 2, 3\}$. We list the permutations of A and assign to each a subscripted Greek letter for a name. The reasons for the choice of names will be clear later. Let

$$\rho_0 = \begin{pmatrix} 1 & 2 & 3 \\ 1 & 2 & 3 \end{pmatrix}, \qquad \mu_1 = \begin{pmatrix} 1 & 2 & 3 \\ 1 & 3 & 2 \end{pmatrix},$$

$$\rho_1 = \begin{pmatrix} 1 & 2 & 3 \\ 2 & 3 & 1 \end{pmatrix}, \qquad \mu_2 = \begin{pmatrix} 1 & 2 & 3 \\ 3 & 2 & 1 \end{pmatrix},$$

$$\rho_2 = \begin{pmatrix} 1 & 2 & 3 \\ 3 & 1 & 2 \end{pmatrix}, \qquad \mu_3 = \begin{pmatrix} 1 & 2 & 3 \\ 2 & 1 & 3 \end{pmatrix},$$

	ρ_0	ρ_1	ρ_2	μ_1	μ_2	μ_3
ρ_0	ρ_0	ρ_1	ρ_2	μ_1	μ_2	μ_3
ρ_1	ρ_1	ρ_2	ρ_0	μ_3	μ_1	μ_2
ρ_2	ρ_2	ρ_0	ρ_1	μ_2	μ_3	μ_1
μ_1	μ_1	μ_2	μ_3	ρ_0	ρ_1	ρ_2
μ_2	μ_2	μ_3	μ_1	ρ_2	ρ_0	ρ_1
μ_3	μ_3	μ_1	μ_2	ρ_1	ρ_2	ρ_0

Table 2.1

The multiplication table for S_3 is shown in Table 2.1. Note that this group is not abelian! It is our first such finite example. We have seen that any group of at most 4 elements is abelian. Later we will see that a group of 5 elements is also abelian. Thus S_3 has minimal order for any nonabelian group. ▲

There is a natural correspondence between the elements of S_3 in Example 4 and the ways in which two copies of an equilateral triangle with vertices 1, 2, and 3 (see Fig. 2.5) can be placed, one covering the other. For this reason, S_3 is also the **group D_3 of symmetries of an equilateral triangle**. Naively, we used ρ_i for *rotations* and μ_i for *mirror images* in bisectors of angles. The notation D_3 stands for the third dihedral group. The nth *dihedral group D_n* is the group of symmetries of the regular n-gon.†

Note that we can consider the elements of S_3 to *act* on the triangle in Fig. 2.5. See the discussion at the start of this section.

EXAMPLE 5 Let us form the dihedral group D_4 of permutations corresponding to the ways that two copies of a square with vertices 1, 2, 3, and 4 can be placed, one covering the other (see Fig. 2.6). D_4 will then be the **group of symmetries of the square**. It is also called the **octic group**. Again we choose seemingly arbitrary notation that we shall explain later. Naively, we are using ρ_i for *rotations,* μ_i for *mirror images* in perpendicular bisectors of sides, and δ_i for *diagonal flips*. There are eight permutations involved here.

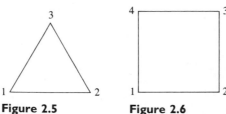

Figure 2.5   **Figure 2.6**

---

† Many people denote the nth dihedral group by D_{2n} rather than by D_n since the order of the group is $2n$.

	ρ_0	ρ_1	ρ_2	ρ_3	μ_1	μ_2	δ_1	δ_2
ρ_0	ρ_0	ρ_1	ρ_2	ρ_3	μ_1	μ_2	δ_1	δ_2
ρ_1	ρ_1	ρ_2	ρ_3	ρ_0	δ_1	δ_2	μ_2	μ_1
ρ_2	ρ_2	ρ_3	ρ_0	ρ_1	μ_2	μ_1	δ_2	δ_1
ρ_3	ρ_3	ρ_0	ρ_1	ρ_2	δ_2	δ_1	μ_1	μ_2
μ_1	μ_1	δ_2	μ_2	δ_1	ρ_0	ρ_2	ρ_3	ρ_1
μ_2	μ_2	δ_1	μ_1	δ_2	ρ_2	ρ_0	ρ_1	ρ_3
δ_1	δ_1	μ_1	δ_2	μ_2	ρ_1	ρ_3	ρ_0	ρ_2
δ_2	δ_2	μ_2	δ_1	μ_1	ρ_3	ρ_1	ρ_2	ρ_0

Table 2.2

Let

$$\rho_0 = \begin{pmatrix} 1 & 2 & 3 & 4 \\ 1 & 2 & 3 & 4 \end{pmatrix}, \qquad \mu_1 = \begin{pmatrix} 1 & 2 & 3 & 4 \\ 2 & 1 & 4 & 3 \end{pmatrix},$$

$$\rho_1 = \begin{pmatrix} 1 & 2 & 3 & 4 \\ 2 & 3 & 4 & 1 \end{pmatrix}, \qquad \mu_2 = \begin{pmatrix} 1 & 2 & 3 & 4 \\ 4 & 3 & 2 & 1 \end{pmatrix},$$

$$\rho_2 = \begin{pmatrix} 1 & 2 & 3 & 4 \\ 3 & 4 & 1 & 2 \end{pmatrix}, \qquad \delta_1 = \begin{pmatrix} 1 & 2 & 3 & 4 \\ 3 & 2 & 1 & 4 \end{pmatrix},$$

$$\rho_3 = \begin{pmatrix} 1 & 2 & 3 & 4 \\ 4 & 1 & 2 & 3 \end{pmatrix}, \qquad \delta_2 = \begin{pmatrix} 1 & 2 & 3 & 4 \\ 1 & 4 & 3 & 2 \end{pmatrix}.$$

The table for D_4 is given in Table 2.2. Note that D_4 is again nonabelian. This group is simply beautiful. It will provide us with nice examples for many concepts we will introduce in group theory. Look at the lovely symmetries in that table! Finally, we give in Fig. 2.7 the lattice diagram for the subgroups of D_4. Look at the lovely symmetries in that diagram! ▲

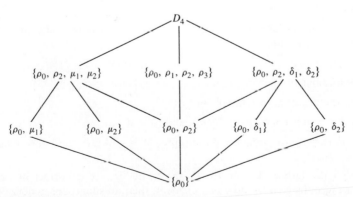

Figure 2.7 Lattice diagram for D_4.

Exercises 2.1

Computation

In Exercises 1 through 5, compute the indicated product involving the following permutations in S_6:

$$\sigma = \begin{pmatrix} 1 & 2 & 3 & 4 & 5 & 6 \\ 3 & 1 & 4 & 5 & 6 & 2 \end{pmatrix},$$

$$\tau = \begin{pmatrix} 1 & 2 & 3 & 4 & 5 & 6 \\ 2 & 4 & 1 & 3 & 6 & 5 \end{pmatrix},$$

$$\mu = \begin{pmatrix} 1 & 2 & 3 & 4 & 5 & 6 \\ 5 & 2 & 4 & 3 & 1 & 6 \end{pmatrix}.$$

1. $\tau\sigma$ **2.** $\tau^2\sigma$ **3.** $\mu\sigma^2$ **4.** $\sigma^{-2}\tau$ **5.** $\sigma^{-1}\tau\sigma$

In Exercises 6 through 9, compute the expression for the permutations σ, τ, and μ defined prior to Exercise 1.

6. $|\langle\sigma\rangle|$ **7.** $|\langle\tau^2\rangle|$ **8.** σ^{100} **9.** μ^{100}

Let A be a set and let $\sigma \in S_A$. For a fixed $a \in A$, the set

$$\mathcal{O}_{a,\sigma} = \{\sigma^n(a) \mid n \in \mathbb{Z}\}$$

is the **orbit** of a **under** σ. In Exercises 10 through 12, find the orbit of 1 under the permutation defined prior to Exercise 1.

10. σ **11.** τ **12.** μ

13. In Table 2.1, we used $\rho_0, \rho_1, \rho_2, \mu_1, \mu_2, \mu_3$ as the names of the 6 elements of S_3. Some authors use the notations $\varepsilon, \rho, \rho^2, \phi, \rho\phi, \rho^2\phi$ for these elements, where their ε is our identity ρ_0, their ρ is our ρ_1, and their ϕ is our μ_1. Verify *geometrically* that their six expressions do give all of S_3.

14. With reference to Exercise 13, give a similar alternative labeling for the 8 elements of D_4 in Table 2.2.

15. Find the number of elements in the set $\{\sigma \in S_4 \mid \sigma(3) = 3\}$.

16. Find the number of elements in the set $\{\sigma \in S_5 \mid \sigma(2) = 5\}$.

17. Consider the group S_3 of Example 4.

a. Find the cyclic subgroups $\langle\rho_1\rangle$, $\langle\rho_2\rangle$, and $\langle\mu_1\rangle$ of S_3.

b. Find *all* subgroups, proper and improper, of S_3 and give the lattice diagram for them.

18. Verify that the lattice diagram for D_4 shown in Fig. 2.7 is correct by finding all (cyclic) subgroups generated by one element, then all subgroups generated by the two elements, etc.

19. Give the multiplication table for the cyclic subgroup of S_5 generated by

$$\rho = \begin{pmatrix} 1 & 2 & 3 & 4 & 5 \\ 2 & 4 & 5 & 1 & 3 \end{pmatrix}.$$

There will be 6 elements. Let them be $\rho, \rho^2, \rho^3, \rho^4, \rho^5$, and $\rho^0 = \rho^6$. Is this group isomorphic to S_3?

20. a. Verify that the six matrices

$$\begin{bmatrix} 1 & 0 & 0 \\ 0 & 1 & 0 \\ 0 & 0 & 1 \end{bmatrix}, \begin{bmatrix} 0 & 1 & 0 \\ 0 & 0 & 1 \\ 1 & 0 & 0 \end{bmatrix}, \begin{bmatrix} 0 & 0 & 1 \\ 1 & 0 & 0 \\ 0 & 1 & 0 \end{bmatrix}, \begin{bmatrix} 1 & 0 & 0 \\ 0 & 0 & 1 \\ 0 & 1 & 0 \end{bmatrix}, \begin{bmatrix} 0 & 0 & 1 \\ 0 & 1 & 0 \\ 1 & 0 & 0 \end{bmatrix},$$

and

$$\begin{bmatrix} 0 & 1 & 0 \\ 1 & 0 & 0 \\ 0 & 0 & 1 \end{bmatrix}$$

form a group under matrix multiplication. [*Hint:* Don't try to compute all products of these matrices. Instead, think how the column vector $\begin{bmatrix} 1 \\ 2 \\ 3 \end{bmatrix}$ is transformed by multiplying it on the left by each of the six matrices.]

b. What group discussed in this section is structurally the same as (isomorphic to) this group of six matrices?

21. After working Exercise 20, write down eight matrices that form a group under permutation multiplication that is isomorphic to D_4.

In this section we discussed the group of symmetries of an equilateral triangle and of a square. In Exercises 22 through 25, give a group that we have discussed in the text that has the same structure (is isomorphic to) the group of symmetries of the indicated figure. You may want to label some special points on the figure, write some permutations corresponding to symmetries, and compute some products of permutations.

22. The figure in Fig. 2.8(a)　　　　　**23.** The figure in Fig. 2.8(b)

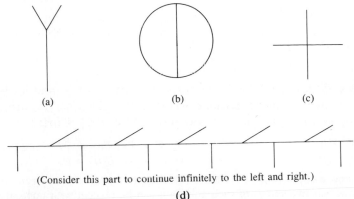

(a)　　　　　　　　　　(b)　　　　　　　　　　(c)

(Consider this part to continue infinitely to the left and right.)

(d)

Figure 2.8

24. The figure in Fig. 2.8(c)   **25.** The figure in Fig. 2.8(d)

Concepts

In Exercises 26 through 30, determine whether the given function is a permutation of $\mathbb{R}$.

26. $f_1:\mathbb{R} \to \mathbb{R}$ defined by $f_1(x) = x + 1$

27. $f_2:\mathbb{R} \to \mathbb{R}$ defined by $f_2(x) = x^2$

28. $f_3:\mathbb{R} \to \mathbb{R}$ defined by $f_3(x) = -x^3$

29. $f_4:\mathbb{R} \to \mathbb{R}$ defined by $f_4(x) = e^x$

30. $f_5:\mathbb{R} \to \mathbb{R}$ defined by $f_5(x) = x^3 - x^2 - 2x$

31. Students often misunderstand the concept of a one-to-one function (mapping). I think I know the reason. You see, a mapping $\phi: A \to B$ has a *direction* associated with it, from A to B. It seems reasonable to expect a one-to-one mapping simply to be a mapping that carries one point of A into one point of B, in the direction indicated by the arrow. But of course, *every* mapping of A into B does this, and Definition 2.2 did not say that at all. With this unfortunate situation in mind, make as good a pedagogical case as you can for calling the functions described in Definition 2.2 *two-to-two functions* instead. (Unfortunately, it is almost impossible to get widely used terminology changed.)

32. Mark each of the following true or false.

_____ a. Every permutation is a one-to-one function.

_____ b. Every function is a permutation if and only if it is one to one.

_____ c. Every function from a finite set onto itself must be one to one.

_____ d. The text still has given no example of a group that is nonabelian.

_____ e. Every subgroup of an abelian group is abelian.

_____ f. Every element of a group generates a cyclic subgroup of the group.

_____ g. The symmetric group S_{10} has 10 elements.

_____ h. The symmetric group S_3 is cyclic.

_____ i. S_n is not cyclic for any n.

_____ j. Every group is isomorphic to some group of permutations.

33. Show by an example that every proper subgroup of a nonabelian group may be abelian.

34. Let A be a nonempty set. What type of algebraic structure mentioned previously in the text is given by the set of *all* functions mapping A into itself under function composition?

Theory

In Exercises 35 through 38, let A be a set, B a subset of A, and let b be one particular element of B. Determine whether the given set is sure to be a subgroup of S_A under the induced operation. Here $\sigma[B] = \{\sigma(x) \mid x \in B\}$.

35. $\{\sigma \in S_A \mid \sigma(b) = b\}$ **36.** $\{\sigma \in S_A \mid \sigma(b) \in B\}$

37. $\{\sigma \in S_A \mid \sigma[B] \subseteq B\}$ **38.** $\{\sigma \in S_A \mid \sigma[B] = B\}$

39. In analogy with Examples 4 and 5, consider a regular plane n-gon for $n \geq 3$. Each way that two copies of such an n-gon can be placed, with one covering the other, corresponds to a certain permutation of the vertices. The set of these permutations is a group, the *n*th **dihedral group** D_n, under permutation multiplication. Find the order of this group D_n. Argue *geometrically* that this group has a subgroup having just half as many elements as the whole group has.

40. Consider a cube that exactly fills a certain cubical box. As in Examples 4 and 5, the ways in which the cube can be placed into the box correspond to a certain group of permutations of the vertices of the cube. This group is the **group of rigid motions** (or **rotations**) **of the cube.** (It should not be confused with the *group of symmetries of the figure, which* will be discussed in the exercises of Section 2.2.) How many elements does this group have? Argue *geometrically* that this group has at least three different subgroups of order 4 and at least four different subgroups of order 3.

41. Show that S_n is a nonabelian group for $n \geq 3$.

42. Strengthening Exercise 41, show that if $n \geq 3$, then the only element of σ of S_n satisfying $\sigma\gamma = \gamma\sigma$ for all $\gamma \in S_n$ is $\sigma = \iota$, the identity permutation.

The following exercise is of a set-theoretic nature. You are asked to prove something we glossed over as intuitive in the text.

43. Let $\phi: A \to B$. A map $\phi^{-1}: B \to A$ is called an **inverse of** ϕ if $\phi^{-1}(\phi(x)) = x$ for all $x \in A$ and $\phi(\phi^{-1}(y)) = y$ for all $y \in B$.

    a. Show that ϕ is one to one and onto B if and only if it has an inverse.

    b. Show that the inverse of a one-to-one map ϕ of A onto B is unique.

44. Orbits were defined before Exercise 10. Let $a, b \in A$ and $\sim \in S_A$. Show that if $\mathcal{O}_{a,\sigma}$ and $\mathcal{O}_{b,\sigma}$ have an element in common, then $\mathcal{O}_{a,\sigma} = \mathcal{O}_{b,\sigma}$.

45. If A is a set, then a subgroup H of S_A is **transitive on** A if for each $a, b \in A$ there exists $\sigma \in H$ such that $\sigma(a) = b$. Show that if A is a nonempty finite set, then there exists a finite cyclic subgroup H of S_A with $|H| = |A|$ that is transitive on A.

46. Referring to the definition before Exercise 10 and to Exercise 45, show that for $\sigma \in S_A$, $\langle\sigma\rangle$ is transitive on A if and only if $\mathcal{O}_{a,\sigma} = A$ for some $a \in A$.

47. (See the warning on page 93.) Let G be a group with binary operation $*$. Let G' be the same set as G, and define a binary operation $*'$ on G' by $x *' y = y * x$ for all $x, y \in G'$.

    a. (Intuitive argument that G' under $*'$ is a group.) Suppose the front wall of your classroom were made of transparent glass, and that all possible products $a * b = c$ and all possible instances $a * (b * c) = (a * b) * c$ of the associative property for G under $*$ were written on the wall with a magic marker. What would a person see when looking at the other side of the wall from the next room in front of yours?

    b. Show from the mathematical definition of $*'$ that G' is a group under $*'$.

Cayley Digraphs

Cayley digraphs were discussed in the exercise set for Section 1.4.

48. Indicate schematically a digraph for D_n using a generating set consisting of a rotation through $2\pi/n$ radians and a reflection (mirror image).

Automata

Automata were discussed in the exercise set for Section 1.1.

Consider a finite-state machine with input alphabet $x_1, x_2, \ldots, x_r$ and internal states $s_0, s_1, s_2, \ldots, s_n$. Let X^+ be the monoid of all input strings $\mathbf{x}$ on the input alphabet under string concatenation, which we now denote by juxtaposition rather than by $*$. Let S be the set of $n + 1$ internal states. Each $\mathbf{x}$ in X^+ determines a function $T_{\mathbf{x}} : S \to S$ defined by $T_{\mathbf{x}}(s_i) = s_j$ where s_j is the internal state in which the machine finds itself if the input string $\mathbf{x}$ was entered while the machine was in state s_i. Such a function $T_{\mathbf{x}}$ is a *state transition function* of the machine.

49. For the parity-check machine in Fig. 1.2 of the exercises in Section 1.1, describe the state transition function $T_{\mathbf{x}}$ for the given input string $\mathbf{x}$ by computing $T_{\mathbf{x}}(s_0)$ and $T_{\mathbf{x}}(s_1)$.

     a. $\mathbf{x} = 0$ b. $\mathbf{x} = 1$ c. $\mathbf{x} = 11101$ d. $\mathbf{x} = 010100$

     Do all possible different state transition functions for this machine appear in your answers to (a)–(d)? If not, give the missing ones. Does your answer here change if the empty string ϵ is included?

50. Repeat Exercise 49 for the three-state machine in Fig. 1.3 of Section 1.1 for the input strings given here, computing $T_{\mathbf{x}}(s_2)$ also.

     a. $\mathbf{x} = 0110$ b. $\mathbf{x} = 0110111$ c. $\mathbf{x} = 1101$

     Do all possible different state transition functions for this machine appear in your answers to (a)–(c)? If not, give the missing ones. Does your answer here change if the empty string ϵ is included?

51. For a machine that has $n + 1$ states, what is the maximum possible number of different state transition functions?

52. Show that the composition $T_{\mathbf{x}} \circ T_{\mathbf{y}}$ of two state transition functions is again a state transition function by exhibiting an input string whose transition function is $T_{\mathbf{x}} \circ T_{\mathbf{y}}$. [*Hint:* Remember that the inputs in a string are applied to the machine in left-to-right order.]

The preceding exercise shows that the set of state transition functions for a machine is closed under function composition. Since function composition is associative, we see that this set under function composition is a semigroup, called the *semigroup of the machine*. If the empty string ϵ is included in X^+, then the semigroup has an identity ϵ, so it is actually a monoid.

53. Give the table for the monoid of the parity-check machine in Exercise 49, labeling the transition functions you found there $T_{\mathbf{x}}$ where $\mathbf{x}$ is as short a string as you can find that produces that transition function. Include the transition function T_{ϵ} for the empty string. Is this monoid a group? Why?

54. Give the table for the monoid of the machine in Exercise 50, labeling the transition functions you found there $T_{\mathbf{x}}$ where $\mathbf{x}$ is as short a string as you can find

that produces that transition function. Include the transition function T_ϵ for the empty string. Is this monoid a group? Why?

Every finite group G (actually, every finite semigroup) defines a simple machine in the following way. Let the elements of G serve as both the input alphabet and the internal states. View the product $g_r g_i = g_s$ in G as saying that when the machine is in state g_r, the input g_i causes it to go to state g_s. Note that we write the input g_i on the right of the state g_r so that if we apply an input string $g_i g_j$ to g_r, the action of the inputs will be in left-to-right order, namely, $g_r(g_i g_j) = (g_r g_i)g_j$. A table for the group thus becomes an input table for the machine, where we think of the list of group elements at the left of the table as the list of states and the list across the top as the list of inputs. We introduced this notion of an input table prior to Exercise 39 of Section 1.1. The machine defined in this fashion is called the *automaton of the group (or semigroup)*. We can then draw a diagram for our group by drawing a state diagram for this machine. A little thought shows that such a state diagram amounts to a Cayley digraph for the group formed by using the entire group as a generating set. Usually a Cayley digraph using as few generators as possible gives the most enlightening diagram of the group.

55. Give the state diagram of the automaton of the group $\mathbb{Z}_2$.

56. Give the state diagram of the automaton of the Klein 4-group shown in Table 1.14.

57. Show that each state transition function of the automaton of a finite group G is a permutation of G.

58. Complete the following statement with what you think is an accurate and interesting conclusion.

Let G be a finite group. Then the semigroup of the automaton of G is _____.

2.2

Orbits, Cycles, and the Alternating Groups

Orbits

Each permutation σ of a set A determines a natural partition of A into cells with the property that $a, b \in A$ are in the same cell if and only if $b = \sigma^n(a)$ for some $n \in \mathbb{Z}$. We establish this partition using an appropriate equivalence relation:

For $a, b \in A$, let $a \sim b$ if and only if $b = \sigma^n(a)$ for some $n \in \mathbb{Z}$. (1)

We now check that $\sim$ defined by (1) is indeed an equivalence relation.

Reflexive   Clearly $a \sim a$ since $a = \iota(a) = \sigma^0(a)$.

Symmetric   If $a \sim b$ then $b = \sigma^n(a)$ for some $n \in \mathbb{Z}$. But then $a = \sigma^{-n}(b)$ and $-n \in \mathbb{Z}$, so $b \sim a$.

Transitive   Suppose $a \sim b$ and $b \sim c$, then $b = \sigma^n(a)$ and $c = \sigma^m(b)$ for some $n, m \in \mathbb{Z}$. Substituting, we find that $c = \sigma^m(\sigma^n(a)) = \sigma^{n+m}(a)$, so $a \sim c$.

DEFINITION 2.5 (Orbits of σ)   Let σ be a permutation of a set A. The equivalence classes in A determined by the equivalence relation (1) are the **orbits of σ**.

EXAMPLE 1   Since the identity permutation ι of A leaves each element of A fixed, the orbits of ι are the one-element subsets of A. ▲

EXAMPLE 2   Find the orbits of the permutation

$$\sigma = \begin{pmatrix} 1 & 2 & 3 & 4 & 5 & 6 & 7 & 8 \\ 3 & 8 & 6 & 7 & 4 & 1 & 5 & 2 \end{pmatrix}$$

in S_8.

Solution   To find the orbit containing 1, we apply σ repeatedly, obtaining symbolically

$$1 \xrightarrow{\sigma} 3 \xrightarrow{\sigma} 6 \xrightarrow{\sigma} 1 \xrightarrow{\sigma} 3 \xrightarrow{\sigma} 6 \xrightarrow{\sigma} 1 \xrightarrow{\sigma} 3 \xrightarrow{\sigma} \cdots.$$

Since σ^{-1} would simply reverse the directions of the arrows in this chain, we see that the orbit containing 1 is $\{1, 3, 6\}$. We now choose an integer from 1 to 8 not in $\{1, 3, 6\}$, say 2, and similarly find that the orbit containing 2 is $\{2, 8\}$. Finally, we find that the orbit containing 4 is $\{4, 7, 5\}$. Since these three orbits include all integers from 1 to 8, we see that the complete list of orbits of σ is

$$\{1, 3, 6\}, \quad \{2, 8\}, \quad \{4, 5, 7\}. \ ▲$$

Cycles

For the remainder of this section, we consider just permutations of a finite set A of n elements. We may as well suppose that $A = \{1, 2, 3, \ldots, n\}$ and that we are dealing with elements of the symmetric group S_n.
    Refer back to Example 2. The orbits of

$$\sigma = \begin{pmatrix} 1 & 2 & 3 & 4 & 5 & 6 & 7 & 8 \\ 3 & 8 & 6 & 7 & 4 & 1 & 5 & 2 \end{pmatrix} \tag{2}$$

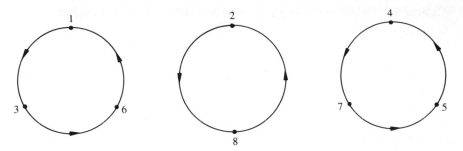

Figure 2.9

are indicated graphically in Fig. 2.9. That is, σ acts on each integer from 1 to 8 on one of the circles by carrying it into the next integer on the circle traveled counterclockwise, in the direction of the arrow. For example, the leftmost circle indicates that $\sigma(1) = 3$, $\sigma(3) = 6$, and $\sigma(6) = 1$. Figure 2.9 is a nice way to visualize the structure of the permutation σ.

Each individual circle in Fig. 2.9 also defines, by itself, a permutation in S_8. For example, the leftmost circle corresponds to the permutation

$$\mu = \begin{pmatrix} 1 & 2 & 3 & 4 & 5 & 6 & 7 & 8 \\ 3 & 2 & 6 & 4 & 5 & 1 & 7 & 8 \end{pmatrix} \tag{3}$$

that acts on 1, 3, and 6 just as σ does, but leaves the remaining integers 2, 4, 5, 7, and 8 fixed. Such a permutation, described graphically by a single circle, is called a *cycle* (for circle). We consider the identity permutation to be a cycle since it can be represented by a circle having only the integer 1, as shown in Fig. 2.10. We now define the term *cycle* in a mathematically precise way.

DEFINITION 2.6 (Cycle) A permutation $\sigma \in S_n$ is a **cycle** if it has at most one orbit containing more than one element. The **length** of a cycle is the number of elements in its largest orbit.

To avoid the cumbersome notation, as in Eq. (3), for a cycle, we introduce

Figure 2.10

a single-row *cyclic notation*. In cyclic notation, the cycle in (3) becomes

$$\mu = (1, 3, 6).$$

We understand by this notation that μ carries the first number 1 into the second number 3, the second number 3 into the next number 6. etc., until finally the last number 6 is carried into the first number 1. An integer not appearing in this notation for μ is understood to be left fixed by μ. Of course, the set on which μ acts, which is $\{1, 2, 3, 4, 5, 6, 7, 8\}$ in our example, must be made clear by the context.

EXAMPLE 3   Working within S_5, we see that

$$(1, 3, 5, 4) = \begin{pmatrix} 1 & 2 & 3 & 4 & 5 \\ 3 & 2 & 5 & 1 & 4 \end{pmatrix}.$$

Observe that

$$(1, 3, 5, 4) = (3, 5, 4, 1) = (5, 4, 1, 3) = (4, 1, 3, 5). \quad \blacktriangle$$

Of course, since cycles are special types of permutations, they can be multiplied just as any two permutations. The product of two cycles need not again be a cycle, however.

Using cyclic notation, we see that the permutation σ in Eq. (2) can be written as a product of cycles:

$$\sigma = \begin{pmatrix} 1 & 2 & 3 & 4 & 5 & 6 & 7 & 8 \\ 3 & 8 & 6 & 7 & 4 & 1 & 5 & 2 \end{pmatrix} = (1, 3, 6)(2, 8)(4, 7, 5). \tag{4}$$

These cycles are **disjoint**, meaning that any integer is moved by at most one of these cycles; thus no one number appears in the notations of two different cycles. Equation (4) exhibits σ in terms of its orbits, and is a one-line description of Fig. 2.9. Every permutation in S_n can be expressed in a similar fashion as a product of the disjoint cycles corresponding to its orbits. We state this as a theorem and write out the proof.

THEOREM 2.2   Every permutation σ of a finite set is a product of disjoint cycles.

PROOF   Let $B_1, B_2, \ldots, B_r$ be the orbits of σ, and let μ_i be the cycle defined by

$$\mu_i(x) = \begin{cases} \sigma(x) & \text{for } x \in B_i \\ x & \text{otherwise.} \end{cases}$$

Clearly $\sigma = \mu_1 \mu_2 \cdots \mu_r$. Since the equivalence-class orbits $B_1, B_2, \ldots, B_r$ are disjoint, the cycles $\mu_1, \mu_2, \ldots, \mu_r$ are disjoint also. $\blacklozenge$

While permutation multiplication in general is not commutative, it is readily seen that *multiplication of disjoint cycles is commutative*. Since the orbits of a permutation are unique, the representation of a permutation as a product of disjoint cycles, none of which is the identity permutation, is unique up to the order of the factors.

EXAMPLE 4   Consider the permutation

$$\begin{pmatrix} 1 & 2 & 3 & 4 & 5 & 6 \\ 6 & 5 & 2 & 4 & 3 & 1 \end{pmatrix}.$$

Let us write it as a product of disjoint cycles. First, 1 is moved to 6 and then 6 to 1, giving the cycle $(1, 6)$. Then 2 is moved to 5, which is moved to 3, which is moved to 2, or $(2, 5, 3)$. This takes care of all elements but 4, which is left fixed. Thus

$$\begin{pmatrix} 1 & 2 & 3 & 4 & 5 & 6 \\ 6 & 5 & 2 & 4 & 3 & 1 \end{pmatrix} = (1, 6)(2, 5, 3).$$

Multiplication of *disjoint* cycles is commutative, so the order of the factors $(1, 6)$ and $(2, 5, 3)$ is not important. ▲

You should practice multiplying permutations in cyclic notation where the cycles may or may not be disjoint. We give an example and provide further practice in Exercises 7 through 9.

EXAMPLE 5   Let $(1, 4, 5, 6)$ and $(2, 1, 5)$ be cycles in S_6. Then

$$(1, 4, 5, 6)(2, 1, 5) = \begin{pmatrix} 1 & 2 & 3 & 4 & 5 & 6 \\ 6 & 4 & 3 & 5 & 2 & 1 \end{pmatrix}$$

and

$$(2, 1, 5)(1, 4, 5, 6) = \begin{pmatrix} 1 & 2 & 3 & 4 & 5 & 6 \\ 4 & 1 & 3 & 2 & 6 & 5 \end{pmatrix}.$$

Neither of these permutations is a cycle. ▲

Even and Odd Permutations

It seems reasonable that every reordering of the sequence $1, 2, \ldots, n$ can be achieved by repeated interchange of positions of pairs of numbers. We discuss this a bit more formally.

DEFINITION 2.7 (Transposition)   A cycle of length 2 is a **transposition**.

Thus a transposition leaves all elements but two fixed, and maps each of these onto the other. A computation shows that

$$(a_1, a_2, \ldots, a_n) = (a_1, a_n)(a_1, a_{n-1}) \cdots (a_1, a_3)(a_1, a_2).$$

Therefore any cycle is a product of transpositions. We then have the following as a corollary to Theorem 2.2.

COROLLARY　Any permutation of a finite set of at least two elements is a product of transpositions.

Naively, this corollary just states that any rearrangement of n objects can be achieved by successively interchanging pairs of them.

EXAMPLE 6　Following the remarks prior to the corollary, we see that $(1, 6)(2, 5, 3)$ is the product $(1, 6)(2, 3)(2, 5)$ of transpositions. ▲

EXAMPLE 7　In S_n for $n \geq 2$, the identity permutation is the product $(1, 2)(1, 2)$ of transpositions. ▲

We have seen that every permutation of a finite set with at least two elements is a product of transpositions. The transpositions may not be disjoint, and a representation of the permutation in this way is not unique. For example, we can always insert at the beginning the transposition (a, b) twice, since $(a, b)(a, b)$ is the identity permutation. What is true is that the number of transpositions used to represent a given permutation must either always be even or always be odd. This is an important fact, and the usual proof of it, which may be found in the first edition of this text, involves a rather artificial construction. In the third edition, we gave a proof published in 1971 by William I. Miller. In the fourth edition, we gave a proof based on a suggestion from David M. Bloom. Since then, we have received two more suggestions, one from Eric Wilson for avoiding the mathematical induction argument in the proof by Bloom, and another proof from David M. Berman, published in 1978, that uses mathematical induction but makes no use of orbits. We have chosen to keep a proof involving orbits since we wish to encourage understanding of this important concept, and present here the proof by Bloom with the change suggested by Wilson. Exercise 28 outlines Berman's proof for students to fill in. We start with a lemma.

LEMMA　Let $\sigma \in S_n$ and let τ be a transposition in S_n. The number of orbits of σ and the number of orbits of $\tau\sigma$ differ by 1.

PROOF　Let $\tau = (i, j)$. We shall show that if i and j are in different orbits of σ, then τ essentially creates a bridge between them so that $\tau\sigma$ has one fewer orbits than σ. On the other hand, if i and j are in the same orbit of σ, we will show that τ will cut it into two orbits in $\tau\sigma$, which will then have one more orbit than σ. Of course multiplication of σ by τ will not affect orbits that do not contain i or j.

Suppose first that i and j are in different orbits of σ. By Theorem 2.2, we can express σ as a product of r disjoint cycles

$$\sigma = \underbrace{(u, \ldots, a, i, \ldots, v)}_{\mu_1}\underbrace{(x, \ldots, b, j, \ldots, y)}_{\mu_2}\mu_3 \cdots \mu_r,$$

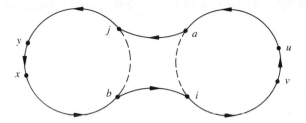

Figure 2.11

where we have written out the cycles μ_1 and μ_2 that contain i and j. Figure 2.11 shows the cycles μ_1 and μ_2, and shows how these two orbits are joined into a single orbit in

$$\tau\mu_1\mu_2 = (i, j)(u, \ldots, a, i, \ldots, v)(x, \ldots, b, j, \ldots, y). \qquad (5)$$

This is easily checked from Eq. (5) by permutation multiplication. (We should also consider cases where one or both of the cycles of σ containing i and j are of length 1. We ask you to draw figures like Fig. 2.11 to check these cases, in Exercise 17.)

Suppose now that i and j are in the same orbit of σ, and let the decomposition of σ into disjoint cycles be

$$\sigma = \underbrace{(x, \ldots, a, i, \ldots, b, j, \ldots, y)}_{\mu_1}\mu_2 \cdots \mu_r.$$

Figure 2.12 shows the cycle μ_1 of σ and indicates the way in which the orbit of μ_1 is split into two orbits in

$$\tau\mu_1 = (i, j)(x, \ldots, a, i, \ldots, b, j, \ldots, y). \qquad (6)$$

This can be verified by performing the permutation multiplication $\tau\sigma$ in Eq. (6). ◆

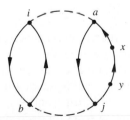

Figure 2.12

THEOREM 2.3 No permutation in S_n can be expressed both as a product

of an even number of transpositions and as a product of an odd number of transpositions.

PROOF   Let $\sigma \in S_n$ and suppose that σ can be expressed as a product of both an even number of transpositions and an odd number of transpositions. Now each transposition is its own inverse, so the inverse of a product of transpositions is the product of the same transpositions in the opposite order. Consequently, σ^{-1} can also be expressed as a product of both an odd number of transpositions and an even number of transpositions. Consider a product $\sigma\sigma^{-1} = \iota$, where we express σ as an even number of transpositions and σ^{-1} as an odd number of transpositions, so that we have an expression

$$\tau_{2m+1}\tau_{2m} \cdots \tau_3\tau_2\tau_1 = \iota$$

for the identity permutation ι as the product of an odd number of transpositions τ_i. Writing this equation in the form $\tau_{2m+1}\tau_{2m} \cdots \tau_3\tau_2\tau_1\iota = \iota$ and applying the lemma repeatedly, we see that the number of orbits of ι must differ from the number of orbits of ι by an odd number, which is impossible. (Note that the number of orbits of ι is n.) Thus we have a contradiction to our assumption that σ can be expressed as both a product of an even number of transpositions and a product of an odd number of transpositions. ◆

DEFINITION 2.8 (Even or Odd Permutation)   A permutation of a finite set is **even** or **odd** according to whether it can be expressed as a product of an even number of transpositions or the product of an odd number of transpositions, respectively.

EXAMPLE 8   The identity permutation ι in S_n is an even permutation since we have $\iota = (1, 2)(1, 2)$. If $n = 1$ so that we cannot form this product, we define ι to be even. On the other hand, the permutation $(1, 4, 5, 6)(2, 1, 5)$ in S_6 can be written as

$$(1, 4, 5, 6)(2, 1, 5) = (1, 6)(1, 5)(1, 4)(2, 5)(2, 1)$$

which has five transpositions, so this is an odd permutation. ▲

For students familiar with the theory of determinants, we mention that Theorem 2.3 follows at once from the fact if A is an $n \times n$ matrix, then interchanging any two rows of A produces a matrix whose determinant is $-\det(A)$. Since a permutation of the rows of the identity matrix I_n cannot produce a matrix with determinant both 1 and -1, such a permutation cannot be achieved both by interchanging rows an even number of times and an odd number of times.

The Alternating Groups

We claim that for $n \geq 2$, the number of even permutations in S_n is the same as the number of odd permutations; that is, S_n is split equally and both numbers are $(n!)/2$. To show this, let A_n be the set of even permutations in S_n and let B_n be the set of odd permutations for $n \geq 2$. We proceed to define a one-to-one function from A_n onto B_n. This is exactly what is needed to show that A_n and B_n have the same number of elements.

Let τ be any fixed transposition in S_n; it exists since $n \geq 2$. We may as well suppose that $\tau = (1, 2)$. We define a function

$$\lambda_\tau : A_n \to B_n$$

by

$$\lambda_\tau(\sigma) = \tau\sigma,$$

that is, $\sigma \in A_n$ is mapped into $(1, 2)\sigma$ by λ_τ. Observe that since σ is even, the permutation $(1, 2)\sigma$ appears as a product of a $(1 + \text{even number})$, or odd number, of transpositions, so $(1, 2)\sigma$ is indeed in B_n. If for σ and $\mu \in a_n$ it is true that $\sigma\lambda_\tau = \mu\lambda_\tau$, then

$$(1, 2)\sigma = (1, 2)\mu,$$

and since S_n is a group, we have $\sigma = \mu$. Thus λ_τ is a one-to-one function. Finally,

$$\tau = (1, 2) = \tau^{-1},$$

so if $\rho \in B_n$, then

$$\tau^{-1}\rho \in A_n,$$

and

$$\lambda_\tau(\tau^{-1}\rho) = \tau(\tau^{-1}\rho) = \rho.$$

Thus λ_τ is onto B_n. Hence the number of elements in A_n is the same as the number in B_n since there is a one-to-one correspondence between the elements of the sets.

Note that the product of two even permutations is again even. Also since $n \geq 2$, S_n has the transposition $(1, 2)$ and $\iota = (1, 2)(1, 2)$ is an even permutation. Finally, note that if σ is expressed as a product of transpositions, the product of the same transpositions taken in just the opposite order is σ^{-1}. Thus if σ is an even permutation, σ^{-1} must also be even. Referring to Theorem 1.4, we see that we have proved the following statement.

If $n \geq 2$, then the collection of all even permutations of $\{1, 2, 3, \ldots, n\}$ forms a subgroup of order $n!/2$ of the symmetric group S_n.

DEFINITION 2.9 (Alternating Group) The subgroup of S_n consisting of the even permutations of n letters is the **alternating group** A_n **on** n **letters**.

Both S_n and A_n are very important groups. We will show in Chapter 2 that every finite group is structurally identical to some subgroup of S_n for some n. Chapter 9 will show that there are no formulas involving radicals for solution of polynomial equations of degree n for $n \geq 5$. This fact is actually due to the structure of A_n, surprising as that may seem!

Exercises 2.2

Computations

In Exercises 1 through 6, find all orbits of the given permutation.

1. $\begin{pmatrix} 1 & 2 & 3 & 4 & 5 & 6 \\ 5 & 1 & 3 & 6 & 2 & 4 \end{pmatrix}$

2. $\begin{pmatrix} 1 & 2 & 3 & 4 & 5 & 6 & 7 & 8 \\ 5 & 6 & 2 & 4 & 8 & 3 & 1 & 7 \end{pmatrix}$

3. $\begin{pmatrix} 1 & 2 & 3 & 4 & 5 & 6 & 7 & 8 \\ 2 & 3 & 5 & 1 & 4 & 6 & 8 & 7 \end{pmatrix}$

4. $\sigma : \mathbb{Z} \to \mathbb{Z}$ where $\sigma(n) = n + 1$

5. $\sigma : \mathbb{Z} \to \mathbb{Z}$ where $\sigma(n) = n + 2$

6. $\sigma : \mathbb{Z} \to \mathbb{Z}$ where $\sigma(n) = n - 3$

In Exercises 7 through 9, compute the indicated product of cycles that are permutations of $\{1, 2, 3, 4, 5, 6, 7, 8\}$.

7. $(1, 4, 5)(7, 8)(2, 5, 7)$

8. $(1, 3, 2, 7)(4, 8, 6)$

9. $(1, 2)(4, 7, 8)(2, 1)(7, 2, 8, 1, 5)$

In Exercises 10 through 12, express the permutation of $\{1, 2, 3, 4, 5, 6, 7, 8\}$ as a product of disjoint cycles, and then as a product of transpositions.

10. $\begin{pmatrix} 1 & 2 & 3 & 4 & 5 & 6 & 7 & 8 \\ 8 & 2 & 6 & 3 & 7 & 4 & 5 & 1 \end{pmatrix}$

11. $\begin{pmatrix} 1 & 2 & 3 & 4 & 5 & 6 & 7 & 8 \\ 3 & 6 & 4 & 1 & 8 & 2 & 5 & 7 \end{pmatrix}$

12. $\begin{pmatrix} 1 & 2 & 3 & 4 & 5 & 6 & 7 & 8 \\ 3 & 1 & 4 & 7 & 2 & 5 & 8 & 6 \end{pmatrix}$

Concepts

13. Mark each of the following true or false.

_____ a. Every permutation is a cycle.

_____ b. Every cycle is a permutation.

_____ c. The definition of even and odd permutations could have been given equally well before Theorem 2.3.

_____ d. Every nontrivial subgroup H of S_9 containing some odd permutation contains a transposition.

_____ e. A_5 has 120 elements.

_____ f. S_n is not cyclic for any $n \geq 1$.

_____ g. A_3 is a commutative group.

_____ h. S_7 is isomorphic to the subgroup of all those elements of S_8 that leave the number 8 fixed.

_____ i. S_7 is isomorphic to the subgroup of all those elements of S_8 that leave the number 5 fixed.

_____ j. The odd permutations in S_8 form a subgroup of S_8.

14. Which of the permutations in S_3 of Example 4 of Section 2.1 are even permutations? Give the table for the alternating group A_3.

15. Recall that element a of a group G with identity e has order $r > 0$ if $a^r = e$ and no smaller positive power of a is the identity. Consider the group S_8.

   a. What is the order of the cyclic $(1, 4, 5, 7)$?

   b. State a theorem suggested by part (a).

   c. What is the order of $\sigma = (4, 5)(2, 3, 7)$? of $\tau = (1, 4)(3, 5, 7, 8)$?

   d. Find the order of each of the permutations given in Exercises 10 through 12 by looking at its decomposition into a product of disjoint cycles.

   e. State a theorem suggested by parts (c) and (d). [*Hint:* The important words you are looking for are *least common multiple.*]

Theory

16. Prove the following about S_n if $n \geq 2$.

   a. Every permutation in S_n can be written as a product of at most $n - 1$ transpositions.

   b. Every permutation in S_n that is not a cycle can be written as a product of at most $n - 2$ transpositions.

   c. Every odd permutation in S_n can be written as a product of $2n + 3$ transpositions, and every even permutation as a product of $2n + 8$ transpositions.

17. a. Draw a figure like Fig. 2.11 to illustrate that if i and j are in different orbits of σ and $\sigma(i) = i$, then the number of orbits of $(i, j)\sigma$ is one less than the number of orbits of σ.

   b. Repeat part (a) if $\sigma(j) = j$ also.

18. Show that for every subgroup H of S_n for $n \geq 2$, either all the permutations in H are even or exactly half of them are even.

19. Let σ be a permutation of a set A. We shall say "σ **moves** $a \in A$" if $\sigma(a) \neq a$. If A is a finite set, how many elements are moved by a cycle $\sigma \in S_A$ of length n?

20. Let A be an infinite set. Let H be the set of all $\sigma \in S_A$ that move (see Exercise 19) only a finite number of elements of A. Show that H is a subgroup of S_A.

21. Let A be an infinite set. Let K be the set of all $\sigma \in S_A$ that move (see Exercise 19) at most 50 elements of A. Is K a subgroup of S_A? Why?

22. Consider S_n for a fixed $n \geq 2$ and let σ be a fixed odd permutation. Show that every odd permutation in S_n is a product of σ and some permutation in A_n.

23. Show that if σ is a cycle of odd length, then σ^2 is a cycle.

24. Following the line of thought opened by Exercise 23, complete the following with a condition involving n and r so that the resulting statement is a theorem:

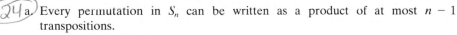

        If σ is a cycle of length n, then σ^r is also a cycle if and only if . . .

25. Let G be a group and let a be a fixed element of G. Show that the map $\lambda_a : G \to G$, given by $\lambda_a(g) = ag$ for $g \in G$, is a permutation of the set G.

26. Referring to Exercise 25 show that $H = \{\lambda_a \mid a \in G\}$ is a subgroup of S_G, the group of all permutations of G.

27. Referring to Exercise 45 of Section 2.1, show that H of Exercise 26 is transitive on the set G. [*Hint:* This is an immediate corollary of one of the theorems in Section 1.2.]

28. (This exercise gives in outline form a proof by David M. Berman of Theorem 2.3.) Consider the symmetric group S_n. We wish to show that no permutation can be written as a product of both an odd number and an even number of transpositions.

   a. Use the equation $\iota = \sigma\sigma^{-1}$ to show that it suffices to prove that the identity permutation can't be written as a product of both an odd number and an even number of transpositions.

   b. Let a *switch* be a transposition $(i, i + 1)$ of two consecutive numbers. Show that the transposition $(i, i + k)$ can be expressed as a product of $2k - 1$ switches.

   c. Conclude from (b) that we need only prove that ι can't be written as a product of an odd number of switches.

   d. Proceeding by induction on n, show that this is the case for $n = 2$.

   e. We take as induction hypothesis that the identity permutation of S_n can't be expressed as an odd number of switches. Consider the identity permutation of S_{n+1} expressed as a product of switches. Show that the number of switches involving the number $n + 1$ must be even.

   f. Show that the switches not involving $n + 1$ must constitute the identity permutation of S_n. [*Hint:* What effect does a switch involving $n + 1$ have on the order among themselves of the numbers $1, 2, \ldots, n$?]

   g. Deduce that there is an even number of switches in the expression for the identity permutation in S_{n+1} as a product of switches.

Cayley Digraphs

29. Figure 2.13 shows a digraph for the alternating group A_4 using the set $S = \{(1, 2, 3), (1, 2)(3, 4)\}$ of generators. Continue labeling the other nine vertices with the elements of A_4, expressed as a product of disjoint cycles.

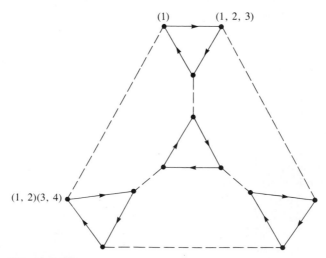

Figure 2.13

Plane Isometries

Consider the Euclidean plane $\mathbb{R}^2$. An **isometry of** $\mathbb{R}^2$ is a permutation $\phi:\mathbb{R}^2 \to \mathbb{R}^2$ that preserves distance, so that the distance between points P and Q is the same as the distance between the points $\phi(P)$ and $\phi(Q)$ for all points P and Q in $\mathbb{R}^2$. If ψ is also an isometry of $\mathbb{R}^2$, then the distance between $\psi(\phi(P))$ and $\psi(\phi(Q))$ must be the same as the distance between $\phi(P)$ and $\phi(Q)$, which in turn is the distance between P and Q, showing that the composition of two isometries is again an isometry. Since the identity map is an isometry and the inverse of an isometry is an isometry, we see that the isometries of $\mathbb{R}^2$ form a subgroup of the group of all permutations of $\mathbb{R}^2$.

Given any subset S of $\mathbb{R}^2$, the isometries of $\mathbb{R}^2$ that carry S onto itself form a subgroup of the group of isometries. This subgroup is the **group of symmetries of** S **in** $\mathbb{R}^2$. In Section 2.1 we gave tables for the group of symmetries of an equilateral triangle and for the group of symmetries of a square in $\mathbb{R}^2$.

Everything we have defined in the two preceding paragraphs could equally well have been done for n-dimensional Euclidean space $\mathbb{R}^n$, but we will concern ourselves chiefly with plane isometries here.

It can be proved that every isometry of the plane is one of just four types (see Artin [5]). We will list the types and show, for each type, a labeled figure that can be carried into itself by an isometry of that type. In each of Figs. 2.14, 2.16, and 2.17, consider the line with spikes shown to be extended infinitely to the left and to the right. We also give an example of each type in terms of coordinates.

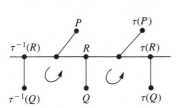

Figure 2.14   Translation τ.

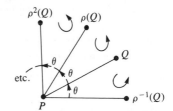

Figure 2.15   Rotation ρ.

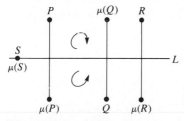

Figure 2.16   Reflection μ.

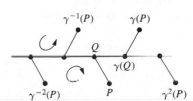

Figure 2.17   Glide reflection γ.

translation τ: Slide every point the same distance in the same direction. See Fig. 2.14. (*Example:* $\tau(x, y) = (x, y) + (2, -3) = (x + 2, y - 3)$.)

rotation ρ: Rotate the plane about a point P through an angle θ. See Fig. 2.15. (*Example:* $\rho(x, y) = (-y, x)$ is a rotation through 90° counterclockwise about the origin $(0, 0)$.)

reflection μ: Map each point into its mirror image (μ for mirror) across a line L, each point of which is left fixed by μ. See Fig. 2.16. The line L is the *axis of reflection.* (*Example:* $\mu(x, y) = (y, x)$ is a reflection across the line $y = x$.)

glide reflection γ: The product of a translation and a reflection across a line mapped into itself by the translation. See Fig. 2.17. (*Example:* $\gamma(x, y) = (x + 4, -y)$ is a glide reflection along the x-axis.)

Notice the little curved arrow that is carried into another curved arrow in each of Figs. 2.14–2.17. For the translation and rotation, the counterclockwise directions of the curved arrows remain the same, but for the reflection and glide reflection, the counterclockwise arrow is mapped into a clockwise arrow. We say that translations and rotations *preserve orientation*, while the reflection and glide reflection *reverse orientation*. We do not classify the identity isometry as any definite one of the four types listed; it could equally well be considered to be a translation by the zero vector or a rotation about any point through an angle of 0°. We always consider a glide reflection to be the product of a reflection and a translation that is different from the identity isometry.

30. This exercise shows that the group of symmetries of a certain type of geometric figure may depend on the dimension of the space in which we consider the figure to lie.

   a. Describe all symmetries of a point in the real line $\mathbb{R}$; that is, describe all isometries of $\mathbb{R}$ that leave one point fixed.

   b. Describe all symmetries (translations, reflections, etc.) of a point in the plane $\mathbb{R}^2$.

   c. Describe all symmetries of a line segment in $\mathbb{R}$.

   d. Describe all symmetries of a line segment in $\mathbb{R}^2$.

   e. Describe some symmetries of a line segment in $\mathbb{R}^3$.

31. Let P stand for an orientation preserving plane isometry and R for an orientation reversing one. Fill in the table with P or R to denote the orientation preserving or reversing property of a product.

	P	R
P		
R		

32. Fill in the table to give *all* possible types of plane isometries given by a product of two types. For example, a product of two rotations may be a rotation, or it may be another type. Fill in the box corresponding to $\rho\rho$ with both letters. Use your answer to Exercise 31 to eliminate some types. Eliminate the identity from consideration.

	τ	ρ	μ	γ
τ				
ρ				
μ				
γ				

33. Draw a plane figure that has a one-element group as its group of symmetries in $\mathbb{R}^2$.

34. Draw a plane figure that has a two-element group as its group of symmetries in $\mathbb{R}^2$.

35. Draw a plane figure that has a three-element group as its group of symmetries in $\mathbb{R}^2$.

36. Draw a plane figure that has a four-element group isomorphic to $\mathbb{Z}_4$ as its group of symmetries in $\mathbb{R}^2$.

37. Draw a plane figure that has a four-element group isomorphic to the Klein 4-group V as its group of symmetries in $\mathbb{R}^2$.

38. For each of the four types of plane isometries (other than the identity), give the possibilities for the order of an isometry of that type in the group of plane isometries.

39. A plane isometry ϕ has a *fixed point* if there exists a point P in the plane such that $\phi(P) = P$. Which of the four types of plane isometries (other than the identity) can have a fixed point?

40. Referring to Exercise 39, which types of plane isometries, if any, have exactly one fixed point?

41. Referring to Exercise 39, which types of plane isometries, if any, have exactly two fixed points?

42. Referring to Exercise 39, which types of plane isometries, if any, have an infinite number of fixed points?

43. Argue geometrically that a plane isometry that leaves three noncolinear points fixed must be the identity map.

44. Using Exercise 43, show algebraically that if two plane isometries ϕ and ψ agree on three noncolinear points, that is, if $\phi(P_i) = \psi(P_i)$ for noncolinear points P_1, P_2, and P_3, then ϕ and ψ are the same map.

45. Do the rotations, together with the identity map, form a subgroup of the group of plane isometries? Why?

46. Do the translations, together with the identity map, form a subgroup of the group of plane isometries? Why?

47. Do the rotations about one particular point P, together with the identity map, form a subgroup of the group of plane isometries? Why?

48. Do the reflections across one particular line L, together with the identity map, form a subgroup of the group of plane isometries? Why?

49. Do the glide reflections, together with the identity map, form a subgroup of the group of plane isometries? Why?

50. Which of the four types of plane isometries can be elements of a *finite* subgroup of the group of plane isometries?

2.3

Cosets and the Theorem of Lagrange

You may have noticed that the order of a subgroup H of a finite group G seems always to be a divisor of the order of G. This is the theorem of Lagrange. We shall prove it by exhibiting a partition of G into cells, all having the same size as H. Thus if there are r such cells, we will have

$$r(\text{order of } H) = (\text{order of } G)$$

from which the theorem follows immediately. The cells in the partition will be called *cosets of H*, and they are important in their own right. In Chapter 3, we will see that if H satisfies a certain property, then each coset can be regarded as an element of a group in a very natural way. We give some indication of such *coset groups* in this section, to help you develop a feel for the topic.

Cosets

Let H be a subgroup of a group G, which may be of finite or infinite order. We exhibit two partitions of G by defining two equivalence relations, $\sim_L$ and $\sim_R$ on G.

THEOREM 2.4 Let H be a subgroup of G. Let the relation $\sim_L$ be defined on G by

$$a \sim_L b \qquad \text{if and only if} \qquad a^{-1}b \in H.$$

Let $\sim_R$ be defined by

$$a \sim_R b \qquad \text{if and only if} \qquad ab^{-1} \in H.$$

Then $\sim_L$ and $\sim_R$ are both equivalence relations on G.

PROOF   We show that $\sim_L$ is an equivalence relation, and leave the proof for $\sim_R$ to Exercise 21. When reading the proof, notice how we must constantly make use of the fact that H is a *subgroup* of G.

Reflexive Let $a \in G$. Then $a^{-1}a = e$ and $e \in H$ since H is a subgroup. Thus $a \sim_L a$.

Symmetric Suppose $a \sim_L b$. Then $a^{-1}b \in H$. Since H is a subgroup, $(a^{-1}b)^{-1}$ is in H and $(a^{-1}b)^{-1} = b^{-1}a$, so $b^{-1}a$ is in H and $b \sim_L a$.

Transitive Let $a \sim_L b$ and $b \sim_L c$. Then $a^{-1}b \in H$ and $b^{-1}c \in H$. Since H is a subgroup, $(a^{-1}b)(b^{-1}c) = a^{-1}c$ is in H, so $a \sim_L c$. ◆

The equivalence relation $\sim_L$ in Theorem 2.4 defines a partition of G, as described in Theorem 0.1. Let's see what the cells in this partition look like. Suppose $a \in G$. The cell containing a consists of all $x \in G$ such that $a \sim_L x$, which means all $x \in G$ such that $a^{-1}x \in H$. Now $a^{-1}x \in H$ if and only if $a^{-1}x = h$ for some $h \in H$, or equivalently, if and only if $x = ah$ for some $h \in H$. Thus the cell containing a is $\{ah \mid h \in H\}$, which we denote by aH. If we go through the same reasoning for the equivalence relation $\sim_R$ defined by H, we find the cell in this partition containing $a \in G$ is $Ha = \{ha \mid h \in H\}$. Since G need not be abelian, we have no reason to expect aH and Ha to be the same subset of G. We give a formal definition.

DEFINITION 2.10 (Cosets) Let H be a subgroup of a group G. The subset $aH = \{ah \mid h \in H\}$ of G is the **left coset** of H containing a, while $Ha = \{ha \mid h \in H\}$ is the **right coset** of H containing a.

EXAMPLE 1 Exhibit the left cosets and the right cosets of the subgroup $3\mathbb{Z}$ of $\mathbb{Z}$.

Solution Our notation here is additive, so the left coset of $3\mathbb{Z}$ containing m is $m + 3\mathbb{Z}$. Taking $m = 0$, we see that

$$3\mathbb{Z} = \{\ldots, -9, -6, -3, 0, 3, 6, 9, \ldots\}$$

is itself one of its left cosets, the coset containing 0. To find another left coset, we select an element of $\mathbb{Z}$ not in $3\mathbb{Z}$, say 1, and find the left coset containing it. We have

$$1 + 3\mathbb{Z} = \{\ldots, -8, -5, -2, 1, 4, 7, 10, \ldots\}.$$

These two left cosets, $3\mathbb{Z}$ and $1 + 3\mathbb{Z}$, do not yet exhaust $\mathbb{Z}$. For example, 2 is in neither of them. The left coset containing 2 is

$$2 + 3\mathbb{Z} = \{\ldots, -7, -4, -1, 2, 5, 8, 11, \ldots\}.$$

It is clear that these three left cosets we have found do exhaust $\mathbb{Z}$, so they constitute the partition of $\mathbb{Z}$ into left cosets of $3\mathbb{Z}$.

Since $\mathbb{Z}$ is abelian, the left coset $m + 3\mathbb{Z}$ and the right coset $3\mathbb{Z} + m$ are the same, so the partition of $\mathbb{Z}$ into right cosets is the same. ▲

We observe two things from Example 1.

For an abelian subgroup H of G, the partition of G into left cosets of H and the partition into right cosets are the same.

Also, looking back at Definition 0.5, we see that the equivalence relation $\sim_R$ for the subgroup $n\mathbb{Z}$ of $\mathbb{Z}$ is the same as the relation of congruence modulo n. Recall that $h \equiv k \pmod{n}$ in $\mathbb{Z}$ if $h - k$ is divisible by n. This is the same as saying that $h + (-k)$ is in $n\mathbb{Z}$, which is relation $\sim_R$ of Theorem 2.4 in additive notation. Thus the partition of $\mathbb{Z}$ into cosets of $n\mathbb{Z}$ is the partition of $\mathbb{Z}$ into residue classes modulo n. For that reason, we often refer to the cells of this partition as *cosets modulo* $n\mathbb{Z}$. Note that we do not have to specify *left* or *right* cosets since they are the same for this abelian group $\mathbb{Z}$.

EXAMPLE 2   The group $\mathbb{Z}_6$ is abelian. Find the partition of $\mathbb{Z}_6$ into cosets of the subgroup $H = \{0, 3\}$.

Solution One coset is $\{0, 3\}$ itself. The coset containing 1 is $1 + \{0, 3\} = \{1, 4\}$. The coset containing 2 is $2 + \{0, 3\} = \{2, 5\}$. Since $\{0, 3\}$, $\{1, 4\}$, and $\{2, 5\}$ exhaust all of $\mathbb{Z}_6$, these are all the cosets. ▲

We point out a fascinating thing that we will develop in detail in Chapter 3. Referring back to Example 2, Table 2.3 gives the binary operation for $\mathbb{Z}_6$ but with elements listed in the order they appear in the cosets $\{0, 3\}$, $\{1, 4\}$, $\{2, 5\}$. We shaded the table according to these cosets.

Suppose we denote these cosets by LT(light), MD(medium), and DK(dark) according to their shading. Table 2.3 then defines a binary operation on these shadings, as shown in Table 2.4. Note that if we replace

$+_6$	0	3	1	4	2	5
0	0	3	1	4	2	5
3	3	0	4	1	5	2
1	1	4	2	5	3	0
4	4	1	5	2	0	3
2	2	5	3	0	4	1
5	5	2	0	3	1	4

Table 2.3

	LT	MD	DK
LT	LT	MD	DK
MD	MD	DK	LT
DK	DK	LT	MD

Table 2.4

LT by 0, MD by 1, and DK by 2 in Table 2.4, we obtain the table for $\mathbb{Z}_3$. Thus the table of shadings forms a group! We will see in Chapter 3 that for a partition of an *abelian* group into cosets of a subgroup, reordering the group table according to the elements in the cosets always gives rise to such a *coset group*.

EXAMPLE 3   Table 2.5 again shows Table 2.1 for the symmetric group S_3 on three letters. Let H be the subgroup $\langle \mu_1 \rangle = \{\rho_0, \mu_1\}$ of S_3. Find the partitions of S_3 into left cosets of H, and the partition into right cosets of H.

Solution   For the partition into left cosets, we have

$$H = \{\rho_0, \mu_1\},$$
$$\rho_1 H = \{\rho_1 \rho_0, \rho_1 \mu_1\} = \{\rho_1, \mu_3\},$$
$$\rho_2 H = \{\rho_2 \rho_0, \rho_2 \mu_1\} = \{\rho_2, \mu_2\}.$$

The partition into right cosets is

$$H = \{\rho_0, \mu_1\},$$
$$H\rho_1 = \{\rho_0 \rho_1, \mu_1 \rho_1\} = \{\rho_1, \mu_2\},$$
$$H\rho_2 = \{\rho_0 \rho_2, \mu_1 \rho_2\} = \{\rho_2, \mu_3\}.$$

The partition into left cosets of H is different from the partition into right

	ρ_0	ρ_1	ρ_2	μ_1	μ_2	μ_3
ρ_0	ρ_0	ρ_1	ρ_2	μ_1	μ_2	μ_3
ρ_1	ρ_1	ρ_2	ρ_0	μ_3	μ_1	μ_2
ρ_2	ρ_2	ρ_0	ρ_1	μ_2	μ_3	μ_1
μ_1	μ_1	μ_2	μ_3	ρ_0	ρ_1	ρ_2
μ_2	μ_2	μ_3	μ_1	ρ_2	ρ_0	ρ_1
μ_3	μ_3	μ_1	μ_2	ρ_1	ρ_2	ρ_0

Table 2.5

	ρ_0	μ_1	ρ_1	μ_3	ρ_2	μ_2
ρ_0	ρ_0	μ_1	ρ_1	μ_3	ρ_2	μ_2
μ_1	μ_1	ρ_0	μ_2	ρ_2	μ_3	ρ_1
ρ_1	ρ_1	μ_3	ρ_2	μ_2	ρ_0	μ_1
μ_3	μ_3	ρ_1	μ_1	ρ_0	μ_2	ρ_2
ρ_2	ρ_2	μ_2	ρ_0	μ_1	ρ_1	μ_3
μ_2	μ_2	ρ_2	μ_3	ρ_1	μ_1	ρ_0

Table 2.6

cosets. For example, the left coset containing ρ_1 is $\{\rho_1, \mu_3\}$, while the right coset containing ρ_1 is $\{\rho_1, \mu_2\}$. This does not surprise us since the group S_3 is not abelian. ▲

Referring to Example 3, Table 2.6 gives permutation multiplication in S_3. The elements are listed in the order they appear in the left cosets $\{\rho_0, \mu_1\}$, $\{\rho_1, \mu_3\}$, $\{\rho_2, \mu_2\}$ found in that example. Again, we have shaded the table light, medium, and dark according to the coset to which the element belongs. Note the difference between this table and Table 2.3. This time, the body of the table does not split up into 2 by 2 blocks opposite and under the shaded cosets at the left and top, as in Table 2.3, and we don't get a coset group.

Table 2.5 is shaded according to the two left cosets of the subgroup $\langle \rho_1 \rangle = \{\rho_0, \rho_1, \rho_2\}$ of S_3. These are also the two right cosets, even though S_3 is not abelian. From Table 2.5 it is clear that we do have a coset group isomorphic to $\mathbb{Z}_2$ in this case. We will see in Chapter 3 that the left cosets of a subgroup H of a group G give rise to a coset group precisely when the partition of G into left cosets of H is the same as the partition into right cosets of H. In such a case, we may simply speak of the *cosets of H,* omitting the adjective left or right. We discuss coset groups in detail in Chapter 3, but we think it will be easier for you to understand them then if you experiment a bit with them now. Some of the exercises in this section are designed for such experimentation.

The Theorem of Lagrange

Let H be a subgroup of a group G. We claim that every left coset and every right coset of H have the same number of elements as H. We show this by exhibiting a *one-to-one* map of H *onto* a left coset gH of H for a fixed element g of G. If H is of finite order, this will show that gH has the same number of elements as H. If H is infinite, the existence of such a map is taken as the *definition* for equality of the size of H and the size of gH.

Our choice for a one-to-one map $\phi : H \rightarrow gH$ is the natural one. Let $\phi(h) = gh$ for each $h \in H$. This map is onto gH by the definition of gH as

$\{gh \mid h \in H\}$. To show that it is one to one, suppose that $\phi(h_1) = \phi(h_2)$ for h_1 and h_2 in H. Then $gh_1 = gh_2$ and by the cancellation law in the group G, we have $h_1 = h_2$. This shows that ϕ is one to one.

Of course, a similar one-to-one map of H onto the right coset Hg can be constructed. (See Exercise 22.) We summarize as follows:

Every coset (left or right) of a subgroup H of a group G has the same number of elements as H.

We can now prove the theorem of Lagrange.

THEOREM 2.5 (Theorem of Lagrange)   Let H be a subgroup of a finite group G. Then the order of H is a divisor of the order of G.

PROOF   Let n be the order of G, and let H have order m. The preceding statement set off by rules shows that every coset of H also has m elements. Let r be the number of cells in the partition of G into left cosets of H. Then $n = rm$, so m is indeed a divisor of n. ◆

Note that this elegant and important theorem comes from the simple counting of cosets and the number of elements in each coset. *Never underestimate results that count something*! We continue to derive consequences of Theorem 2.5, which should be regarded as a counting theorem.

COROLLARY   Every group of prime order is cyclic.

PROOF   Let G be of prime order p, and let a be an element of G different from the identity. Then the cyclic subgroup $\langle a \rangle$ of G generated by a has at least two elements, a and e. But by Theorem 2.5, the order $m \geq 2$ of $\langle a \rangle$ must divide the prime p. Thus we must have $m = p$ and $\langle a \rangle = G$, so G is cyclic. ◆

Since every cyclic group of order p is isomorphic to $\mathbb{Z}_p$, we see that *there is essentially only one group structure of a given prime order p*. Now doesn't this elegant result follow easily from the theorem of Lagrange, a *counting* theorem? *Never underestimate a theorem that counts something*. Proving the preceding corollary is a favorite examination question.

THEOREM 2.6   The order of an element of a finite group divides the order of the group.

PROOF   Remembering that the order of an element is the same as the order of the cyclic subgroup generated by the element, we see that this theorem follows directly from Theorem 2.5. ◆

DEFINITION 2.11 (Index of H in G)   Let H be a subgroup of a group G. The number of left cosets of H in G is the **index** $(G:H)$ **of H in** G.

The index $(G:H)$ just defined may be finite or infinite. If G is finite, then obviously $(G:H)$ is finite and $(G:H) = |G| / |H|$, since every coset of H contains $|H|$ elements. Exercise 30 shows the index $(G:H)$ could be equally well defined as the number of right cosets of H in G. we state a basic theorem concerning indices of subgroups, and leave the proof to the exercises (see Exercise 33).

THEOREM 2.7   Suppose H and K are subgroups of a group G such that $K \le H \le G$, and suppose $(H:K)$ and $(G:H)$ are both finite. Then $(G:K)$ is finite, and $(G:K) = (G:H)(H:K)$.

Theorem 2.5 shows that if there is a subgroup H of a finite group G, then the order of H divides the order of G. Is the converse true? That is, if G is a group of order n, and m divides n, is there always a subgroup of order m? We will see in the next section that this is true for abelian groups. However, A_4 can be shown to have no subgroup of order 6, which gives a counterexample for nonabelian groups.

Exercises 2.3

Computations

1. Find all cosets of the subgroup $4\mathbb{Z}$ of $\mathbb{Z}$.

2. Find all cosets of the subgroup $4\mathbb{Z}$ of $2\mathbb{Z}$.

3. Find all cosets of the subgroup $\langle 2 \rangle$ of $\mathbb{Z}_{12}$.

4. Find all cosets of the subgroup $\langle 4 \rangle$ of $\mathbb{Z}_{12}$.

5. Find all cosets of the subgroup $\langle 18 \rangle$ of $\mathbb{Z}_{36}$.

6. Find all left cosets of the subgroup $\{\rho_0, \mu_2\}$ of the group D_4 given by Table 2.2.

7. Repeat the preceding exercise, but find the right cosets this time. Are they the same as the left cosets?

8. Rewrite Table 2.2 in the order exhibited by the left cosets in Exercise 6. Do you seem to get a coset group of order 4? If so, is it isomorphic to $\mathbb{Z}_4$ or to the Klein 4-group V?

9. Repeat Exercise 6 for the subgroup $\{\rho_0, \rho_2\}$ of D_4.

10. Repeat the preceding exercise, but find the right cosets this time. Are they the same as the left coset?

11. Rewrite Table 2.2 in the order exhibited by the left cosets in Exercise 9. Do you seem to get a coset group of order 4? If so, is it isomorphic to $\mathbb{Z}_4$ or to the Klein 4-group V?

12. Find the index of $\langle 3 \rangle$ in the group $\mathbb{Z}_{24}$.

13. Find the index of $\langle \mu_1 \rangle$ in the group S_3, using the notation of Example 3.

14. Find the index of $\langle \mu_3 \rangle$ in the group D_4 given in Table 2.2.

Concepts

15. Mark each of the following true or false.
_____ a. Every subgroup of every group has left cosets.
_____ b. The number of left cosets of a subgroup of a finite group divides the order of the group.
_____ c. Every group of prime order is abelian.
_____ d. One cannot have left cosets of a finite subgroup of an infinite group.
_____ e. A subgroup of a group is a left coset of itself.
_____ f. Only subgroups of finite groups can have left cosets.
_____ g. A_n is of index 2 in S_n for $n > 1$.
_____ h. The theorem of Lagrange is a nice result.
_____ i. Every finite group contains an element of every order that divides the order of the group.
_____ j. Every finite cyclic group contains an element of every order that divides the order of the group.

In Exercises 16 through 20, give an example of the desired subgroup and group if possible. If impossible, say why it is impossible.

16. A subgroup of an abelian group G whose left cosets and right cosets give different partitions of G

17. A subgroup of a group G whose left cosets give a partition of G into just one cell

18. A subgroup of a group of order 6 whose left cosets give a partition of the group into 6 cells

19. A subgroup of a group of order 6 whose left cosets give a partition of the group into 12 cells

20. A subgroup of a group of order 6 whose left cosets give a partition of the group into 4 cells

Theory

21. Prove that the relation $\sim_R$ of Theorem 2.4 is an equivalence relation.

22. Let H be a subgroup of a group G and let $g \in G$. Define a one-to-one map of H onto Hg. Prove that your map is one to one and is onto Hg.

23. Let H be a subgroup of a group G such that $g^{-1}hg \in H$ for all $g \in G$ and all $h \in H$. Show that every left coset gH is the same as the right coset Hg.

24. Let H be a subgroup of a group G. Prove that if the partition of G into left cosets of H is the same as the partition into right cosets of H, then $g^{-1}hg \in H$ for all $g \in G$ and all $h \in H$. (Note that this is the converse of Exercise 23.)

Let H be a subgroup of a group G and let $a, b \in G$. In Exercises 25 through 28 prove the statement or give a counterexample.

25. If $aH = bH$, then $Ha = Hb$.

26. If $Ha = Hb$, then $b \in Ha$.

27. If $aH = bH$, then $Ha^{-1} = Hb^{-1}$.

28. If $aH = bH$, then $a^2H = b^2H$.

29. Let G be a group of order pq, where p and q are prime numbers. Show that every proper subgroup of G is cyclic.

30. Show that there are the same number of left as right cosets of a subgroup H of a group G; that is, exhibit a one-to-one map of the collection of left cosets onto the collection of right cosets. (Note that this result is obvious by counting for finite groups. Your proof must hold for any group.)

31. Exercise 23 of Section 1.2 showed that every finite group of even order $2n$ contains an element of order 2. Using the theorem of Lagrange, show that if n is odd, then an abelian group of order $2n$ contains precisely one element of order 2.

32. Show that a group with at least two elements but with no proper nontrivial subgroups must be finite and of prime order.

33. Prove Theorem 2.7 [*Hint:* Let $\{a_iH \mid i = 1, \ldots, r\}$ be the collection of distinct left cosets of H in G and $\{b_jK \mid j = 1, \ldots, s\}$ be the collection of distinct left cosets of K in H. Show that

$$\{(a_ib_j)K \mid i = 1, \ldots, r; \; j = 1, \ldots, s\}$$

is the collection of distinct left cosets of K in G.]

34. Show that if H is a subgroup of index 2 in a finite group G, then every left coset of H is also a right coset of H.

35. Show that if a group G with identity e has finite order n, then $a^n = e$ for all $a \in G$.

36. Show that every left coset of the subgroup $\mathbb{Z}$ of the additive group of real numbers contains exactly one representative x in $\mathbb{R}$ such that $0 \le x < 1$.

37. Show that the function *sine* assigns the same value to each representative of any fixed left coset of the subgroup $\langle 2\pi \rangle$ of the additive group $\mathbb{R}$ of real numbers. (Thus *sine* induces a well-defined function on the set of cosets; the value of the function on a coset is obtained when we choose a representative x of the coset and compute $\sin x$.)

38. Let H and K be subgroups of a group G. Define $\sim$ on G by $a \sim b$ if and only if $a = hbk$ for some $h \in H$ and some $k \in K$.

    a. Prove that $\sim$ is an equivalence relation on G.

    b. Describe the elements in the equivalence class containing $a \in G$. (These equivalence classes are called **double cosets**.)

39. Let S_A be the group of all permutations of the set A, and let c be one particular element of A.

    a. Show that $\{\sigma \in S_A \mid \sigma(c) = c\}$ is a subgroup $S_{c,c}$ of S_A.

    b. Let $d \ne c$ be another particular element of A, Is $S_{c,d} = \{\sigma \in S_A \mid \sigma(c) = d\}$ a subgroup of S_A? Why or why not?

    c. Characterize the set $S_{c,d}$ of part (b) in terms of the subgroup $S_{c,c}$ of part (a).

40. Show that a finite cyclic group of order n has exactly one subgroup of each order d dividing n, and that these are all the subgroups it has.

41. The **Euler phi-function** is defined for positive integers n by $\varphi(n) = s$, where s is

the number of positive integers less than or equal to n that are relatively prime to n. Use Exercise 40 to show that

$$n = \sum_{d|n} \varphi(d),$$

the sum being taken over all positive integers d dividing n. [*Hint:* Note that the number of generators of $\mathbb{Z}_d$ is $\varphi(d)$ by the corollary of Theorem 1.9.]

42. Let G be a finite group. Show that if for each positive integer m the number of solutions x of the equation $x^m = e$ in G is at most m, then G is cyclic. [*Hint:* Use Theorem 2.6 and Exercise 41 to show that G must contain an element of order $n = |G|$.]

2.4

Direct Products and Finitely Generated Abelian Groups

Direct Products

Let us take a moment to review our present stockpile of groups. Starting with finite groups, we have the cyclic group $\mathbb{Z}_n$, the symmetric group S_n, and the alternating group A_n for each positive integer n. We also have the octic group D_4 of Example 5 of Section 2.1 and the Klein 4-group V. Of course we know that subgroups of these groups exist. Turning to infinite groups, we have groups consisting of sets of numbers under the usual addition or multiplication, as, for example, $\mathbb{Z}$ and $\mathbb{R}$ under addition. We also have the group S_A of all permutations of an infinite set A, as well as various groups formed from matrices.

One purpose of this section is to show a way to use known groups as building blocks to form more groups. The Klein 4-group will be recovered in this way from the cyclic groups. Employing this procedure with the cyclic groups gives us a large class of abelian groups that can be shown to include all possible structure types for a finite abelian group.

DEFINITION 2.12 (Cartesian Product) The **Cartesian product of sets** $S_1, S_2, \ldots, S_n$ is the set of all ordered n-tuples $(a_1, a_2, \ldots, a_n)$, where $a_i \in S_i$. The Cartesian product is denoted by either

$$S_1 \times S_2 \times \cdots \times S_n$$

or by

$$\prod_{i=1}^{n} S_i.$$

We could also define the Cartesian product of an infinite number of sets, but the definition is considerably more sophisticated and we shall not need it.

Now let $G_1, G_2, \ldots, G_n$ be groups, and let us use multiplicative notation for all the group operations. Regarding the G_i as sets, we can form $\prod_{i=1}^{n} G_i$. Let us show that we can make $\prod_{i=1}^{n} G_i$ into a group by means of a binary operation of *multiplication by components*. Note again that we are being sloppy when we use the same notation for a group as for the set of elements of the group.

THEOREM 2.8 Let $G_1, G_2, \ldots, G_n$ be groups. For $(a_1, a_2, \ldots, a_n)$ and $(b_1, b_2, \ldots, b_n)$ in $\prod_{i=1}^{n} G_i$, define $(a_1, a_2, \ldots, a_n)(b_1, b_2, \ldots, b_n)$ to be $(a_1 b_1, a_2 b_2, \ldots, a_n b_n)$. Then $\prod_{i=1}^{n} G_i$ is a group, the **direct product of the groups** G_i, under this binary operation.

PROOF Note that since $a_i \in G_i$, $b_i \in G_i$, and G_i is a group, we have $a_i b_i \in G_i$. Thus the definition of the binary operation on $\prod_{i=1}^{n} G_i$ given in the statement of the theorem makes sense; that is, $\prod_{i=1}^{n} G_i$ is closed under the binary operation.

The associative law in $\prod_{i=1}^{n} G_i$ is thrown back onto the associative law in each component as follows:

$$
(a_1, a_2, \ldots, a_n)[(b_1, b_2, \ldots, b_n)(c_1, c_2, \ldots, c_n)]
$$
$$
= (a_1, a_2, \ldots, a_n)(b_1 c_1, b_2 c_2, \ldots, b_n c_n)
$$
$$
= (a_1(b_1 c_1), a_2(b_2 c_2), \ldots, a_n(b_n c_n))
$$
$$
= ((a_1, b_1)c_1, (a_2 b_2)c_2, \ldots, (a_n b_n)c_n)
$$
$$
= (a_1 b_1, a_2 b_2, \ldots, a_n b_n)(c_1, c_2, \ldots, c_n)
$$
$$
= [(a_1, a_2, \ldots, a_n)(b_1, b_2, \ldots, b_n)](c_1, c_2, \ldots, c_n).
$$

If e_i is the identity element in G_i, then clearly, with multiplication by components, $(e_1, e_2, \ldots, e_n)$ is an identity in $\prod_{i=1}^{n} G_i$. An inverse of $(a_1, a_2, \ldots, a_n)$ is $(a_1^{-1}, a_2^{-1}, \ldots, a_n^{-1})$; just compute the product by components. Hence $\prod_{i=1}^{n} G_i$ is a group. ◆

In the event that the operation of each G_i is commutative, we sometimes use additive notation in $\prod_{i=1}^{n} G_i$ and refer to $\prod_{i=1}^{n} G_i$ as the **direct sum of the groups** G_i. The notation $\bigoplus_{i=1}^{n} G_i$ is sometimes used in this case in place of $\prod_{i=1}^{n} G_i$, especially with abelian groups with operation $+$. The direct sum of abelian groups $G_1, G_2, \ldots, G_n$ may be written $G_1 \oplus G_2 \oplus \cdots \oplus G_n$. We leave to Exercise 42 the proof that a direct product of abelian groups is again abelian.

It is quickly seen that if the set S_i has r_i elements for $i = 1, \ldots, n$, then $\prod_{i=1}^{n} S_i$ has $r_1 r_2 \cdots r_n$ elements, for in an n-tuple, there are r_1 choices for the first component from S_1, and for each of these there are r_2 choices for the next component from S_2, and so on.

EXAMPLE 1   Consider the group $\mathbb{Z}_2 \times \mathbb{Z}_3$, which has $2 \cdot 3 = 6$ elements, namely $(0,0)$, $(0,1)$, $(0,2)$, $(1,0)$, $(1,1)$, and $(1,2)$. We claim that $\mathbb{Z}_2 \times \mathbb{Z}_3$ is cyclic. It is only necessary to find a generator. Let us try $(1,1)$. Here the operations in $\mathbb{Z}_2$ and $\mathbb{Z}_3$ are written additively, so we do the same in the direct product $\mathbb{Z}_2 \times \mathbb{Z}_3$.

$$(1,1) = (1,1)$$
$$2(1,1) = (1,1) + (1,1) = (0,2)$$
$$3(1,1) = (1,1) + (1,1) + (1,1) = (1,0)$$
$$4(1,1) = 3(1,1) + (1,1) = (1,0) + (1,1) = (0,1)$$
$$5(1,1) = 4(1,1) + (1,1) = (0,1) + (1,1) = (1,2)$$
$$6(1,1) = 5(1,1) + (1,1) = (1,2) + (1,1) = (0,0)$$

Thus $(1,1)$ generates all of $\mathbb{Z}_2 \times \mathbb{Z}_3$. Since there is essentially only one cyclic group structure of a given order, we see that $\mathbb{Z}_2 \times \mathbb{Z}_3$ is isomorphic to $\mathbb{Z}_6$. ▲

EXAMPLE 2   Consider $\mathbb{Z}_3 \times \mathbb{Z}_3$. This is a group of nine elements. We claim that $\mathbb{Z}_3 \times \mathbb{Z}_3$ is *not* cyclic. Since the addition is by components, and since in $\mathbb{Z}_3$ every element added to itself three times gives the identity, the same is true in $\mathbb{Z}_3 \times \mathbb{Z}_3$. Thus no element can generate the group, for a generator added to itself successively could only give the identity after nine summands. We have found another group structure of order 9. A similar argument shows that $\mathbb{Z}_2 \times \mathbb{Z}_2$ is not cyclic. Thus $\mathbb{Z}_2 \times \mathbb{Z}_2$ must be isomorphic to the Klein 4-group. ▲

The preceding examples illustrate the following theorem:

THEOREM 2.9   The group $\mathbb{Z}_m \times \mathbb{Z}_n$ is isomorphic to $\mathbb{Z}_{mn}$ if and only if m and n are relatively prime, that is, the gcd of m and n is 1.

PROOF   Consider the cyclic subgroup of $\mathbb{Z}_m \times \mathbb{Z}_n$ generated by $(1,1)$ as described by Theorem 1.5. As our previous work has shown, the order of this cyclic subgroup is the smallest power of $(1,1)$ that gives the identity $(0,0)$. Here taking a power of $(1,1)$ in our additive notation will involve adding $(1,1)$ to itself repeatedly. Under addition by components, the first component $1 \in \mathbb{Z}_m$ yields 0 only after m summands, $2m$ summands, and so on, and the second component $1 \in \mathbb{Z}_n$ yields 0 only after n summands, $2n$ summands, and so on. For them to yield 0 simultaneously, the number of summands must be a multiple of both m and n. The smallest number that is a multiple of both m and n will be mn if and only if the gcd of m and n is 1; in this case $(1,1)$ generates a cyclic subgroup of order mn, which is the order of the whole group. This shows that $\mathbb{Z}_m \times \mathbb{Z}_n$ is cyclic of order mn, and hence isomorphic to $\mathbb{Z}_{mn}$ if m and n are relatively prime.

For the converse, suppose that the gcd of m and n is $d > 1$. Then mn/d is divisible by both m and n. Consequently, for any (r, s) in $\mathbb{Z}_m \times \mathbb{Z}_n$, we have

$$\underbrace{(r, s) + (r, s) + \cdots + (r, s)}_{mn/d \text{ summands}} = (0, 0).$$

Hence no element (r, s) in $\mathbb{Z}_m \times \mathbb{Z}_n$ can generate the entire group, so $\mathbb{Z}_m \times \mathbb{Z}_n$ is not cyclic and therefore not isomorphic to $\mathbb{Z}_{mn}$. ◆

This theorem can be extended to a product of more than two factors by similar arguments. We state this as a corollary without going through the details of the proof.

COROLLARY The group $\prod_{i=1}^{n} \mathbb{Z}_{m_i}$ is cyclic and isomorphic to $\mathbb{Z}_{m_1 m_2 \cdots m_n}$ if and only if the numbers m_i for $i = 1, \ldots, n$ are such that the gcd of any two of them is 1.

EXAMPLE 3 The preceding corollary shows that if n is written as a product of powers of distinct prime numbers, as in

$$n = (p_1)^{n_1}(p_2)^{n_2} \cdots (p_r)^{n_r},$$

then $\mathbb{Z}_n$ is isomorphic to

$$\mathbb{Z}_{(p_1)^{n_1}} \times \mathbb{Z}_{(p_2)^{n_2}} \times \cdots \times \mathbb{Z}_{(p_r)^{n_r}}.$$

In particular, $\mathbb{Z}_{72}$ is isomorphic to $\mathbb{Z}_8 \times \mathbb{Z}_9$. ▲

We remark that changing the order of the factors in a direct product yields a group isomorphic to the original one. The names of elements have simply been changed via a permutation of the components in some n-tuple.

Exercise 47 of Section 1.4 asked you to define the least common multiple of two positive integers r and s as a generator of a certain cyclic group. It is straightforward to prove that the subset of $\mathbb{Z}$ consisting of all integers that are multiples of both r and s is a subgroup of $\mathbb{Z}$, and hence is a cyclic group. Likewise, the set of all common multiples of n positive integers $r_1, r_2, \ldots, r_n$ is a subgroup of $\mathbb{Z}$ and hence is cyclic.

DEFINITION 2.13 (Least Common Multiple) Let $r_1, r_2, \ldots, r_n$ be positive integers. Their **least common multiple** (abbreviated lcm) is the positive generator of the cyclic group of all common multiples of the r_i, that is, the cyclic group of all integers divisible by each r_i for $i = 1, 2, \ldots, n$.

From Definition 2.13 and our work on cyclic groups, we see that the lcm of $r_1, r_2, \ldots, r_n$ is the smallest positive integer that is a multiple of each r_i for $i = 1, 2, \ldots, n$, hence the name *least common multiple*.

THEOREM 2.10 Let $(a_1, a_2, \ldots, a_n) \in \prod_{i=1}^n G_i$. If a_i is of finite order r_i in G_i, then the order of $(a_1, a_2, \ldots, a_n)$ in $\prod_{i=1}^n G_i$ is equal to the least common multiple of all the r_i.

PROOF This follows by a repetition of the argument used in the proof of Theorem 2.9. For a power of $(a_1, a_2, \ldots, a_n)$ to give $(e_1, e_2, \ldots, e_n)$, the power must simultaneously be a multiple of r_1 so that this power of the first component a_1 will yield e_1, a multiple of r_2, so that this power of the second component a_2 will yield e_2, and so on. ◆

EXAMPLE 4 Find the order of $(8, 4, 10)$ in the group $\mathbb{Z}_{12} \times \mathbb{Z}_{60} \times \mathbb{Z}_{24}$.

Solution Since the gcd of 8 and 12 is 4, we see that 8 is of order $\frac{12}{4} = 3$ in $\mathbb{Z}_{12}$. (See Theorem 1.9.) Similarly, we find that 4 is of order 15 in $\mathbb{Z}_{60}$ and 10 is of order 12 in $\mathbb{Z}_{24}$. The lcm of 3, 15, and 12 is $3 \cdot 5 \cdot 4 = 60$, so $(8, 4, 10)$ is of order 60 in $\mathbb{Z}_{12} \times \mathbb{Z}_{60} \times \mathbb{Z}_{24}$. ▲

EXAMPLE 5 The group $\mathbb{Z} \times \mathbb{Z}_2$ is generated by the elements $(1, 0)$ and $(0, 1)$. More generally, the direct product of n cyclic groups, each of which is either $\mathbb{Z}$ or $\mathbb{Z}_m$ for some positive integer m, is generated by the n n-tuples

$$(1, 0, 0, \ldots, 0), \quad (0, 1, 0, \ldots, 0), \quad (0, 0, 1, \ldots, 0), \quad \ldots, \quad (0, 0, 0, \ldots, 1).$$

Such a direct product might also be generated by fewer elements. For example, $\mathbb{Z}_3 \times \mathbb{Z}_4 \times \mathbb{Z}_{35}$ is generated by the single element $(1, 1, 1)$. ▲

Note that if $\prod_{i=1}^n G_i$ is the direct product of groups G_i, then the subset

$$\bar{G}_i = \{(e_1, e_2, \ldots, e_{i-1}, a_i, e_{i+1}, \ldots, e_n) \mid a_i \in G_i\},$$

that is, the set of all n-tuples with the identity elements in all places but the ith, is a subgroup of $\prod_{i=1}^n G_i$. It is also clear that this subgroup $\bar{G}_i$ is naturally isomorphic to G_i; just rename

$$(e_1, e_2, \ldots, e_{i-1}, a_i, e_{i+1}, \ldots, e_n) \text{ by } a_i.$$

The group G_i is mirrored in the ith component of the elements of $\bar{G}_i$, and the e_j in the other components just ride along. We consider $\prod_{i=1}^n G_i$ to be the *internal direct product* of these subgroups $\bar{G}_i$. The direct product given by Theorem 2.8 is called the *external direct product* of the groups G_i. The terms *internal* and *external*, as applied to a direct product of groups, just reflect whether or not (respectively) we are regarding the component groups as subgroups of the product group. We shall usually omit the words *external* and *internal* and just say *direct product*. Which term we mean will be clear from the context.

The Structure of Finitely Generated Abelian Groups

Some theorems of abstract algebra are easy to understand and use, although their proofs may be quite technical and time-consuming to present. This is one section in the text where we explain the meaning and significance of a theorem but defer its proof to a later chapter. The meaning of any theorem whose proof we defer is well within our understanding, and we feel we should be acquainted with it. It would be impossible for us to meet some of these fascinating facts in a one-semester course if we were to insist on wading through complete proofs of all theorems. The theorem that we now state gives us complete structural information about all sufficiently small abelian groups, in particular, about all finite abelian groups.

THEOREM 2.11 (Fundamental Theorem of Finitely Generated Abelian Groups)   Every finitely generated abelian group G is isomorphic

---

◆ **H I S T O R I C A L   N O T E** ◆

In his *Disquisitiones Arithmeticae* Carl Gauss demonstrated various results in what is today the theory of abelian groups in the context of number theory. Not only did he deal extensively with equivalence classes of quadratic forms, but he also considered residue classes modulo a given integer. Although he noted that results in these two areas were similar, he did not attempt to develop an abstract theory of abelian groups.

In the 1840s, Ernst Kummer in dealing with ideal complex numbers noted that his results were in many respects analogous to those of Gauss. (See the Historical Note in Section 6.1.) But it was Kummer's student Leopold Kronecker (see the Historical Note in Section 8.1) who finally realized that an abstract theory could be developed out of the analogies. As he wrote in 1870, "these principles [from the work of Gauss and Kummer] belong to a more general, abstract realm of ideas. It is therefore appropriate to free their development from all unimportant restrictions, so that one can spare oneself from the necessity of repeating the same argument in different cases. This advantage already appears in the development itself, and the presentation gains in simplicity, if it is given in the most general admissible manner, since the most important features stand out with clarity." Kronecker then proceeded to develop the basic principles of the theory of finite abelian groups and was able to state and prove a version of Theorem 2.11 restricted to finite groups.

to a direct product of cyclic groups in the form

$$\mathbb{Z}_{(p_1)^{r_1}} \times \mathbb{Z}_{(p_2)^{r_2}} \times \cdots \times \mathbb{Z}_{(p_n)^{r_n}} \times \mathbb{Z} \times \mathbb{Z} \times \cdots \times \mathbb{Z},$$

where the p_i are primes, not necessarily distinct. The direct product is unique except for possible rearrangement of the factors; that is, the number (**Betti number** of G) of factors of $\mathbb{Z}$ is unique and the prime powers $(p_i)^{r_i}$ are unique.

PROOF A complete proof can be found in Section 4.4. ◆

EXAMPLE 6 Find all abelian groups, up to isomorphism, of order 360. The phrase *up to isomorphism* signifies that any abelian group of order 360 should be structurally identical (isomorphic) to one of the groups of order 360 exhibited.

Solution We make use of Theorem 2.11. Since our groups are to be of the finite order 360, no factors $\mathbb{Z}$ will appear in the direct product shown in the statement of the theorem.

First we express 360 as a product of prime powers $2^3 3^2 5$. Then using Theorem 2.11, we get as possibilities

1. $\mathbb{Z}_2 \times \mathbb{Z}_2 \times \mathbb{Z}_2 \times \mathbb{Z}_3 \times \mathbb{Z}_3 \times \mathbb{Z}_5$
2. $\mathbb{Z}_2 \times \mathbb{Z}_4 \times \mathbb{Z}_3 \times \mathbb{Z}_3 \times \mathbb{Z}_5$
3. $\mathbb{Z}_2 \times \mathbb{Z}_2 \times \mathbb{Z}_2 \times \mathbb{Z}_9 \times \mathbb{Z}_5$
4. $\mathbb{Z}_2 \times \mathbb{Z}_4 \times \mathbb{Z}_9 \times \mathbb{Z}_5$
5. $\mathbb{Z}_8 \times \mathbb{Z}_3 \times \mathbb{Z}_3 \times \mathbb{Z}_5$
6. $\mathbb{Z}_8 \times \mathbb{Z}_9 \times \mathbb{Z}_5$

Thus there are six different abelian groups (up to isomorphism) of order 360. ▲

Applications

We conclude this section with a sampling of the many theorems we could now prove regarding abelian groups.

DEFINITION 2.14 (Decomposable Group) A group G is **decomposable** if it is isomorphic to a direct product of two proper nontrivial subgroups. Otherwise G is **indecomposable**.

THEOREM 2.12 The finite indecomposable abelian groups are exactly the cyclic groups with order a power of a prime.

PROOF Let G be a finite indecomposable abelian group. Then by Theorem 2.11, G is isomorphic to a direct product of cyclic groups of prime power order. Since G is indecomposable, this direct product must consist of just one cyclic group of prime power order.

Conversely, let p be a prime. Then $\mathbb{Z}_{p^r}$ is indecomposable, for if $\mathbb{Z}_{p^r}$

were isomorphic to $\mathbb{Z}_{p^i} \times \mathbb{Z}_{p^j}$, where $i + j = r$, then every element would have an order at most $p^{\max(i,j)} < p^r$. ◆

THEOREM 2.13 If m divides the order of a finite abelian group G, then G has a subgroup of order m.

PROOF By Theorem 2.11, we can think of G as being

$$\mathbb{Z}_{(p_1)^{r_1}} \times \mathbb{Z}_{(p_2)^{r_2}} \times \cdots \times \mathbb{Z}_{(p_n)^{r_n}},$$

where not all primes p_i need be distinct. Since $(p_1)^{r_1}(p_2)^{r_2} \cdots (p_n)^{r_n}$ is the order of G, then m must be of the form $(p_1)^{s_1}(p_2)^{s_2} \cdots (p_n)^{s_n}$, where $0 \leq s_i \leq r_i$. By Theorem 1.9, $(p_i)^{r_i - s_i}$ generates a cyclic subgroup of $\mathbb{Z}_{(p_i)^{r_i}}$ of order equal to the quotient of $(p_i)^{r_i}$ by the gcd of $(p_i)^{r_i}$ and $(p_i)^{r_i - s_i}$. But the gcd of $(p_i)^{r_i}$ and $(p_i)^{r_i - s_i}$ is $(p_i)^{r_i - s_i}$. Thus $(p_i)^{r_i - s_i}$ generates a cyclic subgroup $\mathbb{Z}_{(p_i)^{r_i}}$ of order

$$[(p_i)^{r_i}]/[(p_i)^{r_i - s_i}] = (p_i)^{s_i}.$$

Recalling that $\langle a \rangle$ denotes the cyclic subgroup generated by a, we see that

$$\langle (p_1)^{r_1 - s_1} \rangle \times \langle (p_2)^{r_2 - s_2} \rangle \times \cdots \times \langle (p_n)^{r_n - s_n} \rangle$$

is the required subgroup of order m. ◆

THEOREM 2.14 If m is a square free integer, that is, m is not divisible by the square of any prime, then every abelian group of order m is cyclic.

PROOF Let G be an abelian group of square free order m. Then by Theorem 2.11 G is isomorphic to

$$\mathbb{Z}_{(p_1)^{r_1}} \times \mathbb{Z}_{(p_2)^{r_2}} \times \cdots \times \mathbb{Z}_{(p_n)^{r_n}},$$

where $m = (p_1)^{r_1}(p_2)^{r_2} \cdots (p_n)^{r_n}$. Since m is square free, we must have all $r_i = 1$ and all p_i distinct primes. The corollary of Theorem 2.9 then shows that G is isomorphic to $\mathbb{Z}_{p_1 p_2 \cdots p_n}$, so G is cyclic. ◆

Exercises 2.4

Computations

1. List the elements of $\mathbb{Z}_2 \times \mathbb{Z}_4$. Find the order of each of the elements. Is this group cyclic?

2. Repeat Exercise 1 for the group $\mathbb{Z}_3 \times \mathbb{Z}_4$.

In Exercises 3 through 7, find the order of the given element of the direct product.   *lcm then gcd*

3. $(2, 6)$ in $\mathbb{Z}_4 \times \mathbb{Z}_{12}$

4. $(2, 3)$ in $\mathbb{Z}_6 \times \mathbb{Z}_{15}$

5. $(8, 10)$ in $\mathbb{Z}_{12} \times \mathbb{Z}_{18}$

6. $(3, 10, 9)$ in $\mathbb{Z}_4 \times \mathbb{Z}_{12} \times \mathbb{Z}_{15}$

7. $(3, 6, 12, 16)$ in $\mathbb{Z}_4 \times \mathbb{Z}_{12} \times \mathbb{Z}_{20} \times \mathbb{Z}_{24}$

8. What is the largest order among the orders of all the cyclic subgroups of $\mathbb{Z}_6 \times \mathbb{Z}_8$? of $\mathbb{Z}_{12} \times \mathbb{Z}_{15}$?

9. Find all proper nontrivial subgroups of $\mathbb{Z}_2 \times \mathbb{Z}_2$.

10. Find all proper nontrivial subgroups of $\mathbb{Z}_2 \times \mathbb{Z}_2 \times \mathbb{Z}_2$.

11. Find all subgroups of $\mathbb{Z}_2 \times \mathbb{Z}_4$ of order 4.

12. Find all subgroups of $\mathbb{Z}_2 \times \mathbb{Z}_2 \times \mathbb{Z}_4$ that are isomorphic to the Klein 4-group.

13. Disregarding the order of the factors, write direct products of two or more groups of the form $\mathbb{Z}_n$ so that the resulting product is isomorphic to $\mathbb{Z}_{60}$ in as many ways as possible.

14. Fill in the blanks.

   a. The cyclic subgroup of $\mathbb{Z}_{24}$ generated by 18 has order _____.

   b. $\mathbb{Z}_3 \times \mathbb{Z}_4$ is of order _____.

   c. The element $(4, 2)$ of $\mathbb{Z}_{12} \times \mathbb{Z}_8$ has order _____.

   d. The Klein 4-group is isomorphic to $\mathbb{Z}\_\_\_ \times \mathbb{Z}\_\_\_$.

   e. $\mathbb{Z}_2 \times \mathbb{Z} \times \mathbb{Z}_4$ has _____ elements of finite order.

In Exercises 15 through 20, find the subgroup generated by the given subset of the group. Then find all left cosets of the subgroup.

15. $\{2, 3\}$ in $\mathbb{Z}_{12}$

16. $\{4, 6\}$ in $\mathbb{Z}_{12}$

17. $\{8, 6, 10\}$ in $\mathbb{Z}_{18}$

18. $\{\rho_2, \mu_1\}$ in D_4 using Table 2.2.

19. $\{\mu_1, \delta_2\}$ in D_4 using Table 2.2.

20. $\{(4, 2), (2, 3)\}$ in $\mathbb{Z}_6 \times \mathbb{Z}_4$.

In Exercises 21 through 25, proceed as in Example 6 to find all groups, up to isomorphism, of the given order.

21. Order 8   **22.** Order 16   **23.** Order 32

24. Order 720   **25.** Order 1089

26. How many abelian groups (up to isomorphism) are there of order 24? of order 25? of order (24)(25)?

27. Following the idea suggested in Exercise 26, let m and n be relatively prime positive integers. Show that if there are (up to isomorphism) r abelian groups of order m and s of order n, then there are (up to isomorphism) rs abelian groups of order mn.

28. Use Exercise 27 to determine the number of abelian groups (up to isomorphism) of order $(10)^5$.

Concepts

29. Mark each of the following true or false.

_____ a. If G_1 and G_2 are any groups, then $G_1 \times G_2$ is always isomorphic to $G_2 \times G_1$.

_____ b. Computation in an external direct product of groups is easy if you know how to compute in each component group.

_____ c. Groups of finite order must be used to form an external direct product.

_____ d. A group of prime order could not be the internal direct product of two proper nontrivial subgroups.

_____ e. $\mathbb{Z}_2 \times \mathbb{Z}_4$ is isomorphic to $\mathbb{Z}_8$.

_____ f. $\mathbb{Z}_2 \times \mathbb{Z}_4$ is isomorphic to S_8.

_____ g. $\mathbb{Z}_3 \times \mathbb{Z}_8$ is isomorphic to S_4.

_____ h. Every element in $\mathbb{Z}_4 \times \mathbb{Z}_8$ has order 8.

_____ i. The order of $\mathbb{Z}_{12} \times \mathbb{Z}_{15}$ is 60.

_____ j. $\mathbb{Z}_m \times \mathbb{Z}_n$ has mn elements whether m and n are relatively prime or not.

30. Give an example illustrating that not every nontrivial abelian group is the internal direct product of two proper nontrivial subgroups.

31. a. How many subgroups of $\mathbb{Z}_5 \times \mathbb{Z}_6$ are isomorphic to $\mathbb{Z}_5 \times \mathbb{Z}_6$?

  b. How many subgroups of $\mathbb{Z} \times \mathbb{Z}$ are isomorphic to $\mathbb{Z} \times \mathbb{Z}$?

32. Give an example of a nontrivial group that is not of prime order and is not the internal direct product of two nontrivial subgroups. S_3

33. Mark each of the following true or false.

_____ a. Every abelian group of prime order is cyclic.

_____ b. Every abelian group of prime power order is cyclic.

_____ c. $\mathbb{Z}_8$ is generated by $\{4, 6\}$.

_____ d. $\mathbb{Z}_8$ is generated by $\{4, 5, 6\}$.

_____ e. All finite abelian groups are classified up to isomorphism by Theorem 2.11.

_____ f. Any two finitely generated abelian groups with the same Betti number are isomorphic.

_____ g. Every abelian group of order divisible by 5 contains a cyclic subgroup of order 5.

_____ h. Every abelian group of order divisible by 4 contains a cyclic subgroup of order 4.

_____ i. Every abelian group of order divisible by 6 contains a cyclic subgroup
of order 6.

_____ j. Every finite abelian group has a Betti number of 0.

34. Let p and q be distinct prime numbers. How does the number (up to isomorphism) of abelian groups of order p^r compare with the number (up to isomorphism) of abelian groups of order q^r?

35. Let G be an abelian group of order 72.

a. Can you say how many subgroups of order 8 G has? Why?

b. Can you say how many subgroups of order 4 G has? Why?

36. Let G be an abelian group. Show that the elements of finite order in G form a subgroup. This subgroup is called the **torsion subgroup** of G.

Exercises 37 through 40 deal with the concept of the torsion subgroup just defined.

37. Find the order of the torsion subgroup of $\mathbb{Z}_4 \times \mathbb{Z} \times \mathbb{Z}_3$; of $\mathbb{Z}_{12} \times \mathbb{Z} \times \mathbb{Z}_{12}$.

38. Find the torsion subgroup of the multiplicative group $\mathbb{R}^*$ of nonzero real numbers.

39. Find the torsion subgroup T of the multiplicative group $\mathbb{C}^*$ of nonzero complex numbers.

40. An abelian group is **torsion free** if e is the only element of finite order. Use Theorem 2.11 to show that every finitely generated abelian group is the internal direct product of its torsion subgroup and of a torsion-free subgroup. (Note that $\{e\}$ may be the torsion subgroup, and is also torsion free.)

41. The part of the decomposition of G in Theorem 2.11 corresponding to the subgroups of prime-power order can also be written in the form $\mathbb{Z}_{m_1} \times \mathbb{Z}_{m_2} \times \cdots \times \mathbb{Z}_{m_r}$, where m_i divides m_{i+1} for $i = 1, 2, \ldots, r-1$. The numbers m_i can be shown to be unique, and are the **torsion coefficients** of G.

a. Find the torsion coefficients of $\mathbb{Z}_4 \times \mathbb{Z}_9$.

b. Find the torsion coefficients of $\mathbb{Z}_6 \times \mathbb{Z}_{12} \times \mathbb{Z}_{20}$.

c. Describe an algorithm to find the torsion coefficients of a direct product of cyclic groups.

Theory

42. Prove that a direct product of abelian groups is abelian.

43. Let G be an abelian group. Let H be the subset of G consisting of the identity e together with all elements of G of order 2. Show that H is a subgroup of G.

44. Following up the idea of Exercise 43 determine whether H will always be a subgroup for every abelian group G if H consists of the identity e together with all elements of G of order 3; of order 4. For what positive integers n will H always be a subgroup for every abelian group G, if H consists of the identity e together with all elements of G of order n? Compare with Exercise 43 of Section 1.3.

45. Find a counterexample for Exercise 43 with the hypothesis that G is abelian omitted.

Let H and K be subgroups of a group G. Exercises 46 and 47 ask you to establish necessary and sufficient criteria for G to appear as the internal direct product of H and K.

46. Let H and K be groups and let $G = H \times K$. Recall that both H and K appear as subgroups of G in a natural way. Show that these subgroups H (actually $H \times \{e\}$) and K (actually $\{e\} \times K$) have the following properties.

i. Every element of G is of the form hk for some $h \in H$ and $k \in K$.

ii. $hk = kh$ for all $h \in H$ and $k \in K$.

iii. $H \cap K = \{e\}$.

47. Let H and K be subgroups of a group G satisfying the three properties listed in the preceding exercise. Show that for each $g \in G$, the expression $g = hk$ for $h \in H$ and $k \in K$ is unique. Then let each g be renamed (h, k). Show that, under this renaming, G becomes structurally identical (isomorphic) to $H \times K$. [*Hint:* You must show that the group operation on $H \times K$ corresponds to the operation on G via this renaming. That is, if g_1 is renamed (h_1, k_1) and g_2 is renamed (h_2, k_2), you must show that $g_1 g_2$ is renamed $(h_1 h_2, k_1 k_2)$.]

48. Show that a finite abelian group is not cyclic if and only if it contains a subgroup isomorphic to $\mathbb{Z}_p \times \mathbb{Z}_p$ for some prime p.

49. Prove that if a finite abelian group has order a power of a prime p, then the order of every element in the group is a power of p. Can the hypothesis of commutativity be dropped? Why?

50. Let G, H, and K be finitely generated abelian groups. Show that if $G \times K$ is isomorphic to $H \times K$, then $G \simeq H$.

51. Show that S_n is generated by $\{(1, 2), (1, 2, 3, \ldots, n)\}$. [*Hint:* Show that as r varies, $(1, 2, 3, \ldots, n)^r (1, 2)(1, 2, 3, \ldots, n)^{n-r}$ gives all the transpositions $(1, 2)$, $(2, 3)$, $(3, 4), \ldots, (n - 1, n)$, $(n, 1)$. Then show that any transposition is a product of some of these transpositions and use the corollary of Theorem 2.2.]

Cayley Digraphs

52. Indicate schematically a digraph for $\mathbb{Z}_m \times \mathbb{Z}_n$ for the generating set $S = \{(1, 0), (0, 1)\}$.

53. Consider Cayley digraphs with two arc types, a solid one with an arrow and a dashed one with no arrow, and consisting of two regular n-gons, for $n \geq 3$, with solid arc sides, one inside the other, with dashed arcs joining the vertices of the outer n-gon to the inner one. Figure 1.11(b) shows such a Cayley diagram with $n = 3$, and Fig. 1.12(b) shows one with $n = 4$. The arrows on the outer n-gon may have the same (clockwise or counterclockwise) direction as those on the inner n-gon, or they may have the opposite direction. Let G be a group with such a Cayley digraph.

a. Under what circumstances will G be abelian?

b. If G is abelian, to what familiar group is it isomorphic?

c. If G is abelian, under what circumstances is it cyclic?

d. If G is not abelian, to what group we have discussed is it isomorphic?

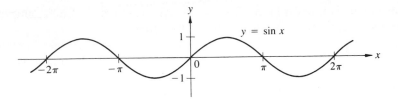

Figure 2.18

Periodic Functions

Consider the function f of one real variable given by $f(x) = \sin x$. It has the well-known graph shown in Fig. 2.18. Now

$$\sin x = \sin(x + 2\pi n) \quad \text{for every integer } n.$$

We say that the function *sine* of one real variable is *invariant* under a *translation* $x' = x + c$ of its domain by an element c of the infinite cyclic subgroup $\langle 2\pi \rangle$ of the group $\mathbb{R}$ under addition. Such a function of one real variable that is invariant under such a translation of its domain is a *periodic function.*

A function of two real variables has as its domain the Euclidean plane. A *translation of the plane* is a mapping τ of the plane into itself of the form $\tau(x, y) = (x + a, y + b)$ for some real numbers a and b. Note that we may write $(x + a, y + b)$ as $(x, y) + (a, b)$ using vector addition. The vector (a, b) is the *translation vector.* A function $f(x, y)$ may be invariant under such a translation of the plane, meaning that we may have $f(x, y) = f(x + a, y + b)$ for all $x, y \in \mathbb{R}$. A *doubly periodic function* is one that is invariant under elements of the group under vector addition generated by *two* translation vectors in different (not opposite) directions, as in Fig. 2.19. The

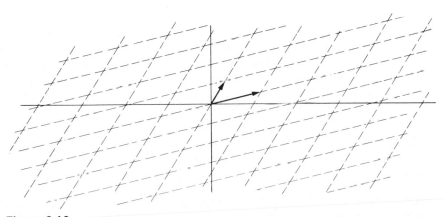

Figure 2.19

group this time is isomorphic to $\mathbb{Z} \times \mathbb{Z}$. We ask you to give some examples of doubly periodic functions in the exercises that follow.

More generally, mathematical analysis defines an *automorphic function* to be one that is invariant under a discrete group of transformations of its domain. The term *discrete* applied to a group means roughly that we have an idea of distance between elements of the group and that each element of the group is at the center of a small circle (or interval, or sphere) of positive radius that contains no other element of the group. Thus $\mathbb{Z}$ is a discrete group; given $n \in \mathbb{Z}$, there is no other integer in the interval $[n - 1/2, n + 1/2]$. However, $\mathbb{R}$ under addition is not discrete, for every interval of positive length having center at 0 contains other real numbers.

54. Give an example of a nonconstant periodic function of one variable that is invariant under the group of translations generated by τ where $\tau(x) = x + 1$.

55. Repeat Exercise 54 if $\tau(x) = x - \sqrt{3}$.

56. Give an example of a nonconstant doubly periodic function of two variables that is invariant under the group of translations generated by $\tau_1(x, y) = (x + 1, y)$ and $\tau_2(x, y) = (x, y + 1)$.

57. Repeat Exercise 56 if $\tau_1(x, y) = (x, y - \sqrt{5})$ and $\tau_2(x, y) = (x - 3, y)$.

58. Find generators τ_1 and τ_2 for the group of translations under which the doubly periodic function $\sin(2x + 3y)$ is invariant.

59. Give an example of a nonconstant automorphic function of two variables that is invariant under the group (under function composition) of rotations of the plane about the origin generated by a rotation through $30°$.

60. Repeat Exercise 59 if the rotations are about the point $(\sqrt{3}, -5)$.

Plane Isometries

We continue the discussion of plane isometries that appeared in the exercise set for Section 2.2. Recall that a plane isometry is a distance-preserving permutation of the points in the plane, and that every plane isometry can be shown to be either a translation, a rotation, a reflection, or a glide reflection. Of these four types, the translation and rotation preserve orientation, while the reflection and glide reflection reverse orientation. The set of all plane isometries forms a group under permutation multiplication (function composition). The theorem that follows describes the possible structures of finite subgroups of the full isometry group.

THEOREM 2.15   Every finite group G of isometries of the plane is isomorphic to either $\mathbb{Z}_n$ or to a dihedral group D_n for some positive integer n.

PROOF   First we show that there is a point P in the plane that is left fixed by every isometry in G. This can be done in the following way, using coordinates in the plane. Suppose $G = \{\phi_1, \phi_2, \ldots, \phi_m\}$ and let

$$(x_i, y_i) = \phi_i(0, 0).$$

Then the point

$$P = (\bar{x}, \bar{y}) = \left(\frac{x_1 + x_2 + \cdots + x_m}{m}, \frac{y_1 + y_2 + \cdots + y_m}{m} \right)$$

is the *centroid* of the set $S = \{(x_i, y_i) \mid i = 1, 2, \ldots, m\}$. The isometries in G permute the points in S among themselves, since if $\phi_i\phi_j = \phi_k$ then $\phi_i(x_j, y_j) = \phi_i[\phi_j(0, 0)] = \phi_k(0, 0) = (x_k, y_k)$. The centroid of a set of points is uniquely determined by its distances from the points, and since each isometry in G just permutes the set S, it must leave the centroid $(\bar{x}, \bar{y})$ fixed. Thus G consists of the identity, rotations about P, and reflections across a line through P.

The orientation-preserving isometries in G form a subgroup H of G which is either all of G or of order $m/2$. This can be shown in the same way that we showed that the even permutations are a subgroup of S_n containing just half the elements of S_n. (See Exercise 61.) Of course H consists of the identity and the rotations in G. If we choose a rotation in G that rotates the plane through as small an angle $\theta > 0$ as possible, it can be shown to generate the subgroup H. (See Exercise 62.) This shows that if $H = G$, then G is cyclic of order m and thus isomorphic to $\mathbb{Z}_m$. Suppose $H \neq G$ so that G contains some reflections. Let $H = \{\iota, \rho_1, \ldots, \rho_{n-1}\}$ with $n = m/2$. If μ is a reflection in G then the coset $H\mu$ consists of all n of the reflections in G.

Consider now a regular n-gon in the plane having P as its center and with a vertex lying on the line through P left fixed by μ. The elements of H rotate this n-gon through all positions, and the elements of $H\mu$ first reflect in an axis through a vertex, effectively turning the n-gon over, and then rotate through all positions. Thus the action of G on this n-gon is the action of D_n, so G is isomorphic to D_n. ◆

The preceding theorem gives the complete story about finite plane isometry groups. We turn now to some infinite groups of plane isometries that arise naturally in decorating and art. Among these are the *discrete frieze groups*. A discrete frieze consists of a pattern of finite width and height that is repeated endlessly in both directions along its baseline to form a strip of infinite length but finite height; think of it as a decorative border strip that goes around a room next to the ceiling on wallpaper. We consider those symmetries that carry each basic pattern onto itself or onto another instance of the pattern in the frieze. All discrete frieze groups are infinite and have a subgroup isomorphic to $\mathbb{Z}$ generated by the translation that slides the frieze lengthwise until the basic pattern is superimposed on the position of its next neighbor pattern in that direction. As a simple example of a discrete frieze, consider integral signs spaced equal distances apart and continuing infinitely to the left and right, indicated schematically like this.

Let us consider the integral signs to be one unit apart. The symmetry group of this frieze is generated by a translation τ sliding the plane one unit to the right, and by a rotation ρ of 180° about a point in the center of some integral sign. There are no horizontal or vertical reflections, and no glide reflections. This frieze group is nonabelian; we can check that $\tau\rho = \rho\tau^{-1}$. The n-th dihedral group D_n is generated by two elements that don't commute, a rotation ρ_1 through $360/n°$ of order n and a reflection μ of order 2 satisfying $\rho_1\mu = \mu\rho_1^{-1}$. Thus it is natural to use the notation D_∞ for this nonabelian frieze group generated by τ of infinite order and ρ of order 2.

As another example, consider the frieze given by an infinite string of D's.

$$\cdots \text{D D D D D D D D D D} \cdots$$

Its group is generated by a translation τ one step to the right and by a vertical reflection μ across a horizontal line cutting through the middle of all the D's. We can check that these group generators commute this time, that is, $\tau\mu = \mu\tau$, so this frieze group is abelian and is isomorphic to $\mathbb{Z} \times \mathbb{Z}_2$.

It can be shown that if we classify such discrete friezes only by whether or not their groups contain a

rotation   horizontal axis reflection

vertical axis reflection   nontrivial glide reflection

then there are a total of seven possibilities. A *nontrivial glide reflection* in a symmetry group is one that is not equal to a product of a translation in that group and a reflection in that group. The group for the string of D's above contains glide reflections across the horizontal line through the centers of the D's, but the translation component of each glide reflection is also in the group so they are all considered trivial glide reflections in that group. The frieze group for

contains a nontrivial glide reflection whose translation component is not an

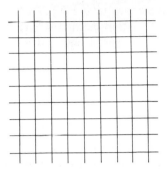

Figure 2.20

element of the group. The exercises exhibit the seven possible cases, and ask you to tell, for each case, which of the four types of isometries displayed above appear in the symmetry group. We do not obtain seven different group structures. Each of the groups obtained can be shown to be isomorphic to one of

$$\mathbb{Z}, \qquad D_\infty, \qquad \mathbb{Z} \times \mathbb{Z}_2 \qquad, \quad \text{or} \quad D_\infty \times \mathbb{Z}_2.$$

Equally interesting is the study of symmetries when a pattern in the shape of a square, parallelogram, rhombus, or hexagon is repeated by translations along *two nonparallel vector directions* to fill the entire plane, like patterns that appear on wallpaper. These groups are called the *wallpaper groups* or the *plane crystallographic groups*. While a frieze could not be carried into itself by a rotation through a positive angle less than 180°, it is possible to have rotations of 60°, 90°, 120°, and 180° for some of these plane-filling patterns. Figure 2.20 provides an illustration where the pattern consists of a square. We are interested in the group of plane isometries that carry this square onto itself or onto another square. Generators for this group are given by two translations (one sliding a square to the next neighbor to the right and one to the next above), by a rotation about the center of a square through 90°, and by a reflection in a vertical (or horizontal) line along the edges of the square. The one reflection is all that is needed to "turn the plane over"; a diagonal reflection can also be used. After being turned over, the translations and rotations can be used again. The isometry group for this *periodic pattern* in the plane surely contains a subgroup isomorphic to $\mathbb{Z} \times \mathbb{Z}$ generated by the unit translations to the right and upward, and a subgroup isomorphic to D_4 generated by those isometries that carry one square (it can be any square) into itself.

If we consider the plane to be filled with parallelograms as in Fig. 2.19 (think of solid-line sides and ignore the arrows and coordinate axes), we do not get all the types of isometries that we did for Fig. 2.20. The symmetry group this time is generated by the translations indicated by the arrows and a rotation through 180° about any vertex of a parallelogram.

It can be shown that there are 17 different types of wallpaper patterns when they are classified according to the types of rotations, reflections, and nontrivial glide reflections that they admit. We refer you to Gallian [9] for pictures of these 17 possibilities and a chart to help you identify them. The exercises illustrate a few of them. The situation in space is more complicated; it can be shown that there are 230 three-dimensional crystallographic groups. The final exercise we give involves rotations in space.

M. C. Escher (1898–1973) was an artist whose work included plane-filling patterns. The exercises include reproductions of four of his works of this type.

61. Completing a detail of the proof of Theorem 2.15, let G be a finite group of plane isometries. Show that the rotations in G, together with the identity isometry, form a subgroup H of G, and that either $H = G$ or $|G| = 2\,|H|$. [*Hint:* Use the same method that we used to show that $|S_n| = 2\,|A_n|$.]

62. Completing a detail in the proof of Theorem 2.15, let G be a finite group consisting of the identity isometry and rotations about one point P in the plane. Show that G is cyclic, generated by the rotation in G that turns the plane counterclockwise about P through the smallest angle $\theta > 0$. [*Hint:* Follow the idea of the proof that a subgroup of a cyclic group is cyclic.]

Exercises 63 through 69 illustrate the seven different types of friezes when they are classified according to their symmetries. Imagine the figure shown to be continued infinitely to the right and left. The symmetry group of a frieze always contains translations. For each of these exercises answer these questions about the symmetry group of the frieze.

a. Does the group contain a rotation?

b. Does the group contain a reflection across a horizontal line?

c. Does the group contain a reflection across a vertical line?

d. Does the group contain a nontrivial glide reflection?

e. To which of the possible groups $\mathbb{Z}$, D_∞, $\mathbb{Z} \times \mathbb{Z}_2$, or $D_\infty \times \mathbb{Z}_2$ do you think the symmetry group of the frieze is isomorphic?

63. F F F F F F F F F F F F F

64. T T T T T T T T T T

65. E E E E E E E E E E E E

66. Z Z Z Z Z Z Z Z Z Z Z

67. H H H H H H H H H H

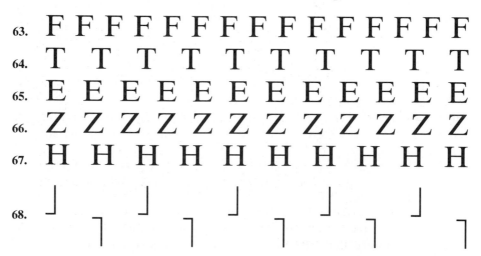

68.

69.

Exercises 70 through 76 describe a pattern to be used to fill the plane by translation in the two directions given by the specified vectors. Answer these questions in each case.

    a. Does the symmetry group contain any rotations? If so, through what possible angles θ where $0 < \theta \leq 180°$?

    b. Does the symmetry group contain any reflections?

    c. Does the symmetry group contain any nontrivial glide reflections?

70. A square with horizontal and vertical edges using translation directions given by vectors $(1, 0)$ and $(0, 1)$.

71. A square as in Exercise 70 using translation directions given by vectors $(1, 1/2)$ and $(0, 1)$.

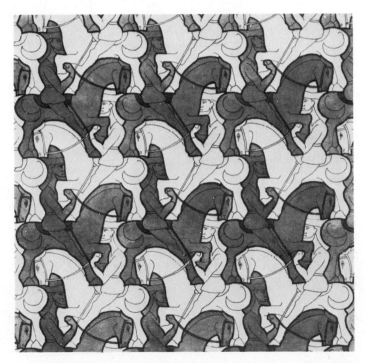

Figure 2.21 The Study of Regular Division of the Plane with Horsemen (© 1946 M. C. Escher Foundation–Baarn–Holland. All rights reserved.)

72. A square as in Exercise 70 with the letter L at its center using translation directions given by vectors $(1, 0)$ and $(0, 1)$.

73. A square as in Exercise 70 with the letter E at its center using translation directions given by vectors $(1, 0)$ and $(0, 1)$.

74. A square as in Exercise 70 with the letter H at its center using translation directions given by vectors $(1, 0)$ and $(0, 1)$.

75. A regular hexagon with a vertex at the top using translation directions given by vectors $(1, 0)$ and $(1, \sqrt{3})$.

76. A regular hexagon with a vertex at the top containing an equilateral triangle with vertex at the top and centroid at the center of the hexagon, using translation directions given by vectors $(1, 0)$ and $(1, \sqrt{3})$.

Figure 2.22 The Study of Regular Division of the Plane with Imaginary Human Figures (© 1936 M. C. Escher Foundation–Baarn–Holland. All rights reserved.)

Exercises 77 through 80 are concerned with art works of M. C. Escher. Neglect the shading in the figures and assume the markings in each human figure, reptile, or horseman are the same, even though they may be invisible due to shading. Answer the same questions (a), (b), and (c) that were asked for Exercises 70 through 76, and also answer this part (d).

  d. Assuming horizontal and vertical coordinate axes with equal scales as usual, give vectors in the two nonparallel directions of vectors that generate the translation subgroup. Don't concern yourself with the length of these vectors.

77. *The Study of Regular Division of the Plane with Horsemen* in Fig. 2.21.

78. *The Study of Regular Division of the Plane with Imaginary Human Figures* in Fig. 2.22.

79. *The Study of Regular Division of the Plane with Reptiles* in Fig. 2.23.

80. *The Study of Regular Division of the Plane with Human Figures* in Fig. 2.24.

81. Show that the rotations of a cube in space form a group isomorphic to S_4. [*Hint:* A rotation of the cube permutes the diagonals through the center of the cube.]

Figure 2.23 The Study of Regular Division of the Plane with Reptiles (© 1939 M. C. Escher Foundation–Baarn–Holland. All rights reserved.)

Figure 2.24 The Study of Regular Division of the Plane with Human Figures (© 1936 M. C. Escher Foundation–Baarn–Holland. All rights reserved.)

2.5

Binary Linear Codes[†]

We are not concerned here with secret codes. Rather, we discuss the problem of encoding information for transmission so that errors occurring during transmission or reception have a good change of being detected, and perhaps even being corrected by an appropriate decoding procedure. The diagram

message → ⟦encode⟧ → ⟦transmit⟧ → ⟦receive⟧ → ⟦decode⟧ → message

---

[†] This section is not used in the remainder of the text.

shows the steps with which we are concerned. Errors could be caused at any stage of the process by equipment malfunction, human error, lightning, sunspots, cross-talk interference, etc.

Numerical Representation of Information

All information can be reduced to sequences of numbers. For example, we could number the letters of our alphabet and represent every word in our language as a finite sequence of numbers. We concentrate on how to encode numbers to detect and possibly correct errors.

We are accustomed to expressing numbers in decimal (base 10) notation, using as *alphabet* the set $\{0, 1, 2, 3, 4, 5, 6, 7, 8, 9\}$. However, they also can be expressed using any integer base greater than or equal to 2. A computer works in binary (base 2) notation which uses the smaller alphabet $\{0, 1\}$; the number 1 can be represented by the presence of an electric charge or by current flowing, while the absence of a charge or current can represent 0. The ASCII code of 256 characters that is widely used by computer programmers includes all the characters that we customarily find on typewriter keyboards, such as

$$A\,a\,B\,b\,Z\,z\,1\,2\,3\,4\,5\,6\,7\,8\,9\,0\,,\,;\,?\,*\,\&\,\#\,!\,+\,-\,/\,\acute{}\,".$$

The 256 characters are assigned numbers from 0 to 255. For example, S is assigned the number 83 (decimal) which is 1010011 (binary) since, reading 1010011 from left to right, we see that

$$1(2^6) + 0(2^5) + 1(2^4) + 0(2^3) + 0(2^2) + 1(2^1) + 1(2^0) = 83.$$

The ASCII code number for the character 7 is 55 (decimal) which is 110111 (binary). Since $2^8 = 256$, each character in the ASCII code can be represented by a sequence of eight 0's or 1's; the S is represented by 01010011 and 7 by 00110111. This discussion makes it clear that all information can be encoded using just the *binary alphabet* $\mathbb{B} = \{0, 1\}$.

Message Words and Code Words

An algebraist refers to a sequence of characters from some alphabet, like 01010011 using the alphabet $\mathbb{B}$ or the sequence *glypt* using our usual letter alphabet, as a *word*; the computer scientist refers to it as a *string*. As we discuss encoding words so that transmission errors can be detected, it is convenient to use as small an alphabet as possible, so we restrict ourselves to the **binary words** using the alphabet $\mathbb{B} = \{0, 1\}$. In Chapters 5 and 8, we will mention other alphabets that are used. Rather than give examples using words of eight characters as in the ASCII code, let us use words of just four characters; the 16 possible words are

0000	0001	0010	0011	0100	0101	0110	0111

1000	1001	1010	1011	1100	1101	1110	1111.

An error in transmitting such a word occurs when a 1 is changed to a 0 or vice versa during transmission. The next two examples exhibit an inefficient way and an efficient way of detecting a transmission error in which *only one* interchange of the characters 0 and 1 occurs. In each example, a binary *message word* is encoded to form the *code word* to be transmitted.

EXAMPLE 1   Suppose we wish to send as the binary message word 1011. To detect any single-error transmission, we could send each character twice, that is, we could *encode* 1011 as 11001111 when we send it. If a single error is made in transmission of the code word 11001111 and the recipient knows the encoding scheme, then the error will be detected. For example, if the received code word is 11001011, the recipient knows there is an error since the 5th and 6th characters are different. Of course, the recipient does not know whether 0 or 1 is the correct character. But note that not all double-error transmissions can be detected. For example, if the received code word is 11000011, the recipient perceives no error, and obtains 1001 upon decoding, which was not the message sent. ▲

One problem with encoding a word by repeating every character as in Example 1 is that the code word is twice as long as the original message word. There is a lot of *redundancy*. The next example shows how we can more efficiently achieve the goal of warning the receiver whenever a single error has been committed.

EXAMPLE 2   Suppose again that we wish to transmit a four-character word on the alphabet $\mathbb{B}$. Let us denote the word symbolically by $x_1x_2x_3x_4$ where each x_i is either 0 or 1. We make use of modulo 2 arithmetic, as in $\mathbb{Z}_2$, where

$$0 + 0 = 0, \quad 1 + 0 = 0 + 1 = 1, \quad \text{and} \quad 1 + 1 = 0, \quad \text{(modulo 2 sums)}$$

while $1 \cdot 0 = 0 \cdot 1 = 0$ and $1 \cdot 1 = 1$. We append to the word $x_1x_2x_3x_4$ the modulo 2 sum

$$x_5 = x_1 + x_2 + x_3 + x_4 \tag{1}$$

This amounts to appending the character 0 if the message word contains an even number of characters 1, and appending a 1 if the number of 1's in the word is odd. Note that the result is a five-character code word $x_1x_2x_3x_4x_5$

definitely containing an even number of 1's. Thus we have

$$x_1 + x_2 + x_3 + x_4 + x_5 = 0 \qquad \text{(modulo 2)}.$$

If the message word is 1011 as in the preceding example, then the encoded word is 10111. If a single error is made in transmitting this code word, changing a single 0 to 1 or a single 1 to 0, then the modulo 2 sum of the five characters will be 1 rather than 0, and the recipient will recognize that there was an error in transmitting the code word. ▲

Example 2 attained the goal of recognizing single-error transmission with less redundancy than in Example 1. In Example 2 we used just one extra character, while in Example 1 we used four extra characters. However, using the scheme in Example 1, we were able to identify which character of the message is affected, while the technique in Example 2 just shows that at least one error was made.

Computers use the scheme in Example 2 when storing the ASCII code number of one of the 256 ASCII characters. An extra 0 or 1 is appended to the binary form of the code number, so that the number of 1's in the augmented word is even. When the encoded ASCII character is retrieved, the computer checks that the number of 1's is indeed even. If it is not, it can try to read that binary word again. The user may be warned that there is a PARITY-CHECK problem if the computer is not successful.

Terminology

Equation (1) in Example 2 is known as a **parity-check equation**. In general, starting with a **message word** $x_1 x_2 \cdots x_k$ of k characters, we encode it to become a **code word** $x_1 x_2 \cdots x_k \cdots x_n$ of n characters. The first k characters are the **information portion** of the encoded word, and the final $n - k$ characters are the **redundancy portion** or **parity-check portion**.

We introduce more notation and terminology to make our discussion easier. Let $\mathbb{B}^n$ be the set of all binary words of n consecutive 0's or 1's. A **binary code** C is any subset of $\mathbb{B}^n$. We can identify a vector of n components with each word in $\mathbb{B}^n$, namely, the vector whose ith component is the ith character in the word. For example, we can identify the word 1101 and the row vector $(1, 1, 0, 1)$. It is convenient to denote the set of all of these row vectors with n components by $\mathbb{B}^n$ also. This notation is similar to the notation $\mathbb{R}^n$ for all n-component vectors of real numbers. On occasion, we may find it convenient to use column vectors rather than row vectors.

The **length** of a word u in $\mathbb{B}^n$ is n, the number of its components. The **Hamming weight** $wt(u)$ of u is the number of components that are 1. Given two binary words u and v in $\mathbb{B}^n$, the **distance between them**, denoted by $d(u, v)$, is the number of components in which the entries in u and v are different, so that one word has a 0 where the other has a 1.

EXAMPLE 3 Consider   the   binary   words $u = 11010011$ and
$v = 01110111$. Both words have length 8. Also $wt(u) = 5$, while $wt(v) = 6$.
The associated vectors differ in the first, third, and sixth components, so
$d(u, v) = 3$. ▲

We can define addition on the set $\mathbb{B}^n$ by adding modulo 2 the characters in
the   corresponding   positions.   Remembering   that $1 + 1 = 0$,   we   add
0011101010 and 1010110001 as follows.

$$
\begin{array}{r}
0011101010 \\
+\ \ 1010110001 \\
\hline
1001011011
\end{array}
$$

We refer to this operation as **word addition**. Word subtraction is similarly
defined. Note that $0 - 1 = 1$. Exercise 17 shows that $\mathbb{B}^n$ with this operation
of word addition is a group. A **binary group code** is a subgroup C of $\mathbb{B}^n$. The
order of C must be a divisor of the order 2^n of $\mathbb{B}^n$, so the order of C must be
2^k for some integer k where $0 \le k \le n$. We refer to such a code as an (n, k)
binary group code.

For those of you who have studied linear algebra, we mention that $\mathbb{B}^n$ can
be viewed as an n-dimensional vector space over the *field of scalars* $\mathbb{B}$, just as
$\mathbb{R}^n$ is an n-dimensional vector space over the field of scalars $\mathbb{R}$. If C is an
(n, k) binary group code, then C can be viewed as a k-dimensional subspace
of $\mathbb{B}^n$. Namely, thinking in terms of the row vectors with n components that
we identify with the words in C, we see that C not only satisfies the additive
group properties, but also the scalar multiplicative properties required for a
subspace, where the only scalars are, of course, 0 and 1. For this reason, C is
often referred to as an (n, k) **linear code**.

Encoding to Enable Correcting a Single-Error Transmission

We now show how, using more than one parity-check equation, we can not
only detect but actually correct a single-error transmission of a code word.
This method of encoding was developed by Richard Hamming in 1948.

Suppose we wish to encode the sixteen message words

0000	0001	0010	0011	0100	0101	0110	0111
1000	1001	1010	1011	1100	1101	1110	1111

in $\mathbb{B}^4$ so that any single-error transmission of a code word can not only be
detected but also be corrected. The basic idea is simple. We append to the
message word $x_1 x_2 x_3 x_4$ some additional binary characters given by parity-
check equations like the equation $x_5 = x_1 + x_2 + x_3 + x_4$ in Example 2, and
try to design the equations so that the minimum distance between the sixteen
code words created will be at least 3. Now with a single-error transmission of
a code word, the distance from the received word and that code word

---

Richard Hamming (b. 1915) had his interest in the question of coding stimulated in 1947 when he was using an early Bell System relay computer on weekends only (because he did not have priority use of the machine). During the week, the machine sounded an alarm when it discovered an error so that an operator could attempt to correct it. On weekends, however, the machine was unattended and would dump any problem in which it discovered an error and proceed to the next one. Hamming's frustration with this behavior of the machine grew when errors cost him two consecutive weekends of work. He decided that if the machine could discover errors—it used a fairly simple error-detecting code—there must be a way for it to correct them and proceed with the solution. He therefore worked on this idea for the next year and discovered several different methods of creating error-correcting codes. Because of patent considerations, Hamming did not publish his solutions until 1950. A brief description of his $(7, 4)$ code, however, appeared in a paper of Claude Shannon (b. 1916) in 1948.

Hamming in fact developed some of the parity-check ideas indicated in the text as well as the geometric model in which the distance between code words is the number of coordinates in which they differ. He also in essence realized that the set of actual code words embedded in $\mathbb{B}^7$ was a four-dimensional subspace of that space.

---

will be 1. If we can make our code words all at least three units apart, the correct code word (before transmission) will be the unique code word at distance 1 from the received word.

In order to detect the error in a single-error transmission of a code word, including not only message word characters but also the redundant parity-check characters, we need to have each component x_i in the code word appear at least once in *some* parity-check equation. Note that in Example 1, each component x_1, x_2, x_3, x_4, and x_5 appears in the parity-check equation $x_5 = x_1 + x_2 + x_3 + x_4$. The parity-check equations

$$x_5 = x_1 + x_2 + x_3, \qquad x_6 = x_1 + x_3 + x_4, \qquad \text{and} \qquad x_7 = x_2 + x_3 + x_4, \quad (2)$$

which we will show accomplish our goal, also satisfy this condition.

Let us see how to get a distance of at least 3 between each pair of the sixteen code words. Of course, the distance between any two of the original sixteen 4-character message words is at least 1 since they are all different. Suppose now that two message words u and v differ in just one component, say x_2. A parity-check equation containing x_2 then yields a different character for u than for v. This shows that if each x_i in our original message word appears in at least two parity-check equations, then any message words at a distance of 1 are encoded into code words of distance at least 3. Note that the three parity-check Eqs. (2) satisfy this condition. It remains to ensure that two message words at a distance of 2 are encoded to increase this distance by at least one. Suppose two message words u and v differ in only the ith and jth components. Now a parity-check equation containing both x_i and x_j will create the same parity-check character for u as for v. Thus for each such combination i, j of positions in our message word, we need some parity-check equation either to contain x_i but not x_j or to contain x_j but not x_i. We see that this condition is satisfied by the three parity-check Eqs. (2) for all possible combinations i, j, namely,

$$1, 2 \quad 1, 3 \quad 1, 4 \quad 2, 3 \quad 2, 4 \quad \text{and} \quad 3, 4.$$

Thus the Eqs. (2) accomplish our goal. The sixteen binary words of length 7, obtained by encoding the sixteen binary words

$$0000 \quad 0001 \quad 0010 \quad 0011 \quad 0100 \quad 0101 \quad 0110 \quad 0111$$

$$1000 \quad 1001 \quad 1010 \quad 1011 \quad 1100 \quad 1101 \quad 1110 \quad 1111$$

of length 4 using the parity-check Eqs. (2) form a subset of $\mathbb{B}^7$ called the Hamming $(7, 4)$ code. In Exercise 18, we ask you to show that the Hamming $(7, 4)$ code is a group code.

We can encode each of the sixteen binary words of length 4 to form the Hamming $(7, 4)$ code by multiplying the vector (x_1, x_2, x_3, x_4) form of the word on the right by a 4×7 matrix G, called the **standard generator matrix**; namely, we compute

$$[x_1, x_2, x_3, x_4] \begin{bmatrix} 1 & 0 & 0 & 0 & 1 & 1 & 0 \\ 0 & 1 & 0 & 0 & 1 & 0 & 1 \\ 0 & 0 & 1 & 0 & 1 & 1 & 1 \\ 0 & 0 & 0 & 1 & 0 & 1 & 1 \end{bmatrix}.$$
$$G$$

To see this, note that the first four columns of G give the 4×4 identity matrix I_4, so the first four entries in the encoded word will yield precisely the message word $x_1 x_2 x_3 x_4$. In columns 5, 6, and 7, we put the coefficients of x_1, x_2, x_3, and x_4 as they appear in the parity-check Eqs. (2) defining x_5, x_6, and x_7 respectively. Table 2.7 shows the sixteen message words and the code words obtained using this generator matrix G. Note that the message words 0011 and 0111 that are at distance 1 have been encoded into 0011100 and 0111001 that are at distance 3. Also, the message words 0101 and 0011 that are at distance 2 have been encoded into 0101110 and 0011100 that are at distance 3.

Decoding a received word w by selecting a code word at minimum distance from w (in some codes more than one code word might be at the minimum distance) is known as **nearest-neighbor decoding**. If transmission errors are

Message	Code Word
0000	0000000
0001	0001011
0010	0010111
0011	0011100
0100	0100101
0101	0101110
0110	0110010
0111	0111001
1000	1000110
1001	1001101
1010	1010001
1011	1011010
1100	1100011
1101	1101000
1110	1110100
1111	1111111

Table 2.7

independent from each other, it can be shown that this is equivalent to *maximum-likelihood decoding*.

EXAMPLE 4   Suppose the Hamming $(7, 4)$ code shown in Table 2.7 is used. If the received word is 1011010, then the decoded message word consists of the first four characters 1011 since 1011010 is a code word. However, suppose the word 0110101 is received. This is not a code word. The closest code word is 0100101, which is at distance 1 from 0110101. Thus we decode 0110101 as 0100, which differs from the first four characters of the received word. On the other hand, if we receive the non codeword 1100111, we decode it as 1100 since the closest code word to 1100111 is 1100011.   ▲

Note in Example 4 that if the code word 0001011 is transmitted and is received as 0011001, with two errors, then the recipient knows that an error was made since 0011001 is not a code word. However, nearest-neighbor decoding yields the code word 0111001 corresponding to a message word 0111 rather than the intended 0001. When retransmission is practical, it may be better to request it when an error is detected rather than to blindly use nearest-neighbor decoding.

Of course, if errors are generated independently and transmission is of high quality, it should be much less likely that a word would be transmitted with two errors than with one error. If the probability of an error in transmission of a single character is p and errors are generated independently, then probability theory shows that in transmitting a word of length n,

the probability of no error is $(1 - p)^n$,

the probability of exactly one error is $np(1 - p)^{n-1}$,

the probability of exactly two errors is $\dfrac{n(n-1)}{2}p^2(1-p)^{n-2}$.

For example, if $p = 0.0001$ so that we can expect about one character out of every 10,000 to be changed and if the length of the word is $n = 10$, then the probabilities of no error, one error, and two errors respectively are approximately 0.999, 0.000999, and 0.0000004.

Parity-Check Matrix Decoding

You can imagine that if we encoded all of the 256 ASCII characters in an $(n, 8)$ linear code and used nearest-neighbor decoding, it would be a job to pore over the list of 256 encoded characters to determine the nearest neighbor to a received code word. There is an easier way, which we illustrate using the Hamming $(7, 4)$ code developed before Example 4. The parity-check equations for that code are

$$x_5 = x_1 + x_2 + x_3, \quad x_6 = x_1 + x_3 + x_4, \quad \text{and} \quad x_7 = x_2 + x_3 + x_4.$$

Let us again concern ourselves with detecting and correcting just single-error transmissions. If these parity-check equations hold for the received word, then no such single error occurred. Suppose, on the other hand, that the first two equations fail and the last one holds. The only character appearing in both of the first two equations but not in the last is x_1, so we could just change the character x_1 from 0 to 1, or vice versa, to decode. Note that each of x_1, x_2, and x_4 is omitted just once but in different equations, x_3 is the only character that appears in all three equations, and each of x_5, x_6, and x_7 appears just once but in different equations. This allows us to identify the incorrect character in a single-error transmission easily. We can be even more systematic. Let us rewrite the equations as

$$x_1 + x_2 + x_3 + x_5 = 0, \qquad x_1 + x_3 + x_4 + x_6 = 0,$$

and

$$x_2 + x_3 + x_4 + x_7 = 0.$$

We form the **parity-check matrix** H whose ith row contains the seven coefficients of $x_1, x_2, x_3, x_4, x_5, x_6,$ and x_7 in the ith equation, namely,

$$H = \begin{bmatrix} 1 & 1 & 1 & 0 & 1 & 0 & 0 \\ 1 & 0 & 1 & 1 & 0 & 1 & 0 \\ 0 & 1 & 1 & 1 & 0 & 0 & 1 \end{bmatrix}.$$

Let w be a received word, written as a column vector. Exercise 29 shows that w is a code word if and only if Hw is the zero column vector, where we are always using modulo 2 arithmetic. If w resulted from a single-error transmission in which the character in the jth position was changed, then Hw will have 1 in its ith component if and only if x_j appears in the ith parity-check equation, so that the column vector Hw will be the jth column of H. Thus we can decode a received word w as follows in the Hamming $(7, 4)$ code of Example 4, and be confident of detecting and correcting any single-error transmission.

Parity-Check Matrix Decoding

1. Compute Hw.
2. If Hw is the zero vector, decode as the first four characters of w.
3. If Hw is the jth column of H then

   a. if $j > 4$ then decode as the first four characters of w,

   b. otherwise decode as the first four characters with the jth character changed.
4. If neither No. 2 nor No. 3 occurs, then more than one error was made; ask for retransmission.

EXAMPLE 5   Suppose the Hamming $(7, 4)$ code shown in Table 2.7 is used and the word $w = 0110101$ is received. We compute that

$$
Hw = \begin{bmatrix} 1 & 1 & 1 & 0 & 1 & 0 & 0 \\ 1 & 0 & 1 & 1 & 0 & 1 & 0 \\ 0 & 1 & 1 & 1 & 0 & 0 & 1 \end{bmatrix} \begin{bmatrix} 0 \\ 1 \\ 1 \\ 0 \\ 1 \\ 0 \\ 1 \end{bmatrix} = \begin{bmatrix} 1 \\ 1 \\ 1 \end{bmatrix}.
$$

Since this is the third column of H, we change the third character in the message portion 0110 and decode as 0100. Note that this is what we obtained in Example 4 where we decoded this word using Table 2.7. ▲

Just as after Example 4, we point out that if two errors are made in transmission, the preceding outline may lead to incorrect decoding. If the code word $v = 0001011$ is transmitted and received as $w = 0011001$ with two errors, then

$$
Hw = \begin{bmatrix} 1 \\ 0 \\ 1 \end{bmatrix},
$$

which is column 2 of the matrix H. Thus decoding by the steps above leads to the incorrect message word 0111. Note that No. 4 above does not say that if more than one error was made, then neither No. 2 nor No. 3 occurs.

Coset Decoding

We conclude with a brief discussion of *coset decoding* for a group code C in $\mathbb{B}^n$ with minimum distance 3 between code words and a parity-check matrix H. We concern ourselves with detecting and correcting a single-error transmission. Consider the n words

$$e_1 = 100 \cdots 00, \qquad e_2 = 010 \cdots 00, \qquad \cdots, \qquad e_n = 000 \cdots 01$$

of weight 1 in $\mathbb{B}^n$. Then $n + 1$ cosets

$$C, \quad e_1 + C, \quad e_2 + C, \quad \cdots \quad e_n + C$$

contain all words of distance at most 1 from some code word in C. For each word w in $\mathbb{B}^n$, we can write w as a column vector and compute the vector Hw. The word form of the vector Hw is called the **syndrome** of w. Since Hc is the zero vector for every $c \in C$, we see that $H(e_i + c) = H(e_i) + H(c) = H(e_i)$ for all $c \in C$. That is, elements in the same coset have the same syndrome. Conversely, if $H(w) = H(v)$ then $H(w - v) = H(w) - H(v) = 0$, the zero vector. Exercise 29 shows that this implies that $w - v = 0$, so $w = v$. Summarizing:

Elements of $\mathbb{B}^n$ are in the same coset of C if and only if they have the same syndrome.

Thus the coset of C containing a word w can be identified by the syndrome of the word.

If $Hw = He_i$, then $w = e_i + c$ for some $c \in C$, and, of course, c is the correct decoding for w using nearest-neighbor decoding, since the distance from c to w is 1. Knowing w and e_i and that $w = e_i + c$ we obtain at once $c = w - e_i = w + e_i$.

Coset Decoding with Syndromes

1. Make a table containing the words of weight at most 1 in $\mathbb{B}^n$ and their syndromes.
2. If the syndrome of w is the zero vector, decode as the first k characters of w.
3. If the syndrome of w is the syndrome of e_i, then decode as the first k characters of the word $w + e_i$.
4. If neither No. 2 nor No. 3 occurs, then more than one error was made; ask for retransmission.

Word	Syndrome
0000000	000
1000000	110
0100000	101
0010000	111
0001000	011
0000100	101
0000010	010
0000001	001

Table 2.8

We illustrate using the Hamming $(7, 4)$ code we exhibited, where the parity-check matrix is

$$H = \begin{bmatrix} 1 & 1 & 1 & 0 & 1 & 0 & 0 \\ 1 & 0 & 1 & 1 & 0 & 1 & 0 \\ 0 & 1 & 1 & 1 & 0 & 0 & 1 \end{bmatrix}.$$

Table 2.8 shows all words of weight at most 1 in $\mathbb{B}^7$ and their syndromes. Note that the nonzero syndromes correspond to the columns of H. Suppose that the received word is $w = 0110101$. Computing Hw, we find the syndrome of w is 111. Table 2.8 shows that this is the syndrome of 0010000, so we decode w by computing $w + e_3 = 0110101 + 0010000 = 0100101$ and taking the first four characters 0100 as the decoded message. Note that this agrees with our decoding of w in Example 4. Of course if Hw is a syndrome not in our list, then w is at distance at least 2 from any code word.

As shown in Exercises 23 and 24, counting some of the cosets of a binary group code C enables us to see how many parity-check equations we at least have to use to enable detecting and correcting single-error transmissions, double-error transmissions, etc.

Our discussion in this section barely scratches the surface of a very active branch of research known as information theory. A standard reference for algebraic coding is F. J. MacWilliams and N. J. A. Sloane, *The Theory of Error-Correcting Codes,* Parts I, II, Amsterdam: North-Holland, 1977.

Exercises 2.5

1. Let 0 stand for a space and let the binary numbers 1 through 15 stand for the letters A B C D E F G H I J K L M N O in that order. Using Table 2.7 for the Hamming $(7, 4)$ code, encode the message A GOOD DOG.

2. With the same understanding as in the preceding exercise, use nearest-neighbor decoding to decode this received message.

0111001   1111111   1010100   0101110   0000000   1100110   1111111

1101000   1101110

In Exercises 3 through 11, consider the $(6, 3)$ linear code C with standard generator matrix

$$G = \begin{bmatrix} 1 & 0 & 0 & 1 & 1 & 0 \\ 0 & 1 & 0 & 1 & 0 & 1 \\ 0 & 0 & 1 & 0 & 1 & 1 \end{bmatrix}.$$

3. Give the parity-check equations for this code.

4. List the code words in C.

5. How many errors can be always be detected using this code?

6. How many errors can always be corrected using this code?

7. Show that there is just one coset of C in $\mathbb{B}^6$ each of whose elements has distance at least 2 from every code word. List the elements in this coset.

8. Assuming that the given word was received, decode it using nearest-neighbor decoding, using your list of code words in Exercise 4. (Recall that in case more than one code word is at minimum distance from the received word, a code word of minimum distance is selected arbitrarily.)

   a. 110111   b. 001011   c. 111011   d. 101010   e. 100101

9. Give the parity-check matrix for this code.

10. Use the parity-check matrix to decode the received words in Exercise 8.

11. List the seven words of weight at most 1 in $\mathbb{B}^6$ and write the syndrome of each of the words next to it. Using that list, execute coset decoding by syndromes of the received words in Exercise 8.

12. Let $u = 1101010111$ and $v = 0111001110$. Find

   a. $wt(u)$ b. $wt(v)$ c. $u + v$ d. $d(u, v)$

13. Show that for word addition of binary words u and v of the same length, we have $u + v = u - v$.

14. If a binary code word u is transmitted and the received word is w, then the sum $u + w$ given by word addition modulo 2 is called the *error pattern*. Explain why this is a descriptive name for this sum.

15. Show that for two binary words of the same length, we have $d(u, v) = wt(u - v)$.

16. Prove the following properties of the distance function for binary words u, v, and w of the same length.

   a. $d(u, v) = 0$ if and only if $u = v$

   b. $d(u, v) = d(v, u)$ (*symmetry*)

   c. $d(u, w) \leq d(u, v) + d(v, w)$ (*triangle inequality*)

   d. $d(u, v) = d(u + w, v + w)$ (*invariance under translation*)

17. Show that $\mathbb{B}^n$ with the operation of word addition is a group.

18. Recall that we call a subset C of $\mathbb{B}^n$ a *group code* if C is a subgroup of $\mathbb{B}^n$. Show that the Hamming $(7, 4)$ code is a group code. [*Hint:* To show closure under word addition, use the fact that the words in the Hamming $(7, 4)$ code can be formed from those in $\mathbb{B}^4$ by multiplying by a generator matrix.]

19. Show that in a binary group code C, the minimum distance between code words is equal to the minimum weight of the nonzero code words.

20. Suppose that you want to be able to recognize that a received word is incorrect when m or fewer of its characters were changed during transmission. What must be the minimum distance between code words to accomplish this?

21. Suppose that you want to be able to find a unique nearest-neighbor for a received word that was transmitted with m or fewer of its characters changed. What must be the minimum distance between code words to accomplish this?

22. Show that if the minimum nonzero weight of code words in a group code C is at least $2t + 1$, then the code can detect any $2t$ errors and correct any t errors. (Compare the result stated in this exercise with your answers to the two preceding ones.)

23. Show that if the minimum distance between the words in an (n, k) binary group code C is at least 3, we must have
$$2^{n-k} \geq 1 + n.$$
[*Hint:* Using cosets, count the number of words in $\mathbb{B}^n$ at distance at most 1 from the 2^k words in C.]

24. Show that if the minimum distance between the words in an (n, k) binary group code C is at least 5, then we must have
$$2^{n-k} \geq 1 + n + \frac{n(n-1)}{2}.$$
[*Hint:* Using cosets, count the number of words in $\mathbb{B}^n$ at distance at most 2 from the 2^k words in C.]

25. Using the formulas in Exercises 23 and 24, find a lower bound for the number of parity-check equations necessary to encode the 2^k words in $\mathbb{B}^k$ so that the minimum distance between different code words is at least m for the given values of m and k. (Note that $k = 8$ would allow us to encode all the ASCII characters, and that $m = 5$ would allow us to detect and correct all single-error and double-error transmissions using nearest-neighbor decoding.)

a. $k = 2$, $m = 3$ b. $k = 4$, $m = 3$ c. $k = 8$, $m = 3$

d. $k = 2$, $m = 5$ e. $k = 4$, $m = 5$ f. $k = 8$, $m = 5$

26. Find parity-check equations to use to encode the 32 words in $\mathbb{B}^5$ into an $(n, 5)$ linear code that can be used to detect and correct any single-error transmission of a code word. (Recall that each character x_i must appear in two parity-check equations and that for each pair x_i, x_j some equation must contain one of them but not the other.) Try to make the number of parity-check equations as small as possible: See Exercise 23. Give the standard generator matrix for your code.

27. The 256 ASCII characters are numbered from 0 to 255, and thus can be represented by the 256 binary words in $\mathbb{B}^8$. Find $n - 8$ parity-check equations that

can be used to form an $(n, 8)$ linear code that can be used to detect and correct any single-error transmission of a code word. Try to make n the value found in part (c) of Exercise 25.

28. Let C be a binary group code.

    a. Show that the words in C of even weight form a subgroup of C.

    b. Show that if some word in C has odd weight, then half the words in C have odd weight and half have even weight.

29. (Uses linear algebra) Let C be an (n, k) linear code with parity-check matrix H. We know that $Hc = 0$ for all $c \in C$. Show conversely that if $w \in \mathbb{B}$ and $Hw = 0$, then $w \in C$. [*Hint:* Use the rank equation

$$(\text{number of columns of } H) = \text{rank}(H) + \text{nullity}(H).]$$

◆

CHAPTER THREE

HOMOMORPHISMS AND FACTOR GROUPS

◆

Chapters 1 and 2 introduced the notion of a group and discussed various types of groups. We now turn to relationships between groups. One such relationship, isomorphism, was used in an informal and intuitive way in Chapters 1 and 2.

A relationship between groups G and G' is generally exhibited in terms of a structure-relating map $\phi: G \to G'$. Section 3.1 introduces such maps, which are called *homomorphisms,* and discusses some of their properties. Section 3.2 provides a careful treatment of the notion of *isomorphic groups* and a discussion of how to decide whether two groups are isomorphic.

We will deal with cosets of a subgroup again when we discuss homomorphisms. In Section 2.3 we mentioned that we can sometimes form a group whose elements are cosets of a subgroup H of a group G. We will see that this notion is closely related to the idea of a homomorphism, and Sections 3.3 and 3.4 provide discussion of such groups of cosets, which we will call *factor groups.*

In Section 3.5 we discuss repeated factoring of a group into irreducible components, which is somewhat like factoring a positive integer into a product of prime integers. The notion of a *solvable group* introduced here is used in the final section of the text, where we discuss formulas involving radicals for solving a polynomial equation. The chapter closes with a discussion of group action on a set and an application to counting.

3.1

Homomorphisms
Structure-Relating Maps

Let G and G' be groups. We are interested in a map $\phi: G \to G'$ that relates the group structure of G to the group structure of G'. Such a map often gives us information about the structure of G' from known structural properties of $G,$ or information about the structure of G from known structural properties of G'. Now group structure is completely determined by the binary operation on the group. We define such a structure-relating map for groups, and then point out how the binary operations of G and G' are related by such a map.

DEFINITION 3.1 (Homomorphism)   A map ϕ of a group G into a group G' is a **homomorphism** if

$$\phi(ab) = \phi(a)\phi(b) \tag{1}$$

for all $a, b \in G$.

Let us examine the idea behind the requirement (1) for a homomorphism $\phi: G \to G'$. In Eq. (1), the product ab on the left-hand side takes place in G, while the product $\phi(a)\phi(b)$ on the right-hand side takes place in G'. Thus Eq. (1) gives a relation between these binary operations, and hence between the two group structures.

For any groups G and G', there is always at least one homomorphism $\phi: G \to G'$, namely the **trivial homomorphism** defined by $\phi(g) = e'$ for all $g \in G$, where e' is the identity in G'. Equation (1) then reduces to the true equation $e' = e'e'$. No information about the structure of G or G' can be gained from the other group using this trivial homomorphism. We give an example illustrating how a homomorphism ϕ mapping G onto G' may give structural information about G'.

EXAMPLE I   Let $\phi: G \to G'$ be a group homomorphism of G onto G'. We claim that if G is abelian, then G' must be abelian. Let $a', b' \in G'$. We must show that $a'b' = b'a'$. Since ϕ is onto G', there exist $a, b \in G$ such that $\phi(a) = a'$ and $\phi(b) = b'$. Since G is abelian, we have $ab = ba$. Using property (1), we have $a'b' = \phi(a)\phi(b) = \phi(ab) = \phi(ba) = \phi(b)\phi(a) = b'a'$, so G' is indeed abelian. ▲

Example 10 will give an illustration showing how information about G' may give information about G via a homomorphism $\phi: G \to G'$. We now give examples of homomorphisms for specific groups.

EXAMPLE 2   Let S_n be the symmetric group on n letters, and let $\phi: S_n \to \mathbb{Z}_2$ be defined by

$$\phi(\sigma) = \begin{cases} 0 & \text{if } \sigma \text{ is an even permutation,} \\ 1 & \text{if } \sigma \text{ is an odd permutation.} \end{cases}$$

Show that ϕ is a homomorphism.

Solution   We must show that $\phi(\sigma\mu) = \phi(\sigma) + \phi(\mu)$ for all choices of σ, $\mu \in S_n$. Note that the operation on the right-hand side of this equation is written additively since it takes place in the group $\mathbb{Z}_2$. Verifying this equation amounts to checking just four cases:

σ odd and μ odd,

σ odd and μ even,

σ even and μ odd,

σ even and μ even.

Checking the first case, if σ and μ can both be written as a product of an odd number of transpositions, then $\sigma\mu$ can be written as the product of an even number of transpositions. Thus $\phi(\sigma\mu) = 0$ and $\phi(\sigma) + \phi(\mu) = 1 + 1 = 0$ in $\mathbb{Z}_2$. The other cases can be checked similarly. ▲

EXAMPLE 3 (Evaluation Homomorphism) Let F be the additive group of all functions mapping $\mathbb{R}$ into $\mathbb{R}$, let $\mathbb{R}$ be the additive group of real numbers, and let c be any real number. Let $\phi_c : F \to \mathbb{R}$ be the evaluation homomorphism defined by $\phi_c(f) = f(c)$ for $f \in F$. Recall that, by definition, the sum of two functions f and g is the function $f + g$ whose value at x is $f(x) + g(x)$. Thus we have

$$\phi_c(f + g) = (f + g)(c) = f(c) + g(c) = \phi_c(f) + \phi_c(g),$$

and Eq. (1) is satisfied, so we have a homomorphism. ▲

EXAMPLE 4 Let $\mathbb{R}^n$ be the additive group of column vectors with n real number components. (This group is of course isomorphic to the direct product of $\mathbb{R}$ under addition with itself for n factors.) Let A be an $m \times n$ matrix of real numbers. Let $\phi : \mathbb{R}^n \to \mathbb{R}^m$ be defined by $\phi(\mathbf{v}) = A\mathbf{v}$ for each column vector $\mathbf{v} \in \mathbb{R}^n$. Then ϕ is a homomorphism, since for $\mathbf{v}, \mathbf{w} \in \mathbb{R}^n$, matrix algebra shows that $\phi(\mathbf{v} + \mathbf{w}) = A(\mathbf{v} + \mathbf{w}) = A\mathbf{v} + A\mathbf{w} = \phi(\mathbf{v}) + \phi(\mathbf{w})$. In linear algebra, such a map computed by multiplying a column vector on the left by a matrix A is known as a **linear transformation.** ▲

EXAMPLE 5 Let $GL(n, \mathbb{R})$ be the multiplicative group of all invertible $n \times n$ matrices. Recall that a matrix A is invertible if and only if its determinant, $\det(A)$, is nonzero. Recall also that for matrices $A, B \in GL(n, \mathbb{R})$ we have

$$\det(AB) = \det(A)\det(B).$$

This means that det is a homomorphism mapping $GL(n, \mathbb{R})$ into the multiplicative group $\mathbb{R}^*$ of nonzero real numbers. ▲

Homomorphisms of a group G into itself are often useful for studying the structure of G. Our next example gives a nontrivial homomorphism of a group into itself.

EXAMPLE 6   Let $r \in \mathbb{Z}$ and let $\phi_r : \mathbb{Z} \to \mathbb{Z}$ be defined by $\phi_r(n) = rn$ for all $n \in \mathbb{Z}$. For all $m, n \in \mathbb{Z}$, we have $\phi_r(m + n) = r(m + n) = rm + rn = \phi_r(m) + \phi_r(n)$ so ϕ_r is a homomorphism. Note that ϕ_0 is the trivial homomorphism, ϕ_1 is the identity map, and ϕ_{-1} maps $\mathbb{Z}$ *onto* $\mathbb{Z}$. For all other r in $\mathbb{Z}$, the map ϕ_r is not onto $\mathbb{Z}$. ▲

EXAMPLE 7   Let $G = G_1 \times G_2 \times \cdots \times G_i \times \cdots \times G_n$ be a direct product of groups. The **projection map** $\pi_i : G \to G_i$ which is defined by $\pi_i(g_1, g_2, \ldots, g_i, \ldots, g_n) = g_i$ is a homomorphism for each $i = 1, 2, \ldots, n$. This follows immediately from the fact that the binary operation of G coincides in the ith component with the binary operation in G_i. ▲

EXAMPLE 8   Let F be the additive group of continuous functions with domain $[0, 1]$ and let $\mathbb{R}$ be the additive group of real numbers. The map $\sigma : F \to \mathbb{R}$ defined by $\sigma(f) = \int_0^1 f(x)\,dx$ for $f \in F$ is a homomorphism, for

$$\sigma(f + g) = \int_0^1 (f + g)(x)\,dx = \int_0^1 [f(x) + g(x)]\,dx$$

$$= \int_0^1 f(x)\,dx + \int_0^1 g(x)\,dx = \sigma(f) + \sigma(g)$$

for all $f, g \in F$. ▲

EXAMPLE 9 (Reduction Modulo n)   Let γ be the natural map of $\mathbb{Z}$ into $\mathbb{Z}_n$ given by $\gamma(m) = r$, where r is the remainder given by the division algorithm when m is divided by n. Show that γ is a homomorphism.

Solution   We need to show that

$$\gamma(s + t) = \gamma(s) + \gamma(t)$$

for $s, t \in \mathbb{Z}$. Using the division algorithm, we let

$$s = q_1 n + r_1 \tag{2}$$

and

$$t = q_2 n + r_2 \tag{3}$$

where $0 \le r_i < n$ for $i = 1, 2$. If

$$r_1 + r_2 = q_3 n + r_3 \tag{4}$$

for $0 \leq r_3 < n$, then adding (2) and (3) we see that

$$s + t = (q_1 + q_2 + q_3)n + r_3,$$

so that $\gamma(s + t) = r_3$.

From Eqs. (2) and (3) we see that $\gamma(s) = r_1$ and $\gamma(t) = r_2$. Equation (4) shows that the sum $r_1 + r_2$ in $\mathbb{Z}_n$ is equal to r_3 also.

Consequently $\gamma(s + t) = \gamma(s) + \gamma(t)$, so we do indeed have a homomorphism. ▲

All the homomorphisms in the preceding three examples are many-to-one maps. That is, different points of the domain of the maps may be carried into the same point. Consider, for illustration, the homomorphism $\pi_1 : \mathbb{Z}_2 \times \mathbb{Z}_4 \to \mathbb{Z}_2$ in Example 7. We have

$$\pi_1(0, 0) = \pi_1(0, 1) = \pi_1(0, 2) = \pi_1(0, 3) = 0,$$

so four elements in $\mathbb{Z}_2 \times \mathbb{Z}_4$ are mapped into 0 in $\mathbb{Z}_2$ by π_1.

Properties of Homomorphisms

We turn to structural features of G and G' that are *preserved* by a homomorphism $\phi : G \to G'$. First we give a set-theoretic definition. Note the use of square brackets when we apply a function to a *subset* of its domain.

DEFINITION 3.2 (Image and Inverse Image) Let ϕ be a mapping of a set X into a set Y, and let $A \subseteq X$ and $B \subseteq Y$. The **image** $\phi[A]$ **of A in Y under** ϕ is $\{\phi(a) \mid a \in A\}$. The set $\phi[X]$ is sometimes called the **range of** ϕ. The **inverse image** $\phi^{-1}[B]$ **of B in X** is $\{x \in X \mid \phi(x) \in B\}$.

THEOREM 3.1 Let ϕ be a homomorphism of a group G into a group ✳ G'.

1. If e is the identity in G, then $\phi(e)$ is the identity e' in G'.
2. If $a \in G$, then $\phi(a^{-1}) = \phi(a)^{-1}$.
3. If H is a subgroup of G, then $\phi[H]$ is a subgroup of G'.
4. If K' is a subgroup of G', then $\phi^{-1}[K']$ is a subgroup of G.

Loosely speaking, ϕ preserves the identity, inverses, and subgroups. ↞

PROOF Let ϕ be a homomorphism of G into G'. Then

$$\phi(a) = \phi(ae) = \phi(a)\phi(e).$$

Multiplying by $\phi(a)^{-1}$, we see that $e' = \phi(e)$. Thus $\phi(e)$ must be the identity e' in G'. The equation

$$e' = \phi(e) = \phi(aa^{-1}) = \phi(a)\phi(a^{-1})$$

shows that $\phi(a^{-1}) = \phi(a)^{-1}$.

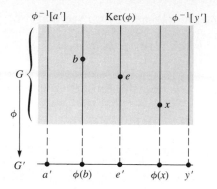

Figure 3.1   Fibres in G under $\phi: G \to G'$.

Turning to No. 3, let H be a subgroup of G, and let $\phi(a)$ and $\phi(b)$ be any two elements in $\phi[H]$. Then $\phi(a)\phi(b) = \phi(ab)$, so we see that $\phi(a)\phi(b) \in \phi[H]$; thus, $\phi[H]$ is closed under the operation of G'. The fact that $e' = \phi(e)$ and $\phi(a^{-1}) = \phi(a)^{-1}$ completes the proof that $\phi[H]$ is a subgroup of G'.

Going the other way for No. 4, let K' be a subgroup of G'. Suppose a and b are in $\phi^{-1}[K']$. Then $\phi(a)\phi(b) \in K'$ since K' is a subgroup. The equation $\phi(ab) = \phi(a)\phi(b)$ shows that $ab \in \phi^{-1}[K']$. Thus $\phi^{-1}[K']$ is closed under the binary operation in G. Also K' must contain the identity $e' = \phi(e)$, so $e \in \phi^{-1}[K']$. If $a \in \phi^{-1}[K']$, then $\phi(a) \in K'$, so $\phi(a)^{-1} \in K'$. But $\phi(a)^{-1} = \phi(a^{-1})$, so we must have $a^{-1} \in \phi^{-1}[K']$. Hence $\phi^{-1}[K']$ is a subgroup of G. ◆

Let $\phi: G \to G'$ be a homomorphism, and let $a' \in G'$. We will often have occasion to refer to the inverse image $\phi^{-1}[\{a'\}]$ of the set $\{a'\}$ containing only one element. To avoid proliferation of symbols, we will denote this inverse image by $\phi^{-1}[a']$, omitting the braces. Borrowing terminology from topology, this inverse image $\phi^{-1}[a']$ of a single element is often called the **fibre over** a' **under** ϕ. We may visualize fibres as shown in Fig. 3.1 where we think of the group G' as being composed of points on the horizontal line at the bottom. The group G consists of points on the vertical line segments, and each such vertical line segment is a fibre under the homomorphism ϕ that projects points on the segment straight down to one point on G'.

EXAMPLE 10   Equation 6 of Section 0.4 shows that $|z_1 z_2| = |z_1| |z_2|$ for complex numbers z_1 and z_2. This means that the absolute value function $|\ |$ is a homomorphism of the group $\mathbb{C}^*$ of nonzero complex numbers under multiplication onto the group $\mathbb{R}^+$ of positive real numbers under multiplication. Since $\{1\}$ is a subgroup of $\mathbb{R}^+$, Theorem 3.1 shows at once that the complex numbers of magnitude 1 form a subgroup of $\mathbb{C}^*$. Recall that the complex numbers can be viewed as filling the coordinate plane, and that the magnitude of a complex number is its distance from the origin. Consequently, the fibres of $\mathbb{C}^*$ under the homomorphism $|\ |$ are circles with center at the origin. ▲

Since $\{e'\}$ is a subgroup of G', Theorem 3.1 shows that the fibre $\phi^{-1}[e']$ under a homomorphism $\phi: G \rightarrow G'$ is a subgroup of G. Theorem 3.2 below will show that the fibres of G under ϕ are precisely the cosets (left and right cosets are the same) of this group. Thus the fibres of G under ϕ form a partition of G, as indicated schematically by Fig. 3.1. This subgroup $\phi^{-1}[e']$ plays such an important role for the homomorphism ϕ that it is given a special name as shown in Fig. 3.1.

DEFINITION 3.3 (Kernel) Let $\phi: G \rightarrow G'$ be a homomorphism of groups. The subgroup $\phi^{-1}[e'] = \{x \in G \mid \phi(x) = e'\}$ is the **kernel of** ϕ, denoted by $\text{Ker}(\phi)$.

Recall that a group homomorphism $\phi: G \rightarrow G'$ may map many different elements of G into a single element of G'. Intuitively, we think of such a homomorphism as *collapsing* each fibre of G into a single element of G'. All elements of $\text{Ker}(\phi)$ are collapsed into the identity e' of G', so the size of $\text{Ker}(\phi)$ measures the amount of collapsing into e'. The fact is that the size of $\text{Ker}(\phi)$ actually measures the amount of collapsing by ϕ into *every* point of the image $\phi[G]$ in G'. Our next theorem demonstrates this by showing that the fibre of G mapped into a single element of $\phi[G]$ is actually a coset of the subgroup $\text{Ker}(\phi)$ in G. We know that all such cosets have the same number of elements as $\text{Ker}(\phi)$ by the argument just before Theorem 2.5.

THEOREM 3.2 Let $\phi: G \rightarrow G'$ be a group homomorphism, and let $H = \text{Ker}(\phi)$. Let $a \in G$. Then the set

$$\phi^{-1}[\phi(a)] = \{x \in G \mid \phi(x) = \phi(a)\}$$

is the left coset aH of H, and is also the right coset Ha of H. Consequently, the two partitions of G into left cosets and into right cosets of H are the same.

PROOF We want to show that

$$\{x \in G \mid \phi(x) = \phi(a)\} = aH.$$

There is a standard way to show that two sets are equal; show that each is a subset of the other.

Suppose that $\phi(x) = \phi(a)$. Then

$$\phi(a)^{-1}\phi(x) = e',$$

where e' is the identity of G'. By Theorem 3.1, we know that $\phi(a)^{-1} = \phi(a^{-1})$, so we have

$$\phi(a^{-1})\phi(x) = e'.$$

Since ϕ is a homomorphism, we have

$$\phi(a^{-1})\phi(x) = \phi(a^{-1}x), \qquad \text{so} \qquad \phi(a^{-1}x) = e'.$$

But this shows that $a^{-1}x$ is in $H = \text{Ker}(\phi)$, so $a^{-1}x = h$ for some $h \in H$, and $x = ah \in aH$. This shows that

$$\{x \in G \mid \phi(x) = \phi(a)\} \subseteq aH.$$

To show containment in the other direction, let $y \in aH$, so that $y = ah$ for some $h \in H$. Then

$$\phi(y) = \phi(ah) = \phi(a)\phi(h) = \phi(a)e' = \phi(a),$$

so that $y \in \{x \in G \mid \phi(x) = \phi(a)\}$.

We leave the similar demonstration that $\{x \in G \mid \phi(x) = \phi(a)\} = Ha$ to Exercise 42. ◆

We give an illustration of Theorem 3.2 from calculus.

EXAMPLE 11   Let D be the additive group of all differentiable functions mapping $\mathbb{R}$ into $\mathbb{R}$, and let F be the additive group of all functions mapping $\mathbb{R}$ into $\mathbb{R}$. Then differentiation gives us a map $\phi:D \to F$, where $\phi(f) = f'$ for $f \in F$. We easily see that ϕ is a homomorphism, for $\phi(f + g) = (f + g)' = f' + g' = \phi(f) + \phi(g)$; the derivative of a sum is the sum of the derivatives.

Now $\text{Ker}(\phi)$ consists of all functions f such that $f' = 0$, the zero constant function. Thus $\text{Ker}(\phi)$ consists of all constant functions, which form a subgroup C of F. Let us find all functions in G mapped into x^2 by ϕ, that is, all functions whose derivative is x^2. Now we know that $x^3/3$ is one such function. By Theorem 3.2, all such functions form the coset $x^3/3 + C$. Doesn't this look familiar? ▲

We will often use the following corollary of Theorem 3.2.

COROLLARY   A group homomorphism $\phi: G \to G'$ is a one-to-one map if and only if $Ker(\phi) = \{e\}$.

PROOF   If $Ker(\phi) = \{e\}$, then for every $a \in G$, the elements mapped into $\phi(a)$ are precisely the elements of the left coset $a\{e\} = \{a\}$, which shows that ϕ is one to one.

Conversely, suppose ϕ is one to one. Now by Theorem 3.1, we know that $\phi(e) = e'$, the identity of G'. Since ϕ is one to one, we see that e is the only element mapped into e' by ϕ, so $Ker(\phi) = \{e\}$. ◆

Theorem 3.2 shows that the kernel of a group homomorphism $\phi: G \to G'$ is a subgroup H of G whose left and right cosets coincide, so that $gH = Hg$ for all $g \in G$. We will see in Section 3.3 that when left and right cosets coincide, we can form a coset group, as discussed intuitively in Section 2.3. Furthermore, we will see that H then appears as the kernel of a homomorphism of G into this coset group in a very natural way. Such subgroups H whose left and right cosets coincide are very useful in studying a group, and are given a special name. We will work with them a lot in Section 3.3.

DEFINITION 3.4 (Normal Subgroup)   A subgroup H of a group G is **normal** if its left and right cosets coincide, that is, if $gH = Hg$ for all $g \in G$.

Note that all subgroups of abelian groups are normal. Also, Theorem 3.2 shows that the kernel of a homomorphism $\phi = G \to G'$ is a normal subgroup of G.

For any group homomorphism $\phi: G \to G'$, two things are of primary

---

◆   **HISTORICAL NOTE**   ◆

Normal subgroups were introduced by Evariste Galois in 1831 as a tool for deciding whether a given equation was solvable by radicals. Galois noted that a subgroup H of a group G of permutations induced two decompositions of G into what we call left cosets and right cosets. If the two decompositions coincide, that is, if the left cosets are the same as the right cosets, Galois called the decomposition "proper." Thus a subgroup giving a proper decomposition is what we call a normal subgroup. Galois stated that if the group of permutations of the roots of an equation has a proper decomposition, then one can solve the given equation if one can first solve an equation corresponding to the subgroup H and then an equation corresponding to the cosets.

Camille Jordan, in his commentaries on Galois's work in 1865 and 1869, elaborated on these ideas considerably. He also defined normal subgroups, though without using the term, essentially as on this page and likewise gave the first definition of a simple group (page 201).

importance: the *kernel of* ϕ, and the *image* $\phi[G]$ *of G in G'*. We have indicated the importance of Ker(ϕ). Sections 3.2 and 3.3 will indicate the importance of the image $\phi[G]$. Exercise 35 asks us to show that if $|G|$ is finite, then $|\phi[G]|$ is finite and is a divisor of $|G|$.

We mention in closing some terminology found in the literature related to Ker(ϕ) and to $\phi[G]$. A map $\mu:A \to B$ that is *one to one* is often called an **injection**. A homomorphism $\phi:G \to G'$ that is *one to one* is often called a **monomorphism**; this is the case if and only if Ker(ϕ) = $\{e\}$. A map of A *onto* B is often called a **surjection**. A homomorphism that maps G onto G' is often called an **epimorphism**; this is the case if and only if $\phi[G] = G'$. We will not use these terms, but rather will stick with *one to one* and *onto*, which we have been using.

Exercises 3.1

Computations

In Exercises 1 through 15, determine whether the given map ϕ is a homomorphism. [*Hint:* The straightforward way to proceed is to check whether $\phi(ab) = \phi(a)\phi(b)$ for all *a* and *b* in the domain of ϕ. However, if we should happen to notice that $\phi^{-1}[e']$ is not a subgroup whose left and right cosets coincide, or that ϕ does not satisfy the properties given in Exercise 35 or 36 for finite groups, then we can say at once that ϕ is not a homomorphism.]

1. Let $\phi:\mathbb{Z} \to \mathbb{R}$ under addition be given by $\phi(n) = n$.

2. Let $\phi:\mathbb{R} \to \mathbb{Z}$ under addition be given by $\phi(x)$ = the greatest integer $\leq x$.

3. Let $\phi:\mathbb{R}^* \to \mathbb{R}^*$ under multiplication be given by $\phi(x) = |x|$.

4. Let $\phi:\mathbb{Z}_6 \to \mathbb{Z}_2$ be given by $\phi(x)$ = the remainder of *x* when divided by 2, as in the division algorithm.

5. Let $\phi:\mathbb{Z}_9 \to \mathbb{Z}_2$ be given by $\phi(x)$ = the remainder of *x* when divided by 2, as in the division algorithm.

6. Let $\phi:\mathbb{R} \to \mathbb{R}^*$, where $\mathbb{R}$ is additive and $\mathbb{R}^*$ is multiplicative, be given by $\phi(x) = 2^x$.

7. Let $\phi_i:G_i \to G_1 \times G_2 \times \cdots \times G_i \times \cdots \times G_r$ be given by $\phi_i(g_i) = (e_1, e_2, \ldots, g_i, \ldots, e_r)$, where $g_i \in G_i$ and e_j is the identity of G_j. This is an **injection map**. Compare with Example 7.

8. Let G be any group and let $\phi:G \to G$ be given by $\phi(g) = g^{-1}$ for $g \in G$.

9. Let *F* be the additive group of functions mapping $\mathbb{R}$ into $\mathbb{R}$ having derivatives of all orders. Let $\phi:F \to F$ be given by $\phi(f) = f''$, the second derivative of *f*.

10. Let *F* be the additive group of all continuous functions mapping $\mathbb{R}$ into $\mathbb{R}$. Let $\mathbb{R}$

be the additive group of real numbers, and let $\phi : F \to \mathbb{R}$ be given by

$$\phi(f) = \int_0^4 f(x)\, dx.$$

11. Let F be the additive group of all functions mapping $\mathbb{R}$ into $\mathbb{R}$, and let $\phi : F \to F$ be given by $\phi(f) = 3f$.

12. Let M_n be the additive group of all $n \times n$ matrices with real entries, and let $\mathbb{R}$ be the additive group of real numbers. Let $\phi(A) = \det(A)$, the determinant of A, for $A \in M_n$.

13. Let M_n and $\mathbb{R}$ be as in Exercise 12. Let $\phi(A) = \text{tr}(A)$ for $A \in M_n$, where the **trace** $\text{tr}(A)$ is the sum of the elements on the main diagonal of A, from the upper-left to the lower-right corner.

14. Let $GL(n, \mathbb{R})$ be the multiplicative group of invertible $n \times n$ matrices, and let $\mathbb{R}$ be the additive group of real numbers. Let $\phi : GL(n, \mathbb{R}) \to \mathbb{R}$ be given by $\phi(A) = \text{tr}(A)$, where $\text{tr}(A)$ is defined in Exercise 13.

15. Let F be the multiplicative group of all continuous functions mapping $\mathbb{R}$ into $\mathbb{R}$ that are nonzero at every $x \in \mathbb{R}$. Let $\mathbb{R}^*$ be the multiplicative group of nonzero real numbers. Let $\phi : F \to \mathbb{R}^*$ be given by $\phi(f) = \int_0^1 f(x)\, dx$.

16. Describe the kernel of the homomorphism $\phi : S_3 \to \mathbb{Z}_2$ presented in Example **2.**

17. Let G be a group. If $\phi : \mathbb{Z} \times \mathbb{Z} \to G$ is a homomorphism and $\phi(1, 0) = h$ while $\phi(0, 1) = k$, find $\phi(m, n)$.

18. How many homomorphisms are there of $\mathbb{Z}$ onto $\mathbb{Z}$?

19. How many homomorphisms are there of $\mathbb{Z}$ into $\mathbb{Z}$?

20. How many homomorphisms are there of $\mathbb{Z}$ into $\mathbb{Z}_2$?

21. Let G be a group, and let $g \in G$. Let $\phi_g : G \to G$ be defined by $\phi_g(x) = gx$ for $x \in G$. For which $g \in G$ is ϕ_g a homomorphism?

22. Let G be a group, and let $g \in G$. Let $\phi_g : G \to G$ be defined by $\phi_g(x) = gxg^{-1}$ for $x \in G$. For which $g \in G$ is ϕ_g a homomorphism?

Concepts

23. Mark each of the following true or false.
    ___T___ a. A_n is a normal subgroup of S_n.
    ___T___ b. For any two groups G and G', there exists a homomorphism of G into G'.
    ___F___ c. Every homomorphism is a one-to-one map.
    ___T___ d. A homomorphism is one to one if and only if the kernel consists of the group of the identity element alone.
    ___F___ e. The image of a group of 6 elements under some homomorphism may have 4 elements.
    ___F___ f. The image of a group of 6 elements under a homomorphism may have 12 elements.
    ___T___ g. There is a homomorphism of some group of 6 elements into some group of 12 elements.
    ___T___ h. There is a homomorphism of some group of 6 elements into some group of 10 elements.
    ___F___ i. A homomorphism may have an empty kernel.

   <u>F</u>   j. It is not possible to have a homomorphism of some infinite group into some finite group.

In Exercises 24 through 34, give an example of a nontrivial homomorphism ϕ for the given groups, if an example exists. If no such homomorphism exists, explain why that is so.

24. $\phi : \mathbb{Z}_{12} \to \mathbb{Z}_5$ **25.** $\phi : \mathbb{Z}_{12} \to \mathbb{Z}_4$

26. $\phi : \mathbb{Z}_2 \times \mathbb{Z}_4 \to \mathbb{Z}_2 \times \mathbb{Z}_5$ **27.** $\phi : \mathbb{Z}_3 \to \mathbb{Z}$

28. $\phi : \mathbb{Z}_3 \to S_3$ **29.** $\phi : \mathbb{Z} \to S_3$

30. $\phi : \mathbb{Z} \times \mathbb{Z} \to 2\mathbb{Z}$ **31.** $\phi : 2\mathbb{Z} \to \mathbb{Z} \times \mathbb{Z}$

32. $\phi : D_4 \to S_3$ **33.** $\phi : S_3 \to S_4$

34. $\phi : S_4 \to S_3$

Theory

35. Let $\phi : G \to G'$ be a group homomorphism. Show that if $|G|$ is finite, then $|\phi[G]|$ is finite and is a divisor of $|G|$.

36. Let $\phi : G \to G'$ be a group homomorphism. Show that if $|G'|$ is finite, then $|\phi[G]|$ is finite and is a divisor of $|G'|$.

37. Let a group G be generated by $\{a_i \mid i \in I\}$, where I is some indexing set and $a_i \in G$. Let $\phi : G \to G'$ be a homomorphism of G into a group G'. Show that the value of ϕ on every element of G is completely determined by the values $\phi(a_i)$. Thus, for example, a homomorphism of a cyclic group is completely determined by the value of the homomorphism on a generator of the group. [*Hint:* Use Theorem 1.11 and, of course, the definition of a homomorphism (3.1).]

38. Show that any group homomorphism $\phi : G \to G'$ where $|G|$ is a prime must either be the trivial homomorphism or a one-to-one map.

39. The **sign of an even permutation** is $+1$ and the **sign of an odd permutation** is -1. Observe that the map $\mathrm{sgn}_n : S_n \to \{1, -1\}$ defined by

$$\mathrm{sgn}_n(\sigma) = \text{sign of } \sigma$$

is a homomorphism of S_n onto the multiplicative group $\{1, -1\}$. What is the kernel? Compare with Example 1.

40. Show that if G, G', and G'' are groups and if $\phi : G \to G'$ and $\gamma : G' \to G''$ are homomorphisms, then the composite map $\gamma\phi : G \to G''$ is a homomorphism.

41. Let G be any group and let a be any element of G. Let $\phi : \mathbb{Z} \to G$ be defined by $\phi(n) = a^n$. Show that ϕ is a homomorphism. Describe the image and the possibilities for the kernel of ϕ.

42. Let $\phi : G \to G'$ be a homomorphism with kernel H and let $a \in G$. Show that $\{x \in G \mid \phi(x) = \phi(a)\} = Ha$.

43. Let G be a group. Let $h, k \in G$ and let $\phi : \mathbb{Z} \times \mathbb{Z} \to G$ be defined by $\phi(m, n) = h^m k^n$. Give a necessary and sufficient condition, involving h and k, for ϕ to be a homomorphism. Prove your condition.

44. Find a necessary and sufficient condition on G such that the map ϕ described in the preceding exercise is a homomorphism for *all* choices of $h, k \in G$.

3.2

Isomorphism and Cayley's Theorem
Definition and Elementary Properties

We now come to the business of making more mathematically precise the idea that two groups, G and G', are structurally the same or *isomorphic*. We have tried to impart the feeling that G and G' are isomorphic if the groups are identical except for the names of the elements and operations. Thus we should be able to obtain G' from G by renaming an element x in G with the name of a certain element x' in G'. That is, each $x \in G$ is assigned its counterpart $x' \in G'$. This is really nothing but a *function* ϕ with domain G. Clearly two different elements x and y in G should have two different counterparts $x' = \phi(x)$ and $y' = \phi(y)$ in G'; that is, the function ϕ must be one to one. Also every element of G' must be the counterpart of some element of G; that is, the function ϕ must be onto G'. This renames the elements. Finally, if the groups are to be structurally the same and if for the moment we denote the group operation of G by $*$ and that of G' by $*'$, then the counterpart of $x * y$ should be $x' *' y'$; that is, $\phi(x * y)$ should be $\phi(x) *' \phi(y)$. Usually we drop the notations $*$ and $*'$ for the operations and use multiplicative notations, writing

$$\phi(xy) = \phi(x)\phi(y). \tag{1}$$

Using our intuitive idea of groups G and G' being structurally identical if they differ only in the names of their elements, we have arrived at the condition that there should be a *one-to-one homomorphism of G onto G'*. (A one-to-one and onto map is sometimes called a **bijection**.)

DEFINITION 3.5 (Isomorphism) An **isomorphism** $\phi : G \to G'$ is a homomorphism that is one to one and onto G'. The usual notation is $G \simeq G'$.

We see from Theorem 3.1 that if $\phi : G \to G'$ is an isomorphism, then $\phi(e) = e'$, the identity of G'. Theorem 3.1 also shows us that if a, a^{-1} are inverse elements of G, then $\phi(a), \phi(a)^{-1}$ are inverse elements of G'. That is, $\phi(a^{-1}) = \phi(a)^{-1}$. Of course, both of these observations are intuitively clear since G' is structurally identical with G and the isomorphism ϕ can be regarded as renaming the elements of G with the names of the elements of G'.

We digress for a moment to discuss an **inverse map**. Let $\phi : A \to B$ be a

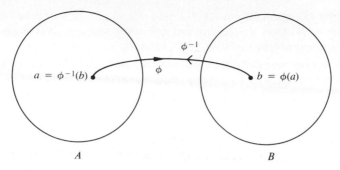

Figure 3.2

one-to-one map of the set A *onto* the set B. Then there is a natural map $\phi^{-1}:B \to A$ which is also one to one and is onto A. We may think of ϕ^{-1} as simply reversing the direction of the solid arrow in Fig. 3.2, where we have used the solid arrowhead for ϕ and the open arrowhead for ϕ^{-1}.

We call ϕ^{-1} the **inverse map to** ϕ. This inverse is not well defined with domain B unless ϕ is one to one and onto B. Figure 3.3 indicates the difficulty if ϕ is not one to one. If $\phi(a_1) = \phi(a_2) = b$, then in Fig. 3.3 we have no way of deciding between a_1 and a_2 as a possible value for an inverse map at b. Recall that a function must assign just *one* element of the range to each element of the domain. Note also that if $b' \in B$ has no ϕ-arrow coming into it, as in Fig. 3.3, that is, if $\phi(a) \neq b'$ for all $a \in A$, then we have no way to select a value for an inverse function at b'. Thus $\phi:A \to B$ must be *one to one* and *onto* B in order for $\phi^{-1}:B \to A$ to exist. When ϕ^{-1} does exist, it is clear that

$$\phi^{-1}\big(\phi(a)\big) = a \quad \text{and} \quad \phi\big(\phi^{-1}(b)\big) = b$$

for all $a \in A$ and all $b \in B$.

Recall that for any function $\phi:A \to B$, we abbreviate the notation $\phi^{-1}[\{b\}] = \{x \in A \mid \phi(x) = b\}$ by $\phi^{-1}[b]$ to avoid clutter in notation. We use square brackets rather than parentheses in this notation to emphasize that ϕ

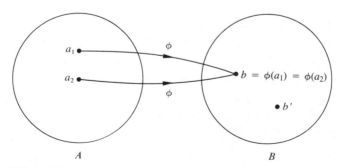

Figure 3.3

may not be one to one or onto B, that is, that the function ϕ may not be
invertible. We will only use the notation $\phi^{-1}(b)$, with the parentheses, when
ϕ is one to one and onto B, so that the inverse function ϕ^{-1} exists.

We now return to isomorphisms. The relation $G \simeq G'$ is an equivalence
relation for any collection of groups. We state this as a theorem, and indicate
the steps of the proof, leaving execution of the steps to the exercises.

THEOREM 3.3   Let $\mathcal{G}$ be any collection of groups, and define $G \simeq G'$
for G and G' in $\mathcal{G}$ if there exists an isomorphism $\phi : G \rightarrow G'$. Then $\simeq$ is
an equivalence relation.

PROOF

Reflexive $(G \simeq G)$ Show in Exercise 15 that $\iota : G \rightarrow G$, where $\iota(g) = g$
for all $g \in G$, is an isomorphism. (Intuitively, we know G has the same
algebraic structure as itself. Every element keeps its own name.)

Symmetric (If $G \simeq G'$, then $G' \simeq G$)   Show in Exercise 16 that if
$\phi : G \rightarrow G'$ is an isomorphism, then $\phi^{-1} : G' \rightarrow G$ is an isomorphism.
(Intuitively, given a structure-preserving way of renaming elements of G'
by names of elements in G, just translate names in the other direction to
rename elements of G by names of elements of G'.)

Transitive (If $G \simeq G'$ and $G' \simeq G''$, then $G \simeq G''$)   Show in Exercise 17
that if $\phi : G \rightarrow G'$ and $\gamma : G' \rightarrow G''$ are isomorphisms, then $\gamma \phi : G \rightarrow G''$ is
an isomorphism. (Intuitively, we rename elements of G' with names from
G using ϕ, and then we rename elements of G'' with the new names in G'
using γ. Algebraic structure is preserved with each renaming.) $\blacklozenge$

Since $\simeq$ is an equivalence relation, every collection of groups can be
partitioned by $\simeq$ into cells such that any two groups in the same cell are
isomorphic, while groups in different cells are not isomorphic. Exercise 11
illustrates such a partition.

How to Show That Groups Are Isomorphic

A few students have trouble understanding and using the concept of
isomorphism. We introduced it informally several sections before this one,
hoping to create an accurate intuitive understanding upon which to build. We
now give an outline showing how to proceed from the definition to show that
two groups, G and G', are isomorphic.

Step 1   *Define the function ϕ that gives the isomorphism of G with G'.* Now
this means that we have to describe, in some fashion, what $\phi(x)$ is to be in
G' for every $x \in G$.

Step 2   *Show that ϕ is a one-to-one function.*

Step 3   *Show that ϕ is onto G'.*

Step 4 *Show that* $\phi(xy) = \phi(x)\phi(y)$ *for all* $x, y \in G$. This is just a question of computation. Compute both sides of the equation and see whether they are the same.

We illustrate this technique with an example.

EXAMPLE I   Let us show that $\mathbb{R}$ under addition is isomorphic to $\mathbb{R}^+$ under multiplication.

Step 1 For $x \in \mathbb{R}$, define $\phi(x) = e^x$. This gives a mapping $\phi : \mathbb{R} \to \mathbb{R}^+$. (Recall that the logarithm function converts multiplication to addition so the exponential function converts addition to multiplication.)

Step 2 If $\phi(x) = \phi(y)$, then $e^x = e^y$, so $x = y$. Thus ϕ is one to one.

Step 3 If $r \in \mathbb{R}^+$, then

$$\phi(\ln r) = e^{\ln r} = r,$$

where $(\ln r) \in \mathbb{R}$. Thus ϕ is onto $\mathbb{R}^+$.

Step 4 For $x, y \in \mathbb{R}$, we have

$$\phi(x + y) = e^{x+y} = e^x e^y = \phi(x)\phi(y). \ \blacktriangle$$

We illustrate this technique again in a theorem.

THEOREM 3.4   Any infinite cyclic group G is isomorphic to the group $\mathbb{Z}$ of integers under addition.

PROOF   We suppose that G has a generator a and use multiplicative notation for the operation in G. Thus

$$G = \{a^n \mid n \in \mathbb{Z}\}.$$

Our Case 1 discussion of the classification of cyclic groups in Section 1.4 showed that the elements a^n of G are all distinct; that is, $a^n \neq a^m$ if $n \neq m$.

Step 1 In this case a natural choice for $\phi : G \to \mathbb{Z}$ is given by $\phi(a^n) = n$ for all $a^n \in G$.

Step 2 If $\phi(a^n) = \phi(a^m)$, then $n = m$ and $a^n = a^m$. Thus ϕ is one to one.

Step 3 For any $n \in \mathbb{Z}$, the element $a^n \in G$ is mapped onto n by ϕ. Thus ϕ is onto $\mathbb{Z}$.

Step 4 Now $\phi(a^n a^m) = \phi(a^{n+m}) = n + m$. (Note that the binary operation here was in G.) It remains to compute $\phi(a^n) + \phi(a^m)$, the +

appearing because the operation in $\mathbb{Z}$ is addition. However, $\phi(a^n) + \phi(a^m)$ is again $n + m$. Thus $\phi(a^n a^m) = \phi(a^n) + \phi(a^m)$. ◆

When we say, "There are precisely m groups of order n, up to isomorphism," we mean that there exist m groups $G_1, G_2, \ldots, G_m$ of order n, no two of which are isomorphic, and that every group of order n is isomorphic to one of them. For example, we have seen that any two groups of order 3 are isomorphic. We express this by saying that *there is only one group of order 3 up to isomorphism.*

EXAMPLE 2 There is only one group of order 1, one of order 2, and one of order 3 up to isomorphism. We saw in Example 5 of Section 1.3 that there are exactly two different groups of order 4 up to isomorphism, the group $\mathbb{Z}_4$ and the Klein 4-group V. There are at least two different groups of order 6 up to isomorphism, namely $\mathbb{Z}_6$ and S_3. ▲

How to Show That Groups Are Not Isomorphic

We turn now to the reverse question, namely:

How do we demonstrate that two groups G and G' are not isomorphic, if this is the case?

This would mean that there is no one-to-one function ϕ from G onto G' with the property $\phi(xy) = \phi(x)\phi(y)$. In general, it is clearly not feasible to try every possible one-to-one function to find out whether it has the above property, except in the case where there are *no* one-to-one functions. This is the case, for example, if G and G' are of finite order and have different numbers of elements.

EXAMPLE 3 $\mathbb{Z}_4$ and S_6 are *not* isomorphic. There is no one-to-one function from $\mathbb{Z}_4$ onto S_6. ▲

For two groups of infinite order, it is not always clear whether there are any one-to-one onto functions. For example, you may think that $\mathbb{Q}$ has "more" elements than $\mathbb{Z}$, but your instructor can show you in five minutes (ask him or her to!) that *there are lots of one-to-one functions from $\mathbb{Z}$ onto $\mathbb{Q}$.* However, it is true that $\mathbb{R}$ *has too many elements to be put into a one-to-one correspondence with* $\mathbb{Z}$. This will take your instructor only another five minutes to show you.

EXAMPLE 4 $\mathbb{Z}$ under addition is *not* isomorphic with $\mathbb{R}$ under addition, for there is no one-to-one function from $\mathbb{Z}$ onto $\mathbb{R}$. ▲

In the event that there are one-to-one mappings of G onto G', we usually show that the groups are not isomorphic (if this is the case) by showing that *one group has some structural property that the other does not possess.* A **structural property of a group** is one that must be shared by any isomorphic group. It must not depend on the names or some other nonstructural characteristics of the elements. The following are examples of some structural properties and some nonstructural properties of groups.

Possible Structural Properties	Possible Nonstructural Properties
1. The group is cyclic	a. The group contains 5.
2. The group is abelian.	b. All elements of the group are numbers.
3. The group has order 8.	c. The group operation is called "composition."
4. The group is finite.	d. The elements of the group are permutations.
5. The group has exactly two elements of order 6.	e. The group operation is denoted by juxtaposition.
6. The equation $x^2 = a$ has a solution for each element a in the group.	f. The group contains no matrices.

Of course there are many other possible structural properties we could have listed. The fact that each of Properties 1 through 6 is indeed structural is a little theorem about isomorphic groups. We ask you to prove a couple of such theorems in the exercises. (See Exercises 19 and 20.) We shall regard them as obviously structural in our work in the text.

EXAMPLE 5 We can't say that $\mathbb{Z}$ and $3\mathbb{Z}$ under addition are not isomorphic because $17 \in \mathbb{Z}$ and $17 \notin 3\mathbb{Z}$. These are *not* structural properties but rather have to do with just the names of the elements. Actually $\mathbb{Z}$ and $3\mathbb{Z}$ are isomorphic. The map $\phi : \mathbb{Z} \to 3\mathbb{Z}$, where $\phi(n) = 3n$ is an isomorphism. ▲

EXAMPLE 6 We can't say that $\mathbb{Z}$ and $\mathbb{Q}$, both under addition, are not isomorphic because $\frac{1}{2} \in \mathbb{Q}$ and $\frac{1}{2} \notin \mathbb{Z}$. But we can say that they are not isomorphic because $\mathbb{Z}$ is cyclic and $\mathbb{Q}$ is not. ▲

EXAMPLE 7 The group $\mathbb{Q}^*$ of nonzero elements of $\mathbb{Q}$ under multiplication is not isomorphic to the group $\mathbb{R}^*$ of nonzero elements of $\mathbb{R}$ under multiplication. One argument is that there is no one-to-one correspondence between them. Another is that every element in $\mathbb{R}^*$ is the cube of some element of $\mathbb{R}^*$, that is, for $a \in \mathbb{R}^*$, the equation $x^3 = a$ has a solution in $\mathbb{R}^*$.

This is not true for $\mathbb{Q}^*$; for example, our work in Section 5.6 will show that the equation $x^3 = 2$ has no solution in $\mathbb{Q}^*$. ▲

EXAMPLE 8 The group $\mathbb{R}^*$ of nonzero real numbers under multiplication is not isomorphic to the group $\mathbb{C}^*$ of nonzero complex numbers under multiplication. Every element of $\mathbb{R}^*$ generates an infinite cyclic subgroup except for 1 and -1, which generate subgroups of orders 1 and 2, respectively. But in $\mathbb{C}^*$, i generates the cyclic subgroup $\{i, -1, -i, 1\}$ of order 4. Or, for another argument, the equation $x^2 = a$ has a solution x in $\mathbb{C}^*$ for every $a \in \mathbb{C}^*$, but $x^2 = -1$ has no solution in $\mathbb{R}^*$. ▲

EXAMPLE 9 The group $\mathbb{R}^*$ of nonzero real numbers under multiplication is not isomorphic to the group $\mathbb{R}$ of real numbers under addition. An equation $x + x = a$ always has a solution in $\langle \mathbb{R}, + \rangle$ for every $a \in \mathbb{R}$, but the corresponding equation $x \cdot x = a$ does not always have a solution in $\langle \mathbb{R}^*, \cdot \rangle$, for example if $a = -1$. ▲

Cayley's Theorem

Look at any group table in the text. Note how each row of the table gives a permutation of the set of elements of the group, as listed at the top of the table. Similarly, each column of the table gives a permutation of the group set, as listed at the left of the table. In view of these observations, it is not surprising that at least every finite group G is isomorphic to a subgroup of the group S_G of all permutations of G. The same is true for infinite groups; Cayley's theorem states that *every* group is isomorphic to some group consisting of permutations under permutation multiplication. This is a nice and intriguing result, and is a classic of group theory. At first glance, the theorem might seem to be a tool to answer *all* questions about groups. What it really shows is the generality of groups of permutations. Examining subgroups of all permutation groups S_A for sets A of all sizes would be a tremendous task. Cayley's theorem does show that if a counterexample exists to some conjecture we have made about groups, then some group of permutations will provide the counterexample. For example, we raised the question in Section 2.3 as to whether a converse of the theorem of Lagrange holds. That is, if d is a divisor of the order n of a finite group G, does G have a subgroup of order d? We will show in the next section that A_4 has no subgroup of order 6. This counterexample is provided by a group of permutations; it is fortunate that it is provided by a subgroup of a small symmetric group such as S_4.

To make it easier to follow the proof of Cayley's theorem we outline the steps to be taken. Starting with any given group G, we shall proceed as follows:

Step I Find a set G' of permutations that is a candidate for forming a group under permutation multiplication isomorphic to G.

Step 2 Prove that G' is a group under permutation multiplication.

Step 3 Define a mapping $\phi: G \to G'$ and show that ϕ is an isomorphism of G with G'.

 THEOREM 3.5 (Cayley's Theorem)   Every group is isomorphic to a group of permutations.

PROOF   Let G be a given group.

Step I   Our first task is to find a set G' of permutations that is a candidate to form a group isomorphic to G. Think of G as just a set, and let S_G be the group of all permutations of G given by Theorem 2.1. (Note that in

---

◆   **H I S T O R I C A L   N O T E**   ◆

Arthur Cayley (1821–1895) gave an abstract-sounding definition of a group in a paper of 1854: "A set of symbols $1, \alpha, \beta, \ldots$, all of them different and such that the product of any two of them (no matter in what order) or the product of any one of them into itself, belongs to the set, is said to be a group." He then proceeded to define a group table and note that every line and column of the table "will contain all the symbols $1, \alpha, \beta, \ldots$." Cayley's symbols, however, always represented operations on sets; it does not seem that he was aware of any other kind of group. He noted, for instance, that the four matrix operations 1, α = inversion, β = transposition, and $\gamma = \alpha\beta$, form, abstractly, the noncyclic group of four elements, In any case, his definition went unnoticed for a quarter of a century.

This paper of 1854 was one of about 300 written during the fourteen years Cayley was practicing law, being unable to find a suitable teaching post. In 1863 he finally became a professor at Cambridge. In 1878 he returned to the theory of groups by publishing four papers, in one of which he stated Theorem 3.5 of this text; his "proof" was simply to notice from the group table that multiplication by any group element permuted the group elements. However, he wrote, "this does not in any wise show that the best or the easiest mode of treating the general problem [of finding all groups of a given order] is thus to regard it as a problem of [permutations]. It seems clear that the better course is to consider the general problem in itself."

The papers of 1878, unlike the earlier one, found a receptive audience; in fact, they were an important influence on Walter Van Dyck's 1882 axiomatic definition of an abstract group, the definition that led to the development of abstract group theory.

the finite case if G has n elements, S_G has $n!$ elements. Thus in general, S_G is clearly too big to be isomorphic to G.) We define a certain subset of S_G. For $a \in G$, let λ_a be the mapping of G into G given by

$$\lambda_a(x) = ax$$

for $x \in G$. (We can think of λ_a as meaning *left multiplication by a.*) If $\lambda_a(x) = \lambda_a(y)$, then $ax = ay$ so $x = y$ by Theorem 1.1. Thus λ_a is a one-to-one function. Also, if $y \in G$, then

$$\lambda_a(a^{-1}y) = a(a^{-1}y) = y,$$

so λ_a maps G onto G. Since $\lambda_a : G \to G$ is both one to one and onto G, λ_a is a permutation of G; that is, $\lambda_a \in S_G$. Let

$$G' = \{\lambda_a \mid a \in G\}.$$

Step 2 We claim that G' is a subgroup of S_G. We must show that G' is closed under permutation multiplication, contains the identity permutation, and contains an inverse for each of its elements. First, we claim that

$$\lambda_a \lambda_b = \lambda_{ab}.$$

To show that these functions are the same, we must show that they have the same action on each $x \in G$. Now

$$(\lambda_a \lambda_b)(x) = \lambda_a(\lambda_b(x)) = \lambda_a(bx) = a(bx) = (ab)x = \lambda_{ab}(x).$$

Thus $\lambda_a \lambda_b = \lambda_{ab}$, so G' is closed under multiplication. Clearly for all $x \in G$,

$$\lambda_e(x) = ex = x,$$

where e is the identity element of G, so λ_e is the identity permutation ι in S_G and is in G'. Since $\lambda_a \lambda_b = \lambda_{ab}$, we have

$$\lambda_a \lambda_{a^{-1}} = \lambda_{aa^{-1}} = \lambda_e,$$

and also

$$\lambda_{a^{-1}} \lambda_a = \lambda_e.$$

Hence

$$(\lambda_a)^{-1} = \lambda_{a^{-1}},$$

so $(\lambda_a)^{-1} \in G'$. Thus G' is a subgroup of S_A.

Step 3 It remains for us to prove now that G is isomorphic to this group

G' that we have described. Define $\phi : G \to G'$ by

$$\phi(a) = \lambda_a$$

for $a \in G$. If $\phi(a) = \phi(b)$, then λ_a and λ_b must be the same permutation of G. In particular.

$$\lambda_a(e) = \lambda_b(e),$$

so $ae = be$ and $a = b$. Thus ϕ is one to one. It is immediate that ϕ is onto G' by the definition of G'. Finally, $\phi(ab) = \lambda_{ab}$, While

$$\phi(a)\phi(b) = \lambda_a\lambda_b.$$

But we saw in Step 2 that λ_{ab} and $\lambda_a\lambda_b$ are the same permutation of G. Thus

$$\phi(ab) = \phi(a)\phi(b). \ \blacklozenge$$

For the proof of the theorem, we could have considered equally well the permutations ρ_a of G defined by

$$\rho_a(x) = xa$$

for $x \in G$. (We can think of ρ_a as meaning *right multiplication by a*.) Exercise 26 shows that these permutations form a subgroup G'' of S_G, again isomorphic to G, but under the map $\mu : G \to G''$ defined by

$$\mu(a) = \rho_{a^{-1}}.$$

DEFINITION 3.6 (Regular Representations) The group G' in the proof of Theorem 3.5 is the **left regular representation** of G, and the group G'' in the preceding comment is the **right regular representation of** G.

EXAMPLE 10 Let us compute the right regular representations of the group given by the group table Table 3.1. By "compute" we mean give

	e	a	b
e	e	a	b
a	a	b	e
b	b	e	a

Table 3.1

	ρ_e	ρ_a	ρ_b
ρ_e	ρ_e	ρ_a	ρ_b
ρ_a	ρ_a	ρ_b	ρ_e
ρ_b	ρ_b	ρ_e	ρ_a

Table 3.2

the elements of the right regular representation and the group table. Here the elements are

$$\rho_e = \begin{pmatrix} e & a & b \\ e & a & b \end{pmatrix}, \qquad \rho_a = \begin{pmatrix} e & a & b \\ a & b & e \end{pmatrix}, \qquad \text{and} \qquad \rho_b = \begin{pmatrix} e & a & b \\ b & e & a \end{pmatrix}.$$

The table for this representation is just like the original table with x renamed ρ_x, as seen in Table 3.2. *This "renaming" is the basic idea of an isomorphism.* For example,

$$\rho_a\rho_b = \begin{pmatrix} e & a & b \\ a & b & e \end{pmatrix}\begin{pmatrix} e & a & b \\ b & e & a \end{pmatrix} = \begin{pmatrix} e & a & b \\ e & a & b \end{pmatrix} = \rho_e. \ \blacktriangle$$

For a finite group given by a group table, ρ_a is the permutation of the elements corresponding to their order in the column under a at the very top, and λ_a is the permutation corresponding to the order of the elements in the row opposite a at the extreme left. The notations ρ_a and λ_a were chosen to suggest right and left multiplication by a, respectively.

Exercises 3.2

Computations

In Exercises 1 through 3, describe all isomorphisms mapping the first group given onto the second group given.

1. The Klein 4-group V of Table 1.14 onto $\mathbb{Z}_2 \times \mathbb{Z}_2$

2. $\mathbb{Z}_2 \times \mathbb{Z}_3$ onto $\mathbb{Z}_6$

3. $\mathbb{Z}_2 \times \mathbb{Z}_5$ onto $\mathbb{Z}_{10}$

An isomorphism of a group with itself is an **automorphism of the group**. In Exercises 4 through 8, find the number of automorphisms of the given group.

4. $\mathbb{Z}_2$ **5.** $\mathbb{Z}_6$ **6.** $\mathbb{Z}_8$ **7.** $\mathbb{Z}$ **8.** $\mathbb{Z}_{12}$

9. Compute the left regular representation of $\mathbb{Z}_4$. Compute the right regular representation of S_3 using the notation of Example 4, Section 2.1.

10. Give two arguments showing that $\mathbb{Z}_4$ is not isomorphic to the Klein 4-group V of Example 5, Section 1.3.

11. Partition the following collection of groups into subcollections of isomorphic groups, as discussed following Theorem 3.3. Here a * superscript means all nonzero elements of the set.

$\mathbb{Z}$ under addition	S_2
$\mathbb{Z}_6$	$\mathbb{R}^*$ under multiplication
$\mathbb{Z}_2$	$\mathbb{R}^+$ under multiplication
S_6	$\mathbb{Q}^*$ under multiplication
$17\mathbb{Z}$ under addition	$\mathbb{C}^*$ under multiplication
$\mathbb{Q}$ under addition	The subgroup $\langle \pi \rangle$ of $\mathbb{R}^*$ under multiplication
$3\mathbb{Z}$ under addition	The subgroup G of S_8 generated by $(1,3,4)(2,6)$
$\mathbb{R}$ under addition	

Concepts

12. Mark each of the following true or false.
_____ a. Any two groups of order 3 are isomorphic.
_____ b. There is, up to isomorphism, only one cyclic group of a given finite order.
_____ c. Any two finite groups with the same number of elements are isomorphic.
_____ d. Every isomorphism is a one-to-one function.
_____ e. Every one-to-one function between groups is an isomorphism.
_____ f. The property of being cyclic (or not being cyclic, as the case may be) is a structural property of a group.
_____ g. A structural property of a group must be shared by every isomorphic group.
_____ h. An abelian group can't be isomorphic to a nonabelian group.
_____ i. An additive group can't be isomorphic to a multiplicative group.
_____ j. $\mathbb{R}$ under addition is isomorphic to a group of permutations.

13. Show that the multiplicative group U of all complex numbers $\cos \theta + i \sin \theta$ of magnitude 1 is not isomorphic to $\mathbb{R}$ under addition.

14. Show that the multiplicative group U of all complex numbers $\cos \theta + i \sin \theta$ of magnitude 1 is not isomorphic to $\mathbb{R}^*$ under multiplication.

Theory

15. Show that the relation $\simeq$ defined in Theorem 3.3 is reflexive.

16. Show that the relation $\simeq$ defined in Theorem 3.3 is symmetric.

17. Show that the relation $\simeq$ defined in Theorem 3.3 is transitive.

18. Let G be a cyclic group with generator a, and let G' be a group isomorphic to G. If $\phi : G \to G'$ is an isomorphism, show that, for every $x \in G$, $\phi(x)$ is completely determined by the value $\phi(a)$.

19. Let G be an abelian group. Prove that being abelian is a structural property of G by showing that if G' is isomorphic to G, then G' is abelian also.

20. Let G be a cyclic group. Prove that the property of being cyclic is a structural property of G. (See Exercise 19.)

21. Let $\langle G, \cdot \rangle$ be a group. Consider the binary operation $*$ on the set G defined by

$$a * b = b \cdot a$$

for $a, b \in G$. Show that $\langle G, * \rangle$ is a group and that $\langle G, * \rangle$ is actually isomorphic to $\langle G, \cdot \rangle$. [*Hint:* Consider the map ϕ with $\phi(a) = a^{-1}$ for $a \in G$.]

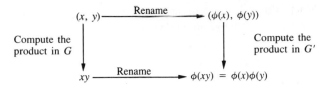

Figure 3.4

Comments: This is one instance where the notations $\langle G, \cdot \rangle$ and $\langle G, * \rangle$ are *very* handy. See the discussion following the definition of a group. Note that if G is finite, then we obtain the group table for $\langle G, * \rangle$ from the group table for $\langle G, \cdot \rangle$ by reading this table from the top to the left instead of from the left to the top.

22. Prove in a manner similar to that used in Theorem 3.4 that every finite cyclic group of order n is isomorphic to $\mathbb{Z}_n$.

23. Let G be a group and let g be a fixed element of G. Show that the map i_g, such that $i_g(x) = gxg^{-1}$ for $x \in G$, is an isomorphism of G with itself, that is, an automorphism of G (see the definition before Exercise 4).

24. Let $\langle S, * \rangle$ be the group of all real numbers except -1 under the operation $*$ defined by $a * b = a + b + ab$ (see Exercise 16 of Section 1.2). Show that $\langle S, * \rangle$ is isomorphic to the group $\mathbb{R}^*$ of nonzero real numbers under multiplication. Actually define an isomorphism $\phi : \mathbb{R}^* \to S$.

25. Let ϕ be an isomorphism of a group G with a group G'. If for $x \in G$ we view $\phi(x)$ as a new name for x or consider $\phi(x)$ as x renamed, then the condition $\phi(xy) = \phi(x)\phi(y)$ corresponds to the statement that the diagram in Fig. 3.4 is commutative. By "the diagram is commutative" we mean that starting in the upper left corner and following the path to the lower right corner, given by (arrow down) (arrow across), gives the same result as following the path (arrow across) (arrow down). Illustrating with the isomorphism ϕ of the answer to Exercise 24, if we consider $\phi(x)$ as x renamed for $x \in \mathbb{R}^*$, we get the diagram in Fig. 3.5 for $x = 2$ and $y = 5$.

a. Starting with the group $\mathbb{R}^*$ under multiplication, suppose x is renamed $x - 4$ for $x \in \mathbb{R}^*$. Let S_1 be the set of all real numbers except -4. Define $*_1$ on S_1 such that $\langle S_1, *_1 \rangle$ is isomorphic to $\mathbb{R}^*$ under multiplication by means of this renaming.

b. Repeat part (a) if $x \in \mathbb{R}^*$ is renamed $x - t$ for fixed $t \in \mathbb{R}$. First determine the required set S_2.

c. Repeat part (a) if $x \in \mathbb{R}^*$ is renamed $x^3 + 1$. First determine the required set S_3.

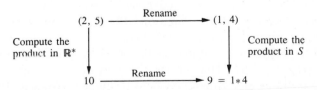

Figure 3.5

26. Let G be a group. Prove that the permutations $\rho_a : G \to G$, where $\rho_a(x) = xa$ for $a \in G$ and $x \in G$, do form a group G'' isomorphic to G.

3.3

Factor Groups

Factor Groups from Homomorphisms

Let H be a subgroup of a finite group G. Suppose we write a table for the group operation of G, listing elements at the top and at the left as they occur in the left cosets of H. We illustrated this in Section 2.3. The body of the table may or may not break up into blocks corresponding to the cosets, giving a group operation on the cosets. See Tables 2.3 and 2.6 in Section 2.3. We start this section with an intuitive argument that if H is the kernel of a group homomorphism $\phi : G \to G'$, then the left cosets of H will be elements of a group whose binary operation is derived from the group operation of G. After our intuitive argument, we will state carefully in a theorem what we seem to have discovered and provide a proof. This will illustrate a fairly common procedure in developing mathematics: we get an intuitive idea of a result, and then we try to write it up precisely and prove it.

Let $\phi : G \to G'$ be a group homomorphism with kernel H. (We are not assuming here that G is finite.) Theorem 3.2 shows that for each $a \in G$, the fibre $\phi^{-1}[\phi(a)] = \{x \in G \mid \phi(x) = \phi(a)\}$ is the left coset aH of H and is also the right coset Ha of H. Since these left and right cosets of H coincide, we will simply refer to them as cosets of H.

Now $\phi[G]$ is a group by Thoerem 3.1. We associate with each $y \in \phi[G]$ the coset $\phi^{-1}[y] = \{x \in G \mid \phi(x) = y\}$. By renaming $y \in \phi[G]$ by the name of the associated coset, that is, by $\phi^{-1}[y]$, we can consider the cosets to form a group. This group will be isomorphic to $\phi[G]$ since it is just $\phi[G]$ renamed.

In summary, the cosets of the kernel of a group homomorphism $\phi : G \to G'$ form a group isomorphic to the subgroup $\phi[G]$ of G'. The binary operation on the cosets can be computed in terms of the group operation of G'. This group of cosets is the **factor group of G modulo H**, and is denoted by G/H.

It would be nice to be able to compute a product $(aH)(bH)$ of cosets in G/H using the binary operation of G, rather than having to make the transition to G'. In our discussion above, we associated the coset aH with the element $\phi(a)$ of $\phi[G]$, and associated bH with $\phi(b)$. Under the renaming of $\phi[G]$ above, we should consider $(aH)(bH)$ to be the coset associated with $\phi(a)\phi(b)$. But $\phi(a)\phi(b) = \phi(ab)$, so the product coset is $(ab)H$. Thus we have $(aH)(bH) = (ab)H$. This means that we might be able to compute products of cosets by choosing an element from each coset, multiplying these elements in G, and finding the coset containing their product.

We are now ready to try to write this up nicely in a theorem, and to give a careful proof.

THEOREM 3.6 Let $\phi : G \to G'$ be a group homomorphism with kernel H. Then the cosets of H form a group, G/H, whose binary operation defines the product $(aH)(bH)$ of two cosets by choosing elements a and b from the cosets, and letting $(aH)(bH) = (ab)H$. Also, the map $\mu : G/H \to \phi[G]$ defined by $\mu(aH) = \phi(a)$ is an isomorphism.

PROOF The first thing we have to worry about is the definition $(aH)(bH) = (ab)H$ for the product of two cosets in G/H. The product is computed by *choosing an element* from each of the cosets aH and bH, and by finding the coset containing their product ab. Any time the definition of something involves making *choices*, we should show that the end result is independent of the choices made. We say that a thing is **well defined** if it is independent of any choices made in its computation. Thus we start by showing that $(aH)(bH) = (ab)H$ gives a *well-defined* operation on G/H. To that end, suppose that $ah_1 \in aH$ and $bh_2 \in bH$ are two other representative elements from these cosets. We must show that $(ab)H = (ah_1bh_2)H_1$, that is, that ab and ah_1bh_2 lie in the same coset of G/H. By Theorem 3.2, we need only show that $\phi(ab) = \phi(ah_1bh_2)$. Recall that $\phi(h) = e'$, the identity of G', for all $h \in H$. Thus

$$\phi(ah_1bh_2) = \phi(a)\phi(h_1)\phi(b)\phi(h_2) = \phi(a)e'\phi(b)e'$$
$$= \phi(a)\phi(b) = \phi(ab).$$

Thus we do indeed have a well-defined binary operation on G/H.
  We now check the group axioms for G/H.

Associative Now

$$(aH)[(bH)(cH)] = (aH)[(bc)H] = [a(bc)]H$$
$$= [(ab)c]H = [(ab)H](cH)$$
$$= [(aH)(bH)](cH).$$

Thus associativity in G/H follows from associativity in G, since we compute in G/H by choosing representatives from cosets and computing in G.

Identity Note that

$$(aH)(eH) = (ae)H = aH = (ea)H = (eH)(aH),$$

so $H = eH$ acts as identity coset in G/H.

Inverses　Note that

$$(a^{-1}H)(aH) = (a^{-1}a)H = eH = (aa^{-1})H = (aH)(a^{-1}H),$$

so $a^{-1}H$ is the inverse of aH in G/H.

Thus G/H is a group.

We now show that $\mu : G/H \to \phi[G]$, given by $\mu(aH) = \phi(a)$, is an isomorphism. Now μ is defined by *choosing* elements from cosets and computing ϕ of these elements. We must start by showing that this map μ is *well defined*, independent of the choices of elements from cosets. To that end, suppose that ah is another element of the coset aH, where $h \in H$. Then $\phi(ah) = \phi(a)\phi(h) = \phi(a)e' = \phi(a)$, so μ is indeed well defined. We proceed to show it is an isomorphism.

Homomorphism Property　Note that

$$\mu((aH)(bH)) = \mu((ab)H) = \phi(ab) = \phi(a)\phi(b) = \mu(aH)\mu(bH).$$

This demonstrates the homomorphism property.

One to one　Suppose $\mu(aH) = \mu(bH)$. Then $\phi(a) = \phi(b)$, so that aH and bH are the same coset, by Theorem 3.2.

Onto $\phi[G]$　Let $y \in \phi[G]$. Then $y = \phi(x)$ for some $x \in G$, and $\mu(xH) = \phi(x) = y$. Our proof is complete.　◆

EXAMPLE 1　Example 9 of Section 3.1 considered the map $\gamma : \mathbb{Z} \to \mathbb{Z}_n$, where $\gamma(m)$ is the remainder when m is divided by n in accordance with the division algorithm. We know that γ is a homomorphism. Of course, $\mathrm{Ker}(\gamma) = n\mathbb{Z}$. By Theorem 3.6, we see that the factor group $\mathbb{Z}/n\mathbb{Z}$ is isomorphic to $\mathbb{Z}_n$. The cosets of $n\mathbb{Z}$ are the *residue classes modulo n* described in Section 0.2. For example, taking $n = 5$, we see the cosets of $5\mathbb{Z}$ are

$$5\mathbb{Z} = \{\ldots, -10, -5, 0, 5, 10, \ldots\},$$

$$1 + 5\mathbb{Z} = \{\ldots, -9, -4, 1, 6, 11, \ldots\},$$

$$2 + 5\mathbb{Z} = \{\ldots, -8, -3, 2, 7, 12, \ldots\},$$

$$3 + 5\mathbb{Z} = \{\ldots, -7, -2, 3, 8, 13, \ldots\},$$

$$4 + 5\mathbb{Z} = \{\ldots, -6, -1, 4, 9, 14, \ldots\},$$

Note that the isomorphism $\mu : \mathbb{Z}/5\mathbb{Z} \to \mathbb{Z}_5$ of Theorem 3.6 assigns to each coset of $5\mathbb{Z}$ its smallest nonnegative element. That is, $\mu(5\mathbb{Z}) = 0$, $\mu(1 + 5\mathbb{Z}) = 1$, etc.　▲

It is very important that we learn how to compute in a factor group. We may multiply (add) two cosets and choosing *any* two representative elements, multiplying (adding) them and finding the coset in which the resulting product (sum) lies. ◆

EXAMPLE 2 Consider the factor group $\mathbb{Z}/5\mathbb{Z}$ with the cosets shown above. We can add $(2 + 5\mathbb{Z}) + (4 + 5\mathbb{Z})$ by choosing 2 and 4, finding $2 + 4 = 6$, and noticing that 6 is in the coset $1 + 5\mathbb{Z}$. We could equally well add these two cosets by choosing 27 in $2 + 5\mathbb{Z}$ and -16 in $4 + 5\mathbb{Z}$; the sum $27 + (-16) = 11$ is also in the coset $1 + 5\mathbb{Z}$. ▲

The factor groups $\mathbb{Z}/n\mathbb{Z}$ in the preceding example are classics. Recall that we refer to the cosets of $n\mathbb{Z}$ as *residue classes modulo n.* Two integers in the same coset are *congruent modulo n.* This terminology is carried over to other factor groups. A factor group G/H is often called the **factor group of G modulo H**. Elements in the same coset of H are often said to be **congruent modulo H**. By abuse of notation, we may sometimes write $\mathbb{Z}/n\mathbb{Z} = \mathbb{Z}_n$ and think of $\mathbb{Z}_n$ as the additive group of residue classes of $\mathbb{Z}$ modulo $\langle n \rangle$, or abusing notation further, modulo n.

Factor Groups from Normal Subgroups

So far, we have obtained factor groups only from homomorphisms. Let G be a group and let H be a subgroup of G. Now H has both left cosets and right cosets, and in general, a left coset aH need not be the same set as the right coset Ha. Suppose we try to define a binary operation on left cosets by defining

$$(aH)(bH) = (ab)H \tag{1}$$

as in the statement of Theorem 3.6. Equation (1) attempts to define left coset multiplication by choosing representatives a and b from the cosets. As discussed in the proof of Theorem 3.6, Eq. (1) is meaningless unless it gives a *well-defined* operation, independent of the representative elements a and b chosen from the cosets. The theorem that follows shows that Eq. (1) gives a well-defined binary operation if and only if the left and right cosets aH and Ha coincide for all $a \in G$.

THEOREM 3.7 Let H be a subgroup of a group G. Then left coset multiplication is well defined by the equation

$$(aH)(bH) = (ab)H$$

if and only if left and right cosets coincide, so that $aH = Ha$ for all $a \in G$.

PROOF Suppose first that $(aH)(bH) = (ab)H$ does give a well-defined binary operation on left cosets. Let $a \in G$. We want to show that aH and Ha are the same set. We use the standard technique of showing that each is a subset of the other.

Let $x \in aH$. Choosing representatives $x \in aH$ and $a^{-1} \in a^{-1}H$, we have $(xH)(a^{-1}H) = (xa^{-1})H$. On the other hand, choosing representatives $a \in aH$ and $a^{-1} \in a^{-1}H$, we see that $(aH)(a^{-1}H) = eH = H$. Using our assumption that left coset multiplication by representatives is well defined, we must have $xa^{-1} = h \in H$. Then $x = ha$ so $x \in Ha$ and $aH \subseteq Ha$.

Now suppose that $y \in Ha$, so that $y = h_1 a$ for some $h_1 \in H$. Then $y^{-1} = a^{-1}h_1^{-1} \in a^{-1}H$. Since $(a^{-1}H)(aH) = H$ and left coset multiplication is assumed to be well defined, we see that $y^{-1}a = h_2 \in H$ so $y = ah_2^{-1}$. This shows that $Ha \subseteq aH$, and consequently $aH = Ha$. This completes the proof that if left coset multiplication by choosing representatives is well defined, then left and right cosets must coincide.

We turn now to the converse: If left and right cosets coincide, then left coset multiplication by representatives is well defined. Due to our hypothesis, we can simply say *cosets*, omitting *left* and *right*. Suppose we wish to compute $(aH)(bH)$. Choosing $a \in aH$ and $b \in bH$, we obtain the coset$(ab)H$. Choosing different representatives $ah_1 \in aH$ and $bh_2 \in bH$, we obtain the coset ah_1bh_2H. We must show that these are the same coset. Now $h_1b \in Hb = bH$, so $h_1b = bh_3$ for some $h_3 \in H$. Thus

$$(ah_1)(bh_2) = a(h_1b)h_2 = a(bh_3)h_2 = (ab)(h_3h_2)$$

and $(ab)(h_3h_2) \in (ab)H$. Therefore ah_1bh_2 is in $(ab)H$, and the proof is finished. ◆

Theorem 3.7 shows that if left and right cosets of H coincide, then Eq. (1) gives a well-defined binary operation on cosets. We wonder whether the cosets do form a group with such coset multiplication. This is indeed true, and the check of the group axioms is identical to that given in the proof of Theorem 3.6.

COROLLARY   Let H be a subgroup of G whose left and right cosets coincide. Then the cosets of H form a group G/H under the binary operation $(aH)(bH) = (ab)H$.

i.e., normal

DEFINITION 3.7 (Factor Group)   The group G/H in the preceding corollary is the **factor group** (or **quotient group**) of G modulo H.

Recall that a subgroup H of G is **normal** if its left and right cosets coincide (see Definition 3.4). Recall also that the kernel of a homomorphism $\phi : G \rightarrow G'$ is a normal subgroup of G.

EXAMPLE 3   Since $\mathbb{Z}$ is an abelian group, $n\mathbb{Z}$ is a normal subgroup. Theorem 3.7 allows us to construct the factor group $\mathbb{Z}/n\mathbb{Z}$ with no reference to a homomorphism. This is the *elegant* approach to the demonstration that there exists a cyclic group of order n, as opposed to our naive approach in Theorem 1.8 of Section 1.4. ▲

We derive some alternative characterizations of normal subgroups, which often provide us with an easier way to check normality than finding both the left and the right coset decompositions.

Suppose that H is a subgroup of G such that $ghg^{-1} \in H$ for all $g \in G$. Then $gHg^{-1} = \{ghg^{-1} \mid h \in H\} \subseteq H$ for all $g \in G$. We claim that actually $gHg^{-1} = H$. We must show that $H \subseteq gHg^{-1}$ for all $g \in G$. Let $h \in H$. Replacing g by g^{-1} in the relation $ghg^{-1} \in H$, we obtain $g^{-1}h(g^{-1})^{-1} = g^{-1}hg = h_1$ where $h_1 \in H$. Consequently, $h = gh_1g^{-1} \in gHg^{-1}$, and we are done.

Suppose that $gH = Hg$ for all $g \in G$. Then $gh = h_1g$, so $ghg^{-1} \in H$ for all $g \in G$ and all $h \in H$. By the preceding paragraph, this means that $gHg^{-1} = H$ for all $g \in G$. Conversely, if $gHg^{-1} = H$ for all $g \in G$, then $ghg^{-1} = h_1$ so $gh = h_1g \in Hg$, and $gH \subseteq Hg$. But also, $g^{-1}Hg = H$ giving $g^{-1}hg = h_2$, so that $hg = gh_2$ and $Hg \subseteq gH$.

We have shown that the following three conditions are equivalent characterizations for a normal subgroup H of a group G.

1. $ghg^{-1} \in H$ for all $g \in G$ and $h \in H$.
2. $gHg^{-1} = H$ for all $g \in G$.
3. $gH = Hg$ for all $g \in G$.

Condition (2) is often taken as the definition of a normal subgroup H of a group G.

EXAMPLE 4 Every subgroup H of an abelian group G is normal. We need only note that $gh = hg$ for all $h \in H$ and all $g \in G$, so, of course, $ghg^{-1} = h \in H$ for all $g \in G$ and all $h \in H$. ▲

Exercise 22 of Section 3.1 shows that the map $i_g : G \to G$ defined by $i_g(x) = gxg^{-1}$ is a homomorphism of G into itself. We see that $gag^{-1} = gbg^{-1}$ if and only if $a = b$, so i_g is one to one. Since $g(g^{-1}yg)g^{-1} = y$, we see that i_g is onto G, so it is an isomorphism of G with itself.

DEFINITION 3.8 (Automorphism) An isomorphism $\phi : G \to G$ is an **automorphism** of G. The automorphism $i_g : G \to G$ where $i_g(x) = gxg^{-1}$ is the **inner automorphism of G by g**.

The equivalence of Conditions (1) and (3) shows that $gH = Hg$ for all $g \in G$ if and only if $i_g[H] = H$ for all $g \in G$, that is, if and only if H is **invariant** under all inner automorphisms of G. It is important to realize that $i_g[H] = H$ is an equation in sets; we need not have $i_g(h) = h$ for all $h \in H$. That is, i_g may perform a nontrivial *permutation* of the set H. We see that the normal subgroups of a group G are precisely those that are invariant under all inner automorphisms.

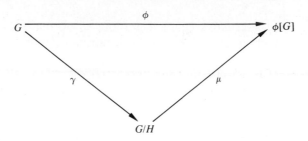

Figure 3.6

Fundamental Homomorphism Theorem

We have seen that every homomorphism $\phi: G \to G'$ gives rise to a natural factor group (Theorem 3.6), namely $G/\mathrm{Ker}(\phi)$. We now show that each factor group G/H gives rise to a natural homomorphism having H as kernel.

THEOREM 3.8   Let H be a normal subgroup of G. Then $\gamma: G \to G/H$ given by $\gamma(x) = xH$ is a homomorphism with kernel H.

PROOF   Let $x, y \in G$. Then

$$\gamma(xy) = (xy)H = (xH)(yH) = \gamma(x)\gamma(y),$$

so γ is a homomorphism. Since $xH = H$ if and only if $x \in H$, we see that the kernel of γ is indeed H. ◆

We have seen in Theorem 3.6 that if $\phi: G \to G'$ is a homomorphism with kernel H, then $\mu: G/H \to \phi[G]$ where $\mu(gH) = \phi(g)$ is an isomorphism. Theorem 3.8 shows that $\gamma: G \to G/H$ defined by $\gamma(g) = gH$ is a homomorphism. Figure 3.6 shows these groups and maps. We see that the homomorphism ϕ can be *factored*, $\phi = \mu\gamma$, where γ is a homomorphism and μ is a one-to-one homomorphism with image $\phi[G]$. We state this as a theorem.

THEOREM 3.9 (Fundamental Homomorphism Theorem)   Let $\phi: G \to G'$ be a group homomorphism with kernel H. Then $\phi[G]$ is a group, and the map $\mu: G/H \to \phi[G]$ given by $\mu(gH) = \phi(g)$ is an isomorphism. If $\gamma: G \to G/H$ is the homomorphism given by $\gamma(g) = gH$, then for each $g \in G$, we have $\phi(g) = \mu\gamma(g)$.

The isomorphism μ in Theorem 3.9 is referred to as a *natural* or *canonical* isomorphism, and the same adjectives are used to describe the homomorphism γ. There may be other isomorphisms and homomorphisms for these same groups, but the maps μ and γ have a special status and are uniquely determined by Theorem 3.9.

In summary, every homomorphism with domain G gives rise to a factor group G/H, and every factor group G/H gives rise to a homomorphism mapping G into G/H. Homomorphisms and factor groups are closely related. We close with an example indicating how useful this relationship can be.

EXAMPLE 5 Classify the group $(\mathbb{Z}_4 \times \mathbb{Z}_2)/(\{0\} \times \mathbb{Z}_2)$ according to the fundamental theorem of finitely generated abelian groups (Theorem 2.11 of Section 2.4).

Solution The projection map $\pi_1 : \mathbb{Z}_4 \times \mathbb{Z}_2 \to \mathbb{Z}_4$ given by $\pi_1(x, y) = x$ is a homomorphism of $\mathbb{Z}_4 \times \mathbb{Z}_2$ onto $\mathbb{Z}_4$ with kernel $\{0\} \times \mathbb{Z}_2$. By Theorem 3.9, we know that the given factor group is isomorphic to $\mathbb{Z}_4$. ▲

Exercises 3.3 3.2

Computations

In Exercises 1 through 8, find the order of the given factor group.

1. $\mathbb{Z}_6/\langle 3 \rangle$

2. $(\mathbb{Z}_4 \times \mathbb{Z}_{12})/(\langle 2 \rangle \times \langle 2 \rangle)$

3. $(\mathbb{Z}_4 \times \mathbb{Z}_2)/\langle (2, 1) \rangle$

4. $(\mathbb{Z}_3 \times \mathbb{Z}_5)/(\{0\} \times \mathbb{Z}_5)$

5. $(\mathbb{Z}_2 \times \mathbb{Z}_4)/\langle (1, 1) \rangle$

6. $(\mathbb{Z}_{12} \times \mathbb{Z}_{18})/\langle (4, 3) \rangle$

7. $(\mathbb{Z}_2 \times S_3)/\langle (1, \rho_1) \rangle$

8. $(\mathbb{Z}_{11} \times \mathbb{Z}_{15})/\langle (1, 1) \rangle$

In Exercises 9 through 15, give the order of the element in the factor group.

9. $5 + \langle 4 \rangle$ in $\mathbb{Z}_{12}/\langle 4 \rangle$

10. $26 + \langle 12 \rangle$ in $\mathbb{Z}_{60}/\langle 12 \rangle$

11. $(2, 1) + \langle (1, 1) \rangle$ in $(\mathbb{Z}_3 \times \mathbb{Z}_6)/\langle (1, 1) \rangle$

12. $(3, 1) + \langle (1, 1) \rangle$ in $(\mathbb{Z}_4 \times \mathbb{Z}_4)/\langle (1, 1) \rangle$

13. $(3, 1) + \langle (0, 2) \rangle$ in $(\mathbb{Z}_4 \times \mathbb{Z}_8)/\langle (0, 2) \rangle$

14. $(3, 3) + \langle (1, 2) \rangle$ in $(\mathbb{Z}_4 \times \mathbb{Z}_8)/\langle (1, 2) \rangle$

15. $(2, 0) + \langle (4, 4) \rangle$ in $(\mathbb{Z}_6 \times \mathbb{Z}_8)/\langle (4, 4) \rangle$

16. Compute $i_{\rho_1}[H]$ for the subgroup $H = \{\rho_0, \mu_1\}$ of the group S_3 of Example 4, Section 2.1.

Concepts

Students often write nonsense when first proving theorems about factor groups. The next two exercises are designed to call attention to one basic type of error.

17. A student is asked to show that if H is a normal subgroup of an abelian group G, then G/H is abelian. The student's proof starts as follows:

We must show that G/H is abelian. Let a and b be two elements of G/H.

    a. Why does the instructor reading this proof expect to find nonsense from here on in the student's paper?

    b. What should the student have written?

    c. Complete the proof.

18. A **torsion group** is a group all of whose elements have finite order. A group is **torsion free** if the identity is the only element of finite order. A student is asked to prove that if G is a torsion group, then so is G/H for every normal subgroup H of G. The student writes

> We must show that each element of G/H is of finite order. Let $x \in G/H$.

Answer the same questions as in Exercise 17.

19. Mark each of the following true or false.

    _____ a. It makes sense to speak of the factor group G/N if and only if N is a normal subgroup of the group G.

    _____ b. Every subgroup of an abelian group G is a normal subgroup of G.

    _____ c. An inner automorphism of an abelian group must be just the identity map.

    _____ d. Every factor group of a finite group is again of finite order.

    _____ e. Every factor group of a torsion group is a torsion group. (See Exercise 18.)

    _____ f. Every factor group of a torsion-free group is torsion free.

    _____ g. Every factor group of an abelian group is abelian.

    _____ h. Every factor group of a nonabelian group is nonabelian.

    _____ i. $\mathbb{Z}/n\mathbb{Z}$ is cyclic of order n.

    _____ j. $\mathbb{R}/n\mathbb{R}$ is cyclic of order n, where $n\mathbb{R} = \{nr \mid r \in \mathbb{R}\}$ and $\mathbb{R}$ is under addition.

Theory

20. Show that A_n is a normal subgroup of S_n and compute S_n/A_n; that is, find a known group to which S_n/A_n is isomorphic.

21. Prove that the torsion subgroup T of an abelian group G is a normal subgroup of G, and that G/T is torsion free. (See Exercises 36 and 40 of Section 2.4.)

22. A subgroup H is **conjugate to a subgroup** K of a group G if there exists an inner automorphism i_g of G such that $i_g[H] = K$. Show that conjugacy is an equivalence relation on the collection of subgroups of G.

23. Characterize the normal subgroups of a group G in terms of their appearance in cells of the partition given by the conjugacy relation in the preceding exercise.

24. Referring to Exercise 22, find all subgroups of S_3 (Example 4 of Section 2.1) that are conjugate to $\{\rho_0, \mu_2\}$.

25. Let H be a normal subgroup of a group G, and let $m = (G:H)$. Show that $a^m \in H$ for every $a \in G$.

26. Show that an intersection of normal subgroups of a group G is again a normal subgroup of G.

27. Show that it makes sense to speak of the smallest normal subgroup of a group G that contains a fixed subset S of G. [*Hint:* Use Exercise 26.]

28. Let G be a group. An element of G that can be expressed in the form $aba^{-1}b^{-1}$ for some $a, b \in G$ is a **commutator** in G. The preceding exercise shows that there is a smallest normal subgroup C of a group G containing all commutators in G; the subgroup C is the **commutator subgroup** of G. Show that G/C is an abelian group.

29. Show that if a finite group G has exactly one subgroup H of a given order, then H is a normal subgroup of G.

30. Show that if H and N are subgroups of a group G, and N is normal in G, then $H \cap N$ is normal in H. Show by an example that $H \cap N$ need not be normal in G.

31. Let G be a group containing at least one subgroup of a fixed finite order s. Show that the intersection of all subgroups of G of order s is a normal subgroup of G. [*Hint:* Use the fact that if H has order s, then so does $x^{-1}Hx$ for all $x \in G$.]

32. a. Show that all automorphisms of a group G form a group under function composition.

    b. Show that the inner automorphisms of a group G form a normal subgroup of the group of all automorphisms of G under function composition. [*Warning:* Be sure to show that the inner automorphisms do form a subgroup.]

33. Show that the set of all $g \in G$ such that $i_g : G \to G$ is the identity inner automorphism i_e is a normal subgroup of a group G.

34. Let G and G' be groups, and let H and H' be normal subgroups of G and G', respectively. Let ϕ be a homomorphism of G into G'. Show that ϕ induces a natural homomorphism $\phi_* : (G/H) \to (G'/H')$ if $\phi[H] \subseteq H'$. (This fact is used constantly in algebraic topology.)

35. Use the properties $\det(AB) = \det(A) \cdot \det(B)$ and $\det(I_n) = 1$ for $n \times n$ matrices to show the following.

    a. The $n \times n$ matrices with determinant 1 form a normal subgroup of $GL(n, \mathbb{R})$.

    b. The $n \times n$ matrices with determinant ± 1 form a normal subgroup of $GL(n, \mathbb{R})$.

3.4

Factor-Group Computations and Simple Groups

Factor groups can be a tough topic for students to grasp. There is nothing like a bit of computation to strengthen understanding in mathematics. We start by attempting to improve our intuition concerning factor groups. Since we will be dealing with normal subgroups throughout this section, we often denote a subgroup of a group G by N rather than by H.

Let N be a normal subgroup of G. In the factor group G/N, the subgroup N acts as identity element. We may regard N as being *collapsed* to a single element, either to 0 in additive notation or to e in multiplicative notation.

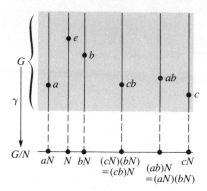

Figure 3.7

This collapsing of N together with the algebraic structure of G require that other subsets of G, namely the cosets of N, also collapse into a single element in the factor group. A visualization of this collapsing is provided by Fig. 3.7. Recall from Theorem 3.8 that $\gamma: G \to G/N$ defined by $\gamma(a) = aN$ for $a \in G$ is a homomorphism of G onto G/N. Figure 3.7 is very similar to Fig. 3.1, but in Fig. 3.7 the image group under the homomorphism is actually formed from G. We can view the "line" G/N at the bottom of the figure as obtained by collapsing to a point each coset of N in another copy of G. Each point of G/N thus corresponds to a whole vertical line segment in the figure, representing a coset of N (fibre under γ) in G. It is crucial to remember that multiplication of cosets in G/N can be computed by multiplying in G, using any representative elements of the cosets as shown in the figure.

Additively, two elements of G will collapse into the same element of G/N if they differ by an element of N. Multiplicatively, a and b collapse if ab^{-1} is in N. The degree of collapsing can vary from nonexistent to catastrophic. We illustrate the two extreme cases by examples.

EXAMPLE 1   The trivial subgroup $N = \{0\}$ of $\mathbb{Z}$ is, of course, a normal subgroup. Compute $\mathbb{Z}/\{0\}$.

Solution   Since $N = \{0\}$ has only one element, every coset of N has only one element. That is, the cosets are of the form $\{m\}$ for $m \in \mathbb{Z}$. There is no collapsing at all, and consequently $\mathbb{Z}/\{0\} \simeq \mathbb{Z}$. Each $m \in \mathbb{Z}$ is simply renamed $\{m\}$ in $\mathbb{Z}/\{0\}$. ▲

EXAMPLE 2   Let n be a positive integer. The set $n\mathbb{R} = \{nr \mid r \in \mathbb{R}\}$ is a subgroup of $\mathbb{R}$ under addition, and it is normal since $\mathbb{R}$ is abelian. Compute $\mathbb{R}/n\mathbb{R}$.

Solution   A bit of thought shows that actually $n\mathbb{R} = \mathbb{R}$, because each $x \in \mathbb{R}$ is of the form $n(x/n)$ and $x/n \in \mathbb{R}$. Thus $\mathbb{R}/n\mathbb{R}$ has only one element, the subgroup $n\mathbb{R}$. The factor group is a trivial group consisting only of the identity. ▲

As illustrated in Examples 1 and 2, for any group G, we have $G/\{e\} \simeq G$ and $G/G \simeq \{e\}$, where $\{e\}$ is the trivial group consisting only of the identity element e. These two extremes of factor groups are of little importance. We would like knowledge of a factor group G/N to give some information about the structure of G. If $N = \{e\}$, the factor group has the same structure as G and we might as well have tried to study G directly. If $N = G$, the factor group has no significant structure to supply information about G. If G is a finite group and $N \neq \{e\}$ is a normal subgroup of G, then G/N is a smaller group than G, and consequently may have a more simple structure than G. The multiplication of cosets in G/N reflects the multiplication in G, since products of cosets can be computed by multiplying in G representative elements of the cosets.

We give two examples showing that even when G/N has order 2, we may be able to deduce some useful results. If G is a finite group and G/N has just two elements, then we must have $|G| = 2|N|$. Note that every subgroup H containing just half the elements of a finite group G must be a normal subgroup since for each element a in G but not in H, both the left coset aH and the right coset Ha must consist of all elements in G that are not in H. Thus the left and right cosets of H coincide and H is a normal subgroup of G.

EXAMPLE 3 Since $|S_n| = 2|A_n|$, we see that A_n is a normal subgroup of S_n, and S_n/A_n has order 2. Let σ be an odd permutation in S_n. If we rename the two elements A_n and σA_n of S_n/A_n by *even* (for A_n) for *odd* (for σA_n), then the multiplication in S_n/A_n shown in Table 3.3 becomes

(even)(even) = even   (odd)(even) = odd

(even)(odd) = odd   (odd)(odd) = even.

Thus the factor group reflects these multiplicative properties for all the permutations in S_n. ▲

Example 3 illustrates that while knowing the product of two cosets in G/N does not tell us what the product of two elements of G is, it may tell us that the product in G of two *types* of elements is itself of a certain type.

	A_n	σA_n
A_n	A_n	σA_n
σA_n	σA_n	A_n

Table 3.3

EXAMPLE 4   (Falsity of the Converse of the Theorem of Lagrange)   The Theorem of Lagrange states if H is a subgroup of a finite group G, then the order of H divides the order of G. We show that it is false that if d divides the order of G, then there exists a subgroup H of G having order d. Namely, we show that A_4, which has order 12, contains no subgroup of order 6.

Suppose that H were a subgroup of A_4 having order 6. As observed before Example 3, it would follow that H would be a normal subgroup of A_4. Then A_4/H would have only two elements, H and σH for some $\sigma \in A_4$ not in H. Since in a group of order 2, the square of each element is the identity, we would have $HH = H$ and $(\sigma H)(\sigma H) = H$. Now computation in a factor group can be achieved by computing with representatives in the original group. Thus, computing in A_4, we find that for each $\alpha \in H$ we must have $\alpha^2 \in H$ and for each $\beta \in \sigma H$ we must have $\beta^2 \in H$. That is, the square of every element in A_4 must be in H. But in A_4, we have

$$(1, 2, 3) = (1, 3, 2)^2 \qquad \text{and} \qquad (1, 3, 2) = (1, 2, 3)^2$$

so $(1, 2, 3)$ and $(1, 3, 2)$ are in H. A similar computation shows that $(1, 2, 4)$, $(1, 4, 2)$, $(1, 3, 4)$, $(1, 4, 3)$, $(2, 3, 4)$, and $(2, 4, 3)$ are all in H. This shows that there must be at least eight elements in H, contradicting the fact that H was supposed to have order 6. ▲

We now turn to several examples that *compute* factor groups. If the group we start with is finitely generated and abelian, then its factor group will be also. *Computing* such a factor group means classifying it according to the fundamental theorem (Theorem 2.11).

EXAMPLE 5   Let us compute the factor group $(\mathbb{Z}_4 \times \mathbb{Z}_6)/\langle(0, 1)\rangle$. Here $\langle(0, 1)\rangle$ is the cyclic subgroup H of $\mathbb{Z}_4 \times \mathbb{Z}_6$ generated by $(0, 1)$. Thus

$$H = \{(0, 0), (0, 1), (0, 2), (0, 3), (0, 4), (0, 5)\}.$$

Since $\mathbb{Z}_4 \times \mathbb{Z}_6$ has 24 elements and H has 6 elements, all cosets of H must have 6 elements, and $(\mathbb{Z}_4 \times \mathbb{Z}_6)/H$ must have order 4. Since $\mathbb{Z}_4 \times \mathbb{Z}_6$ is abelian, so is $(\mathbb{Z}_4 \times \mathbb{Z}_6)/H$ (remember, we compute in a factor group by means of representatives from the original group). In additive notation, the cosets are

$$H = (0, 0) + H, \qquad (1, 0) + H, \qquad (2, 0) + H, \qquad (3, 0) + H.$$

Since we can compute by choosing the representatives $(0, 0)$, $(1, 0)$, $(2, 0)$, and $(3, 0)$, it is clear that $(\mathbb{Z}_4 \times \mathbb{Z}_6)/H$ is isomorphic to $\mathbb{Z}_4$. Note that this is what we would expect, since in a factor group modulo H, everything in H becomes the identity; that is, we are essentially setting everything in H equal to zero. Thus the whole second factor $\mathbb{Z}_6$ of $\mathbb{Z}_4 \times \mathbb{Z}_6$ is collapsed, leaving just the first factor $\mathbb{Z}_4$. ▲

Example 5 is a special case of a general theorem that we now state and prove. We should acquire an intuitive feeling for this theorem in terms of *collapsing one of the factors to the identity.*

THEOREM 3.10 Let $G = H \times K$ be the direct product of groups H and K. Then $\bar{H} = \{(h, e) \mid h \in H\}$ is a normal subgroup of G. Also $G/\bar{H}$ is isomorphic to K in a natural way. Similarly, $G/\bar{K} \simeq H$ in a natural way.

PROOF Consider the homomorphism $\pi_2 : H \times K \to K$, where $\pi_2(h, k) = k$. (See Example 7 of Section 3.1.) Since $\text{Ker}(\pi_2) = \bar{H}$ and π_2 is onto K, Theorem 3.9 tells us that $\bar{H}$ is a normal subgroup of $H \times K$ and that $(H \times K)/\bar{H} \simeq K$. ◆

We continue with additional computations of abelian factor groups. To illustrate how easy it is to compute in a factor group if we can compute in the whole group, we prove the following theorem.

THEOREM 3.11 A factor group of a cyclic group is cyclic.

PROOF Let G be cyclic with generator a, and let N be a normal subgroup of G. We claim the coset aN generates G/N. We must compute all powers of aN. But this amounts to computing, in G, all powers of the representative a and all these powers give all elements in G. Hence the powers of aN certainly give all cosets of N and G/N is cyclic. ◆

EXAMPLE 6 Let us compute the factor group $(\mathbb{Z}_4 \times \mathbb{Z}_6)/\langle(0, 2)\rangle$. Now $(0, 2)$ generates the subgroup

$$H = \{(0, 0), (0, 2), (0, 4)\}$$

of $\mathbb{Z}_4 \times \mathbb{Z}_6$ of order 3. Here the first factor $\mathbb{Z}_4$ of $\mathbb{Z}_4 \times \mathbb{Z}_6$ is left alone. The $\mathbb{Z}_6$ factor, on the other hand, is essentially collapsed by a subgroup of order 3, giving a factor group in the second factor of order 2 that must be isomorphic to $\mathbb{Z}_2$. Thus $(\mathbb{Z}_4 \times \mathbb{Z}_6)/\langle(0, 2)\rangle$ is isomorphic to $\mathbb{Z}_4 \times \mathbb{Z}_2$. ▲

EXAMPLE 7 Let us compute the factor group $(\mathbb{Z}_4 \times \mathbb{Z}_6)/\langle(2, 3)\rangle$. *Be careful!* There is a great temptation to say that we are setting the 2 of $\mathbb{Z}_4$ and the 3 of $\mathbb{Z}_6$ both equal to zero, so that $\mathbb{Z}_4$ is collapsed to a factor group isomorphic to $\mathbb{Z}_2$ and $\mathbb{Z}_6$ to one isomorphic to $\mathbb{Z}_3$, giving a total factor group isomorphic to $\mathbb{Z}_2 \times \mathbb{Z}_3$. *This is wrong!* Note that

$$H = \langle(2, 3)\rangle = \{(0, 0), (2, 3)\}$$

is of order 2, so $(\mathbb{Z}_4 \times \mathbb{Z}_6)/\langle(2, 3)\rangle$ has order 12, not 6. Setting $(2, 3)$ equal to zero does not make $(2, 0)$ and $(0, 3)$ equal to zero individually, so the factors don't collapse separately.

The possible abelian groups of order 12 are $\mathbb{Z}_4 \times \mathbb{Z}_3$ and $\mathbb{Z}_2 \times \mathbb{Z}_2 \times \mathbb{Z}_3$, and we must decide to which one our factor group is isomorphic. These two groups are most easily distinguished in that $\mathbb{Z}_4 \times \mathbb{Z}_3$ has an element of order 4, and $\mathbb{Z}_2 \times \mathbb{Z}_2 \times \mathbb{Z}_3$ does not. We claim that the coset $(1, 0) + H$ is of order 4 in the factor group $(\mathbb{Z}_4 \times \mathbb{Z}_6)/H$. To find the smallest power of a coset giving the identity in a factor group modulo H, we must, by choosing representatives, find the smallest power of a representative that is in the subgroup H. Now,

$$4(1, 0) = (1, 0) + (1, 0) + (1, 0) + (1, 0) = (0, 0)$$

is the first time that $(1, 0)$ added to itself gives an element of H. Thus $(\mathbb{Z}_4 \times \mathbb{Z}_6)/\langle(2, 3)\rangle$ has an element of order 4 and is isomorphic to $\mathbb{Z}_4 \times \mathbb{Z}_3$ or $\mathbb{Z}_{12}$. ▲

EXAMPLE 8   Let us compute (that is, classify as in Theorem 2.11) the group $(\mathbb{Z} \times \mathbb{Z})/\langle(1, 1)\rangle$. We may visualize $\mathbb{Z} \times \mathbb{Z}$ as the points in the plane with both coordinates integers, as indicated by the dots in Fig. 3.8. The subgroup $\langle(1, 1)\rangle$ consists of those points that lie on the 45° line through the origin, indicated in the figure. The coset $(1, 0) + \langle(1, 1)\rangle$ consists of those dots on the 45° line through the point $(1, 0)$, also shown in the figure. Continuing,

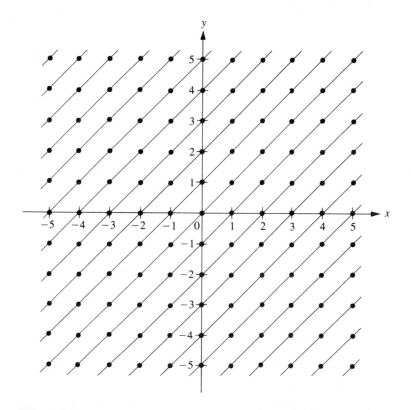

Figure 3.8

we see that each coset consists of those dots lying on one of the 45° lines in the figure. We may choose the representatives

$$\ldots, (-3, 0), (-2, 0), (-1, 0), (0, 0), (1, 0), (2, 0), (3, 0), \ldots$$

of these cosets to compute in the factor group. Since these representatives correspond precisely to the points of $\mathbb{Z}$ on the x-axis, we see that the factor group $(\mathbb{Z} \times \mathbb{Z})/\langle (1, 1) \rangle$ is isomorphic to $\mathbb{Z}$. ▲

Simple Groups

As we mentioned in the preceding section, one feature of a factor group is that it gives crude information about the structure of the whole group. Of course, sometimes there may be no nontrivial proper normal subgroups. For example, the corollary of Theorem 2.5 shows that a group of prime order can have no nontrivial proper subgroups of any sort.

DEFINITION 3.9 (Simple Group)   A group is **simple** if it has no proper nontrivial normal subgroups.

THEOREM 3.12   The alternating group A_n is simple for $n \geq 5$.

PROOF   See Exercise 34. ◆

We will have an important use for Theorem 3.12 in the very last chapter of this text. There are many simple groups other than those given above. For example, A_5 is of order 60 and A_6 is of order 360, and there is a simple group of nonprime order, namely 168, between these orders.

The complete determination and classification of all finite simple groups were recently completed. Hundreds of mathematicians worked on this task from 1950 to 1980. Section 3.5 on series of groups will indicate that a finite group has a sort of factorization into simple groups, where the factors are unique up to order. The situation is similar to the factorization of positive integers into primes. The new knowledge of all finite simple groups can now be used to solve some problems of finite group theory.

We have seen in this text that a finite simple abelian group is isomorphic to $\mathbb{Z}_p$ for some prime p. In 1964, Thompson and Feit [22] proved a longstanding conjecture of Burnside by showing that every finite nonabelian simple group is of even order. Further great strides toward the complete classification were made by Aschbacher in the 1970s. Early in 1980, Griess announced that he had constructed a predicted "monster" simple group of order

$$808, 017, 424, 794, 512, 875, 886, 459, 904, 961,$$

$$710, 757, 005, 754, 368, 000, 000, 000.$$

Aschbacher added the final details of the classification in August 1980. The research papers contributing to the entire classification fill roughly five thousand journal pages.

We turn to the characterization of those normal subgroups N of a group G for which G/N is a simple group. First we state an addendum to Theorem 3.1 on properties of a group homomorphism. The proof is left to Exercises 30 and 31.

THEOREM 3.13 Let $\phi: G \to G'$ be a group homomorphism. If N is a normal subgroup of G, then $\phi[N]$ is a normal subgroup of $\phi[G]$. Also, if N' is a normal subgroup of $\phi[G]$, then $\phi^{-1}[N']$ is a normal subgroup of G.

Theorem 3.13 should be viewed as saying that a homomorphism $\phi: G \to G$ preserves normal subgroups between G and $\phi[G]$. It is important to note that $\phi[N]$ may not be normal in G', even though N is normal in G. For example, $\phi: \mathbb{Z}_2 \to S_3$, where $\phi(0) = \rho_0$ and $\phi(1) = \mu_1$ is a homomorphism, but $\{\rho_0, \mu_1\}$ is not a normal subgroup of S_3.
We can now characterize when G/N is a simple group.

DEFINITION 3.10 (Maximal Normal Subgroup) A **maximal normal subgroup of a group** G is a normal subgroup M not equal to G such that there is no proper normal subgroup N of G properly containing M.

THEOREM 3.14 M is a maximal normal subgroup of G if and only if G/M is simple.

PROOF Let M be a maximal normal subgroup of G. Consider the canonical homomorphism $\gamma: G \to G/M$ given by Theorem 3.8. Now γ^{-1} of any proper normal subgroup of G/M would be a proper normal subgroup of G properly containing M. But M is maximal so this can't happen. Thus G/M must be simple.
  Conversely, Theorem 3.13 shows that if N is a normal subgroup of G properly containing M, then $\gamma[N]$ is normal in G/M. If also $N \neq G$, then

$$\gamma[N] \neq G/M \qquad \text{and} \qquad \gamma[N] \neq \{M\}.$$

Thus, if G/M is simple so that no such $\gamma[N]$ can exist, no such N can exist, and M is maximal. ◆

The Center and Commutator Subgroups

Every nonabelian group G has two important normal subgroups, the *center* $Z(G)$ of G and the *commutator subgroup* C of G. (The letter Z comes from the German word *Zentrum* meaning center.) The center $Z(G)$ is defined by

$$Z(G) = \{z \in G \mid zg = gz \text{ for all } g \in G\}.$$

Exercise 47 of Section 1.3 shows that $Z(G)$ is an abelian subgroup of G. Since for each $g \in G$ and $z \in Z(G)$ we have $gzg^{-1} = zgg^{-1} = ze = z$, we see at once that $Z(G)$ is a normal subgroup of G. If G is abelian, then $Z(G) = G$; in this case, the center is not useful.

EXAMPLE 9 The center of a group G always contains the identity e. It may be that $Z(G) = \{e\}$, in which case we say that **the center of G is trivial**. For example, examination of Table 2.1 in Section 2.1 for the group S_3 shows us that $Z(S_3) = \{\rho_0\}$, so the center of S_3 is trivial. (This is a special case of Exercise 33, which shows that the center of every nonabelian group of order pq for primes p and q is trivial.) Consequently, the center of $S_3 \times \mathbb{Z}_5$ must be $\{\rho_0\} \times \mathbb{Z}_5$, which is isomorphic to $\mathbb{Z}_5$. ▲

Turning to the commutator subgroup, recall that in forming a factor group of G modulo a normal subgroup N, we are essentially putting every element in G that is in N equal to e, for N forms our new identity in the factor group. This indicates another use for factor groups. Suppose, for example, that we are studying the structure of a nonabelian group G. Since Therem 2.11 gives complete information about the structure of all sufficiently small abelian groups, it might be of interest to try to form an abelian group as much like G as possible, *an abelianized version* of G, by starting with G and then requiring that $ab = ba$ for all a and b in our new group structure. To require that $ab = ba$ is to say that $aba^{-1}b^{-1} = e$ in our new group. An element $aba^{-1}b^{-1}$ in a group is a **commutator of the group**. Thus we wish to attempt to form an abelianized version of G by replacing every commutator of G by e. By the first observation of this paragraph, we should then attempt to form the factor group of G modulo the smallest normal subgroup we can find that contains all commutators of G.

THEOREM 3.15 The set of all commutators $aba^{-1}b^{-1}$ of a group G generates a normal subgroup C (the **commutator subgroup**) of G, and G/C is abelian. Furthermore, G/N is abelian if and only if $C \leq N$.

PROOF The commutators certainly generate a subgroup C, we must show that it is normal in G. Note that the inverse $(aba^{-1}b^{-1})^{-1}$ of a commutator is again a commutator, namely $bab^{-1}a^{-1}$. Also $e = eee^{-1}e^{-1}$ is a commutator. Theorem 1.11 then shows that C consists precisely of all finite products of commutators. For $x \in C$, we must show that $g^{-1}xg \in C$ for all $g \in G$, or that if x is a product of commutators, so is $g^{-1}xg$ for all $g \in G$. By inserting $e = gg^{-1}$ between each product of commutators occurring in x, we see that it is sufficient to show for each commutator $cdc^{-1}d^{-1}$ that $g^{-1}(cdc^{-1}d^{-1})g$ is in C. But

$$g^{-1}(cdc^{-1}d^{-1})g = (g^{-1}cdc^{-1}(e)(d^{-1}g)$$
$$= (g^{-1}cdc^{-1})(gd^{-1}dg^{-1})(d^{-1}g)$$
$$= [(g^{-1}c)d(g^{-1}c)^{-1}d^{-1}][dg^{-1}d^{-1}g],$$

which is in C. Thus C is normal in G.

The rest of the theorem is obvious if we have acquired the proper feeling for factor groups. One doesn't visualize in this way, but writing out that G/C is abelian follows from

$$(aC)(bC) = abC = ab(b^{-1}a^{-1}ba)C$$
$$= (abb^{-1}a^{-1})baC = baC = (bC)(aC).$$

Furthermore, if G/N is abelian, then $(a^{-1}N)(b^{-1}N) = (b^{-1}N)(a^{-1}N)$; that is, $aba^{-1}b^{-1}N = N$, so $aba^{-1}b^{-1} \in N$, and $C \leq N$. Finally, if $C \leq N$, then

$$(aN)(bN) = abN = ab(b^{-1}a^{-1}ba)N$$
$$= (abb^{-1}a^{-1})baN = baN = (bN)(aN). \blacklozenge$$

EXAMPLE 10   For the group S_3 in Table 2.1 of Section 2.1, we find that one commutator is $\rho_1\mu_1\rho_1^{-1}\mu_1^{-1} = \rho_1\mu_1\rho_2\mu_1 = \mu_3\mu_2 = \rho_2$. We similarly find that $\rho_2\mu_1\rho_2^{-1}\mu_1^{-1} = \rho_2\mu_1\rho_1\mu_1 = \mu_2\mu_3 = \rho_1$. Thus the commutator subgroup C of S_3 contains A_3. Since A_3 is a normal subgroup of S_3 and S_3/A_3 is abelian, Theorem 3.15 shows that $C = A_3$. ▲

Exercises 3.4

Computations

In Exercises 1 through 12, classify the given group according to the fundamental theorem of finitely generated abelian groups.

1. $(\mathbb{Z}_2 \times \mathbb{Z}_4)/\langle(0, 1)\rangle$

2. $(\mathbb{Z}_2 \times \mathbb{Z}_4)/\langle(0, 2)\rangle$

3. $(\mathbb{Z}_2 \times \mathbb{Z}_4)/\langle(1, 2)\rangle$

4. $(\mathbb{Z}_4 \times \mathbb{Z}_8)/\langle(1, 2)\rangle$

5. $(\mathbb{Z}_4 \times \mathbb{Z}_4 \times \mathbb{Z}_8)/\langle(1, 2, 4)\rangle$

6. $(\mathbb{Z} \times \mathbb{Z})/\langle(0, 1)\rangle$

7. $(\mathbb{Z} \times \mathbb{Z})/\langle(1, 2)\rangle$

8. $(\mathbb{Z} \times \mathbb{Z} \times \mathbb{Z})/\langle(1, 1, 1)\rangle$

9. $(\mathbb{Z} \times \mathbb{Z} \times \mathbb{Z}_4)/\langle(3, 0, 0)\rangle$

10. $(\mathbb{Z} \times \mathbb{Z} \times \mathbb{Z}_8)/\langle(0, 4, 0)\rangle$

11. $(\mathbb{Z} \times \mathbb{Z})/\langle(2, 2)\rangle$

12. $(\mathbb{Z} \times \mathbb{Z} \times \mathbb{Z})/\langle(3, 3, 3)\rangle$

13. Find both the center $Z(D_4)$ and the commutator subgroup C of the group D_4 of symmetries of the square in Table 2.2 of Section 2.1.

14. Describe all subgroups of order ≤ 4 of $\mathbb{Z}_4 \times \mathbb{Z}_4$, and in each case classify the factor group of $\mathbb{Z}_4 \times \mathbb{Z}_4$ modulo the subgroup as in Theorem 2.11. That is, describe the subgroup and say that the factor group of $\mathbb{Z}_4 \times \mathbb{Z}_4$ modulo the subgroup is isomorphic to $\mathbb{Z}_2 \times \mathbb{Z}_4$, or whatever the case may be. [*Hint:* $\mathbb{Z}_4 \times \mathbb{Z}_4$ has six different cyclic subgroups of order 4. Describe them by giving a generator, such as the subgroup $\langle(1, 0)\rangle$. There is one subgroup of order 4 that is isomorphic to the Klein 4-group. There are three subgroups of order 2.]

Concepts

15. Mark each of the following true or false.
    _____ a. Every factor group of a cyclic group is cyclic.
    _____ b. A factor group of a noncyclic group is again noncyclic.

_____ c. $\mathbb{R}/\mathbb{Z}$ under addition has no element of order 2.

_____ d. $\mathbb{R}/\mathbb{Z}$ under addition has elements of order n for all $n \in \mathbb{Z}^+$.

_____ e. $\mathbb{R}/\mathbb{Z}$ under addition has an infinite number of elements of order 4.

_____ f. If the commutator subgroup C of a group G is $\{e\}$, then G is abelian.

_____ g. If G/H is abelian, then the commutator subgroup C of G contains H.

_____ h. The commutator subgroup of a simple group G must be G itself.

_____ i. The commutator subgroup of a nonabelian simple group G must be G itself.

_____ j. All nontrivial finite simple groups have prime order.

In Exercises 16 through 20, let F be the additive group of all functions mapping $\mathbb{R}$ into $\mathbb{R}$, and let F^* be the multiplicative group of all elements of F that do not assume the value 0 at any point of $\mathbb{R}$.

16. Let K be the subgroup of F consisting of the constant functions. Find a subgroup of F to which F/K is isomorphic.

17. Let K^* be the subgroup of F^* consisting of the nonzero constant functions. Find a subgroup of F^* to which F^*/K^* is isomorphic.

18. Let K be the subgroup of continuous functions in F. Can you find an element of F/K having order 2? Why?

19. Let K^* be the subgroup of F^* consisting of the continuous functions in F^*. Can you find an element of F^*/K^* having order 2? Why?

20. Let K^* be as in Exercise 19. Describe all $n \in \mathbb{Z}^+$ such that F^*/K^* has an element of order n.

In Exercises 21 through 23, let U be the multiplicative group $\{z \in \mathbb{C} \mid |z| = 1\}$.

21. Let $z_0 \in U$. Show that $z_0 U = \{z_0 z \mid z \in U\}$ is a subgroup of U, and compute $U/z_0 U$.

22. To what group we have mentioned in the text is $U/\langle -1 \rangle$ isomorphic?

23. Let $z_n = \cos(2\pi/n) + i \sin(2\pi/n)$ where $n \in \mathbb{Z}^+$. To what group we have mentioned is $U/\langle z_n \rangle$ isomorphic?

24. To what group mentioned in the text is the additive group $\mathbb{R}/\mathbb{Z}$ isomorphic?

25. Give an example of a group G having no elements of finite order >1 but having a factor group G/H, all of whose elements are of finite order.

26. Let H and K be subgroups of a group G. Give an example showing that we may have $H \simeq K$ while G/H is not isomorphic to G/K.

27. Describe the center of every simple

a. abelian group   whole group

b. nonabelian group. $\{e\}$ the center is a normal subgroup.

28. Describe the commutator subgroup of every simple

a. abelian group

b. nonabelian group.

Theory

29. Show that if a finite group G contains a proper subgroup of index 2 in G, then G is not simple.

30. Let $\phi: G \to G'$ be a group homomorphism, and let N be a normal subgroup of G. Show that $\phi[N]$ is a normal subgroup of $\phi[G]$.

31. Let $\phi: G \to G'$ be a group homomorphism, and let N' be a normal subgroup of G'. Show that $\phi^{-1}[N']$ is a normal subgroup of $\phi[G]$.

32. Show that if G is nonabelian, then the factor group $G/Z(G)$ is not cyclic. [*Hint:* Show the equivalent contrapositive, namely, that if $G/Z(G)$ is cyclic then G is abelian (and hence $Z(G) = G$).]

33. Using Exercise 32, show that a nonabelian group G of order pq where p and q are primes has a trivial center.

34. Prove that A_n is simple for $n \geq 5$, following the steps and hints given.

    a. Show A_n contains every 3-cycle if $n \geq 3$.

    b. Show A_n is generated by the 3-cycles for $n \geq 3$. [*Hint:* Note that $(a, b)(c, d) = (a, c, b)(a, c, d)$ and $(a, c)(a, b) = (a, b, c)$.]

    c. Let r and s be fixed elements of $\{1, 2, \ldots, n\}$ for $n \geq 3$. Show that A_n is generated by the n "special" 3-cycles of the form (r, s, i) for $1 \leq i \leq n$. [*Hint:* Show every 3-cycle is the product of "special" 3-cycles by computing

$$(r, s, i)^2, \qquad (r, s, j)(r, s, i)^2, \qquad (r, s, j)^2(r, s, i),$$

    and

$$(r, s, i)^2(r, s, k)(r, s, j)^2(r, s, i).$$

    Observe that these products give all possible types of 3-cycles.]

    d. Let N be a normal subgroup of A_n for $n \geq 3$. Show that if N contains a 3-cycle, then $N = A_n$. [*Hint:* Show that $(r, s, i) \in N$ implies that $(r, s, j) \in N$ for $j = 1, 2, \ldots, n$ by computing

$$((r, s)(i, j))(r, s, i)^2((r, s)(i, j))^{-1}.]$$

    e. Let N be a nontrivial normal subgroup of A_n for $n \geq 5$. Show that one of the following cases must hold, and conclude in each case that $N = A_n$.

Case 1 N contains a 3-cycle.

Case 2 N contains a product of disjoint cycles, at least one of which has length greater than 3. [*Hint:* Suppose N contains the disjoint product $\sigma = \mu(a_1, a_2, \ldots, a_r)$. Show $\sigma^{-1}(a_1, a_2, a_3)\sigma(a_1, a_2, a_3)^{-1}$ is in N, and compute it.]

Case 3 N contains a disjoint product of the form $\sigma = \mu(a_4, a_5, a_6)(a_1, a_2, a_3)$. [*Hint:* Show $\sigma^{-1}(a_1, a_2, a_4)\sigma(a_1, a_2, a_4)^{-1}$ is in N, and compute it.]

Case 4 N contains a disjoint product of the form $\sigma = \mu(a_1, a_2, a_3)$ where μ is a product of disjoint 2-cycles. [*Hint:* Show $\sigma^2 \in N$ and compute it.]

Case 5 N contains a disjoint product σ of the form $\sigma = \mu(a_3, a_4)(a_1, a_2)$, where μ is a product of an even number of disjoint 2-cycles. [*Hint:* Show that $\sigma^{-1}(a_1, a_2, a_3)\sigma(a_1, a_2, a_3)^{-1}$ is in N, and compute it to deduce that $\alpha = (a_2, a_4)(a_1, a_3)$ is in N. Using $n \geq 5$ for the first time, find $i \in \{1, 2, \ldots, n\}$, where $i \neq a_1, a_2, a_3, a_4$. Let $\beta = (a_1, a_3, i)$. Show that $\beta^{-1}\alpha\beta\alpha \in N$, and compute it.]

35. Let N be a normal subgroup of G and let H be any subgroup of G. Let

$HN = \{hn \mid h \in H, n \in N\}$. Show that HN is a subgroup of G, and is the smallest subgroup containing both N and H.

36. With reference to the preceding exercise, let M also be a normal subgroup of G. Show that NM is again a normal subgroup of G.

37. Show that if H and K are normal subgroups of a group G such that $H \cap K = \{e\}$, then $hk = kh$ for all $h \in H$ and $k \in K$. [*Hint:* Consider the commutator $hkh^{-1}k^{-1} = (hkh^{-1})k^{-1} = h(kh^{-1}k^{-1})$.]

3.5

Series of Groups[†]
Subnormal and Normal Series

This section is concerned with the notion of a *series* of a group G, which gives insight into the structure of G. The results, presented here without proof, hold for both abelian and nonabelian groups. They are not too important for finitely generated abelian groups because of our strong Theorem 2.11. Many of our illustrations will be taken from abelian groups, however, for ease of computation. The results are proved in the following chapter.

> **DEFINITION 3.11 (Subnormal and Normal Series)** A **subnormal** (or **subinvariant) series of a group** G is a finite sequence $H_0, H_1, \ldots, H_n$ of subgroups of G such that $H_i < H_{i+1}$ and H_i is a normal subgroup of H_{i+1} with $H_0 = \{e\}$ and $H_n = G$. A **normal** (or **invariant) series of** G is a finite sequence $H_0, H_1, \ldots, H_n$ of normal subgroups of G such that $H_i < H_{i+1}$, $H_0 = \{e\}$, and $H_n = G$.

Note that for abelian groups the notions of subnormal and normal series coincide, since every subgroup is normal. A normal series is always subnormal, but the converse need not be true. We defined a subnormal series before a normal series, since the concept of a subnormal series is more important for our work.

EXAMPLE 1 Two examples of normal series of $\mathbb{Z}$ under addition are

$$\{0\} < 8\mathbb{Z} < 4\mathbb{Z} < \mathbb{Z}$$

and

$$\{0\} < 9\mathbb{Z} < \mathbb{Z}. \; \blacktriangle$$

EXAMPLE 2 Consider the group D_4 of symmetries of the square in

---

[†] The only reason that this section appears here rather than in (optional) Chapter 4 is to introduce the idea of a *solvable group*, which we need in Section 9.9 to show the insolvability by radicals of quintic equations.

Example 5 in Section 2.1. The series

$$\{\rho_0\} < \{\rho_0, \mu_1\} < \{\rho_0, \rho_2, \mu_1, \mu_2\} < D_4$$

is a subnormal series, as we could check using Table 2.2 in Section 2.1. It is not a normal series since $\{\rho_0, \mu_1\}$ is not normal in D_4. ▲

DEFINITION 3.12 (Refinement)   A subnormal (normal) series $\{K_j\}$ is a **refinement of a subnormal (normal) series** $\{H_i\}$ of a group G if $\{H_i\} \subseteq \{K_j\}$, that is, if each H_i is one of the K_j.

EXAMPLE 3   The series

$$\{0\} < 72\mathbb{Z} < 24\mathbb{Z} < 8\mathbb{Z} < 4\mathbb{Z} < \mathbb{Z}$$

is a refinement of the series

$$\{0\} < 72\mathbb{Z} < 8\mathbb{Z} < \mathbb{Z}.$$

Two new terms, $4\mathbb{Z}$ and $24\mathbb{Z}$, have been inserted. ▲

Of interest in studying the structure of G are the factor groups H_{i+1}/H_i. These are defined for both normal and subnormal series, since H_i is normal in H_{i+1} in either case.

DEFINITION 3.13 (Isomorphic Series)   Two subnormal (normal) series $\{H_i\}$ and $\{K_j\}$ of the same group G are **isomorphic** if there is a one-to-one correspondence between the collections of factor groups $\{H_{i+1}/H_i\}$ and $\{K_{j+1}/K_j\}$ such that corresponding factor groups are isomorphic.

Clearly, two isomorphic subnormal (normal) series must have the same number of groups.

EXAMPLE 4   The two series of $\mathbb{Z}_{15}$,

$$\{0\} < \langle 5 \rangle < \mathbb{Z}_{15}$$

and

$$\{0\} < \langle 3 \rangle < \mathbb{Z}_{15},$$

are isomorphic. Both $\mathbb{Z}_{15}/\langle 5 \rangle$ and $\langle 3 \rangle/\{0\}$ are isomorphic to $\mathbb{Z}_5$, and $\mathbb{Z}_{15}/\langle 3 \rangle$ is isomorphic to $\langle 5 \rangle/\{0\}$, or to $\mathbb{Z}_3$. ▲

The Jordan–Hölder Theorem

The following theorem is fundamental to the theory of series.

THEOREM 3.16 (Schreier)   Two subnormal (normal) series of a group G have isomorphic refinements.

PROOF   See Section 4.1 for proof of this theorem. ◆

The proof of Theorem 3.16 is really not too difficult. However, we know from experience that some students get lost in the proof and then tend to feel that they can't understand the theorem. We don't prove it in this chapter even though many students could follow it. Let us, however, illustrate this theorem.

EXAMPLE 5 Let us try to find isomorphic refinements of the series

$$\{0\} < 8\mathbb{Z} < 4\mathbb{Z} < \mathbb{Z}$$

and

$$\{0\} < 9\mathbb{Z} < \mathbb{Z}$$

given in Example 1. Consider the refinement

$$\{0\} < 72\mathbb{Z} < 8\mathbb{Z} < 4\mathbb{Z} < \mathbb{Z}$$

of $\{0\} < 8\mathbb{Z} < 4\mathbb{Z} < \mathbb{Z}$ and the refinement

$$\{0\} < 72\mathbb{Z} < 18\mathbb{Z} < 9\mathbb{Z} < \mathbb{Z}$$

of $\{0\} < 9\mathbb{Z} < \mathbb{Z}$. In both cases the refinements have four factor groups isomorphic to $\mathbb{Z}_4, \mathbb{Z}_2, \mathbb{Z}_9$, and $72\mathbb{Z}$ or $\mathbb{Z}$. The *order* in which the factor groups occur is different to be sure. ▲

We now come to the real meat of the theory.

DEFINITION 3.14 (Composition Series) A subnormal series $\{H_i\}$ of a group G is a **composition series** if all the factor groups H_{i+1}/H_i are simple. A normal series $\{H_i\}$ of G is a **principal** or **chief series** if all the factor groups H_{i+1}/H_i are simple.

Note that for abelian groups the concepts of composition and principal series coincide. Also, since every normal series is subnormal, every principal series is a composition series for any group, abelian or not.

EXAMPLE 6 We claim that $\mathbb{Z}$ has no composition (and also no principal) series. For if

$$\{0\} = H_0 < H_1 < \cdots < H_{n-1} < H_n = \mathbb{Z}$$

is a subnormal series, H_1 must be of the form $r\mathbb{Z}$ for some $r \in \mathbb{Z}^+$. But then H_1/H_0 is isomorphic to $r\mathbb{Z}$, which is infinite cyclic with many nontrivial proper normal subgroups, for example $2r\mathbb{Z}$. Thus $\mathbb{Z}$ has no composition (and also no principal) series. ▲

EXAMPLE 7 The series

$$\{e\} < A_n < S_n$$

for $n \geq 5$ is a composition series (and also a principal series) of S_n, since

$A_n/\{e\}$ is isomorphic to A_n, which is simple for $n \geq 5$, and S_n/A_n is isomorphic to $\mathbb{Z}_2$, which is simple. Likewise the two series given in Example 4 are composition series (and also principal series) of $\mathbb{Z}_{15}$. They are isomorphic, as shown in that example. This illustrates our main theorem, which will be stated shortly. ▲

Observe that by Theorem 3.14 H_{i+1}/H_i is simple if and only if H_i is a maximal normal subgroup of H_{i+1}. Thus for a composition series, each H_i must be maximal normal subgroup of H_{i+1}. *To form a composition series of a group G, we just hunt for a maximal normal subgroup H_{n-1} of G, then for a maximal normal subgroup H_{n-2} of H_{n-1}, and so on, If this process terminates in a finite number of steps, we have a composition series.* Note that by Theorem 3.14, a composition series cannot have any further refinement. *To form a principal series, we have to hunt for a maximal normal subgroup H_{n-1} of G, then for a maximal normal subgroup H_{n-2} of H_{n-1} that is also normal in G, and so on.* The main theorem is as follows.

THEOREM 3.17 (Jordan–Hölder Theorem) Any two composition (principal) series of a group G are isomorphic.

PROOF   Let $\{H_i\}$ and $\{K_j\}$ be two composition (principal) series of G. By Theorem 3.16, they have isomorphic refinements. But since all factor groups are already simple, Theorem 3.14 shows that neither series has any further refinement. Thus $\{H_i\}$ and $\{K_j\}$ must already be isomorphic. ◆

◆   H I S T O R I C A L   N O T E   ◆

This first appearance of what became the Jordan–Hölder theorem occurred in 1869 in a commentary on the work of Galois by the brilliant French algebraist Camille Jordan (1838–1922). The context of its appearance is the study of permutation groups associated with the roots of polynomial equations. Jordan asserted that even though the sequence of normal subgroups $G, I, J, \ldots$ of the group of the equation is not necessarily unique, nevertheless the sequence of indices of this composition series is unique. Jordan gave a proof in his monumental 1870 *Treatise on Substitutions and Algebraic Equations*. This latter work, through restricted to what we now call permutation groups, remained the standard treatise on group theory for many years.

The Hölder part of the theorem, that the sequence of factor groups in a composition series is unique up to order, was due to Otto Hölder (1859–1937), who played a very important role in the development of group theory once the completely abstract definition of a group had been given. Among his other contributions, he gave the first abstract definition of a "factor group" and determined the structure of all finite groups of square-free order.

For a finite group, we should regard a composition series as a type of factorization of the group into simple factor groups, analogous to the factorization of a positive integer into primes. In both cases, the factorization is unique, up to the order of the factors.

THEOREM 3.18 If G has a composition (principal) series, and if N is a proper normal subgroup of G, then there exists a composition (principal) series containing N.

PROOF The series

$$\{e\} < N < G$$

is both a subnormal and a normal series. Since G has a composition series $\{H_i\}$, then by Theorem 3.16, there is a refinement of $\{e\} < N < G$ to a subnormal series isomorphic to a refinement of $\{H_i\}$. But as a composition series, $\{H_i\}$ can have no further refinement. Thus $\{e\} < N < G$ can be refined to a subnormal series all of whose factor groups are simple, that is, to a composition series. A similar argument holds if we start with a principal series $\{K_j\}$ of G. ◆

EXAMPLE 8 A composition (and also a principal) series of $\mathbb{Z}_4 \times \mathbb{Z}_9$ containing $\langle(0, 1)\rangle$ is

$$\{(0, 0)\} < \langle(0, 3)\rangle < \langle(0, 1)\rangle < \langle 2\rangle \times \langle 1\rangle < \langle 1\rangle \times \langle 1\rangle = \mathbb{Z}_4 \times \mathbb{Z}_9. \quad ▲$$

The next definition is basic to the last chapter of this text, which deals with solutions of polynomial equations in terms of radicals.

DEFINITION 3.15 (Solvable Group) A group G is **solvable** if it has a composition series $\{H_i\}$ such that all factor groups H_{i+1}/H_i are abelian.

By the Jordan–Hölder theorem, we see that for a solvable group, *every* composition series $\{H_i\}$ must have abelian factor groups H_{i+1}/H_i.

EXAMPLE 9 The group S_3 is solvable, because the composition series

$$\{e\} < A_3 < S_3$$

has factor groups isomorphic to $\mathbb{Z}_3$ and $\mathbb{Z}_2$, which are abelian. The group S_5 is not solvable, for since A_5 is simple, the series

$$\{e\} < A_5 < S_5$$

is a composition series, and $A_5/\{e\}$, which is isomorphic to A_5, is not abelian. *This group A_5 of order 60 can be shown to be the smallest group that is not solvable.* This fact is closely connected with the fact that a polynomial equation of degree 5 is not in general solvable by radicals, but a polynomial equation of degree ≤ 4 is. ▲

The Ascending Central Series

We mention one subnormal series for a group G that can be formed using centers of groups. Recall from the preceding section that the center $Z(G)$ of a group G is defined by

$$Z(G) = \{z \in G \mid zg = gz \text{ for all } g \in G\},$$

and that $Z(G)$ is a normal subgroup of G. If we have the table for a finite group G, it is easy to find the center. An element a will be in the center of G if and only if the elements in the row opposite a at the extreme left are given in the same order as the elements in the column under a at the very top of the table.

Now let G be a group, and let $Z(G)$ be the center of G. Since $Z(G)$ is normal in G, we can form the factor group $G/Z(G)$ and find the center $Z(G/Z(G))$ of this factor group. Since $Z(G/Z(G))$ is normal in $G/Z(G)$, if $\gamma : G \rightarrow G/Z(G)$ is the canonical map, then by Theorem 3.13, $\gamma^{-1}[Z(G/Z(G))]$ is a normal subgroup $Z_1(G)$ of G. We can then form the factor group $G/Z_1(G)$ and find its center, take $(\gamma_1)^{-1}$ of it to get $Z_2(G)$, and so on.

DEFINITION 3.16 (Ascending Central Series)   The series

$$\{e\} \leq Z(G) \leq Z_1(G) \leq Z_2(G) \leq \cdots$$

described in the preceding discussion is the **ascending central series of the group** G.

EXAMPLE 10   The center of S_3 is just the identity $\{\rho_0\}$. Thus the ascending central series of S_3 is

$$\{\rho_0\} \leq \{\rho_0\} \leq \{\rho_0\} \leq \cdots.$$

The center of the group D_4 of symmetries of the square in Example 5 in Section 1.4 is $\{\rho_0, \rho_2\}$. (Do you remember that we said that this group would give us nice examples of many things we discussed?) Since $D_4/\{\rho_0, \rho_2\}$ is of order 4 and hence abelian, its center is all of $D_4/\{\rho_0, \rho_2\}$. Thus the ascending central series of D_4 is

$$\{\rho_0\} \leq \{\rho_0, \rho_2\} \leq D_4 \leq D_4 \leq D_4 \leq \cdots. \quad \blacktriangle$$

Exercises 3.5

Computations

In Exercises 1 through 5, give isomorphic refinements of the two series.

1. $\{0\} < 10\mathbb{Z} < \mathbb{Z}$ and $\{0\} < 25\mathbb{Z} < \mathbb{Z}$

2. $\{0\} < 60\mathbb{Z} < 20\mathbb{Z} < \mathbb{Z}$ and $\{0\} < 245\mathbb{Z} < 49\mathbb{Z} < \mathbb{Z}$

3. $\{0\} < \langle 3 \rangle < \mathbb{Z}_{24}$ and $\{0\} < \langle 8 \rangle < \mathbb{Z}_{24}$

4. $\{0\} < \langle 18 \rangle < \langle 3 \rangle < \mathbb{Z}_{72}$ and $\{0\} < \langle 24 \rangle < \langle 12 \rangle < \mathbb{Z}_{72}$

5. $\{(0,0)\} < (60\mathbb{Z}) \times \mathbb{Z} < (10\mathbb{Z}) \times \mathbb{Z} < \mathbb{Z} \times \mathbb{Z}$
   and $\{(0,0)\} < \mathbb{Z} \times (80\mathbb{Z}) < \mathbb{Z} \times (20\mathbb{Z}) < \mathbb{Z} \times \mathbb{Z}$

6. Find all composition series of $\mathbb{Z}_{60}$ and show that they are isomorphic.

7. Find all composition series of $\mathbb{Z}_{48}$ and show that they are isomorphic.

8. Find all composition series of $\mathbb{Z}_5 \times \mathbb{Z}_5$.

9. Find all composition series of $S_3 \times \mathbb{Z}_2$.

10. Find all composition series of $\mathbb{Z}_2 \times \mathbb{Z}_5 \times \mathbb{Z}_7$.

11. Find the center of $S_3 \times \mathbb{Z}_4$.

12. Find the center of $S_3 \times D_4$.

13. Find the ascending central series of $S_3 \times \mathbb{Z}_4$.

14. Find the ascending central series of $S_3 \times D_4$.

Concepts

15. Mark each of the following true or false.
    _____ a. Every normal series is also subnormal.
    _____ b. Every subnormal series is also normal.
    _____ c. Every principal series is a composition series.
    _____ d. Every composition series is a principal series.
    _____ e. Every abelian group has exactly one composition series.
    _____ f. Every finite group has a composition series.
    _____ g. A group is solvable if and only if it has a composition series with simple factor groups.
    _____ h. S_7 is a solvable group.
    _____ i. The Jordan–Hölder theorem has some similarity with the Fundamental Theorem of Arithmetic, which states that every positive integer greater than 1 can be factored into a product of primes uniquely up to order.
    _____ j. Every finite group of prime order is solvable.

16. Find a composition series of $S_3 \times S_3$. Is $S_3 \times S_3$ solvable?

17. Is the group D_4 of symmetries of the square in Example 5 of Section 1.4 solvable?

Theory

18. Show that if
$$H_0 = \{e\} < H_1 < H_2 < \cdots < H_n = G$$
is a subnormal (normal) series for a group G, and if H_{i+1}/H_i is of finite order s_{i+1}, then G is of finite order $s_1 s_2 \cdots s_n$.

19. Show that an infinite abelian group can have no composition series. [*Hint:* Use

Exercise 18, together with the fact that an infinite abelian group always has a proper normal subgroup.]

20. Show that a finite direct product of solvable groups is solvable.

3.6

Group Action on a Set†

We have seen examples of how groups may *act on things*, like the group of symmetries of a triangle or of a square, the group of rotations of a cube, the general linear group acting on $\mathbb{R}^n$, and so on. In this section, we give the general notion of group action on a set. The next section will give an application to counting.

The Notion of a Group Action

We are now familiar with functions and Cartesian products, so we can take a more sophisticated view of a binary operation on a set S than we did in Section 1.1. A **binary operation on** S is a function mapping $S \times S$ into S. If we denote the function by $*$, then $*(s_1, s_2) = s_3$ is more conventionally expressed as $s_1 * s_2 = s_3$. The function $*$ gives us a rule for "multiplying" any element of S by an element of S to yield again an element of S.

More generally, for any sets A, B, and C, we can view a map $*: A \times B \to C$ as defining a "multiplication," where any element a of A times any element b of B has as value some element c of C. Of course, we write $a * b = c$, or simply $ab = c$. In this chapter, we will be concerned with the case where X is a set, G is a group, and we have a map $*: G \times X \to X$. We shall write $*(g, x)$ as $g * x$ or gx.

DEFINITION 3.17 (Group Action)   Let X be a set and G a group. An **action of** G **on** X is a map $*: G \times X \to X$ such that

1. $ex = x$ for all $x \in X$,
2. $(g_1 g_2)(x) = g_1(g_2 x)$ for all $x \in X$ and all $g_1, g_2 \in G$.

Under these conditions, X is a G-**set**.

EXAMPLE I   Let X be any set, and let H be a subgroup of the group S_X of all permutations of X. Then X is an H-set, where the action of $\sigma \in H$ on X is its action as an element of S_X, so that $\sigma x = \sigma(x)$ for all $x \in X$. Condition 2 is a consequence of the definition of permutation multiplication as function composition, and Condition 1 is immediate from the definition of the identity permutation as the identity function. Note that, in particular, $\{1, 2, 3, \ldots, n\}$ is an S_n set. ▲

---

† This section is used only in the following one and in Section 4.2.

Our next theorem will show that for every G-set X and each $g \in G$, the map $\sigma_g : X \to X$ defined by $\sigma_g(x) = gx$ is a permutation of X, and that there is a homomorphism $\phi : G \to S_X$ such that the action of G on X is essentially the Example 1 action of the image subgroup $H = \phi[G]$ of S_X on X. So actions of subgroups of S_X on X describe all possible group actions on X. When studying the set X, actions using subgroups of S_X suffice. However, sometimes a set X is used to study G via a group action of G on X; this occurs in Section 4.2. Thus we need the more general concept given by Definition 3.17.

THEOREM 3.19   Let X be a G-set. For each $g \in G$, the function $\sigma_g : X \to X$ defined by $\sigma_g(x) = gx$ for $x \in X$ is a permutation of X. Furthermore, the map $\phi : G \to S_X$ defined by $\phi(g) = \sigma_g$ is a homomorphism with the property that $\phi(g)(x) = gx$.

PROOF   To show that σ_g is a permutation of X, we must show that it is a one-to-one map of X onto itself. suppose that $\sigma_g(x_1) = \sigma_g(x_2)$ for $x_1, x_2 \in X$. Then $gx_1 = gx_2$. Consequently, $g^{-1}(gx_1) = g^{-1}(gx_2)$. Using Condition 2 in Definition 3.17, we see that $(g^{-1}g)x_1 = (g^{-1}g)x_2$, so $ex_1 = ex_2$. Condition 1 of the definition then yields $x_1 = x_2$, so σ_g is one to one. The two conditions of the definition show that for $x \in X$, we have $\sigma_g(g^{-1}x) = g(g^{-1})x = (gg^{-1})x = ex = x$, so σ_g maps X onto X. Thus σ_g is indeed a permutation.

To show that $\phi : G \to S_X$ defined by $\phi(g) = \sigma_g$ is a homomorphism, we must show that $\phi(g_1 g_2) = \phi(g_1)\phi(g_2)$ for all $g_1, g_2 \in G$. We show that equality of these two permutations in S_X by showing they both carry an $x \in X$ into the same element. Using the two conditions in Definition 3.17 and the rule for function composition, we obtain

$$\phi(g_1 g_2)(x) = \sigma_{g_1 g_2}(x) = (g_1 g_2)x = g_1(g_2 x) = g_1 \sigma_{g_2}(x) = \sigma_{g_1}(\sigma_{g_2}(x))$$
$$= (\sigma_{g_1} \circ \sigma_{g_2})(x) = (\sigma_{g_1}\sigma_{g_2})(x) = (\phi(g_1)\phi(g_2))(x).$$

Thus ϕ is a homomorphism. The stated property of ϕ follows at once since by our definitions, we have $\phi(g)(x) = \sigma_g(x) = gx$.   ◆

It follows from the preceding theorem and Theorem 3.9 that if X is a G-set, then the subset of G leaving every element of X fixed is a normal subgroup N of G, and we can regard X as a G/N-set where the action of a coset gN on X is given by $(gN)x = gx$ for each $x \in X$. If $N = \{e\}$, then the identity of G is the only element that leaves every $x \in X$ fixed; we then say that G **acts faithfully** on X. A group G is **transitive** on a G-set X if for each $x_1, x_2 \in X$, there exists $g \in G$ such that $gx_1 = x_2$. Note that G is transitive on X if and only if the subgroup $\phi[G]$ of S_X is transitive on X, as defined in Exercise 45 of Section 2.1.

We continue with more examples of G-sets.

EXAMPLE 2   Every group G is itself a G-set, where the action on $g_2 \in G$ by $g_1 \in G$ is given by left multiplication. That is, $*(g_1, g_2) = g_1 g_2$. If H is a subgroup of G, we can also regard G as an H-set, where $*(h, g) = hg$. ▲

EXAMPLE 3   Let H be a subgroup of G. Then G is an H-set under conjugation where $*(h, g) = hgh^{-1}$ for $g \in G$ and $h \in H$. Condition 1 is obvious, and for Condition 2 note that

$$*(h_1 h_2, g) = (h_1 h_2)g(h_1 h_2)^{-1} = h_1(h_2 g h_2^{-1})h_1^{-1} = *(h_1, *(h_2, g)).$$

We always write this action of H on G by conjugation as hgh^{-1}. The abbreviation hg described before the definition would cause terrible confusion with the group operation of G. We will see in Section 4.2 that this action of H on G is very important in studying the structure of the group G. ▲

EXAMPLE 4   For students who have studied vector spaces with real (or complex) scalars, we mention that the axioms $(rs)\mathbf{v} = r(s\mathbf{v})$ and $1\mathbf{v} = \mathbf{v}$ for scalars r and s and a vector $\mathbf{v}$ show that the set of vectors is an $\mathbb{R}^*$-set (or a $\mathbb{C}^*$-set) for the multiplicative group of nonzero scalars. ▲

EXAMPLE 5   Let H be a subgroup of G, and let L_H be the set of all left cosets of H. Then L_H is a G-set, where the action of $g \in G$ on the left coset xH is given by $g(xH) = (gx)H$. A series of exercises shows that every G-set is isomorphic to one that may be formed using these left coset G-sets as building blocks. (See Exercises 12 through 15.) ▲

EXAMPLE 6   Let G be the group $D_4 = \{\rho_0, \rho_1, \rho_2, \rho_3, \mu_1, \mu_2, \delta_1, \delta_2\}$ of symmetries of the square, described in Example 5 of Section 2.1. In Fig. 3.9

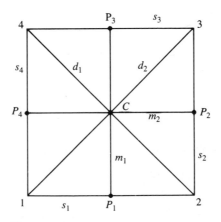

Figure 3.9

	1	2	3	4	s_1	s_2	s_3	s_4	m_1	m_2	d_1	d_2	C	P_1	P_2	P_3	P_4
ρ_0	1	2	3	4	s_1	s_2	s_3	s_4	m_1	m_2	d_1	d_2	C	P_1	P_2	P_3	P_4
ρ_1	2	3	4	1	s_2	s_3	s_4	s_1	m_2	m_1	d_2	d_1	C	P_2	P_3	P_4	P_1
ρ_2	3	4	1	2	s_3	s_4	s_1	s_2	m_1	m_2	d_1	d_2	C	P_3	P_4	P_1	P_2
ρ_3	4	1	2	3	s_4	s_1	s_2	s_3	m_2	m_1	d_2	d_1	C	P_4	P_1	P_2	P_3
μ_1	2	1	4	3	s_1	s_4	s_3	s_2	m_1	m_2	d_2	d_1	C	P_1	P_4	P_3	P_2
μ_2	4	3	2	1	s_3	s_2	s_1	s_4	m_1	m_2	d_2	d_1	C	P_3	P_2	P_1	P_4
δ_1	3	2	1	4	s_2	s_1	s_4	s_3	m_2	m_1	d_1	d_2	C	P_2	P_1	P_4	P_3
δ_2	1	4	3	2	s_4	s_3	s_2	s_1	m_2	m_1	d_1	d_2	C	P_4	P_3	P_2	P_1

Table 3.4

we show the square with vertices 1, 2, 3, 4 as in Fig. 2.6. We also label the sides s_1, s_2, s_3, s_4, the diagonals d_1 and d_2, vertical and horizontal axes m_1 and m_2, the center point C, and midpoints P_i of the sides s_i. Recall that ρ_i corresponds to rotating the square counterclockwise through $\pi/2$ radians, μ_i corresponds to flipping on the axis m_i, and δ_i to flipping on the diagonal d_i. We let

$$X = \{1, 2, 3, 4, s_1, s_2, s_3, s_4, m_1, m_2, d_1, d_2, C, P_1, P_2, P_3, P_4\}.$$

Then X can be regarded as a D_4-set in a natural way. Table 3.4 describes completely the action of D_4 on X and is given to provide geometric illustrations of ideas to be introduced. We should be sure that we understand how this table is formed before continuing. ▲

Isotropy Subgroups

Let X be a G-set. Let $x \in X$ and $g \in G$. It will be important to know when $gx = x$. We let

$$X_g = \{x \in X \mid gx = x\} \qquad \text{and} \qquad G_x = \{g \in G \mid gx = x\}.$$

EXAMPLE 7   For the D_4-set X in Example 6, we have

$$X_{\rho_0} = X, \qquad X_{\rho_1} = \{C\}, \qquad X_{\mu_1} = \{s_1, s_3, m_1, m_2, C, P_1, P_3\}$$

Also, with $G = D_4$,

$$G_1 = \{\rho_0, \delta_2\}, \qquad G_{s_3} = \{\rho_0, \mu_1\}, \qquad G_{d_1} = \{\rho_0, \rho_2, \delta_1, \delta_2\}.$$

We leave the computation of the other X_σ and G_x to Exercises 1 and 2. ▲

Note that the subsets G_x given in the preceding example were, in each case, subgroups of G. This is true in general.

THEOREM 3.20   Let X be a G-set. Then G_x is a subgroup of G for each $x \in X$.

PROOF Let $x \in X$ and let $g_1, g_2 \in G_x$. Then $g_1 x = x$ and $g_2 x = x$. Consequently, $(g_1 g_2) x = g_1 (g_2 x) = g_1 x = x$, so $g_1 g_2 \in G_x$, and G_x is closed under the induced operation of G. Of course $ex = x$, so $e \in G_x$. If $g \in G_x$, then $gx = x$, so $x = ex = (g^{-1}g)x = g^{-1}(gx) = g^{-1}x$, and consequently $g^{-1} \in G_x$. Thus G_x is a subgroup of G. ◆

DEFINITION 3.18 (Isotropy Subgroup) Let X be a G-set and let $x \in X$. The subgroup G_x is the **isotropy subgroup of** x.

Orbits

For the D_4-set X of Example 6 with action table in Table 3.4, the elements in the subset $\{1, 2, 3, 4\}$ are carried into elements of this same subset under action by D_4. Furthermore, each of the elements 1, 2, 3, and 4 is carried into all the other elements of the subset by the various elements of D_4. We proceed to show that every G-set X can be partitioned into subsets of this type.

THEOREM 3.21 Let X be a G-set. For $x_1, x_2 \in X$, let $x_1 \sim x_2$ if and only if there exists $g \in G$ such that $gx_1 = x_2$. Then $\sim$ is an equivalence relation on S.

PROOF For each $x \in X$, we have $ex = x$, so $x \sim x$ and $\sim$ is reflexive.

Suppose $x_1 \sim x_2$, so $gx_1 = x_2$ for some $g \in G$. Then $g^{-1}x_2 = g^{-1}(gx_1) = (g^{-1}g)x_1 = ex_1 = x_1$, so $x_2 \sim x_1$, and $\sim$ is symmetric.

Finally, if $x_1 \sim x_2$ and $x_2 \sim x_3$, then $g_1 x_1 = x_2$ and $g_2 x_2 = x_3$ for some $g_1, g_2 \in G$. Then $(g_2 g_1)x_1 = g_2(g_1 x_1) = g_2 x_2 = x_3$, so $x_1 \sim x_3$ and $\sim$ is transitive. ◆

DEFINITION 3.19 (Orbit) Let X be a G-set. Each cell in the partition of the equivalence relation described in Theorem 3.21 is an **orbit in** X **under** G. If $x \in X$, the cell containing x is the **orbit of** x. We let this cell be Gx.

The relationship between the orbits in X and the group structure of G lies at the heart of the applications that appear in the next section and Section 4.2. The following theorem gives this relationship. Recall that for a set X, we use $|X|$ for the number of elements in X, and $(G:H)$ is the index of a subgroup H in a group G.

THEOREM 3.22 Let X be a G-set and let $x \in X$. Then $|Gx| = (G:G_x)$.

PROOF We define a one-to-one map ψ from Gx onto the collection of left cosets of G_x in G. Let $x_1 \in Gx$. Then there exists $g_1 \in G$ such that $g_1 x = x_1$. We define $\psi(x_1)$ to be the left coset $g_1 G_x$ of G_x. We must show that this map ψ is well defined, independent of the choice of $g_1 \in G$ such that $g_1 x = x_1$. Suppose also that $g_1'x = x_1$. Then $g_1 x = g_1'x$, so $g_1^{-1}(g_1 x) = g_1^{-1}(g_1'x)$, from which we deduce $x = (g_1^{-1}g_1')x$. Therefore

$g_1^{-1}g_1' \in G_x$, so $g_1' \in g_1 G_x$, and $g_1 G_x = g_1' G_x$. Thus the map ψ is well defined.

To show the map ψ is one to one, suppose $x_1, x_2 \in Gx$, and $\psi(x_1) = \psi(x_2)$. Then there exist $g_1, g_2 \in G$ such that $x_1 = g_1 x, x_2 = g_2 x$, and $g_2 \in g_1 G_x$. Then $g_2 = g_1 g$ for some $g \in G_x$, so $x_2 = g_2 x = g_1(gx) = g_1 x = x_1$. Thus ψ is one to one.

Finally, we show that each left coset of G_x in G is of the form $\psi(x_1)$ for some $x_1 \in Gx$, Let $g_1 G_x$ be a left coset. Then if $g_1 x = x_1$, we have $g_1 G_x = \psi(x_1)$. Thus ψ maps Gx onto the collection of right cosets. ◆

EXAMPLE 8 Let X be the D_4-set in Example 6, with action table given by Table 3.4. With $G = D_4$, we have $G1 = \{1, 2, 3, 4\}$ and $G_1 = \{\rho_0, \delta_2\}$. Since $|G| = 8$, we have $|G1| = (G:G_1) = 4$. ▲

We should remember not only the cardinality equation in Theorem 3.22 but also that the *elements of G carrying x into $g_1 x$ are precisely the elements of the left coset $g_1 G_x$*. Namely, if $g \in G_x$, then $(g_1 g)x = g_1(gx) = g_1 x$. On the other hand, if $g_2 x = g_1 x$, then $g_1^{-1}(g_2 x) = x$ so $(g_1^{-1} g_2)x = x$. Thus $g_1^{-1} g_2 \in G_x$ so $g_2 \in g_1 G_x$.

Exercises 3.6

Computations

In Exercises 1 through 3, let

$$X = \{1, 2, 3, 4, s_1, s_2, s_3, s_4, m_1, m_2, d_1, d_2, C, P_1, P_2, P_3, P_4\}$$

be the D_4-set of Example 6 with action table in Table 3.4. Find the following, where $G = D_4$.

1. The fixed sets X_σ for each $\sigma \in D_4$, that is, $X_{\rho_0}, X_{\rho_1}, \ldots, X_{\delta_2}$

2. The isotropy subgroups G_x for each $x \in X$, that is, $G_1, G_2, \ldots, G_{P_3}, G_{P_4}$

3. The orbits in X under D_4

Concepts

4. Let X be a G-set and let $S \subseteq X$. If $Gs \subseteq S$ for all $s \in S$, then S is a **sub-G-set**. Characterize a sub-G-set of a G-set X in terms of orbits in X under G.

5. Characterize a transitive G-set in terms of its orbits.

6. Mark each of the following true or false.

———   a. Every G-set is also a group.
———   b. Each element of a G-set is left fixed by the identity of G.
———   c. If every element of a G-set is left fixed by the same element g of G, then g must be the identity e.
———   d. Let X be a G-set with $x_1, x_2 \in X$ and $g \in G$. If $gx_1 = gx_2$, then $x_1 = x_2$.
———   e. Let X be a G-set with $x \in X$ and $g_1, g_2 \in G$. If $g_1 x = g_2 x$, then $g_1 = g_2$.
———   f. Each orbit of a G-set X is a transitive sub-G-set.
———   g. Let X be a G-set and let $H \leq G$. Then X can be regarded in a natural way as an H set.
———   h. With reference to (g), the orbits in X under H are the same as the orbits in X under G.
———   i. If X is a G-set, then each element of G acts as a permutation of X.
———   j. Let X be a G-set and let $x \in X$. If G is finite, then $|G| = |Gx| \cdot |G_x|$.

7. Let X and Y be G-sets with the *same* group G. An **isomorphism** between G-sets X and Y is a map $\phi : X \to Y$ that is one to one, onto Y, and satisfies $g\phi(x) = \phi(gx)$ for all $x \in X$ and $g \in G$. Two G-sets are **isomorphic** if such an isomorphism between them exists. Let X be the D_4-set of Example 6.

   a. Find two distinct orbits of X that are isomorphic sub-D_4-sets.

   b. Show that the orbits $\{1, 2, 3, 4\}$ and $\{s_1, s_2, s_3, s_4\}$ are not isomorphic sub-D_4-sets. [*Hint:* Find an element of G that acts in an essentially different fashion on the two orbits.]

   c. Are the orbits you gave for your answer to part (a) the only two different isomorphic sub-D_4-sets of X?

8. Let X be the D_4-set in Example 6.

   a. Does D_4 act faithfully on X?

   c. Find all orbits in X on which D_4 acts faithfully as a sub-D_4-set.

Theory

9. Let X be a G-set. Show that G acts faithfully on X if and only if no two distinct elements of G have the same action on each element of X.

10. Let X be a G-set and let $Y \subseteq X$. Let $G_Y = \{g \in G \mid gy = y \text{ for all } y \in Y\}$. Show G_Y is a subgroup of G, generalizing Theorem 3.20.

11. Let G be the additive group of real numbers. Let the action of $\theta \in G$ on the real plane $\mathbb{R}^2$ be given by rotating the plane counterclockwise about the origin through θ radians. Let P be a point other than the origin in the plane.

   a. Show $\mathbb{R}^2$ is a G-set.

   b. Describe geometrically the orbit containing P.

   c. Find the group G_P.

Exercises 12 through 15 shows how all possible G-sets, up to isomorphism, can be formed from a group G.

12. Let $\{X_i \mid i \in I\}$ be a disjoint collection of sets, so $X_i \cap X_j = \emptyset$ for $i \neq j$. Let each X_i be a G-set for the same group G.

a. Show that $\bigcup_{i \in I} X_i$ can be viewed in a natural way as a *G*-set, the **union** of the *G*-sets X_i.

b. Show that every *G*-set X is the union of its orbits.

13. Let X be a transitive *G*-set, and let $x_0 \in X$. Show that X is isomorphic to the *G*-set L of all cosets of G_{x_0}, described in Example 5. [*Hint:* For $x \in X$, suppose $x = gx_0$, and define $\phi: X \to L$ by $\phi(x) = gG_{x_0}$. Be sure to show ϕ is well defined!]

14. Let X_i for $i \in I$ be *G*-sets for the same group G, and suppose the sets X_i are not necessarily disjoint. Let $X_i' = \{(x, i) \mid x \in X_i\}$ for each $i \in I$. Then the sets X_i' are disjoint, and each can still be regarded as a *G*-set in an obvious way. (The elements of X_i have simply been tagged by i to distinguish them from the elements of X_j for $i \neq j$.) The *G*-set $\bigcup_{i \in I} X_i'$ is the **disjoint union** of the *G*-sets X_i. Using Exercises 12 and 13, show that every *G*-set is isomorphic to a disjoint union of left coset *G*-sets, as described in Example 5.

15. The preceding exercises show that every *G*-set is isomorphic to a disjoint union of left costs *G*-sets. The question then arises whether left coset *G*-sets of distinct subgroups H and K of G can themselves be isomorphic. Note that the map defined in the hint of Exercise 13 depends on the choice of x_0 as "base point." If x_0 is replaced by $g_0 x_0$ and if $G_{x_0} \neq G_{g_0 x_0}$, then the collections L_H of left cosets of $H = G_{x_0}$ and L_K of left cosets of $K = G_{g_0 x_0}$ form distinct *G*-sets that must be isomorphic, since both L_H and L_K are isomorphic to X.

a. Let X be a transitive *G*-set and let $x_0 \in X$ and $g_0 \in G$. If $H = G_{x_0}$, describe $K = G_{g_0 x_0}$ in terms of H and g_0.

b. Based on (a), conjecture conditions on subgroups H and K of G such that the left coset *G*-sets of H and K are isomorphic.

c. Prove your conjecture in (b).

16. Up to isomorphism, how many transitive $\mathbb{Z}_4$ sets X are there? (Use the preceding exercises.) Give an example of each isomorphism type, listing an action table of each as in Table 3.4. Take lowercase names *a*, *b*, *c*, and so on for the elements in the set X.

17. Repeat Exercise 16 for the group $\mathbb{Z}_6$.

18. Repeat Exercise 16 for the group S_3. List the elements of S_3 in the order ι, $(1\,2\,3)$, $(1\,3\,2)$, $(2\,3)$, $(1\,3)$, $(1\,2)$.

3.7

Applications of *G*-Sets to Counting[†]

This section presents an application of our work with *G*-sets to counting. Suppose, for example, we wish to count how many distinguishable ways the six faces of a cube can be marked with from one to six dots to form a die. The standard die is marked so that when placed on a table with the 1 on the

---

[†] This section is not used in the remainder of the text.

bottom and the 2 toward the front, the 6 is on top, the 3 on the left, the 4 on the right, and the 5 on the back. Of course, other ways of marking the cube to give a distinguishably different die are possible.

Let us distinguish between the faces of the cube for the moment and call them the bottom, top, left, right, front, and back. Then the bottom can have any one of six marks from one dot to six dots, the top any one of the five remaining marks, and so on. There are $6! = 720$ ways the cube faces can be marked in all. Some markings yield the same die as others, in the sense that one marking can be carried into another by a rotation of the marked cube. For example, if the standard die described above is rotated 90° counterclockwise as we look down on it, then 3 will be on the front face rather than 2, but it is the same die.

There are 24 possible positions of a cube on a table, for any one of six faces can be placed down, and then any one of four to the front, giving $6 \cdot 4 = 24$ possible positions. Any position can be achieved from any other by a rotation of the die. These rotations form a group G, which is isomorphic to a subgroup of S_8 (see Exercise 40 of Section 2.1). We let X be the 720 possible ways of marking the cube and let G act on X by rotation of the cube. We consider two markings to give the same die if one can be carried into the other under action by an element of G, that is, by rotating the cube. In other words, we consider each *orbit* in X under G to correspond to a single die, and different orbits to give different dice. The determination of the number of distinguishable dice thus leads to the question of determining the number of orbits under G in a G-set X.

The following theorem gives a tool for determining the number of orbits in a G-set X under G. Recall that for each $g \in G$ we let X_g be the set of elements of X left fixed by g, so that $X_g = \{x \in X \mid gx = x\}$. Recall also that for each $x \in X$, we let $G_x = \{g \in G \mid gx = x\}$, and Gx is the orbit of x under G.

THEOREM 3.23 (Burnside's Formula)   Let G be a finite group and X a finite G-set. If r is the number of orbits in X under G, then

$$r \cdot |G| = \sum_{g \in G} |X_g|. \tag{1}$$

PROOF   We consider all pairs (g, x) where $gx = x$, and let N be the number of such pairs. For each $g \in G$ there are $|X_g|$ pairs having g as first member. Thus,

$$N = \sum_{g \in G} |X_g|. \tag{2}$$

On the other hand, for each $x \in X$ there are $|G_x|$ pairs having x as second member. Thus we also have

$$N = \sum_{x \in X} |G_x|.$$

By Theorem 3.22 we have $|Gx| = (G:G_x)$. But we know that $(G:G_x) = |G|/|G_x|$, so we obtain $|G_x| = |G|/|Gx|$. Then

$$N = \sum_{x \in X} \frac{|G|}{|Gx|} = |G| \left(\sum_{x \in X} \frac{1}{|Gx|} \right).$$

Now $1/|Gx|$ has the same value for all x in the same orbit, and if we let $\mathcal{O}$ be any orbit, then

$$\sum_{x \in \mathcal{O}} \frac{1}{|Gx|} = 1.$$

Thus we obtain

$$N = |G| \, (\text{number of orbits in } X \text{ under } G) = |G| \cdot r. \tag{3}$$

Comparison of Eq. 2 and Eq. 3 gives Eq. 1. ◆

COROLLARY If G is a finite group and X is a finite G-set, then

$$(\text{number of orbits in } X \text{ under } G) = \frac{1}{|G|} \cdot \sum_{g \in G} |X_g|.$$

PROOF The proof of this corollary follows immediately from the preceding theorem. ◆

Let us continue our computation of the number of distinguishable dice as our first example.

EXAMPLE 1 We let X be the set of 720 different markings of faces of a cube using from one to six dots. Let G be the group of 24 rotations of the cube as discussed above. We saw that the number of distinguishable dice is the number of orbits in X under G. Now $|G| = 24$. For $g \in G$ where $g \neq e$, we have $|X_g| = 0$, because any rotation other than the identity changes any one of the 720 markings into a different one. However, $|X_e| = 720$ since the identity leaves all 720 markings fixed. Then by the corollary to Theorem 3.23,

$$(\text{number of orbits}) = \frac{1}{24} \cdot 720 = 30,$$

so there are 30 distinguishable dice. ▲

Of course the number of distinguishable dice could be counted without using the machinery of the preceding corollary, but by using elementary combinatorics as often taught in a freshman finite math course. In marking a cube to make a die, we can, by rotation if necessary, assume the face marked 1 is down. There are five choices for the top (opposite) face. By rotating the die as we look down on it, any one of the remaining four faces could be brought to the front position, so there are no different choices involved for the front face. But with respect to the number on the front face, there are $3 \cdot 2 \cdot 1$ possibilities for the remaining three side faces. Thus there are $5 \cdot 3 \cdot 2 \cdot 1 = 30$ possibilities in all.

The next two examples appear in some finite math texts and are easy to solve by elementary means. We use the corollary to Theorem 3.23 so that we have more practice thinking in terms of orbits.

EXAMPLE 2   How many distinguishable ways can seven people be seated at a round table, where there is no distinguishable "head" to the table? Of course there are 7! ways to assign prople to the different chairs. We take X to be the 7! possible assignments. A rotation of people achieved by asking each person to move one place to the right results in the same arrangement. Such a rotation generates a cyclic group G of order 7, which we consider to act on X in the obvious way. Again, only the identity e leaves any arrangement fixed, and it leaves all 7! arrangements fixed. By the corollary to Theorem 3.23.

$$\left(\text{number of orbits}\right) = \frac{1}{7} \cdot 7! = 6! = 720. \ \blacktriangle$$

EXAMPLE 3   How many distinguishable necklaces (with no clasp) can be made using seven different-colored beads of the same size? Unlike the table in Example 2, the necklace can be turned over as well as rotated. Thus we consider the full dihedral group D_7 of order $2 \cdot 7 = 14$ as acting on the set X of 7! possibilities. Then the number of distinguishable necklaces is

$$\left(\text{number of orbits}\right) = \frac{1}{14} \cdot 7! = 360. \ \blacktriangle$$

In using the corollary to Theorem 3.23, we have to compute $|G|$ and $|X_g|$ for each $g \in G$. In the examples and the exercises, $|G|$ will pose no real problem. Let us give an example where $|X_g|$ is not as trivial to compute as in the preceding examples. We will continue to assume knowledge of very elementary combinatorics.

EXAMPLE 4   Let us find the number of distinguishable ways the edges of an equilateral triangle can be painted if four different colors of paint are available, assuming only one color is used on each edge, and the same color may be used on different edges.

Of course there are $4^3 = 64$ ways of painting the edges in all, since each of the three edges may be any one of four colors. We consider X to be the set of these 64 possible painted triangles. The group G acting on X is the group of symmetries of the triangle, which is isomorphic to S_3 and which we consider to be S_3. We use the notation for elements in S_3 given in Section 2.1. We need to compute $|X_g|$ for each of the six elements g in S_3.

$|X_{\rho_0}| = 64$ Every painted triangle is left fixed by ρ_0.

$|X_{\rho_1}| = 4$ To be invariant under ρ_1, all edges must be the same color, and there are 4 possible colors.

$|X_{\rho_2}| = 4$ Same reason as for ρ_1.

$|X_{\mu_1}| = 16$ The edges that are interchanged must be the same color (4 possibilities) and the other edge may also be any of the colors (times 4 possibilities).

$|X_{\mu_2}| = |X_{\mu_3}| = 16$ Same reason as for μ_1.

Then

$$\sum_{g \in S_3} |X_g| = 64 + 4 + 4 + 16 + 16 + 16 = 120.$$

Thus

$$(\text{number of orbits}) = \frac{1}{6} \cdot 120 = 20,$$

and there are 20 distinguishable painted triangles. ▲

EXAMPLE 5 We repeat Example 4 with the assumption that a different color is used on each edge. The number of possible ways of painting the edges is then $4 \cdot 3 \cdot 2 = 24$, and we let X be the set of 24 possible painted triangles. Again, the group acting on X can be considered to be S_3. Since all edges are a different color, we see $|X_{\rho_0}| = 24$ while $|X_g| = 0$ for $g \neq \rho_0$. Thus

$$(\text{number of orbits}) = \frac{1}{6} \cdot 24 = 4,$$

so there are four distinguishable triangles. ▲

Exercises 3.7

Computations

In each of the following exercises use the corollary to Theorem 3.23 to work the problem even though the answer might be obtained by more elementary methods.

1. Find the number of orbits in $\{1, 2, 3, 4, 5, 6, 7, 8\}$ under the cyclic subgroup $\langle (1, 3, 5, 6) \rangle$ of S_8.

2. Find the number of orbits in $\{1, 2, 3, 4, 5, 6, 7, 8\}$ under the subgroup of S_8 generated by $(1, 3)$ and $(2, 4, 7)$.

3. Find the number of distinguishable tetrahedral dice that can be made using one, two, three, and four dots on the faces of a regular tetrahedron, rather than a cube.

4. Wooden cubes of the same size are to be painted a different color on each face to make children's blocks. How many distinguishable blocks can be made if eight colors of paint are available?

5. Answer Exercise 4 if colors may be repeated on different faces at will. [*Hint:* The 24 rotations of a cube consist of the identity, 9 that leave a pair of opposite faces invariant, 8 that leave a pair of opposite vertices invariant, and 6 leaving a pair of opposite edges invariant.]

6. Each of the eight corners of a cube is to be tipped with one of four colors, each of which may be used on from one to all eight corners. Find the number of distinguishable markings possible. (See the hint of Exercise 5.)

7. Find the number of distinguishable ways the edges of a square of cardboard can be painted if six colors of paint are available and

   a. no color is used more than once.

   b. the same color can be used on any number of edges.

8. Consider six straight wires of equal lengths with ends soldered together to form edges of a regular tetrahedron. Either a 50-ohm or 100-ohm resistor is to be inserted in the middle of each wire. Assume there are at least six of each type of resistor available. How many essentially different wirings are possible?

9. A rectangular prism 2 ft long with 1-ft square ends is to have each of its six faces painted with one of six possible colors. How many distinguishable painted prisms are possible if

   a. no color is to be repeated on different faces,

   b. each color may be used on any number of faces?

◆

CHAPTER FOUR

ADVANCED GROUP THEORY[†]

◆

The preceding three chapters presented the fundamentals of group theory. This chapter fills in some gaps and gives a brief look at a few more topics in the subject. Section 4.1 proves two useful theorems exhibiting some isomorphisms between factor groups formed from subgroups of a group G. It then uses these theorems to prove the Jordan–Hölder theorem (Theorem 3.17). The Sylow theory in Section 4.2 gives some information on the existence and number of subgroups with prime-power order p^r of a finite group G whose order is divisible by p. Section 4.3 illustrates how Sylow theory can sometimes be used to determine, up to isomorphism, all finite group structures of a certain order, or to show that all groups of some order must contain a proper normal subgroup. Section 4.4 presents the notion of a free abelian group and gives a proof of Theorem 2.11, the fundamental theorem for finitely generated abelian groups. A brief look at free groups appears in Section 4.5. To conclude, Section 4.6 describes how a finitely generated group might be specified in terms of generators and equations that express relationships between those generators.

4.1

Isomorphism Theorems: Proof of the Jordan–Hölder Theorem

The Isomorphism Theorems

There are several theorems concerning isomorphic factor groups that are known as the *isomorphism theorems* of group theory. The first of these is Theorem 3.9, which we restate for easy reference. The theorem is diagrammed in Fig. 4.1.

---

† Topics in this chapter will not be used in subsequent chapters of the text.

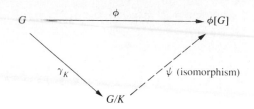

Figure 4.1

THEOREM 4.1 (First Isomorphism Theorem) Let $\phi:G \to G'$ be a homomorphism with kernel K, and let $\gamma_K:G \to G/K$ be the canonical homomorphism. There is a unique isomorphism $\psi:G/K \to \phi[G]$ such that $\phi(x) = \psi(\gamma_K(x))$ for each $x \in G$.

Recall that if H and N are subgroups of a group G, then

$$HN = \{hn \mid h \in H, n \in N\}.$$

We define the **join** $H \vee N$ of H and N as the intersection of all subgroups of G that contain HN; thus $H \vee N$ is the smallest subgroup of G containing HN. Of course $H \vee N$ is also the smallest subgroup of G containing both H and N, since any such subgroup must contain HN. In general, HN need not be a subgroup of G. However, we have the following lemma.

LEMMA 4.1 If N is a normal subgroup of G, and if H is any subgroup of G, then $H \vee N = HN = NH$. Furthermore, if H is also normal in G, then HN is normal in G.

PROOF We show that HN is a subgroup of G, from which $H \vee N = HN$ follows at once. Let $h_1, h_2 \in H$ and $n_1, n_2 \in N$. Since N is a normal subgroup, we have $n_1h_2 = h_2n_3$ for some $n_3 \in N$. Then $(h_1n_1)(h_2n_2) = h_1(n_1h_2)n_2 = h_1(h_2n_3)n_2 = (h_1h_2)(n_3n_2) \in HN$, so HN is closed under the induced operation in G. Clearly $e = ee$ is in HN, For $h \in H$ and $n \in N$, we have $(hn)^{-1} = n^{-1}h^{-1} = h^{-1}n_4$ for some $n_4 \in N$, since N is a normal subgroup. Thus $(hn)^{-1} \in HN$, so $HN \leq G$. A similar argument shows that NH is a subgroup, so $NH = H \vee N = HN$.

Now suppose that H is also normal in G, and let $h \in H$, $n \in N$, and $g \in G$. Then $ghng^{-1} = (ghg^{-1})(gng^{-1}) \in HN$, so HN is indeed normal in G. ◆

We are now ready for the second isomorphism theorem.

THEOREM 4.2 (Second Isomorphism Theorem) Let H be a subgroup of G and let N be a normal subgroup of G. Then $(HN)/N \simeq H/(H \cap N)$.

PROOF Since N is normal in G, we see at once that $H \cap N$ is normal in H (see Exercise 9). Let $h \in H$ and $n \in N$. We attempt to define $\phi:HN \to H/(H \cap N)$ by $\phi(hn) = h(H \cap N)$. We must show ϕ is well defined. Let $h_1 \in H$ and $n_1 \in N$, and suppose $h_1n_1 = hn$. Then

$h^{-1}h_1 = nn_1^{-1}$, so $h^{-1}h_1$ is in both H and N, and thus is in $H \cap N$. Consequently, $h(H \cap N) = h_1(H \cap N)$ in $H/(H \cap N)$. Thus $\phi(h_1n_1) = \phi(hn)$, so ϕ is well defined.

We claim ϕ is a homomorphism onto $H/(H \cap N)$. Let $n_1, n_2 \in N$ and $h_1, h_2 \in H$. As in Lemma 4.1, we can write $n_1h_2 = h_2n_3$ since N is normal in G. Then

$$\phi((h_1n_1)(h_2n_2)) = \phi((h_1h_2)(n_3n_2))$$
$$= h_1h_2(H \cap N)$$
$$= h_1(h \cap N) \cdot h_2(H \cap N)$$
$$= \phi(h_1n_1) \cdot \phi(h_2n_2).$$

so ϕ is a homomorphism. Since $\phi(he) = h(H \cap N)$ for all $h \in H$, we see ϕ is onto $H/(H \cap N)$.

The kernel of ϕ consists of all $hn \in HN$ such that $h \in H \cap N$; this kernel is thus $(H \cap N)N$. Clearly $(H \cap N)N = N$. Thus ϕ is a homomorphism onto $h/(H \cap N)$ with kernel N, so by Theorem 4.1, $HN/N \simeq H/(H \cap N)$. ◆

EXAMPLE I Let $G = \mathbb{Z} \times \mathbb{Z} \times \mathbb{Z}$, $H = \mathbb{Z} \times \mathbb{Z} \times \{0\}$, and $N = \{0\} \times \mathbb{Z} \times \mathbb{Z}$. Then clearly $HN = \mathbb{Z} \times \mathbb{Z} \times \mathbb{Z}$ and $H \cap N = \{0\} \times \mathbb{Z} \times \{0\}$. We have $(HN)/N \simeq \mathbb{Z}$ and also $H/(H \cap N) \simeq \mathbb{Z}$ ▲

If H and K are two normal subgroups of G and $K \leq H$, then H/K is a normal subgroup of G/K. The third isomorphism theorem concerns these groups.

THEOREM 4.3 (Third Isomorphism Theorem) Let H and K be normal subgroups of a group g with $K \leq H$. Then $G/H \simeq (G/K)/(H/K)$.

PROOF Let $\phi: G \rightarrow (G/K)/(H/K)$ be given by $\phi(a) = (aK)(H/K)$ for $a \in G$. Clearly ϕ is onto $(G/K)/(H/K)$, and for $a, b \in G$,

$$\phi(ab) = [(ab)K](H/K) = [(aK)(bK)](II/K)$$
$$= [(aK)(H/K)][(bK)(H/K)]$$
$$= \phi(a)\phi(b),$$

so ϕ is a homomorphism. The kernel consists of those $x \in G$ such that $\phi(x) = H/K$. These x are just the elements of H. Then Theorem 4.1 shows that $G/H \simeq (G/K)/(H/K)$. ◆

A nice way of viewing Theorem 4.3 is to regard the canonical map $\gamma_H: G \rightarrow G/H$ as being factored via a normal subgroup K of G, $K \leq H \leq G$, to give

$$\gamma_H = \gamma_{H/K}\gamma_K,$$

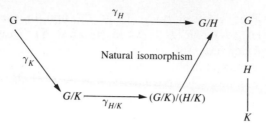

Figure 4.2   **Figure 4.3**

up to a natural isomorphism, as illustrated in Fig. 4.2. Another way of visualizing this theorem is to use the lattice diagram in Fig. 4.3, where each group is a normal subgroup of G and is contained in the one above it. *The larger the normal subgroup, the smaller the factor group.* Thus we can think of G collapsed by H, that is, G/H, as being smaller than G collapsed by K. Theorem 4.3 states that we can collapse G all the way down to G/H in two steps. First collapse to G/K, and then using H/K collapse this to $(G/K)/(H/K)$. The overall result is the same (up to isomorphism) as collapsing G by H.

EXAMPLE 2   Consider $K = 6\mathbb{Z} < H = 2\mathbb{Z} < G = \mathbb{Z}$.   Then $G/H = \mathbb{Z}/2\mathbb{Z} \simeq \mathbb{Z}_2$. Now $G/K = \mathbb{Z}/6\mathbb{Z}$ has elements

$$6\mathbb{Z}, \qquad 1 + 6\mathbb{Z}, \qquad 2 + 6\mathbb{Z}, \qquad 3 + 6\mathbb{Z}, \qquad 4 + 6\mathbb{Z}, \qquad \text{and} \qquad 5 + 6\mathbb{Z}.$$

Of these six cosets, $6\mathbb{Z}$, $2 + 6\mathbb{Z}$, and $4 + 6\mathbb{Z}$ lie in $2\mathbb{Z}/6\mathbb{Z}$. Thus $(\mathbb{Z}/6\mathbb{Z})/(2\mathbb{Z}/6\mathbb{Z})$ has two elements and is isomorphic to $\mathbb{Z}_2$ also. Alternatively, we see that $\mathbb{Z}/6\mathbb{Z} \simeq \mathbb{Z}_6$, and $2\mathbb{Z}/6\mathbb{Z}$ corresponds *under this isomorphism to the* cyclic subgroup $\langle 2 \rangle$ of $\mathbb{Z}_6$. Thus $(\mathbb{Z}/6\mathbb{Z})/(2\mathbb{Z}/6\mathbb{Z}) \simeq \mathbb{Z}_6/\langle 2 \rangle \simeq \mathbb{Z}_2 \simeq \mathbb{Z}/2\mathbb{Z}$.

The Zassenhaus (Butterfly) Lemma

The proof of the Jordan–Hölder theorem follows quite easily from a rather technical lemma developed by Zassenhaus. This lemma (4.2) is also called the butterfly lemma, since Fig. 4.4, which accompanies the lemma, has a butterfly shape.

Let H and K be subgroups of a group G, and let H^* be a normal subgroup of H and K^* be a normal subgroup of K. Applying the first statement in Lemma 4.1 to H^* and $H \cap K$ as subgroups of H, we see that $H^*(H \cap K)$ is a group. Similar arguments show that $H^*(H \cap K^*)$, $K^*(H \cap K)$, and $K^*(H^* \cap K)$ are also groups. It is not hard to show that $H^* \cap K$ is a normal subgroup of $H \cap K$ (see Exercise 10). The same argument using Lemma 4.1 applied to $H^* \cap K$ and $H \cap K^*$ as subgroups of $H \cap K$ shows that $L = (H^* \cap K)(H \cap K^*)$ is a group. Thus we have the lattice of subgroups shown in Fig. 4.4. It is not hard to verify the inclusion relations indicated by the diagram.

Since both $H \cap K^*$ and $H^* \cap K$ are normal subgroups of $H \cap K$, the second statement in Lemma 4.1 shows that $L = (H^* \cap K)(H \cap K^*)$ is a normal subgroup of $H \cap K$. We have denoted this particular normal subgroup relationship by the heavy middle line in Fig. 4.4. We claim the other two heavy lines also indicate normal subgroup relationships, and that the three factor groups given by the three normal subgroup relations are all isomorphic. To show this, we shall define a homomorphism $\phi : H^*(H \cap K) \rightarrow (H \cap K)/L$, and show ϕ is onto $(H \cap K)/L$ with kernel $H^*(H \cap K^*)$. It will then follow at once from Theorem 4.1 that $H^*(H \cap K^*)$ is normal in $H^*(H \cap K)$, and that $H^*(H \cap K)/H^*(H \cap K^*) \simeq (H \cap K)/L$. A similar result for the groups on the right-hand heavy line in Fig. 4.4 then follows by symmetry.

Let $\phi : H^*(H \cap K) \rightarrow (H \cap K)/L$ be defined as follows. For $h \in H^*$ and $x \in H \cap K$, let $\phi(hx) = xL$. We show ϕ is well defined and a homomorphism. Let $h_1, h_2 \in H^*$ and $x_1, x_2 \in H \cap K$. If $h_1 x_1 = h_2 x_2$, then $h_2^{-1} h_1 = x_2 x_1^{-1} \in H^* \cap (H \cap K) = H^* \cap K \subseteq L$, so $x_1 L = x_2 L$. Thus ϕ is well defined. Since H^* is normal in H, there is h_3 in H^* such that $x_1 h_2 = h_3 x_1$. Then

$$\phi((h_1 x_1)(h_2 x_2)) = \phi((h_1 h_3)(x_1 x_2)) = (x_1 x_2)L$$
$$= (x_1 L)(x_2 L) = \phi(h_1 x_1) \cdot \phi(h_2 x_2).$$

Thus ϕ is a homomorphism.

Obviously ϕ is onto $(H \cap K)/L$. Finally if $h \in H^*$ and $x \in H \cap K$, then $\phi(Hx) = xL = L$ if and only if $x \in L$, or if and only if $hx \in H^*L = H^*(H^* \cap K)(H \cap K^*) = H^*(H \cap K^*)$. Thus $\mathrm{Ker}(\phi) = H^*(H \cap K^*)$.

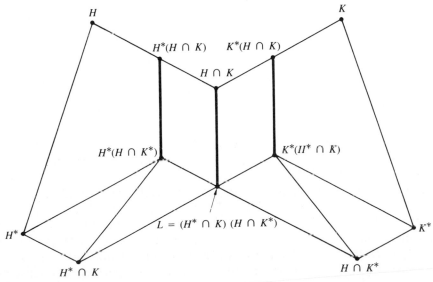

Figure 4.4

We have proved the following lemma.

LEMMA 4.2 (Zassenhaus Lemma) Let H and K be subgroups of a group G and let H^* and K^* be normal subgroups of H and K respectively. Then

1. $H^*(H \cap K^*)$ is a normal subgroup of $H^*(H \cap K)$.
2. $K^*(H^* \cap K)$ is a normal subgroup of $K^*(H \cap K)$.
3. $H^*(H \cap K)/H^*(H \cap K^*) \simeq K^*(H \cap K)/K^*(H^* \cap K)$
$$\simeq (H \cap K)/[(H^* \cap K)(H \cap K^*)].$$

Proof of the Schreier Theorem

Section 3.5 showed that the Jordan–Hölder theorem follows at once from the Schreier theorem (Theorem 3.16). We restate the Schreier theorem here for easy reference and give the proof.

THEOREM 4.4 (Schreier Theorem) Two subnormal (normal) series of a group G have isomorphic refinements.

PROOF Let G be a group and let

$$\{e\} = H_0 < H_1 < H_2 < \cdots < H_n = G \tag{1}$$

and

$$\{e\} = K_0 < K_1 < K_2 < \cdots < K_m = G \tag{2}$$

be two subnormal series for G. For i where $0 \le i \le n - 1$, form the chain of groups

$$H_i = H_i(H_{i+1} \cap K_0) \le H_i(H_{i+1} \cap K_1)$$
$$\le \cdots \le H_i(H_{i+1} \cap K_m) = H_{i+1}.$$

This inserts $m - 1$ not necessarily distinct groups between H_i and H_{i+1}. If we do this for each i where $0 \le i \le n - 1$ and let $H_{i,j} = H_i(H_{i+1} \cap K_j)$, then we obtain the chain of groups

$$\{e\} = H_{0,0} \le H_{0,1} \le H_{0,2} \le \cdots \le H_{0,m-1} \le H_{1,0}$$
$$\le H_{1,1} \le H_{1,2} \le \cdots \le H_{1,m-1} \le H_{2,0}$$
$$\le H_{2,1} \le H_{2,2} \le \cdots \le H_{2,m-1} \le H_{3,0}$$
$$\le \cdots$$
$$\le H_{n-1,1} \le H_{n-1,2} \le \cdots \le H_{n-1,m-1} \le H_{n-1,m}$$
$$= G. \tag{3}$$

This chain (3) contains $nm + 1$ not necessarily distinct groups, and $H_{i,0} = H_i$ for each i. By the Zassenhaus lemma, chain (3) is a subnormal chain, that is, each group is normal in the following group. This chain refines the series (1).

In a symmetric fashion, we set $K_{j,i} = K_j(K_{j+1} \cap H_i)$ for $0 \le j \le m - 1$ and $0 \le i \le n$. This gives a subnormal chain

$$\{e\} = K_{0,0} \le K_{0,1} \le K_{0,2} \le \cdots \le K_{0,n-1} \le K_{1,0}$$
$$\le K_{1,1} \le K_{1,2} \le \cdots \le K_{1,n-1} \le K_{2,0}$$
$$\le K_{2,1} \le K_{2,2} \le \cdots \le K_{2,n-1} \le K_{3,0}$$
$$\le \cdots$$
$$\le K_{m-1,1} \le K_{m-1,2} \le \cdots \le K_{m-1,n-1} \le K_{m-1,n}$$
$$= G. \tag{4}$$

This chain (4) contains $mn + 1$ not necessarily distinct groups, and $K_{j,0} = K_j$ for each j. This chain refines the series (2).

By the Zassenhaus lemma (4.2), we have

$$H_i(H_{i+1} \cap K_{j+1})/H_i(H_{i+1} \cap K_j) \simeq K_j(K_{j+1} \cap H_{i+1})/K_j(K_{j+1} \cap H_i),$$

or

$$H_{i,j+1}/H_{i,j} \simeq K_{j,i+1}/K_{j,i} \tag{5}$$

for $0 \le i \le n - 1$ and $0 \le j \le m - 1$. The isomorphisms of relation (5) give a one-to-one correspondence of isomorphic factor groups between the subnormal chains (3) and (4). To verify this correspondence, note that $H_{i,0} = H_i$ and $H_{i,m} = H_{i+1}$, while $K_{j,0} = K_j$ and $K_{j,n} = K_{j+1}$. Each chain in (3) and (4) contains a rectangular array of mn symbols $\le$. Each $\le$ gives rise to a factor group. The factor groups arising from the rth *row* of $\le$'s in chain (3) correspond to the factor groups arising from the rth *column* of $\le$'s in chain (4). Deleting repeated groups from the chains in (3) and (4), we obtain subnormal series of distinct groups that are isomorphic refinements of chains (1) and (2). This establishes the theorem for subnormal series.

For normal series, where all H_i and K_j are normal in G, we merely observe that all the groups $H_{i,j}$ and $K_{j,i}$ formed above are also normal in G, so the same proof applies. This normality of $H_{i,j}$ and $K_{j,i}$ follows at once from the second assertion in Lemma 4.1 and from the fact that intersections of normal subgroups of a group yield normal subgroups. ◆

Exercises 4.1

Computations

In using the three isomorphism theorems, it is often necessary to know the actual correspondence given by the isomorphism and not just the fact that the groups are isomorphic. The first six exercises give us training for this.

1. Let $\phi : \mathbb{Z}_{12} \to \mathbb{Z}_3$ be the homomorphism such that $\phi(1) = 2$.

   a. Find the kernel K of ϕ.

b. List the cosets in $\mathbb{Z}_{12}/K$, showing the elements in each coset.

c. Give the correspondence between $\mathbb{Z}_{12}/K$ and $\mathbb{Z}_3$ given by the map ψ described in Theorem 4.1.

2. Let $\phi:\mathbb{Z}_{18} \to \mathbb{Z}_{12}$ be the homomorphism where $\phi(1) = 10$.

a. Find the kernel K of ϕ.

b. List the cosets in $\mathbb{Z}_{18}/K$, showing the elements in each coset.

c. Find the group $\phi[\mathbb{Z}_{18}]$.

d. Give the correspondence between $\mathbb{Z}_{18}/K$ and $\phi[\mathbb{Z}_{18}]$ given by the map ψ described in Theorem 4.1.

3. In the group $\mathbb{Z}_{24}$, let $H = \langle 4 \rangle$ and $N = \langle 6 \rangle$.

a. List the elements in HN (which we might write $H + N$ for these additive groups) and in $H \cap N$.

b. List the cosets in HN/N, showing the elements in each coset.

c. List the cosets in $H/(H \cap N)$, showing the elements in each coset.

d. Give the correspondence between HN/N and $H/(H \cap N)$ described in the proof of Theorem 4.2.

4. Repeat Exercise 3 for the group $\mathbb{Z}_{36}$ with $H = \langle 6 \rangle$ and $N = \langle 9 \rangle$.

5. In the group $G = \mathbb{Z}_{24}$, let $H = \langle 4 \rangle$ and $K = \langle 8 \rangle$.

a. List the cosets in G/H, showing the elements in each coset.

b. List the cosets in G/K, showing the elements in each coset.

c. List the cosets in H/K, showing the elements in each coset.

d. List the cosets in $(G/K)/(H/K)$, showing the elements in each coset.

e. Give the correspondence between G/H and $(G/K)/(H/K)$ described in the proof of Theorem 4.3.

6. Repeat Exercise 5 for the group $G = \mathbb{Z}_{36}$ with $H = \langle 9 \rangle$ and $K = \langle 18 \rangle$.

Concepts

7. Let G be $\mathbb{Z}_{36}$. Refer to the proof of Theorem 4.4. Let the subnormal series (1) be

$$\{0\} < \langle 12 \rangle < \langle 3 \rangle < \mathbb{Z}_{36}$$

and let the subnormal series (2) be

$$\{0\} < \langle 18 \rangle < \mathbb{Z}_{36}.$$

Find chains (3) and (4) and exhibit the isomorphic factor groups as described in the proof. Write chains (3) and (4) in the rectangular array shown in the text.

8. Repeat Exercise 7 for the group $\mathbb{Z}_{24}$ with the subnormal series (1)

$$\{0\} < \langle 12 \rangle < \langle 4 \rangle < \mathbb{Z}_{24}$$

and (2)

$$\{0\} < \langle 6 \rangle < \langle 3 \rangle < \mathbb{Z}_{24}.$$

Theory

9. Show that if H and N are subgroups of G, and N is normal in G, then $H \cap N$ is normal in H.

10. Let H^*, H, and K be subgroups of G with H^* normal in H. Show that $H^* \cap K$ is normal in $H \cap K$.

11. Let $H, K,$ and L be normal subgroups of G with $H < K < L$. Let $A = G/H$, $B = K/H$, and $C = L/H$.

    a. Show that B and C are normal subgroups of A, and $B < C$.

    b. To what group is $(A/B)/(C/B)$ isomorphic?

12. Let K and L be normal subgroups of G with $K \vee L = G$, and $K \cap L = \{e\}$. Show that $G/K \simeq L$ and $G/L \simeq K$.

13. Show that a subgroup K of a solvable group G is solvable. [*Hint:* Let $H_0 = \{e\} < H_1 < \cdots < H_n = G$ be a composition series for G. Show that the distinct groups among $K \cap H_i$ for $i = 0, \ldots, n$ form a composition series for K. Observe that
$$(K \cap H_i)/(K \cap H_{i-1}) \simeq [H_{i-1}(K \cap H_i)]/[H_{i-1}],$$
by Theorem 4.2, with $H = K \cap H_i$ and $N = H_{i-1}$, and that $H_{i-1}(K \cap H_i) \leq H_i$.]

14. Let $H_0 = \{e\} < H_1 < \cdots < H_n = G$ be a composition series for a group G. Let N be a normal subgroup of G, and suppose that N is a simple group. Show that the distinct groups among $H_0, H_i N$ for $i = 0, \ldots, n$ also form a composition series for G. [*Hint:* H_i/N is a group by Lemma 4.1. Show that $H_{i-1}N$ is normal in $H_i N$. By Theorem 4.2,
$$(H_i N)/(H_{i-1}N) \simeq H_i/[H_i \cap (H_{i-1}N)],$$
and the latter group is isomorphic to
$$[H_i/H_{i-1}]/[(H_i \cap (H_{i-1}N))/H_{i-1}],$$
by Theorem 4.3. But H_i/H_{i-1} is simple.]

15. Let G be a group, and let $H_0 = \{e\} < H_1 < \cdots < H_n = G$ be a composition series for G. Let N be a normal subgroup of G, and let $\gamma: G \rightarrow G/N$ be the canonical map. Show that the distinct groups among $\gamma[H_i]$ for $i = 0, \ldots, n$ form a composition series for G/N. [*Hint:* Observe that the map
$$\psi: H_i N \rightarrow \gamma(H_i)/\gamma[H_{i-1}]$$
defined by
$$\psi(h_i n) - \gamma(h_i n)\gamma[H_{i-1}]$$
is a homomorphism with kernel $H_{i-1}N$. By Theorem 4.1,
$$\gamma(H_i)/\gamma[H_{i-1}] \sim (H_i N)/(H_{i-1}N).$$
Proceed via Theorem 4.2, as shown in the hint for Exercise 14.]

16. Prove that a homomorphic image of a solvable group is solvable. [*Hint:* Apply Exercise 15 to get a composition series for the homomorphic image. The hints of Exercises 14 and 15 then show how the factor groups of this composition series in the image look.]

4.2

Sylow Theorems

Theorem 2.11 of Section 2.4 gives us complete information about all finite abelian groups. The study of finite nonabelian groups is much more complicated. The Sylow theorems give us some important information about them.

We know the order of a subgroup of a finite group G must divide $|G|$. If G is abelian, then there exist subgroups of every order dividing $|G|$. We showed in Example 4 of Section 3.4 that A_4, which has order 12, has no subgroup of order 6. Thus a nonabelian group G may have no subgroup of some order d dividing $|G|$; the "converse of the theorem of Lagrange" does not hold. The Sylow theorems give a weak converse. Namely, they show that if d is a power of a prime and d divides $|G|$, then G does contain a subgroup of order d. (Note that 6 is not a power of a prime.) The Sylow theorems also give some information concerning the number of such subgroups and their relationship to each other. We will see that these theorems are very useful in studying finite nonabelian groups.

Proofs of the Sylow theorems give us another application of action of a group on a set described in Section 3.6. This time, the set itself is formed from the group; in some instances the set is the group itself, sometimes it is a collection of cosets of a subgroup, and sometimes it is a collection of subgroups.

p-Groups

Section 3.7 gave applications of Burnside's formula that counted the number of orbits in a finite G-set. Most of our results in this section flow from an equation that counts the number of elements in a finite G-set.

Let X be a finite G-set. Recall that for $x \in X$, the orbit of x in X under G is $Gx = \{gx \mid g \in G\}$. Suppose there are r orbits in X under G, and let $\{x_1, x_2, \ldots, x_r\}$ contain one element from each orbit in X. Now every element of X is in precisely one orbit, so

$$|X| = \sum_{i=1}^{r} |Gx_i| . \tag{1}$$

There may be one-element orbits in X. Let $X_G = \{x \in X \mid gx = x \text{ for all } g \in G\}$. Thus X_G is precisely the union of the one-element orbits in X. Let us suppose there are s one-element orbits, where $0 \le s \le r$. Then $|X_G| = s$, and reordering the x_i if necessary, we may rewrite Eq. (1) as

$$|X| = |X_G| + \sum_{i=s+1}^{r} |Gx_i| . \tag{2}$$

Most of the results of this section will flow from Eq. (2), We shall develop Sylow theory as in Hungerford [11], where credit is given to R. J. Nunke for the line of proof. The proof of Theorem 4.6 (Cauchy's theorem) is credited there to J. H. McKay.

Theorem 4.5, which follows, is not quite a counting theorem, but it does have a numerical conclusion. It counts modulo p. The theorem seems to be amazingly powerful. In the rest of the chapter, if we choose the correct set, the correct group action on it, and apply Theorem 4.5, what we want seems to fall right into our lap! Compared with older proofs, the arguments are extremely pretty and elegant.

Throughout this section, p will always be a prime integer.

THEOREM 4.5   Let G be a group of order p^n and let X be a finite G-set. Then $|X| \equiv |X_G| \pmod{p}$.

PROOF   In the notation of Eq. (2), we know that $|Gx_i| = (G:G_{x_i})$ by Theorem 3.22. But $(G:G_{x_i})$ divides $|G|$, and consequently p divides $(G:G_{x_i})$ and thus divides $|Gx_i|$ for $s + 1 \le i \le r$. Equation (2) then shows that $|X| - |X_G|$ is divisible by p, so $|X| \equiv |X_G| \pmod{p}$.   ◆

DEFINITION 4.1 (p-Group)   A group G is a p-**group** if every element in G has order a power of the prime p. A subgroup of a group G is a p-**subgroup of** G if the subgroup is itself a p-group.

Our goal in this chapter is to show that a finite group G has a subgroup of every prime-power order dividing $|G|$. As a first step, we prove Cauchy's theorem, which says that if p divides $|G|$, then G has a subgroup of order p.

THEOREM 4.6 (Cauchy's Theorem)   Let G be a finite group and let p divide $|G|$. Then G has an element of order p and, consequently, a subgroup of order p.

PROOF   We form the set X of all p-tuples $(g_1, g_2, \ldots, g_p)$ of elements of G having the property that the product of the coordinates in G is e. That is,

$$X = \{(g_1, g_2, \ldots, g_p) \mid g_i \in G \quad \text{and} \quad g_1 g_2 \cdots g_p = e\}.$$

We claim p divides $|X|$. In forming a p-tuple in X, we may let $g_1, g_2, \ldots, g_{p-1}$ be any elements of G, and g_p is then uniquely determined as $(g_1 g_2 \cdots g_{p-1})^{-1}$. Thus $|X| = |G|^{p-1}$ and since p divides $|G|$, we see that p divides $|X|$.

Let σ be the cycle $(1, 2, 3, \ldots, p)$ in S_p. We let σ act on X by

$$\sigma(g_1, g_2, \ldots, g_p) = \left(g_{\sigma(1)}, g_{\sigma(2)}, \ldots, g_{\sigma(p)}\right) = (g_2, g_3, \ldots, g_p, g_1).$$

Note that $(g_2, g_3, \ldots, g_p, g_1) \in X$, for $g_1(g_2, g_3 \cdots g_p) = e$ implies that $g_1 = (g_2 g_3 \cdots g_p)^{-1}$, so $(g_2 g_3 \cdots g_p)g_1 = e$ also. Thus σ acts on X, and we consider the subgroup $\langle \sigma \rangle$ of S_p to act on X by iteration in the natural way.

Now $|\langle \sigma \rangle| = p$, so we may apply Theorem 4.5, and we know that $|X| \equiv |X_{\langle \sigma \rangle}| \pmod{p}$. Since p divides $|X|$, it must be that p divides $|X_{\langle \sigma \rangle}|$ also. Let us examine $X_{\langle \sigma \rangle}$. Now $(g_1, g_2, \ldots, g_p)$ is left fixed by σ, and hence by $\langle \sigma \rangle$, if and only if $g_1 = g_2 = \cdots = g_p$. We know at least one element in $X_{\langle \sigma \rangle}$, namely $(e, e, \ldots, e)$. Since p divides $|X_{\langle \sigma \rangle}|$, there must be at least p elements in $X_{\langle \sigma \rangle}$. Hence there exists some element $a \in G$, $a \ne e$, such that $(a, a, \ldots, a) \in X_{\langle \sigma \rangle}$ and $a^p = e$, so a has order p. Of course $\langle a \rangle$ is a subgroup of G of order p.   ◆

COROLLARY   Let G be a finite group. Then G is a p-group if and only if $|G|$ is a power of p.

PROOF   We leave the proof of this corollary to Exercise 10.   ◆

The Sylow Theorems

Let G be a group, and let $\mathscr{S}$ be the collection of all subgroups of G. We make $\mathscr{S}$ into a G-set by letting G act on $\mathscr{S}$ by conjugation. That is, if $H \in \mathscr{S}$ so $H \le G$ and $g \in G$, then g acting on H yields the conjugate subgroup gHg^{-1}. (To avoid confusion we will never write this action as gH.) Now $G_H = \{g \in G \mid gHg^{-1} = H\}$ is a subgroup of G by Theorem 3.20, and H is a normal subgroup of G_H. Since G_H consists of *all* elements of G that leave H invariant under conjugation, G_H is the largest subgroup of G having H as a normal subgroup.

DEFINITION 4.2 (Normalizer)   The subgroup H_G just discussed is the **normalizer of H in** G and we will be denoted $N[H]$ from now on.

LEMMA 4.3   Let H be a p-subgroup of a finite group G. Then

$$(N[H]:H) \equiv (G:H) \,(\mathrm{mod}\, p).$$

PROOF   Let $\mathscr{L}$ be the set of left cosets of H in G, and let H act on $\mathscr{L}$ by left translation, so that $h(xH) = (hx)H$. Then $\mathscr{L}$ becomes an H-set. Note that $|\mathscr{L}| = (G:H)$.

Let us determine $\mathscr{L}_H$, that is, those left cosets that are fixed under action by all elements of H. Now $xH = h(xH)$ if and only if $H = x^{-1}hxH$, or if and only if $x^{-1}hx \in H$. Thus $xH = h(xH)$ for all $h \in H$ if and only if

---

◆   H I S T O R I C A L   N O T E   ◆

The Sylow theorems are due to the Norwegian mathematician Peter Ludvig Mejdell Sylow (1832–1918), who published them in a brief paper in 1872. Sylow stated the theorems in terms of permutation groups (since the abstract definition of a group had not yet been given). Georg Frobenius re-proved the theorems for abstract groups in 1887, even though he noted that in fact every group can be considered as a permutation group (Theorem 3.5 of this text). Sylow himself immediately applied the theorems to the question of solving algebraic equations and showed that any equation whose Galois group is a power of a prime p is solvable by radicals.

Sylow spent most of his professional life as a high school teacher in Halden, Norway, and was only appointed to a position at Christiana University in 1898. He devoted eight years of his life to the project of editing the mathematical works of his countryman Niels Henrik Abel.

$x^{-1}hx = x^{-1}h(x^{-1})^{-1} \in H$ for all $h \in H$, or if and only if $x^{-1} \in N[H]$, or if and only if $x \in N[H]$. Thus the left cosets in $\mathscr{L}_H$ are those contained in $N[H]$. The number of such cosets is $(N[H]:H)$, so $|\mathscr{L}_H| = (N[H]:H)$.

Since H is a p-group, it has order a power of p by the corollary to Theorem 4.6. Theorem 4.5 then tells us that $|\mathscr{L}| \equiv |\mathscr{L}_H| \pmod{p}$, that is, that $(G:H) \equiv (N[H]:H) \pmod{p}$. ◆

COROLLARY Let H be a p-subgroup of a finite group G. If p divides $(G:H)$, then $N[H] \neq H$.

PROOF It follows from Lemma 4.3 that p divides $(N[H]:H)$, which must then be different from 1. Thus $H \neq N[H]$. ◆

We are now ready for the first of the Sylow theorems, which asserts the existence of prime-power subgroups of G for any prime power dividing $|G|$.

THEOREM 4.7 (First Sylow Theorem) Let G be a finite group and let $|G| = p^n m$ where $n \geq 1$ and where p does not divide m. Then

1. G contains a subgroup of order p^i for each i where $1 \leq i \leq n$,
2. Every subgroup H of G of order p^i is a normal subgroup of a subgroup of order p^{i+1} for $1 \leq i < n$.

PROOF

1. We know G contains a subgroup of order p by Cauchy's theorem (Theorem 4.6). We use an induction argument and show that the existence of a subgroup of order p^i for $i < n$ implies the existence of a subgroup of order p^{i+1}. Let H be a subgroup of order p^i. Since $i < n$, we see p divides $(G:H)$. By Lemma 4.3, we then know p divides $(N[H]:H)$. Since H is a normal subgroup of $N[H]$, we can form $N[H]/H$, and we see that p divides $|N[H]/H|$. By Cauchy's theorem, the factor group $N[H]/H$ has a subgroup K which is of order p. If $\gamma:N[H] \to N[H]/H$ is the canonical homomorphism, then $\gamma^{-1}[K] = \{x \in N[H] \mid \gamma(x) \in K\}$ is a subgroup of $N[H]$ and hence of G. This subgroup contains H and is of order p^{i+1}.
2. We repeat the construction in part 1 and note that $H < \gamma^{-1}[K] \leq N[H]$ where $|\gamma^{-1}[K]| = p^{i+1}$. Since H is normal in $N[H]$, it is of course normal in the possibly smaller group $\gamma^{-1}[K]$. ◆

DEFINITION 4.3 (Sylow Subgroup) A **Sylow p-subgroup** P of a group G is a maximal p-subgroup of G, that is, a p-subgroup contained in no larger p-subgroup.

Let G be a finite group, where $|G| = p^n m$ as in Theorem 4.7. The theorem shows that the Sylow p-subgroups of G are precisely those subgroups of order p^n. If P is a Sylow p-subgroup, every conjugate gPg^{-1} of P is also a Sylow p-subgroup. The second Sylow theorem states that every Sylow p-subgroup can be obtained from P in this fashion; that is, any two Sylow p-subgroups are conjugate.

THEOREM 4.8 (Second Sylow Theorem) Let P_1 and P_2 be Sylow p-subgroups of a finite group G. Then P_1 and P_2 are conjugate subgroups of G.

PROOF Here we will let one of the subgroups act on left cosets of the other, and use Theorem 4.5. Let $\mathscr{L}$ be the collection of left cosets of P_1, and let P_2 act on $\mathscr{L}$ by $y(xP_1) = (yx)P_1$ for $y \in P_2$. Then $\mathscr{L}$ is a P_2-set. By Theorem 4.5, $|\mathscr{L}_{P_2}| \equiv |\mathscr{L}| \pmod{p}$, and $|\mathscr{L}| = (G:P_1)$ is not divisible by p, so $|\mathscr{L}_{P_2}| \neq 0$. Let $xP_1 \in \mathscr{L}_{P_2}$. Then $yxP_1 = xP_1$ for all $y \in P_2$, so $x^{-1}yxP_1 = P_1$ for all $y \in P_2$. Thus $x^{-1}yx \in P_1$ for all $y \in P_2$, so $x^{-1}P_2x \leq P_1$. Since $|P_1| = |P_2|$ we must have $P_1 = x^{-1}P_2x$, so P_1 and P_2 are indeed conjugate subgroups. ◆

The final Sylow theorem gives information on the number of Sylow p-subgroups. A few illustrations are given after the theorem, and many more are given in the next section.

THEOREM 4.9 (Third Sylow Theorem) If G is a finite group and p divides $|G|$, then the number of Sylow p-subgroups is congruent to 1 modulo p and divides $|G|$.

PROOF Let P be one Sylow p-subgroup of G. Let $\mathscr{S}$ be the set of all Sylow p-subgroups and let P act on $\mathscr{S}$ by conjugation, so that $x \in P$ carries $T \in \mathscr{S}$ into xTx^{-1}. By Theorem 4.5, $|\mathscr{S}| \equiv |\mathscr{S}_P| \pmod{p}$. Let us find $\mathscr{S}_P$. If $T \in \mathscr{S}_P$, then $xTx^{-1} = T$ for all $x \in P$. Thus $P \leq N[T]$. Of course $T \leq N[T]$ also. Since P and T are both Sylow p-subgroups of G, they are also Sylow p-subgroups of $N[T]$. But then they are conjugate in $N[T]$ by Theorem 4.8. Since T is a normal subgroup of $N[T]$, it is its only conjugate in $N[T]$. Thus $T = P$. Then $\mathscr{S}_P = \{P\}$. Since $|\mathscr{S}| \equiv |\mathscr{S}_P| = 1 \pmod{p}$, we see the number of Sylow p-subgroups is congruent to 1 modulo p.

Now let G act on $\mathscr{S}$ by conjugation. Since all Sylow p-subgroups are conjugate, there is only one orbit in $\mathscr{S}$ under G. If $P \in \mathscr{S}$, then $G_P = N[P]$. Then $|\mathscr{S}| = |\text{orbit of } P| = (G:G_P)$ by Theorem 3.22. But $(G:G_P)$ is a divisor of $|G|$, so the number of Sylow p-subgroups divides $|G|$. ◆

EXAMPLE 1 The Sylow 2-subgroups of S_3 have order 2. The subgroups of order 2 of S_3 in Example 4 of Section 2.1 are

$$\{\rho_0 \ \mu_1\}, \qquad \{\rho_0 \ \mu_2\}, \qquad \{\rho_0 \ \mu_3\}.$$

Note that there are three subgroups and that $3 \equiv 1 \pmod{2}$. Also, 3 divides 6, the order of S_3. We can readily check that

$$i_{\rho_2}\{\rho_0, \mu_1\} = \{\rho_0, \mu_3\} \qquad \text{and} \qquad i_{\rho_1}\{\rho_0, \mu_1\} = \{\rho_0, \mu_2\}$$

where $i_{\rho_j}(x) = \rho_j x \rho_j^{-1}$, illustrating that they are all conjugate. ▲

EXAMPLE 2 Let us use the Sylow theorems to show that no group of order 15 is simple. Let G have order 15. We claim that G has a normal subgroup of order 5. By Theorem 4.7 G has at least one subgroup of order 5, and by Theorem 4.9 the number of such subgroups is congruent to 1 modulo 5 and divides 15. Since 1, 6, and 11 are the only positive numbers less than 15 that are congruent to 1 modulo 5, and since among these only the number 1 divides 15, we see that G has exactly one subgroup P of order 5. But for each $g \in G$, the inner automorphism i_g of G with $i_g(x) = gxg^{-1}$ maps P onto a subgroup gPg^{-1}, again of order 5. Hence we must have $gPg^{-1} = P$ for all $g \in G$, so P is a normal subgroup of G. Therefore, G is not simple. (Example 3 of Section 4.3 will show that G must actually be abelian and therefore must be cyclic.) ◆

We trust that Example 2 gives some inkling of the power of Theorem 4.9. *Never underestimate a theorem that counts something, even modulo p.*

Exercises 4.2

Computations

In Exercises 1 through 4, fill in the blanks.

1. A Sylow 3-subgroup of a group of order 12 has order _____.

2. A Sylow 3-subgroup of a group of order 54 has order _____.

3. A group of order 24 must have either _____ or _____ Sylow 2-subgroups. (Use only the information given in Theorem 4.9.)

4. A group of order $255 = (3)(5)(17)$ must have either _____ or _____ Sylow 3-subgroups and _____ or _____ Sylow 5-subgroups. (Use only the information given in Theorem 4.9.)

5. Find all Sylow 3-subgroups of S_4 and demonstrate that they are all conjugate.

6. Find two Sylow 2-subgroups of S_4 and show that they are conjugate.

Concepts

7. Mark each of the following true or false.
   _____ a. Any two Sylow p-subgroups of a finite group are conjugate.
   _____ b. Theorem 4.9 shows that a group of order 15 has only one Sylow 5-subgroup.
   _____ c. Every Sylow p-subgroup of a finite group has order a power of p.
   _____ d. Every p-subgroup of every finite group is a Sylow p-subgroup.
   _____ e. Every finite abelian group has exactly one Sylow p-subgroup for each prime p dividing the order of G.
   _____ f. The normalizer in G of a subgroup H of G is always a normal subgroup of G.
   _____ g. If H is a subgroup of G, then H is always a normal subgroup of $N[H]$.

    _____   h. A Sylow p-subgroup of a finite group G is normal in G if and only if it is the only Sylow p-subgroup of G.

    _____   i. If G is an abelian group and H is a subgroup of G, then $N[H] = H$.

    _____   j. A group of prime-power order p^n has no Sylow p-subgroup.

Theory

8. Let G be a finite group and let p divide $|G|$. Prove that if G has precisely one proper Sylow p-subgroup, it is a normal subgroup, so G is not simple.

9. Show that every group of order 45 has a normal subgroup of order 9.

10. Prove the corollary of Theorem 4.6.

11. Let G be a finite group and let p divide $|G|$. Let P be a Sylow p-subgroup of G. Show that $N[N[P]] = N[P]$. [*Hint:* Argue that P is the only Sylow p-subgroup of $N[N[P]]$, and use Theorem 4.8.]

12. Let G be a finite group and let p divide $|G|$. Let P be a Sylow p-subgroup of G and let H be any p-subgroup of G. Show there exists $g \in G$ such that $gHg^{-1} \leq P$.

13. Show that every group of order $(35)^3$ has a normal subgroup of order 125.

14. Show that there are no simple groups of order $255 = (3)(5)(17)$.

15. Show that there are no simple groups of order $p'm$, where p is a prime and $m < p$.

16. Let G be a finite group. Regard G as a G-set where G acts on itself by conjugation.

    a. Show that G_G is the center $Z(G)$ of G. (See Section 3.4.)

    b. Use Theorem 4.5 to show that the center of a finite nontrivial p-group is nontrivial.

17. Show that a finite group of order p^n contains *normal* subgroups H_i for $0 \leq i \leq n$ such that $|H_i| = p^i$ and $H_i < H_{i+1}$ for $0 \leq i < n$. [*Hint:* See Exercise 16 and get an idea from Section 3.5.]

18. Show that a normal p-subgroup of a finite group is contained in every Sylow p-subgroup.

4.3

Applications of the Sylow Theory

In this section we give several applications of the Sylow theorems. It is intriguing to see how easily certain facts about groups of particular orders can be deduced. However, we should realize that we are working only with groups of finite order and really making only a small dent in the general problem of determining the structure of all finite groups. If the order of a group has only a few factors, then the techniques illustrated in this chapter may be of some use in determining the structure of the group. This will be demonstrated further in Section 4.6, where we shall show how it is sometimes possible to describe all groups (up to isomorphism) of certain orders, even when some of the groups are not abelian. However, if the order of a finite group is highly composite, that is, has a large number of factors, the problem is in general much harder.

Applications to p-Groups and the Class Equation

THEOREM 4.10 Every group of prime-power order (that is, every finite p-group) is solvable.

PROOF If G has order p^r, it is immediate from Theorem 4.7 that G has a subgroup H_i of order p^i normal in a subgroup H_{i+1} of order p^{i+1} for $1 \le i \le r$. Then

$$\{e\} = H_0 < H_1 < H_2 < \cdots < H_r = G$$

is a composition series, where the factor groups are of order p, and hence abelian and actually cyclic. Thus, G is solvable. ◆

The older proofs of the Sylow theorems used the *class equation*. The line of proof in Section 4.2 avoided explicit mention of the class equation, although Eq. (2) of Section 4.2 is a general form of it. We now develop the classic class equation so you will be familiar with it.

Let X be a finite G-set where G is a finite group. Then Eq. (2) of Section 4.2 tells us that

$$|X| = |X_G| + \sum_{i=s+1}^{r} |Gx_i| \tag{1}$$

where x_i is an element in the ith orbit in X. Consider now the special case of Eq. (1) where $X = G$ and the action of G on G is by conjugation, so $g \in G$ carries $x \in X = G$ into gxg^{-1}. Then

$$X_G = \{x \in G \mid gxg^{-1} = x \text{ for all } g \in G\}$$
$$= \{x \in G \mid xg = gx \text{ for all } g \in G\} = Z(G),$$

the center of G. If we let $c = |Z(G)|$ and $n_i = |Gx_i|$ in Eq. (1), then we obtain

$$|G| = c + n_{c+1} + \cdots + n_r \tag{2}$$

where n_i is the number of elements in the ith orbit of G under conjugation by itself. Note that n_i divides $|G|$ for $c + 1 \le i \le r$ since in Eq. (1) we know $|Gx_i| = (G:G_{x_i})$, which is a divisor of $|G|$.

DEFINITION 4.4 (Class Equation) Equation (2) is the **class equation of** G. Each orbit in G under conjugation by G is a **conjugate class in** G.

EXAMPLE 1 It is readily checked that for S_3 of Example 4, Section 2.1, the conjugate classes are

$$\{\rho_0\}, \qquad \{\rho_1, \rho_2\}, \qquad \{\mu_1, \mu_2, \mu_3\}.$$

The class equation of S_3 is

$$6 = 1 + 2 + 3. \; \blacktriangle$$

For illustration of the use of the class equation, we prove a theorem that Exercise 16(b) in Section 4.2 asked us to prove.

THEOREM 4.11   The center of a nontrivial p-group G is nontrivial.

PROOF   In Eq. (2) for G, each n_i divides $|G|$ for $c + 1 \le i \le r$, so p divides each n_i, and p divides $|G|$. Therefore p divides c. Now $e \in Z(G)$, so $c \ge 1$. Therefore $c \ge p$, and there exists some $a \in Z(G)$ where $a \ne e$. ◆

We turn now to a lemma on direct products that will be used in some of the theorems that follow.

LEMMA 4.4   Let G be a group containing normal subgroups H and K such that $H \cap K = \{e\}$ and $H \vee K = G$. Then G is isomorphic to $H \times K$.

PROOF   We start by showing that $hk = kh$ for $k \in K$ and $h \in H$. Consider the commutator $hkh^{-1}k^{-1} = (hkh^{-1})k^{-1} = h(kh^{-1}k^{-1})$. Since H and K are normal subgroups of G, the two groupings with parentheses show that $hkh^{-1}k^{-1}$ is in both K and H. Since $K \cap H = \{e\}$, we see that $hkh^{-1}k^{-1} = e$, so $hk = kh$.

Let $\phi : H \times K \to G$ be defined by $\phi(h, k) = hk$. Then

$$\phi((h, k)(h', k')) = \phi(hh', kk') = hh'kk'$$
$$= hkh'k' = \phi(h, k)\phi(h', k'),$$

so ϕ is a homomorphism.

If $\phi(h, k) = e$, then $hk = e$, so $h = k^{-1}$, and both h and k are in $H \cap K$, Thus $h = k = e$, so $\mathrm{Ker}(\phi) = \{(e, e)\}$ and ϕ is one to one.

By Lemma 4.1 in Section 4.1, we know that $HK = H \vee K$, and $H \vee K = G$ by hypothesis. Thus ϕ is onto G, and $H \times K \simeq G$. ◆

THEOREM 4.12   For a prime number p, every group G of order p^2 is abelian.

PROOF   If G is not cyclic, then every element except e must be of order p. Let a be such an element. Then the cyclic subgroup $\langle a \rangle$ of order p does not exhaust G. Also let $b \in G$ with $b \notin \langle a \rangle$. Then $\langle a \rangle \cap \langle b \rangle = \{e\}$, since an element c in $\langle a \rangle \cap \langle b \rangle$ with $c \ne e$ would generate both $\langle a \rangle$ and $\langle b \rangle$, giving $\langle a \rangle = \langle b \rangle$, contrary to construction. From Theorem 4.7, $\langle a \rangle$ is normal in some subgroup of order p^2 of G, that is, normal in all of G. Likewise $\langle b \rangle$ is normal in G. Now $\langle a \rangle \vee \langle b \rangle$ is a subgroup of G properly containing $\langle a \rangle$ and of order dividing p^2. Hence $\langle a \rangle \vee \langle b \rangle$ must be all of G. Thus the hypotheses of Lemma 4.4 are satisfied, and G is isomorphic to $\langle a \rangle \times \langle b \rangle$ and therefore abelian. ◆

Further Applications

We turn now to a discussion of whether there exist simple groups of certain orders. We have seen that every group of prime order is simple. We also asserted that A_n is simple for $n \geq 5$ and that A_5 is the smallest simple group that is not of prime order. There was a famous conjecture of Burnside that every finite simple group of nonprime order must be of even order. It was a triumph when this was proved by Thompson and Feit [22].

THEOREM 4.13   If p and q are distinct primes with $p < q$, then every group G of order pq has a single subgroup of order q and this subgroup is normal in G. Hence G is not simple. If q is not congruent to 1 modulo p, then G is abelian and cyclic.

PROOF   Theorems 4.7 and 4.9 tell us that G has a Sylow q-subgroup and that the number of such subgroups is congruent to 1 modulo q and divides pq, and therefore must divide p. Since $p < q$, the only possibility is the number 1. Thus there is only one Sylow q-subgroup Q of G. This group Q must be normal in G, for under an inner automorphism it would be carried into a group of the same order, hence itself. Thus G is not simple.

Likewise, there is a Sylow p-subgroup P of G, and the number of these divides q and is congruent to 1 modulo p. This number must be either 1 or q. If q is not congruent to 1 modulo p, then the number must be 1 and P is normal in G. Let us assume that $q \neq 1 \pmod{p}$. Since every element in Q other than e is of order q and every element in P other than e is of order p, we have $Q \cap P = \{e\}$. Also $Q \vee P$ must be a subgroup of G properly containing Q and of order dividing pq. Hence $Q \vee P = G$ and by Lemma 4.4 is isomorphic to $Q \times P$ or $\mathbb{Z}_q \times \mathbb{Z}_p$. Thus G is abelian and cyclic. ◆

We need another lemma for some of the counting arguments that follow.

LEMMA 4.5   If H and K are finite subgroups of a group G, then

$$|HK| = \frac{(|H|)(|K|)}{|H \cap K|}.$$

PROOF   Recall that $HK = \{hk \mid h \in H, k \in K\}$. Let $|H| = r$, $|K| = s$, and $|H \cap K| = t$. Now HK has at most rs elements. However, it is possible for $h_1 k_1$ to equal $h_2 k_2$, for $h_1, h_2 \in H$ and $k_1, k_2 \in K$; that is, there may be some collapsing. If $h_1 k_1 = h_2 k_2$, then let

$$x = (h_2)^{-1} h_1 = k_2 (k_1)^{-1}.$$

Now $x = (h_2)^{-1} h_1$ shows that $x \in H$, and $x = k_2(k_1)^{-1}$ shows that $x \in K$. Hence $x \in (H \cap K)$, and

$$h_2 = h_1 x^{-1} \quad \text{and} \quad k_2 = xk_1.$$

On the other hand, if for $y \in (H \cap K)$ we let $h_3 = h_1 y^{-1}$ and $k_3 = yk_1$, then clearly $h_3 k_3 = h_1 k_1$, with $h_3 \in H$ and $k_3 \in K$. Thus each element $hk \in HK$ can be represented in the form $h_i k_i$, for $h_i \in H$ and $k_i \in K$, as many times as there are elements of $H \cap K$, that is, t times. Therefore, the number of elements in HK is rs/t. ◆

Lemma 4.5 is another result that counts something, so don't underestimate it. The lemma will be used in the following way: A finite group G can't have subgroups H and K that are too large with intersections that are too small, or the order of HK would have to exceed the order of G, which is impossible. For example, a group of order 24 can't have two subgroups of orders 12 and 8 with an intersection of order 2.

The remainder of this section consists of several examples illustrating techniques of proving that all groups of certain orders are abelian or that they have nontrivial proper normal subgroups, that is, that they are not simple. We will use one fact we mentioned before only in the exercises. *A subgroup H of index 2 in a finite group G is always normal,* for by counting, we see that there are only the left cosets H itself and the coset consisting of all elements in G not in H. The right cosets are the same. Thus every right coset is a left coset, and H is normal in G.

EXAMPLE 2   No group of order p^r for $r > 1$ is simple, where p is a prime. For by Theorem 4.7 such a group G contains a subgroup of order p^{r-1} normal in a subgroup of order p^r, which must be all of G. Thus a group of order 16 is not simple; it has a normal subgroup of order 8. ▲

EXAMPLE 3   Every group of order 15 is cyclic (hence abelian and not simple, since 15 is not a prime). This is because $15 = (5)(3)$, and 5 is not congruent to 1 modulo 3. By Theorem 4.13 we are done. ▲

EXAMPLE 4   No group of order 20 is simple, for such a group G contains Sylow 5-subgroups in number congruent to 1 modulo 5 and a divisor of 20, hence only 1. This Sylow 5-subgroup is then normal, since all conjugates of it must be itself. ▲

EXAMPLE 5   No group of order 30 is simple. We have seen that if there is only one Sylow p-subgroup for some prime p dividing 30, we are done. By Theorem 4.9 the possibilities for the number of Sylow 5-subgroups are 1 to 6, and those for Sylow 3-subgroups are 1 or 10. But if G has six Sylow 5-subgroups, then the intersection of any two is a subgroup of each of order dividing 5, and hence just $\{e\}$. Thus each contains 4 elements of order 5 that are in none of the others. Hence G must contain 24 elements of order 5. Similarly, if G has 10 Sylow 3-subgroups, it has at least 20 elements of order

3. The two types of Sylow subgroups together would require at least 44 elements in G. Thus there is a normal subgroup either of order 5 or of order 3. ▲

EXAMPLE 6   No group of order 48 is simple. Indeed, we shall show that a group G of order 48 has a normal subgroup of either order 16 or order 8. By Theorem 4.9 G has either one or three Sylow 2-subgroups of order 16. If there is only one subgroup of order 16, it is normal in G, by a now familiar argument.

Suppose that there are three subgroups of order 16, and let H and K be two of them. Then $H \cap K$ must be of order 8, for if $H \cap K$ were of order ≤ 4, then by Lemma 4.5 HK would have at least $(16)(16)/4 = 64$ elements, contradicting the fact that G has only 48 elements. Therefore, $H \cap K$ is normal in both H and K (being of index 2, or by Theorem 4.7). Hence the normalizer of $H \cap K$ contains both H and K and must have order a multiple >1 of 16 and a divisor of 48, therefore 48. Thus $H \cap K$ must be normal in G. ▲

EXAMPLE 7   No group of order 36 is simple. Such a group G has either one or four subgroups of order 9. If there is only one such subgroup, it is normal in G. If there are four such subgroups, let H and K be two of them. As in Example 6, $H \cap K$ must have at least 3 elements, or HK would have to have 81 elements, which is impossible. Thus the normalizer of $H \cap K$ has as order a multiple >1 of 9 and a divisor of 36; hence the order must be either 18 or 36. If the order is 18, the normalizer is then of index 2 and therefore is normal in G. If the order is 36, then $H \cap K$ is normal in G. ▲

EXAMPLE 8   Every group of order $255 = (3)(5)(17)$ is abelian (hence cyclic by Theorem 2.11 and not simple, since 255 is not a prime). By Theorem 4.9 such a group G has only one subgroup H of order 17, Then G/H has order 15 and is abelian by Example 3. By Theorem 3.15, Section 3.4, we see that the commutator subgroup C of G is contained in H. Thus as a subgroup of H, C has either order 1 or 17. Theorem 4.9 also shows that G has either 1 or 85 subgroups of order 3 and either 1 or 51 subgroups of order 5. However, 85 subgroups of order 3 would require 170 elements of order 3, and 51 subgroups of order 5 would require 204 elements of order 5 in G; both together would then require 375 elements in G, which is impossible. Hence there is a subgroup K having either order 3 or order 5 and normal in G. Then G/K has either order $(5)(17)$ or order $(3)(17)$, and in either case Theorem 4.13 shows that G/K is abelian. Thus $C \leq K$ and has order either 3, 5, or 1. Since $C \leq H$ showed that C has order 17 or 1, we conclude that C has order 1. Hence $C = \{e\}$, and $G/C \simeq G$ is abelian. Theorem 2.11 then shows that G is cyclic. ▲

Exercises 4.3

Computations

1. Let D_4 be the group of symmetries of the square in Example 5, Section 2.1.

   a. Find the decomposition of D_4 into conjugate classes.

   b. Write the class equation for D_4.

2. By arguments similar to those used in the examples of this section, convince yourself that every nontrivial group of order not a prime and less than 60 contains a nontrivial proper normal subgroup and hence is not simple. You need not write out the details. (The hardest cases were discussed in the examples.)

Concepts

3. Mark each of the following true or false.

   _____ a. Every group of order 159 is cyclic.

   _____ b. Every group of order 102 has a nontrivial proper normal subgroup.

   _____ c. Every solvable group is of prime-power order.

   _____ d. Every group of prime-power order is solvable.

   _____ e. It would become quite tedious to show that no group of nonprime order between 60 and 168 is simple by the methods illustrated in the text.

   _____ f. No group of order 21 is simple.

   _____ g. Every group of 125 elements has at least 5 elements that commute with every element in the group.

   _____ h. Every group of order 42 has a normal subgroup of order 7.

   _____ i. Every group of order 42 has a normal subgroup of order 8.

   _____ j. The only simple groups are the groups $\mathbb{Z}_p$ and A_n, where p is a prime and $n \neq 4$.

Theory

4. Prove that every group of order $(5)(7)(47)$ is abelian and cyclic.

5. Prove that no group of order 96 is simple.

6. Prove that no group of order 160 is simple.

7. This exercise determines the conjugate classes of S_n for every integer $n \geq 1$.

   a. Show that if $\sigma = (a_1, a_2, \ldots, a_m)$ is a cycle in S_n and τ is any element of S_n, then $\tau\sigma\tau^{-1} = (\tau a_1, \tau a_2, \ldots, \tau a_m)$.

   b. Argue from (a) that any two cycles in S_n of the same length are conjugate.

   c. Argue from (a) and (b) that a product of s disjoint cycles in S_n of lengths r_i for $i = 1, 2, \ldots, s$ is conjugate to every other product of s disjoint cycles of lengths r_i in S_n.

   d. Show that the number of conjugate classes in S_n is $p(n)$, where $p(n)$ is the number of ways, neglecting the order of the summands, that n can be expressed as a sum of positive integers. The number $p(n)$ is the **number of partitions of** n.

   e. Compute $p(n)$ for $n = 1, 2, 3, 4, 5, 6, 7$.

8. Find the conjugate classes and the class equation for S_4. [*Hint:* Use Exercise 7.]

9. Find the class equation for S_5 and S_6. [*Hint:* Use Exercise 7.]

10. Show that the number of conjugate classes in S_n is also the number of different abelian groups (up to isomorphism) of order p^n, where p is a prime number. [*Hint:* Use Exercise 7.]

11. Show that if $n > 2$, the center of S_n is the subgroup consisting of the identity permutation only. [*Hint:* Use Exercise 7.]

4.4

Free Abelian Groups

In this section we introduce the concept of free abelian groups and prove some results concerning them. The section concludes with a demonstration of the fundamental theorem of finitely generated abelian groups (Theorem 2.11, Section 2.4).

Free Abelian Groups

We should review the notions of a generating set for a group G and a finitely generated group, as given in Section 1.4. In this section we shall deal exclusively with abelian groups and use additive notations as follows:

0 for the identity, + for the operation,

$$na = \underbrace{a + a + \cdots + a}_{n \text{ summands}}$$

$$-na = \underbrace{(-a) + (-a) + \cdots + (-a)}_{n \text{ summands}} \quad \Bigg\} \quad \text{for } n \in \mathbb{Z}^+ \text{ and } a \in G,$$

$0a = 0$ for the first 0 in $\mathbb{Z}$ and the second in G.

We shall continue to use the symbol $\times$ for direct product of groups rather than change to direct sum notation.

$\{(1, 0), (0, 1)\}$ is a generating set for the group $\mathbb{Z} \times \mathbb{Z}$ since $(n, m) = n(1, 0) + m(0, 1)$ for any (n, m) in $\mathbb{Z} \times \mathbb{Z}$. This generating set has the property that each element of $\mathbb{Z} \times \mathbb{Z}$ can be *uniquely* expressed in the form $n(1, 0) + m(0, 1)$. That is, the coefficients n and m in $\mathbb{Z}$ are unique.

THEOREM 4.14 Let X be a subset of a nonzero abelian group G. The following conditions on X are equivalent.

1. Each nonzero element a in G can be *uniquely* expressed in the form $a = n_1 x_1 + n_2 x_2 + \cdots + n_r x_r$ for $n_i \neq 0$ in $\mathbb{Z}$ and distinct x_i in X.
2. X generates G, and $n_1 x_1 + n_2 x_2 + \cdots + n_r x_r = 0$ for $n_i \in \mathbb{Z}$ and distinct $x_i \in X$ if and only if $n_1 = n_2 - \cdots = n_r = 0$.

PROOF   Suppose Condition 1 is true. Since $G \neq \{0\}$, we have $X \neq \{0\}$. It follows from 1 that $0 \notin X$, for if $x_i = 0$ and $x_j \neq 0$, then $x_j = x_i + x_j$, which would contradict the uniqueness of the expression from x_j. From 1, X generates G, and $n_1 x_1 + n_2 x_2 + \cdots + n_r x_r = 0$ if $n_1 = n_2 = \cdots = n_r = 0$. Suppose that $n_1 x_1 + n_2 x_2 + \cdots + n_r x_r = 0$ with some $n_i \neq 0$; by dropping terms with zero coefficients and renumbering, we can assume all $n_i \neq 0$. Then

$$x_1 = x_1 + \left(n_1 x_1 + n_2 x_2 + \cdots + n_r x_r \right)$$
$$= \left(n_1 + 1 \right) x_1 + n_2 x_2 + \cdots + n_r x_r,$$

which gives two ways of writing $x_1 \neq 0$, contradicting the uniqueness assumption in Condition 1. Thus Condition 1 implies Condition 2.

We now show that Condition 2 implies Condition 1. Let $a \in G$. Since X generates G, we see a can be written in the form $a = n_1 x_1 + n_2 x_2 + \cdots + n_r x_r$. Suppose a has another such expression in terms of elements of X. By using some zero coefficients in the two expressions, we can assume they involve the same elements in X and are of the form

$$a = n_1 x_1 + n_2 x_2 + \cdots + n_r x_r$$
$$a = m_1 x_1 + m_2 x_2 + \cdots + m_r x_r.$$

Subtracting, we obtain

$$0 = \left(n_1 - m_1 \right) x_1 + \left(n_2 - m_2 \right) x_2 + \cdots + \left(n_r - m_r \right) x_r,$$

so $n_i - m_i = 0$ by Condition 2, and $n_i = m_i$ for $i = 1, 2, \ldots, r$. Thus the coefficients are unique. ◆

DEFINITION 4.5 (Free Abelian Group)   An abelian group having a nonempty generating set X satisfying the conditions described in Theorem 4.14 is a **free abelian group**, and X is a **basis** for the group.

EXAMPLE I   The group $\mathbb{Z} \times \mathbb{Z}$ is free abelian and $\{(1,0), (0,1)\}$ is a basis. Similarly, a basis for the free abelian group $\mathbb{Z} \times \mathbb{Z} \times \mathbb{Z}$ is $\{(1,0,0), (0,1,0), (0,0,1)\}$, and so on. Thus finite direct products of the group $\mathbb{Z}$ with itself are free abelian groups. ▲

EXAMPLE 2   The group $\mathbb{Z}_n$ is not free abelian, for $nx = 0$ for every $x \in \mathbb{Z}_n$, and $n \neq 0$, which would contradict Condition 2. ▲

Suppose a free abelian group G has a finite basis $X = \{x_1, x_2, \ldots, x_r\}$. If $a \in G$ and $a \neq 0$, then a has a *unique* expression of the form

$$a = n_1 x_1 + n_2 x_2 + \cdots + n_r x_r \quad \text{for} \quad n_i \in \mathbb{Z}.$$

We define

$$\phi : G \to \underbrace{\mathbb{Z} \times \mathbb{Z} \times \cdots \times \mathbb{Z}}_{r \text{ factors}}$$

by $\phi(a) = (n_1, n_2, \ldots, n_r)$ and $\phi(0) = (0, 0, \ldots, 0)$. It is straightforward to check that ϕ is an isomorphism. We leave the details to the exercises (see Exercise 7) and state the result as a theorem.

THEOREM 4.15   If G is a nonzero free abelian group with a basis of r elements, then G is isomorphic to $\mathbb{Z} \times \mathbb{Z} \times \cdots \times \mathbb{Z}$ for r factors.

It is a fact that any two bases of a free abelian group G contain the same number of elements. We shall prove this only if G has a finite basis, although it is also true if every basis of G is infinite. The proof is really lovely; it gives an easy characterization of the number of elements in a basis in terms of the size of a factor group.

THEOREM 4.16   Let $G \neq \{0\}$ be a free abelian group with a finite basis. Then every basis of G is finite, and all bases have the same number of elements.

PROOF   Let G have a basis $\{x_1, x_2, \ldots, x_r\}$. Then G is isomorphic to $\mathbb{Z} \times \mathbb{Z} \times \cdots \times \mathbb{Z}$ for r factors. Let $2G = \{2g \mid g \in G\}$. It is readily checked that $2G$ is a subgroup of G. Since $G \simeq \mathbb{Z} \times \mathbb{Z} \times \cdots \times \mathbb{Z}$ for r factors, we have

$$G/2G \simeq (\mathbb{Z} \times \mathbb{Z} \times \cdots \times \mathbb{Z})/(2\mathbb{Z} \times 2\mathbb{Z} \times \cdots \times 2\mathbb{Z})$$

$$\simeq \mathbb{Z}_2 \times \mathbb{Z}_2 \times \cdots \times \mathbb{Z}_2$$

for r factors. Thus $|G/2G| = 2^r$, so the number of elements in any finite basis X is $\log_2 |G/2G|$. Thus any two finite bases have the same number of elements.

It remains to show that G cannot also have an infinite basis. Let Y be any basis for G, and let $\{y_1, y_2, \ldots, y_s\}$ be distinct elements in Y. Let H be the subgroup of G generated by $\{y_1, y_2, \ldots, y_s\}$, and let K be the subgroup of G generated by the remaining elements of Y. It is readily checked that $G \simeq H \times K$,   so $G/2G \simeq (H \times K)/(2H \times 2K) \simeq (H/2H) \times (K/2K)$. Since $|H/2H| = 2^s$, we see $|G/2G| \geq 2^s$. Since we have $|G/2G| = 2^r$, we see that $s \leq r$. Then Y cannot be an infinite set, for we could take $s > r$.   ◆

DEFINITION 4.6 (Rank)   If G is a free abelian group, the **rank** of G is the number of elements in a basis for G. (All bases have the same number of elements.)

Proof of the Fundamental Theorem

We shall prove the fundamental theorem (Theorem 2.11, Section 2.4) by showing that any finitely generated abelian group is isomorphic to a factor group of the form

$$(\mathbb{Z} \times \mathbb{Z} \times \cdots \times \mathbb{Z})/(d_1\mathbb{Z} \times d_2\mathbb{Z} \times \cdots \times d_s\mathbb{Z} \times \{0\} \times \cdots \times \{0\}),$$

where both "numerator" and "denominator" have n factors, and d_1 divides d_2, which divides $d_3 \ldots$, which divides d_s. The prime-power decomposition of Theorem 2.11 in Section 2.4 will then follow.

To show that G is isomorphic to such a factor group, we will show that there is a homomorphism of $\mathbb{Z} \times \mathbb{Z} \times \cdots \times \mathbb{Z}$ onto G with kernel of the form $d_1\mathbb{Z} \times d_2\mathbb{Z} \times \cdots \times d_s\mathbb{Z} \times \{0\} \times \cdots \times \{0\}$. The result will then follow by Theorem 3.9, Section 3.3. The theorems that follow give the details of the argument. Our purpose in these introductory paragraphs is to let us see where we are going as we read what follows.

THEOREM 4.17 Let G be a finitely generated abelian group with generating set $\{a_1, a_2, \ldots, a_n\}$. Let

$$\phi: \underbrace{\mathbb{Z} \times \mathbb{Z} \times \cdots \times \mathbb{Z}}_{n \text{ factors}} \to G$$

be defined by $\phi(h_1, h_2, \ldots, h_n) = h_1a_1 + h_2a_2 + \cdots + h_na_n$. Then ϕ is a homomorphism onto G.

PROOF From the meaning of h_ia_i for $h_i \in \mathbb{Z}$ and $a_i \in G$, we see at once that

$$\begin{aligned}
\phi[(h_1, \ldots, h_n) + (k_1, \ldots, k_n)] &= \phi(h_1 + k_1, \ldots, h_n + k_n) \\
&= (h_1 + k_1)a_1 + \cdots + (h_n + k_n)a_n \\
&= (h_1a_1 + k_1a_1) + \cdots + (h_na_n + k_na_n) \\
&= (h_1a_1 + \cdots + h_na_n) \\
&\quad + (k_1a_1 + \cdots + k_na_n) \\
&= \phi(k_1, \ldots, k_n) + \phi(h_1, \ldots, h_n).
\end{aligned}$$

Since $\{a_1, \ldots, a_n\}$ generates G, clearly the homomorphism ϕ is onto G. ◆

We now prove a "replacement property" that makes it possible for us to adjust a basis.

THEOREM 4.18 If $X = \{x_1, \ldots, x_r\}$ is a basis for a free abelian group G and $t \in \mathbb{Z}$, then for $i \neq j$, the set

$$Y = \{x_1, \ldots, x_{j-1}, x_j + tx_i, x_j, x_{j+1}, \ldots, x_r\}$$

is also a basis for G.

PROOF Since $x_j = (-t)x_i + (1)(x_j + tx_i)$, we see that x_j can be recovered from Y, which thus also generates G. Suppose

$$n_1x_1 + \cdots + n_{j-1}x_{j-1} + n_j(x_j + tx_i) + n_{j+1}x_{j+1} + \cdots + n_rx_r = 0.$$

Then

$$n_1x_1 + \cdots + (n_i + n_jt)x_i + \cdots + n_jx_j + \cdots + n_rx_r = 0.$$

and since X is a basis, $n_1 = \cdots = n_i + n_jt = \cdots = n_j = \cdots = n_r = 0$. From $n_j = 0$ and $n_i + n_jt = 0$, it follows that $n_i = 0$ also, so $n_1 = \cdots = n_i = \cdots = n_j = \cdots = n_r = 0$, and Condition 2 of Theorem 4.14 is satisfied. Thus Y is a basis. ◆

EXAMPLE 3   A basis for $\mathbb{Z} \times \mathbb{Z}$ is $\{(1, 0), (0, 1)\}$. Another basis is $\{(1, 0), (4, 1)\}$ for $(4, 1) = 4(1, 0) + (0, 1)$. However, $\{(3, 0), (0, 1)\}$ is not a basis. For example, we can't express $(2, 0)$ in the form $n_1(3, 0) + n_2(0, 1)$ for $n, n_2 \in \mathbb{Z}$. Here $(3, 0) = (1, 0) + 2(1, 0)$, and a multiple of a basis element was added to *itself*, rather than to a *different* basis element. ▲

A free abelian group G of finite rank may have many bases. We show that if $K \leq G$, then K is also free abelian with rank not exceeding that of G. Equally important, there exist bases of G and K nicely related to each other.

THEOREM 4.19   Let G be a nonzero free abelian group of finite rank n, and let K be a nonzero subgroup of G. Then K is free abelian of rank $s \leq n$. Furthermore, there exists a basis $\{x_1, x_2, \ldots, x_n\}$ for G and positive integers $d_1, d_2, \ldots, d_s$ where d_i divides d_{i+1} for $i = 1, \ldots, s - 1$, such that $\{d_1x_1, d_2x_2, \ldots, d_sx_s\}$ is a basis for K.

PROOF   We show that K has a basis of the described form, which will show that K is free abelian of rank at most n. Suppose $Y = \{y_1, \ldots, y_n\}$ is a basis for G. All nonzero elements in K can be expressed in the form

$$k_1 y_1 + \cdots + k_n y_n,$$

where some $|k_i|$ is nonzero. Among *all* bases Y for G, select one Y_1 that yields the minimal such nonzero value $|k_i|$ as all nonzero elements of K are written in terms of the basis elements in Y_1. By renumbering the elements of Y_1 if necessary, we can assume there is $w_1 \in K$ such that

$$w_1 = d_1 y_1 + k_2 y_2 + \cdots + k_n y_n$$

where $d_1 > 0$ and d_1 is the minimal attainable coefficient as just described. Using the division algorithm, we write $k_j = d_1q_j + r_j$ where $0 \leq r_j < d_1$ for $j = 2, \ldots, n$. Then

$$w_1 = d_1(y_1 + q_2 y_2 + \cdots + q_n y_n) + r_2 y_2 + \cdots + r_n y_n. \qquad (1)$$

Now let $x_1 = y_1 + q_2 y_2 + \cdots + q_n y_n$. By Theorem 4.18, $\{x_1, y_2, \ldots, y_n\}$

is also a basis for G. From Eq. (1) and our choice of Y_1 for minimal coefficient d_1, we see that $r_2 = \cdots = r_n = 0$. Thus $d_1 x_1 \in K$.

We now consider bases for G of the form $\{x_1, y_2, \ldots, y_n\}$. Each element of K can be expressed in the form

$$h_1 x_1 + k_2 y_2 + \cdots + k_n y_n.$$

Since $d_1 x_1 \in K$, we can subtract a suitable multiple of $d_1 x_1$ and then using the minimality of d_1 to see that h_1 is a multiple of d_1, we see we actually have $k_2 y_2 + \cdots + k_n y_n$ in K. Among all such bases $\{x_1, y_2, \ldots, y_n\}$, we choose one Y_2 that leads to some $k_i \neq 0$ of minimal magnitude. (It is possible all k_i are always zero. In this case, K is generated by $d_1 x_1$ and we are done.) By renumbering the elements of Y_2 we can assume that there is $w_2 \in K$ such that

$$w_2 = d_2 y_2 + \cdots + k_n y_n$$

where $d_2 < 0$ and d_2 is minimal as just described. Exactly as in the preceding paragraph, we can modify our basis from $Y_2 = \{x_1, y_2, \ldots, y_n\}$ to a basis $\{x_1, x_2, y_3, \ldots, y_n\}$ for G where $d_1 x_1 \in K$ and $d_2 x_2 \in K$. Writing $d_2 = d_1 q + r$ for $0 \leq r < d_1$, we see that $\{x_1 + q x_2, x_2, y_3, \ldots, y_n\}$ is a basis for G, and $d_1 x_1 + d_2 x_2 = d_1 (x_1 + q x_2) + r x_2$ is in K. By our minimal choice of d_1, we see $r = 0$, so d_1 divides d_2.

We now consider all bases of the form $\{x_1, x_2, y_3, \ldots, y_n\}$ for G and examine elements of K of the form $k_3 y_3 + \cdots + k_n y_n$. The pattern is clear. The process continues until we obtain a basis $\{x_1, x_2, \ldots, x_s, y_{s+1}, \ldots, y_n\}$ where the only element of K of the form $k_{s+1} y_{s+1} + \cdots + k_n y_n$ is zero, that is, all k_i are zero. We then let $x_{s+1} = y_{s+1}, \ldots, x_n = y_n$ and obtain a basis for G of the form described in the statement of Theorem 4.19. ◆

THEOREM 4.20   Every finitely generated abelian group is isomorphic to a group of the form

$$\mathbb{Z}_{m_1} \times \mathbb{Z}_{m_2} \times \cdots \times \mathbb{Z}_{m_r} \times \mathbb{Z} \times \mathbb{Z} \times \cdots \times \mathbb{Z},$$

where m_i divides m_{i+1} for $i = 1, \ldots, r - 1$.

PROOF   For the purposes of this proof, it will be convenient to use as notations $\mathbb{Z}/1\mathbb{Z} = \mathbb{Z}/\mathbb{Z} \simeq \mathbb{Z}_1 = \{0\}$. Let G be finitely generated by n elements. Let $F = \mathbb{Z} \times \mathbb{Z} \times \cdots \times \mathbb{Z}$ for n factors. Consider the homomorphism $\phi: F \to G$ of Theorem 4.17, and let K be the kernel of this homomorphism. Then there is a basis for F of the form $\{x_1, \ldots, x_n\}$, where $\{d_1 x_1, \ldots, d_s x_s\}$ is a basis for K and d_i divides d_{i+1} for $i = 1, \ldots, s - 1$. By Theorem 3.9, Section 3.3, G is isomorphic to F/K. But

$$F/K \simeq (\mathbb{Z} \times \mathbb{Z} \times \cdots \times \mathbb{Z})/(d_1 \mathbb{Z} \times d_2 \mathbb{Z} \times \cdots \times d_s \mathbb{Z} \times \{0\} \times \cdots \times \{0\})$$
$$\simeq \mathbb{Z}_{d_1} \times \mathbb{Z}_{d_2} \times \cdots \times \mathbb{Z}_{d_s} \times \mathbb{Z} \times \cdots \times \mathbb{Z}.$$

It is possible that $d_1 = 1$, in which case $\mathbb{Z}_{d_1} = \{0\}$ and can be dropped (up to isomorphism) from this product. Similarly, d_2 may be 1, and so on. We let m_1 be the first $d_i > 1$, m_2 be the next d_i, and so on, and our theorem follows at once. ◆

We have demonstrated the toughest part of Theorem 2.11. Of course a prime-power decomposition exists since we can break the groups $\mathbb{Z}_{m_i}$ into prime-power factors. The only remaining part of Theorem 2.11 concerns the uniqueness of the Betti number, of the torsion coefficients, and of the prime powers. The Betti number appears as the rank of the free abelian group G/T, where T is the torsion subgroup of G. This rank is invariant by Theorem 4.16, which shows the uniqueness of the Betti number. The uniqueness of the torsion coefficients and of prime powers is a bit more difficult to show. We give some exercises that indicate their uniqueness (see Exercises 12 through 20).

Exercises 4.4

Computations

1. Find a basis $\{(a_1, a_2, a_3), (b_1, b_2, b_3), (c_1, c_2, c_3)\}$ for $\mathbb{Z} \times \mathbb{Z} \times \mathbb{Z}$ with all $a_i \neq 0$, all $b_i \neq 0$, and all $c_i \neq 0$. (Many answers are possible.)

2. Is $\{(2, 1), (3, 1)\}$ a basis for $\mathbb{Z} \times \mathbb{Z}$? Prove your assertion.

3. Is $\{(2, 1), (4, 1)\}$ a basis for $\mathbb{Z} \times \mathbb{Z}$? Prove your assertion.

4. Find conditions on $a, b, c, d \in \mathbb{Z}$ for $\{(a, b), (c, d)\}$ to be a basis for $\mathbb{Z} \times \mathbb{Z}$. [*Hint:* Solve $x(a, b) + y(c, d) = (e, f)$ in $\mathbb{R}$, and see when the x and y lie in $\mathbb{Z}$.]

Concepts

5. Show by example that it is possible for a proper subgroup of a free abelian group of finite rank r also to have rank r.

6. Mark each of the following true or false.
   _____ a. Every free abelian group is torsion free.
   _____ b. Every finitely generated torsion-free abelian group is a free abelian group.
   _____ c. There exists a free abelian group of every positive integer rank.
   _____ d. A finitely generated abelian group is free abelian if its Betti number equals the number of elements in some generating set.
   _____ e. If X generates a free abelian group G and $X \subseteq Y \subseteq G$, then Y generates G.
   _____ f. If X is a basis for a free abelian group G and $X \subseteq Y \subseteq G$, then Y is a basis for G.
   _____ g. Every nonzero free abelian group has an infinite number of bases.
   _____ h. Every free abelian group of rank at least 2 has an infinite number of bases.
   _____ i. If K is a nonzero subgroup of a finitely generated free abelian group, then K is free abelian.

——— j. If K is a nonzero subgroup of a finitely generated free abelian group, then G/K is free abelian.

Theory

7. Complete the proof of Theorem 4.15. (See the two sentences preceding the theorem.)

8. Show that a free abelian group contains no nonzero elements of finite order.

9. Show that if G and G' are free abelian groups, then $G \times G'$ is free abelian.

10. Show that free abelian groups of finite rank are precisely the finitely generated abelian groups containing nonzero elements of finite order.

11. Show that $\mathbb{Q}$ under addition is not a free abelian group. [*Hint:* Show that no two distinct rational numbers n/m and r/s could be contained in a set satisfying Condition 2 of Theorem 4.14.]

Exercises 12 through 17 deal with showing the uniqueness of the prime powers appearing in the prime-power decomposition of the torsion subgroup T of a finitely generated abelian group.

12. Let p be a fixed prime. Show that the elements of T having as order some power of p, together with zero, form a subgroup T_p of T.

13. Show that in any prime-power decomposition of T, the subgroup T_p in the preceding exercise is isomorphic to the direct product of those cyclic factors of order some power of the prime p. [This reduces our problem to showing that the group T_p cannot have essentially different decompositions into products of cyclic groups.]

14. Let G be any abelian group and let n be any positive integer. Show that $G[n] = \{x \in G \mid nx = 0\}$ is a subgroup of G. (In multiplicative notation, $G[n] = \{x \in G \mid x^n = e\}$.)

15. Referring to Exercise 14, show that $\mathbb{Z}_{p^r}[p] \simeq \mathbb{Z}_p$ for any $r \geq 1$ and prime p.

16. Using Exercise 15, show that

$$\left(\mathbb{Z}_{p^{r_1}} \times \mathbb{Z}_{p^{r_2}} \times \cdots \times \mathbb{Z}_{p^{r_m}}\right)[p] \simeq \underbrace{\mathbb{Z}_p \times \mathbb{Z}_p \times \cdots \times \mathbb{Z}_p}_{m \text{ factors}}$$

provided each $r_i \geq 1$.

17. Let G be a finitely generated abelian group and T_p the subgroup defined in Exercise 12. Suppose $T_p \simeq \mathbb{Z}_{p^{r_1}} \times \mathbb{Z}_{p^{r_2}} \times \cdots \times \mathbb{Z}_{p^{r_m}} \simeq \mathbb{Z}_{p^{s_1}} \mathbb{Z}_{p^{s_2}} \times \cdots \times \mathbb{Z}_{p^{s_n}}$, where $1 \leq r_1 \leq r_2 \leq \cdots \leq r_m$ and $1 \leq s_1 \leq s_2 \leq \cdots \leq s_n$. We need to show that $m = n$ and $r_i = s_i$ for $i = 1, \ldots, n$ to complete the demonstration of uniqueness of the prime-power decomposition.

   a. Use Exercise 16 to show that $n = m$.

   b. Show $r_1 = s_1$. Suppose $r_i = s_i$ for all $i < j$. Show $r_j = s_j$, which will complete the proof. [*Hint:* Suppose $r_j < s_j$. Consider the subgroup $p^{r_j} T_p = \{p^{r_j} x \mid x \in T_p\}$, and show that this subgroup would then have two prime-power decompositions involving different numbers of nonzero factors. Then argue that this is impossible by part (a) of this exercise.]

Let T be the torsion subgroup of a finitely generated abelian group. Suppose

$T \simeq \mathbb{Z}_{m_1} \times \mathbb{Z}_{m_2} \times \cdots \times \mathbb{Z}_{m_r} \simeq \mathbb{Z}_{n_1} \times \mathbb{Z}_{n_2} \times \cdots \times \mathbb{Z}_{n_s}$, where m_i divides m_{i+1} for $i = 1, \ldots, r - 1$, n_j divides n_{j+1} for $n = 1, \ldots, s - 1$, and $m_1 > 1$ and $n_1 > 1$. We wish to show that $r = s$ and $m_k = n_k$ for $k = 1, \ldots, r$, demonstrating the uniqueness of the torsion coefficients. This is done in Exercises 18 through 20.

18. Indicate how a prime-power decomposition can be obtained from a torsion-coefficient decomposition. (Observe that the preceding exercises show the prime powers obtained are unique.)

19. Argue from Exercise 18 that m_r and n_s can both be characterized as follows. Let $p_1, \ldots, p_t$ be the distinct primes dividing $|T|$, and let $p_1^{h_1}, \ldots, p_t^{h_t}$ be the highest powers of these primes appearing in the (unique) prime-power decomposition. Then $m_r = n_s = p_1^{h_1} p_2^{h_2} \cdots p_t^{h_t}$.

20. Characterize m_{r-1} and n_{s-1}, showing that they are equal, and continue to show $m_{r-i} = n_{s-i}$ for $i = 1, \ldots, r - 1$, and then $r = s$.

4.5

Free Groups

In this section and the next we discuss a portion of group theory that is of great interest not only in algebra but in topology as well. In fact, an excellent and readable discussion of free groups and presentations of group is found in Crowell and Fox [46, Chapters 3 and 4].

Words and Reduced Words

Let A be any (not necessarily finite) set of elements a_i for $i \in I$. We think of A as an **alphabet** and of the a_i as **letters** in the alphabet. Any symbol of the form a_i^n with $n \in \mathbb{Z}$ is a **syllable** and a finite string w of syllables written in juxtaposition is a **word**. We also introduce the **empty word** 1, which has no syllables.

EXAMPLE I Let $A = \{a_1, a_2, a_3\}$. Then

$$a_1 a_3^{-4} a_2^{2} a_3, \qquad a_2^{3} a_2^{-1} a_3 u_1^{2} u_1^{-7}, \qquad \text{and} \qquad a_3^{2}$$

are all words, if we follow the convention of understanding that a_i^1 is the same as a_i. ▲

There are two natural types of modifications of certain words, the **elementary contractions**. The first type consists of replacing an occurrence of $a_i^m a_i^n$ in a word by a_i^{m+n}. The second type consists of replacing an occurrence of a_i^0 in a word by 1, that is, dropping it out of the word. By means of a finite number of elementary contractions, every word can be changed to a **reduced word**, one for which no more elementary contractions are possible. Note that these elementary contractions formally amount to the usual manipulations of integer exponents.

EXAMPLE 2   The reduced form of the word $a_2{}^3a_2{}^{-1}a_3a_1{}^2a_1{}^{-7}$ of Example 1 is $a_2{}^2a_3a_1{}^{-5}$. ▲

It should be said here once and for all that we are going to gloss over several points that some books spend pages proving, usually by complicated induction arguments broken down into many cases. For example, suppose we are given a word and wish to find its reduced form. There may be a variety of elementary contractions that could be performed first. How do we know that the reduced word we end up with is the same no matter the order in which we perform the elementary contractions? The student will probably say this is obvious. Some authors spend considerable effort proving this. The author tends to agree here with the student. Proofs of this sort he regards as tedious, and they have never made him more comfortable about the situation. However, the author is the first to acknowledge that he is not a great mathematician. In deference to the fact that many mathematicians feel that these things do need considerable discussion, we shall mark an occasion when we just state such facts by the phrase, "It would seem obvious that," keeping the quotation marks.

Free Groups

Let the set of all reduced words formed from our alphabet A be $F[A]$. We now make $F[A]$ into a group in a natural way. For w_1 and w_2 in $F[A]$, define $w_1 \cdot w_2$ to be the reduced form of the word obtained by the juxtaposition w_1w_2 of the two words.

EXAMPLE 3   If
$$w_1 = a_2{}^3a_1{}^{-5}a_3{}^2$$
and
$$w_2 = a_3{}^{-2}a_1{}^2a_3a_2{}^{-2},$$
then $w_1 \cdot w_2 = a_2{}^3a_1{}^{-3}a_3a_2{}^{-2}$. ▲

"It would seem obvious that" this operation of multiplication on $F[A]$ is well defined and associative. The empty word 1 acts as an identity element. "It would seem obvious that" given a reduced word $w = F[A]$, if we form the word obtained by first writing the syllables of w in the opposite order and second by replacing each $a_i{}^n$ by $a_i{}^{-n}$, then the resulting word w^{-1} is a reduced word also, and
$$w \cdot w^{-1} = w^{-1} \cdot w = 1.$$

DEFINITION 4.7 (Free Group)   The group $F[A]$ just described is the **free group generated** by A.

Look back at Theorem 1.11 and the definition preceding it to see that the present use of the term *generated* is consistent with the earlier use.

Starting with a group G and a generating set $\{a_i \mid i \in I\}$ which we will abbreviate by $\{a_i\}$, we might ask if G is *free* on $\{a_i\}$, that is, if G is essentially the free group generated by $\{a_i\}$. We define precisely what this is to mean.

DEFINITION 4.8 (Free on A) If G is a group with a set $A = \{a_i\}$ of generators, and if G is isomorphic to $F[A]$ under a map $\phi : G \rightarrow F[A]$ such that $\phi(a_i) = a_i$, then G is **free on** A, and the a_i are **free generators of** G. A group is **free** if it is free on some nonempty set A.

EXAMPLE 4 The only example of a free group that has occurred before is $\mathbb{Z}$, which is free on one generator. Note that, every free group is infinite. ▲

Refer to the literature for proofs of the next three theorems. We will not be using these results. They are stated simply to inform us of these interesting facts.

THEOREM 4.21 If a group G is free on A and also on B, then the sets A and B have the same number of elements; that is, any two sets of free generators of a free group have the same cardinality.

DEFINITION 4.9 (Rank) If G is free on A, the number of elements in A is the **rank of the free group** G.

Actually, the next theorem is quite evident from Theorem 4.21.

THEOREM 4.22 Two free groups are isomorphic if and only if they have the same rank.

THEOREM 4.23 A nontrivial proper subgroup of a free group is free.

EXAMPLE 5 Let $F[\{x, y\}]$ be the free group on $\{x, y\}$. Let

$$y_k = x^k y x^{-k}$$

for $k \geq 0$. The y_k for $k \geq 0$ are free generators for the subgroup of $F[\{x, y\}]$ that they generate. This illustrates that although a subgroup of a free group is free, the rank of the subgroup may be much greater than the rank of the whole group! ▲

Homomorphisms of Free Groups

Our work in this section will be concerned primarily with homomorphisms defined on a free group. The results here are simple and elegant.

THEOREM 4.24 Let G be generated by $A = \{a_i \mid i \in I\}$ and let G' be any group. If a_i' for $i \in I$ are any elements in G', not necessarily distinct, then there is at most one homomorphism $\phi: G \to G'$ such that $\phi(a_i) = a_i'$. If G is free on A, then there is exactly one such homomorphism.

PROOF   Let ϕ be a homomorphism from G into G' such that $\phi(a_i) = a_i'$. Now by Theorem 1.11, for any $x \in G$ we have

$$x = \prod_j a_{i_j}^{n_j}$$

for some finite product of the generators a_i, where the a_{i_j} appearing in the product need not be distinct. Then since ϕ is a homomorphism, we must have

$$\phi(x) = \prod_j \phi(a_{i_j}^{n_j}) = \prod_j (a_{i_j}')^{n_j}.$$

Thus a homomorphism is completely determined by its values on elements of a generating set. This shows that there is at most one homomorphism such that $\phi(a_i) = a_i'$.

Now suppose G is free on A; that is, $G = F[A]$. For

$$x = \prod_j a_{i_j}^{n_j}$$

in G, define $\psi: G \to G'$ by

$$\psi(x) = \prod_j (a_{i_j}')^{n_j}.$$

This map is well defined, since $F[A]$ consists precisely of reduced words; no two different formal products in $F[A]$ are equal. Since the rules for computation involving exponents in G' are formally the same as those involving exponents in G, it is clear that $\psi(xy) = \psi(x)\psi(y)$ for any elements x and y in G, so ψ is indeed a homomorphism. ◆

Perhaps we should have proved the first part of this theorem earlier, rather than having relegated it to the exercises. Note that the theorem states that *a homomorphism of a group is completely determined if we know its value on each element of a generating set*. This was Exercise 37 of Section 3.1. In particular, a homomorphism of a cyclic group is completely determined by its value on any single generator of the group.

THEOREM 4.25   Every group G' is a homomorphic image of a free group G.

PROOF   Let $G' = \{a_i' \mid i \in I\}$, and let $A = \{a_i \mid i \in I\}$ be a set with the same number of elements as G'. Let $G = F[a]$. Then by Theorem 4.24 there exists a homomorphism ψ mapping G into G' such that $\psi(a_i) = a_i'$. Clearly the image of G under ψ is all of G'. ◆

Another Look at Free Abelian Groups

It is important that we do not confuse the notion of a free group with the notion of a free abelian group. A free group on more than one generator is not abelian. In the preceding chapter, we defined a free abelian group as an abelian group that has a basis, that is, a generating set satisfying properties described in Theorem 4.14. There is another approach, via free groups, to free abelian groups. We now describe this approach.

Let $F[A]$ be the free group on the generating set A. We shall write F in place of $F[A]$ for the moment. Note that F is not abelian if A contains more than one element. Let C be the commutator subgroup of F. Then F/C is an abelian group, and it is not hard to show that F/C is free abelian with basis $\{a + C \mid a \in A\}$. If $a + C$ is renamed a, we can view F/C as a free abelian group with basis A. This indicates how a free abelian group having a given set as basis can be constructed. Every free abelian group can be constructed in this fashion, up to isomorphism. That is, if G is free abelian with basis X, form the free group $F[X]$, form the factor group of $F[X]$ modulo its commutator subgroup, and we have a group isomorphic to G.

Theorems 4.21, 4.22, and 4.23 hold for free abelian groups as well as for free groups. In fact, the abelian version of Theorem 4.23 was proved for the finite rank case in Theorem 4.19, Section 4.4. In contrast to Example 5 for free groups, it is true that for a free abelian group the rank of a subgroup is at most the rank of the entire group. Theorem 4.19 also showed this for the finite rank case.

Exercises 4.5

Computations

1. Find the reduced form and the inverse of the reduced form of each of the following words.

   a. $a^2b^{-1}b^3a^3c^{-1}c^4b^{-2}$ 　　　　　　　　 b. $a^2a^{-3}b^3a^4c^4c^2a^{-1}$

2. Compute the products given in parts (a) and (b) of Exercise 1 in the case that $\{a, b, c\}$ is a set of generators forming a basis for a free abelian group. Find the inverse of these products.

3. How many different homomorphisms are there of a free group of rank 2 into

   a. $\mathbb{Z}_4$? 　　　　　　　　 b. $\mathbb{Z}_6$? 　　　　　　　　 c. S_3?

4. How many different homomorphisms are there of a free group of rank 2 onto

   a. $\mathbb{Z}_4$?   b. $\mathbb{Z}_6$?   c. S_3?

5. How many different homomorphisms are there of a free abelian group of rank 2 into

   a. $\mathbb{Z}_4$?   b. $\mathbb{Z}_6$?   c. S_3?

6. How many different homomorphisms are there of a free abelian group of rank 2 onto

   a. $\mathbb{Z}_4$?   b. $\mathbb{Z}_6$?   c. S_3?

Concepts

7. Take one of the instances in this section in which the phrase "It would seem obvious that" was used and discuss your reaction in that instance.

8. Mark each of the following true or false.
   _____ a. Every proper subgroup of a free group is a free group.
   _____ b. Every proper subgroup of every free abelian group is a free group.
   _____ c. A homomorphic image of a free group is a free group.
   _____ d. Every free abelian group has a basis.
   _____ e. The free abelian groups of finite rank are precisely the finitely generated abelian groups.
   _____ f. No free group is abelian.
   _____ g. No free abelian group is free.
   _____ h. No free abelian group of rank >1 is free.
   _____ i. Any two free groups are isomorphic.
   _____ j. Any two free abelian groups of the same rank are isomorphic.

Theory

9. Let G be a finitely generated abelian group with identity 0. A finite set $\{b_1, \ldots, b_n\}$, where $b_i \in G$, is a **basis for** G if $\{b_1, \ldots, b_n\}$ generates G and $\sum_{i=1}^{n} m_i b_i = 0$ if and only if each $m_i b_i = 0$, where $m_i \in \mathbb{Z}$.

   a. Show that $\{2, 3\}$ is not a basis for $\mathbb{Z}_4$. Find a basis for $\mathbb{Z}_4$.

   b. Show that both $\{1\}$ and $\{2, 3\}$ are bases for $\mathbb{Z}_6$. (This shows that for a finitely generated abelian group G with torsion, the number of elements in a basis may vary; that is, it need not be an *invariant* of the group G.)

   c. Is a basis for a free abelian group as we defined it in Section 4.4 a basis in the sense in which it is used in this exercise?

   d. Show that every finite abelian group has a basis $\{b_1, \ldots, b_n\}$, where the order of b_i divides the order of b_{i+1}.

In present-day expositions of algebra, a frequently used technique (particularly by the disciples of N. Bourbaki) for introducing a new algebraic entity is the following:

1. Describe algebraic properties that this algebraic entity is to possess.

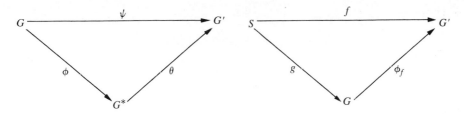

Figure 4.5   **Figure 4.6**

2. Prove that any two algebraic entities with these properties are isomorphic, that is, that these properties characterize the entity.

3. Show that at least one such entity exists.

The next three exercises illustrate this technique for three algebraic entities, each of which we have met before. So that we don't give away their identities, we use fictitious names for them in the first two exercises. The last part of these first two exercises asks us to give the usual name for the entity.

10. Let G be any group. An abelian group G^* is a **blip group of** G if there exists a fixed homomorphism ϕ of G onto G^* such that each homomorphism ψ of G into an abelian group G' can be factored as $\psi = \theta\phi$, where θ is a homomorphism of G^* into G' (see Fig. 4.5).

 a. Show that any two blip groups of G are isomorphic. [*Hint:* Let G_1^* and G_2^* be two blip groups of G. Then each of the fixed homomorphisms $\phi_1: G \to G_1^*$ and $\phi_2: G \to G_2^*$ can be factored via the other blip group according to the definition of a blip group; that is, $\phi_1 = \theta_1\phi_2$ and $\phi_2 = \theta_2\phi_1$. Show that θ_1 is an isomorphism of G_2^* onto G_1^* by showing that both $\theta_1\theta_2$ and $\theta_2\theta_1$ are identity maps.]

 b. Show for every group G that a blip group G^* of G exists.

 c. What concept that we have introduced before corresponds to this idea of a blip group of G?

11. Let S be any set. A group G together with a fixed function $g: S \to G$ constitutes a **blop group on** S if for each group G' and map $f: S \to G'$ there exists a *unique* homomorphism ϕ_f of G into G' such that $f = \phi_f g$ (see Fig. 4.6).

 a. Let S be a fixed set. Show that if both G_1, together with $g_1: S \to G_1$, and G_2, together with $g_2: S \to G_2$, are blop groups on S, then G_1 and G_2 are isomorphic. [*Hint:* Show that g_1 and g_2 are one-to-one maps and that g_1S and g_2S generate G_1 and G_2 respectively. Then proceed in a way analogous to that given by the hint of Exercise 10.]

 b. Let S be a set. Show that a blop group on S exists. You may use any theorems of the text.

 c. What concept that we have introduced before corresponds to this idea of a blop group on S?

12. Characterize a free abelian group by properties in a fashion similar to that used in Exercise 11.

Group Presentations

Definition

Following most of the literature on group presentations, in this chapter we let 1 be the identity of a group. The idea of a *group presentation* is to form a group by giving a set of generators for the group and certain equations or relations that we want the generators to satisfy. We want the group to be as free as it possibly can be on the generators to these relations.

EXAMPLE 1   Suppose G has generators x and y and is *free except for the relation* $xy = yx$, which we may express as $xyx^{-1}y^{-1} = 1$. Note that the condition $xy = yx$ is exactly what is needed to make G commutative, even though $xyx^{-1}y^{-1}$ is just one of the many possible commutators of $F[\{x, y\}]$. Thus G is free abelian on two generators and is isomorphic to $F[\{x, y\}]$ modulo its commutator subgroup. This commutator subgroup of $F[\{x, y\}]$ is the smallest normal subgroup containing $xyx^{-1}y^{-1}$, since any normal subgroup containing $xyx^{-1}y^{-1}$ gives rise to a factor group that is abelian and thus contains the commutator subgroup by Theorem 3.15. ▲

The preceding example illustrates the general situation. Let $F[A]$ be a free group and suppose that we want to form a new group as much like $F[A]$ as it can be, subject to certain equations that we want satisfied. Any equation can be written in a form in which the right-hand side is 1. Thus we can consider the equations to be $r_i = 1$ for $i \in I$, where $r_i \in F[A]$. If we require that $r_i = 1$, then we will have to have

$$x(r_i^n)x^{-1} = 1$$

for any $x \in F[A]$ and $n \in \mathbb{Z}$. Also any product of elements equal to 1 will again have to equal 1. Thus any finite product of the form

$$\prod_j x_j(r_{i_j}^{n_j})x_j^{-1},$$

where the r_{i_j} need not be distinct, will have to equal 1 in the new group. It is readily checked that the set of all these finite products is a normal subgroup R of $F[A]$. Thus any group looking as much as possible like $F[A]$, subject to the requirements $r_i = 1$, also has $r = 1$ for every $r \in R$. But $F[A]/R$ looks like $F[A]$ (remember that we multiply cosets by choosing representatives) except that R has been collapsed to form the identity 1. Hence the group we are after is (at least isomorphic to) $F[A]/R$. We can view this group as described by the generating set A and the set $\{r_i \mid i \in I\}$, which we will abbreviate $\{r_i\}$.

DEFINITION 4.10 (Presentation)   Let A be a set and let $\{r_i\} \subseteq F[A]$. Let R be the least normal subgroup of $F[A]$ containing the r_i. An isomorphism ϕ of $F[A]/R$ onto a group G is a **presentation** of G. The sets A and $\{r_i\}$

give a **group presentation**. The set A is the set of **generators for the presentation** and each r_i is **relator**. Each $r \in R$ is a **consequence of** $\{r_i\}$. An equation $r_i = 1$ is a **relation**. A **finite presentation** is one in which both A and $\{r_i\}$ arc finite sets.

This definition may seem complicated, but it really isn't. In Example 1 $\{x, y\}$ is our set of generators and $xyx^{-1}y^{-1}$ is the only relator. The equation $xyx^{-1}y^{-1} = 1$, or $xy = yx$, is a relation. This was an example of a finite presentation.

If a group presentation has generators x_j and relators r_i, we shall use the notations

$$(x_j:r_i) \qquad \text{or} \qquad (x_j:r_i = 1)$$

to denote the group presentation. We may refer to $F[\{x_j\}]/R$ *as the group with presentation* $(x_j:r_i)$.

Isomorphic Presentations

EXAMPLE 2   Consider the group presentation with

$$A = \{a\} \qquad \text{and} \qquad \{r_i\} = \{a^6\},$$

that is, the presentation

$$(a:a^6 = 1).$$

This group defined by one generator a, with the relation $a^6 = 1$, is isomorphic to $\mathbb{Z}_6$.

Now consider the group defined by two generators a and b, with $a^2 = 1$, $b^3 = 1$, and $ab = ba$, that is, the group with presentation

$$(a, b:a^2, b^3, aba^{-1}b^{-1}).$$

The condition $a^2 = 1$ gives $a^{-1} = a$. Also $b^3 = 1$ gives $b^{-1} = b^2$. Thus every element in this group can be written as a product of nonnegative powers of a and b. The relation $aba^{-1}b^{-1} = 1$, that is, $ab = ba$, allows us to write first all the factors involving a and then the factors involving b. Hence every element of the group is equal to some $a^m b^n$. But then $a^2 = 1$ and $b^3 = 1$ show that there are just six distinct elements,

$$1, b, b^2, a, ab, ab^2.$$

Therefore this presentation also gives a group of order 6 that is abelian, and by Theorem 2.11, Section 2.4, it must again be cyclic and isomorphic to $\mathbb{Z}_6$. ▲

The preceding example illustrates that different presentations may give isomorphic groups. When this happens, we have **isomorphic presentation**. To determine whether two presentations are isomorphic may be very hard. It has recently been shown (see Rabin [23]) that a number of such problems connected with this theory are not generally solvable; that is, there is no *routine* and well-defined way of discovering a solution in all cases. These unsolvable problems include the problem of deciding whether two presentations are isomorphic, whether a group given by a presentation is finite, free, abelian, or trivial, and the famous *word problem* of determining whether a given word w is a consequence of a given set of relators $\{r_i\}$.

The importance of this material is indicated by our Theorem 4.25, Section 4.5, which guarantees that *every group has a presentation*.

EXAMPLE 3   Let us show that

$$(x, y : y^2x = y, \ yx^2y = x)$$

is a presentation of the trivial group of one element. We need only show that x and y are consequences of the relators y^2xy^{-1} and yx^2yx^{-1}, or that $x = 1$ and $y = 1$ can be deduced from $y^2x = y$ and $yx^2y = x$. We illustrate both techniques.

As a consequence of y^2xy^{-1}, we get yx upon conjugation by y^{-1}. From yx we deduce $x^{-1}y^{-1}$, and then $(x^{-1}y^{-1})(yx^2yx^{-1})$ gives xyx^{-1}. Conjugating xyx^{-1} by x^{-1}, we get y. From y we get y^{-1}, and $y^{-1}(yx)$ is x.

Working with relations instead of relators, from $y^2x = y$ we deduce $yx = 1$ upon multiplication by y^{-1} on the left. Then substituting $yx = 1$ into $yx^2y = x$, that is, $(yx)(xy) = x$, we get $xy = x$. Then multiplying by x^{-1} on the left, we have $y = 1$. Substituting this in $yx = 1$, we get $x = 1$.

Both techniques amount to the same work, but it somehow seems more natural to most of us to work with relations. ▲

Applications

We conclude this chapter with two applications.

EXAMPLE 4   Let us determine all groups of order 10 up to isomorphism. We know from Theorem 2.11, Section 2.4 that every abelian group of order 10 is isomorphic to $\mathbb{Z}_{10}$. Suppose that G is nonabelian of order 10. By Sylow theory, G contains a normal subgroup H of order 5, and H must be cyclic. Let a be a generator of H. Then G/H is of order 2 and thus isomorphic to $\mathbb{Z}_2$. If $b \in G$ and $b \notin H$, we must then have $b^2 \in H$. Since every element of H except 1 has order 5, if b^2 were not equal to 1, then b^2 would have order 5, so b would have order 10. This would mean that G would be cyclic, contradicting our assumption that G is not abelian. Thus $b^2 = 1$. Finally, since H is a normal subgroup of G, $bHb^{-1} = H$, so in particular, $bab^{-1} \in H$. Since conjugation by b is an automorphism of H, bab^{-1} must be another element of H of order 5, hence bab^{-1} equals a, a^2, a^3, or a^4. But $bab^{-1} = a$ would give $ba = ab$, and then G would be abelian, since a and b generate G. Thus the possibilities for presentations of G are:

1. $(a, b : a^5 = 1, b^2 = 1, ba = a^2b)$,
2. $(a, b : a^5 = 1, b^2 = 1, ba = a^3b)$,
3. $(a, b : a^5 = 1, b^2 = 1, ba = a^4b)$.

Note that all three of these presentations can give groups of order at most 10, since the last relation $ba = a^ib$ enables us to express every product of a's and b's in G in the form a^sb^t. Then $a^5 = 1$ and $b^2 = 1$ show that the set

$$S = \{a^0b^0, a^1b^0, a^2b^0, a^3b^0, a^4b^0, a^0b^1, a^1b^1, a^2b^1, a^3b^1, a^4b^1\}$$

includes all elements of G.

It is not yet clear that all these elements in S are distinct, so that we have in all three cases a group of order 10. For example, the group presentation

$$(a, b : a^5 = 1, b^2 = 1, ba = a^2b)$$

gives a group in which, using the associative law, we have

$$a = b^2a = (bb)a = b(ba) = b(a^2b) = (ba)(ab)$$
$$= (a^2b)(ab) = a^2(ba)b = a^2(a^2b)b = a^4b^2 = a^4.$$

Thus in this group, $a = a^4$, so $a^3 = 1$, which, together with $a^5 = 1$, yields $a^2 = 1$. But $a^2 = 1$, together with $a^3 = 1$, means that $a = 1$. Hence every element in the group with presentation

$$(a, b : a^5 = 1, b^2 = 1, ba = a^2b)$$

is equal to either 1 or b; that is, this group is isomorphic to $\mathbb{Z}_2$. A similar study of

$$(bb)a = b(ba)$$

for

$$(a, b : a^5 = 1, b^2 = 1, ba = a^3b)$$

shows that $a = a^4$ again, so this also yields a group isomorphic to $\mathbb{Z}_2$.

This leaves just

$$(a, b : a^5 = 1, b^2 = 1, ba = a^4b)$$

as a candidate for a nonabelian group of order 10. In this case, it can be shown that all elements of S are distinct, so this presentation does give a nonabelian group G of order 10. How can we show that all elements in S represent distinct elements of G? The easy way is to observe that we know that there is at least one nonabelian group of order 10, the dihedral group D_5. Since G is the only remaining candidate, we must have $G \approx D_5$. Another attack is as follows. Let us try to make S into a group by defining $(a^sb^t)(a^ub^v)$ to be a^xb^y, where x is the remainder of $s + u(4^t)$ when divided by 5, and y is the remainder of $t + v$ when divided by 2, in the sense of the division algorithm in Section 1.4. In other words, we use the relation $ba = a^4b$ as a guide in defining the product $(a^sb^t)(a^ub^v)$ of two elements of S. We see that a^0b^0 acts as identity, and that given a^ub^v, we can determine t and s

successively by letting
$$t \equiv -v \,(\text{mod}\, 2)$$
and then
$$s = -u(4')(\text{mod}\, 5),$$

giving $a^s b'$, which is a left inverse for $a^u b^v$. We will then have a group structure on S if and only if the associative law holds. Exercise 11 asks us to carry out the straightforward computation for the associative law and to discover a condition for S to be a group under such a definition of multiplication. The criterion of the exercise in this case amounts to the valid congruence

$$4^2 \equiv 1 \,(\text{mod}\, 5).$$

Thus we do get a group of order 10. Note that
$$2^2 \not\equiv 1 \,(\text{mod}\, 5)$$
and
$$3^2 \not\equiv 1 \,(\text{mod}\, 5),$$

so Exercise 11 also shows that
$$\left(a, b : a^5 = 1, b^2 = 1, ba = a^2 b\right)$$
and
$$\left(a, b : a^5 = 1, b^2 = 1, ba = a^3 b\right)$$

do not give groups of order 10. ▲

EXAMPLE 5   Let us determine all groups of order 8 up to isomorphism. We know the three abelian ones:

$$\mathbb{Z}_8, \qquad \mathbb{Z}_2 \times \mathbb{Z}_4, \qquad \mathbb{Z}_2 \times \mathbb{Z}_2 \times \mathbb{Z}_2.$$

Using generators and relations, we shall give presentations of the nonabelian groups.

Let G be nonabelian of order 8. Since G is nonabelian, it has no elements of order 8, so each element but the identity is of order either 2 or 4. If every element were of order 2, then for $a, b \in G$, we would have $(ab)^2 = 1$, that is, $abab = 1$. Then since $a^2 = 1$ and $b^2 = 1$ also, we would have

$$ba = a^2 bab^2 = a(ab)^2 b = ab,$$

contrary to our assumption that G is not abelian. Thus G has an element of order 4.

Let $\langle a \rangle$ be a subgroup of G or order 4. If $b \notin \langle a \rangle$, the cosets $\langle a \rangle$ and $b\langle a \rangle$ exhaust all of G. Hence a and b are generators for G and $a^4 = 1$. Since $\langle a \rangle$ is normal in G (by Sylow theory, or because it is of index 2), $G/\langle a \rangle$ is isomorphic to $\mathbb{Z}_2$ and we have $b^2 \in \langle a \rangle$. If $b^2 = a$ or $b^2 = a^3$, then b would be of order 8. Hence $b^2 = 1$ or $b^2 = a^2$. Finally, since $\langle a \rangle$ is normal, we have $bab^{-1} \in \langle a \rangle$, and since $b\langle a \rangle b^{-1}$ is a subgroup conjugate to $\langle a \rangle$ and hence isomorphic to $\langle a \rangle$, we see that bab^{-1} must be an element of order 4. Thus $bab^{-1} = a$ or $bab^{-1} = a^3$. If bab^{-1} were equal to a, then ba would equal ab, which would make G abelian. Hence $bab^{-1} = a^3$, so $ba = a^3b$. Thus we have two possibilities for G, namely,

$$G_1 : (a, b : a^4 = 1, b^2 = 1, ba = a^3b)$$

and

$$G_2 : (a, b : a^4 = 1, b^2 = a^2, ba = a^3b).$$

Note that $a^{-1} = a^3$, and that b^{-1} is b in G_1 and b^3 in G_2. These facts, along with the relation $ba = a^3b$, enable us to express every element in G_i in the form $a^m b^n$, as in Examples 2 and 4. Since $a^4 = 1$ and either $b^2 = 1$ or $b^2 = a^2$, the possible elements in each group are

$$1, \quad a, \quad a^2, \quad a^3, \quad b, \quad ab, \quad a^2b, \quad a^3b.$$

Thus G_1 and G_2 each have order at most 8. That G_1 is a group of order 8 can be seen from Exercise 11. An argument similar to that used in Exercise 11 shows that G_2 has order 8 also.

Since $ba = a^3b \neq ab$, we see that both G_1 and G_2 are nonabelian. That the two groups are not isomorphic follows from the fact that a computation shows that G_1 has only two elements of order 4, namely, a and a^3. On the other hand, in G_2 all elements but 1 and a^2 are of order 4. We leave the computations of the tables for these groups to Exercise 3. To illustrate, suppose we wish to compute $(a^2b)(a^3b)$. Using $ba = a^3b$ repeatedly, we get

$$(a^2b)(a^3b) = a^2(ba)a^2b = a^5(ba)ab = a^8(ba)b = a^{11}b^2.$$

Then for G_1, we have

$$a^{11}b^2 = a^{11} = a^3,$$

but if we are in G_2, we get

$$a^{11}b^2 = a^{13} = a.$$

The group G_1 is the **octic group** and is nothing more than our old friend the group D_4 of symmetries of the square. The group G_2 is the **quaternion group**; the reason for the name will be explained in Section 5.7. ▲

Exercises 4.6

Computations

1. Give a presentation of $\mathbb{Z}_4$ involving one generator; involving two generators; involving three generators.

2. Give a presentation of S_3 involving three generators.

3. Give the tables for both the octic group

$$(a, b : a^4 = 1, b^2 = 1, ba = a^3b)$$

and the quaternion group

$$(a, b: a^4 = 1, b^2 = a^2, ba = a^3 b).$$

In both cases, write the elements in the order $1, a, a^2, a^3, b, ab, a^2 b, a^3 b$. (Note that we don't have to compute *every* product. We know that these presentations give groups of order 8, and once we have computed enough products the rest are forced so that each row and each column of the table has each element exactly once.)

4. Determine all groups of order 14 up to isomorphism. [*Hint:* Follow the outline of Example 4 and use Exercise 11, part (b).]

5. Determine all groups of order 21 up to isomorphism. [*Hint:* Follow the outline of Example 4 and use Exercise 11, part (*b*). It may seem that there are two presentations giving nonabelian groups. Show that they are isomorphic.]

Concepts

6. Mark each of the following true or false.

_____ a. Every group has a presentation.
_____ b. Every group has many different presentations.
_____ c. Every group has two presentations that are not isomorphic.
_____ d. Every group has a finite presentation.
_____ e. Every group with a finite presentation is of finite order.
_____ f. Every cyclic group has a presentation with just one generator.
_____ g. Every conjugate of a relator is a consequence of the relator.
_____ h. Two presentations with the same number of generators are always isomorphic.
_____ i. In a presentation of an abelian group, the set of consequences of the relators contains the commutator subgroup of the free group on the generators.
_____ j. Every presentation of a free group has 1 as the only relator.

Theory

7. Use the methods of this section and Exercise 11, part (b), to show that there are no nonabelian groups of order 15. (See also Example 3 of Section 4.3.)

8. Show that

$$(a, b: a^3 = 1, b^2 = 1, b^2 = 1, ba = a^2 b)$$

gives a group of order 6. Show that it is nonabelian.

9. Show that the presentation

$$(a, b: a^3 = 1, b^2 = 1, ba = a^2 b)$$

of Exercise 8 gives (up to isomorphism) the only nonabelian group of order 6, and hence gives a group isomorphic to S_3.

10. We showed in Example 4 of Section 3.4 that A_4 has no subgroup of order 6. The preceding exercise shows that such a subgroup of A_4 would have to be isomorphic to either $\mathbb{Z}_6$ or S_3. Show again that this is impossible by considering orders of elements.

11. Let

$$S = \{a^i b^j \mid 0 \le i < m, 0 \le j < n\},$$

that is, S consists of all formal products $a^i b^j$ starting with $a^0 b^0$ and ending with

$a^{m-1}b^{n-1}$. Let r be a positive integer, and define multiplication on S by

$$(a^s b^t)(a^u b^v) = a^x b^y,$$

where x is the remainder of $s + u(r^t)$ when divided by m, and y is the remainder of $t + v$ when divided by n, in the sense of the division algorithm in Section 1.4.

a. Show that a necessary and sufficient condition for the associative law to hold and for S to be a group under this multiplication is that $r^n \equiv 1 \pmod{m}$.

b. Deduce from part (a) that the group presentation

$$(a, b : a^m = 1, b^n = 1, ba = a^r b)$$

gives a group of order mn if and only if $r^n \equiv 1 \pmod{m}$.

12. Show that if $n = pq$, with p and q primes and $q > p$ and $q \equiv 1 \pmod{p}$, then there is exactly one nonabelian group (up to isomorphism) of order n. Assume (as will be proved later) that the $q - 1$ nonzero elements of $\mathbb{Z}_q$ form a cyclic group $\mathbb{Z}_q^*$ under multiplication modulo q. [*Hint:* The solutions of $x^p \equiv 1 \pmod{q}$ form a cyclic subgroup of $\mathbb{Z}_q^*$ with elements $1, r, r^2, \ldots, r^{p-1}$. In the group with presentation $(a, b : a^q = 1, \quad b^p = 1, \quad ba = a^r b)$, we have $bab^{-1} = a^r$, so $b^j a b^{-j} = a^{(r^j)}$. Thus, since b^j generates $\langle b \rangle$ for $j = 1, \ldots, p - 1$, this presentation is isomorphic to

$$(a, b^j : a^q = 1, (b^j)^p = 1, (b^j)a = a^{(r^j)}(b^j)),$$

so all the presentations $(a, b : a^q = 1, b^p = 1, ba = a^{(r^j)}b)$ are isomorphic.]

◆

CHAPTER FIVE

INTRODUCTION TO RINGS AND FIELDS

◆

All our work thus far has been concerned with sets on which a single binary operation has been defined. Our years of work with the integers and real numbers show that a study of sets on which two binary operations have been defined should be of great importance. Algebraic structures of this type are introduced in this chapter. In one sense this chapter seems more intuitive than those that precede it, for the structures studied are closely related to those we have worked with for many years. We will also be concerned with a fundamental problem of algebra: solving a polynomial equation. However, we will be continuing with our axiomatic approach. So, from another viewpoint this study is more complicated than group theory, for we now have two binary operations and more axioms to deal with.

5.1

Rings and Fields

Definitions and Basic Properties

The most general algebraic structure with two binary operations that we shall study is called a *ring*. As Example 1 following Definition 5.1 indicates, we have all worked with rings since grade school.

DEFINITION 5.1 (Ring) A ring $\langle R, +, \cdot \rangle$ is a set R together with two binary operations $+$ and $\cdot$, which we call addition and multiplication, defined on R such that the following axioms are satisfied:

$\mathcal{R}_1$. $\langle R, + \rangle$ is an abelian group.

$\mathcal{R}_2$. Multiplication is associative.

$\mathcal{R}_3$. For all $a, b, c \in R$, the **left distributive law**, $a \cdot (b + c) = (a \cdot b) + (a \cdot c)$ and the **right distributive law** $(a + b) \cdot c = (a \cdot c) + (b \cdot c)$ hold.

The theory of rings grew out of the study of two particular classes of rings, polynomial rings in n variables over the real or complex numbers (Section 5.5) and the "integers" of an algebraic number field (Section 7.3). It was David Hilbert (1862–1943) who first introduced the term *ring,* in connection with the latter example, but it was not until the second decade of the twentieth century that a fully abstract definition appeared. The theory of commutative rings was given a firm axiomatic foundation by Emmy Noether (1882–1935) in her monumental paper "Ideal Theory in Rings," which appeared in 1921. A major concept of this paper is the ascending chain condition for ideals. Noether proved that in any ring in which every ascending chain of ideals has a maximal element (see Section 7.1), every ideal is finitely generated.

Emmy Noether received her doctorate from the University of Erlangen, Germany, in 1907. Hilbert invited her to Göttingen in 1915, but his efforts to secure her a paid position were blocked because of her sex. Hilbert complained, "I do not see that the sex of the candidate is an argument against her admission [to the faculty]. After all, we are a university, not a bathing establishment." Noether was, however, able to lecture under Hilbert's name. Ultimately, after the political changes accompanying the end of the First World War reached Göttingen, she was given in 1923 a paid position at the University. For the next decade, she was very influential in the development of the basic concepts of modern algebra. Along with other Jewish faculty members, however, she was forced to leave Göttingen in 1933. She spent the final two years of her life at Bryn Mawr College near Philadelphia.

EXAMPLE I   We are well aware that axioms $\mathcal{R}_1$, $\mathcal{R}_2$, and $\mathcal{R}_3$ for a ring hold in any subset of the complex numbers that is a group under addition and that is closed under multiplication. For example, $\langle \mathbb{Z}, +, \cdot \rangle$, $\langle \mathbb{Q}, +, \cdot \rangle$, $\langle \mathbb{R}, +, \cdot \rangle$, and $\langle \mathbb{C}, +, \cdot \rangle$ are all rings. ▲

It is customary to denote multiplication in a ring by juxtaposition, using ab in place of $a \cdot b$. We shall also observe the usual convention that multiplication is performed before addition in the absence of parentheses, so the left distributive law, for example, becomes

$$a(b + c) = ab + ac,$$

without the parentheses on the right side of the equation. Also, as a convenience analogous to our notation in group theory, we shall somewhat incorrectly refer to a *ring R* in place of a *ring* $\langle R, +, \cdot \rangle$, provided that no confusion will result. In particular, from now on $\mathbb{Z}$ will always be $\langle \mathbb{Z}, +, \cdot \rangle$, and $\mathbb{Q}$, $\mathbb{R}$, and $\mathbb{C}$ will also be the rings in Example 1. We may on occasion refer to $\langle R, + \rangle$ as *the additive group of the ring R.*

EXAMPLE 2 Let R be any ring and let $M_n(R)$ be the collection of all $n \times n$ matrices having elements of R as entries. The operations of addition and multiplication in R allow us to add and multiply matrices in the usual fashion, explained in Section 0.4. We can quickly check that $\langle M_n(R), + \rangle$ is an abelian group. The associativity of matrix multiplication and the two distributive laws in $M_n(R)$ are more tedious to demonstrate, but straightforward calculations indicate that they follow from the same properties in R. We will assume from now on that we know that $M_n(R)$ is a ring. In particular, we have the rings $M_n(\mathbb{Z})$, $M_n(\mathbb{Q})$, $M_n(\mathbb{R})$, and $M_n(\mathbb{C})$. Note that multiplication is not a commutative operation in any of these rings for $n \geq 2$. ▲

EXAMPLE 3 Let F be the set of all functions $f : \mathbb{R} \to \mathbb{R}$. We know that $\langle F, + \rangle$ is an abelian group under the usual function addition,

$$(f + g)(x) = f(x) + g(x).$$

We define multiplication on R by

$$(fg)(x) = f(x)g(x).$$

That is, fg is the function whose value at x is $f(x)g(x)$. It is readily checked that F is a ring; we leave the demonstration to Exercise 32. We have used this juxtaposition notation $\sigma\mu$ for the composite function $\sigma(\mu(x))$ when discussing permutation multiplication. If we were to use both function multiplication and function composition in F, we would use the notation $f \circ g$ for the composite function. However, we will be using composition of functions almost exclusively with homomorphisms, which we will denote by Greek letters, and the usual product defined in this example chiefly when multiplying polynomial functions $f(x)g(x)$, so no confusion should result. ▲

EXAMPLE 4 Recall that in group theory, $n\mathbb{Z}$ is the cyclic subgroup of $\mathbb{Z}$ under addition consisting of all integer multiples of the integer n. Since $(nr)(ns) = n(nrs)$, we see that $n\mathbb{Z}$ is closed under multiplication. The associative and distributive laws which hold in $\mathbb{Z}$ then assure us that $\langle n\mathbb{Z}, +, \cdot \rangle$ is a ring. From now on in the text, we will consider $n\mathbb{Z}$ to be this ring. ▲

EXAMPLE 5 Consider the cyclic group $\langle \mathbb{Z}_n, + \rangle$. If we define for $a, b \in \mathbb{Z}_n$ the product ab as the remainder of the usual product of integers when divided by n, it can be shown that $\langle \mathbb{Z}_n, +, \cdot \rangle$ is a ring. We shall feel free to use this fact. For example, in $\mathbb{Z}_{10}$ we have $(3)(7) = 1$. This operation on $\mathbb{Z}_n$ is **multiplication modulo** n. We do not check the ring axioms here, for they will follow in Section 6.1 from some of the theory we develop there. From now on, $\mathbb{Z}_n$ will always be the ring $\langle \mathbb{Z}_n, +, \cdot \rangle$. ▲

EXAMPLE 6 If $R_1, R_2, \ldots, R_n$ are rings, we can form the set $R_1 \times R_2 \times \cdots \times R_n$ of all ordered n-tuples $(r_1, r_2, \ldots, r_n)$, where $r_i \in R_i$. Defining addition and multiplication of n-tuples by components (just as for groups), we see at once from the ring axioms in each component that the set of all these n-tuples forms a ring under addition and multiplication by components. The ring $R_1 \times R_2 \times \cdots \times R_n$ is the **direct product** of the rings R_i. ▲

Continuing matters of notation, we shall always let 0 be the additive identity of a ring. The additive inverse of an element a of a ring is $-a$. We shall frequently have occasion to refer to a sum

$$a + a + \cdots + a$$

having n summands. We shall let this sum be $n \cdot a$, always using the dot. However, *$n \cdot a$ is not to be construed as a multiplication of n and a in the ring, for the integer n may not be in the ring at all.* If $n < 0$, we let

$$n \cdot a = (-a) + (-a) + \cdots + (-a)$$

for $|n|$ summands. Finally, we define

$$0 \cdot a = 0$$

for $0 \in \mathbb{Z}$ on the left side of the equation and $0 \in R$ on the right side. Actually the equation $0a = 0$ holds also for $0 \in R$ on both sides. The following theorem proves this and various other elementary but important facts. Note the strong use of the distributive laws in the proof of this theorem. Axiom $\mathcal{R}_1$ for a ring concerns only addition, and axiom $\mathcal{R}_2$ concerns only multiplication. This shows that in order to prove anything that gives a relationship between these two operations, we are going to have to use axiom $\mathcal{R}_3$. For example, the first thing that we will show in Theorem 5.1 is that $0a = 0$ for any element a in a ring R. Now this relation involves both addition and multiplication. The multiplication $0a$ stares us in the face, and 0 is an *additive* concept. Thus we will have to come up with an argument that uses the distributive law to prove this.

THEOREM 5.I If R is a ring with additive identity 0, then for any $a, b \in R$ we have

1. $0a = a0 = 0$,
2. $a(-b) = (-a)b = -(ab)$,
3. $(-a)(-b) = ab$.

PROOF For Property 1, note that by axioms $\mathcal{R}_1$ and $\mathcal{R}_2$,

$$a0 + a0 = a(0 + 0) = a0 = 0 + a0.$$

Then by the cancellation law for the additive group $\langle R, + \rangle$, we have $a0 = 0$. Likewise,

$$0a + 0a = (0 + 0)a = 0a = 0 + 0a$$

implies that $0a = 0$. This proves Property 1.

In order to understand the proof of Property 2, we must remember that, by *definition*, $-(ab)$ is the element that when added to ab gives 0. Thus to show that $a(-b) = -(ab)$, we must show precisely that $a(-b) + ab = 0$. By the left distributive law,

$$a(-b) + ab = a(-b + b) = a0 = 0,$$

since $a0 = 0$ by Property 1. Likewise,

$$(-a)b + ab = (-a + a)b = 0b = 0.$$

For Property 3, note that

$$(-a)(-b) = -(a(-b))$$

by Property 2. Again by Property 2,

$$-(a(-b)) = -(-(ab)),$$

and $-(-(ab))$ is the element that when added to $-(ab)$ gives 0. This is ab by definition of $-(ab)$ and by the uniqueness of an inverse in a group. Thus, $(-a)(-b) = ab$. ◆

It is important that you *understand* the preceding proof. The theorem allows us to use our usual rules for signs.

Homomorphisms and Isomorphisms

From our work in group theory, it is quite clear how a structure-relating map of a ring R into a ring R' should be defined.

DEFINITION 5.2 (Homomorphism) Let R and R' be rings. A map $\phi: R \to R'$ is a **homomorphism** if the following two conditions are satisfied for all $a, b \in R$:

1. $\phi(a + b) = \phi(a) + \phi(b)$.
2. $\phi(ab) = \phi(a)\phi(b)$.

In the preceding definition, Condition 1 is the statement that ϕ is a homomorphism mapping the abelian group $\langle R, + \rangle$ into $\langle R', + \rangle$. Condition 2 requires that ϕ relate the multiplicative structures of the rings R and R' in the same way. Since ϕ is also a group homomorphism, all the results concerning group homomorphisms are valid for the additive structure of the rings. In particular, ϕ is one to one if and only if its **kernel**

$\text{Ker}(\phi) = \{a \in R \mid \phi(a) = 0'\}$ is just the subset $\{0\}$ of R. The homomorphism ϕ of the group $\langle R, + \rangle$ gives rise to a factor group. We expect that a ring homomorphism will give rise to a factor ring. This is indeed the case. We delay discussion of this to Chapter 6, where the treatment will parallel our treatment of factor groups in Chapter 3.

EXAMPLE 7   Let F be the ring of all functions mapping R into R defined in Example 3. For each $a \in R$, we have the **evaluation homomorphism** $\phi_a : F \to R$, where $\phi_a(f) = f(a)$ for $f \in F$. We defined this homomorphism for the group $\langle F, + \rangle$ in Example 2 of Section 3.1, but we did not do much with it in group theory. We will be working a great deal with it in the rest of this text, for finding a real solution of a polynomial equation $p(x) = 0$ amounts precisely to finding $a \in R$ such that $\phi_a(p) = 0$. Much of the remainder of this text deals with solving polynomial equations. We leave the demonstration of the multiplicative homomorphism property 2 for ϕ_a to Exercise 33. ▲

EXAMPLE 8   The map $\phi : \mathbb{Z} \to \mathbb{Z}_n$ where $\phi(a)$ is the remainder of a modulo n is a ring homomorphism for each positive integer n. We know that $\phi(a + b) = \phi(a) + \phi(b)$ by group theory. To show the multiplicative property, write $a = q_1 n + r_1$ and $b = q_2 n + r_2$ according to the division algorithm. Then $ab = n(q_1 q_2 n + r_1 q_2 + q_1 r_2) + r_1 r_2$. Thus $\phi(ab)$ is the remainder of $r_1 r_2$ when divided by n. Since $\phi(a) = r_1$ and $\phi(b) = r_2$, Example 5 shows that $\phi(a)\phi(b)$ is also this same remainder, so $\phi(ab) = \phi(a)\phi(b)$. From group theory, we anticipate that the ring $\mathbb{Z}_n$ might be isomorphic to a factor ring $\mathbb{Z}/n\mathbb{Z}$. This is indeed the case; factor rings will be discussed in Chapter 6. ▲

Based on our work in group theory, we should realize that in the study of any sort of mathematical structure, an idea of basic importance is the concept of two systems being *structurally identical,* that is, one being just like the other except for names. In algebra this concept is always called *isomorphism.* The concept of two rings being just alike except for names of elements leads us, just as it did for groups, to the following definition.

DEFINITION 5.3 (Isomorphism)   An **isomorphism** $\phi : R \to R'$ from a ring R to a ring R' is a homomorphism that is one to one and onto R'. The rings R and R' are then **isomorphic**.

From our work in group theory, we expect that isomorphism gives an equivalence relation on any collection of rings. We need to check that the multiplicative property of an isomorphism is satisfied for the inverse map $\phi^{-1} : R' \to R$ (to complete the symmetry argument). Similarly, we check that if $\mu : R' \to R''$ is also a ring isomorphism, then the multiplicative requirement holds for the composite map $\mu\phi : R \to R''$ (to complete the transitivity argument). We ask you to do this in Exercise 34.

EXAMPLE 9 As abelian groups, $\langle \mathbb{Z}, + \rangle$ and $\langle 2\mathbb{Z}, + \rangle$ are isomorphic under the map $\phi : \mathbb{Z} \to \mathbb{Z}$, with $\phi(x) = 2x$ for $x \in \mathbb{Z}$. Here ϕ is *not* a ring isomorphism, for $\phi(xy) = 2xy$, while $\phi(x)\phi(y) = 2x2y = 4xy$. ▲

Multiplicative Questions; Fields

Many of the rings we have mentioned, such as $\mathbb{Z}$, $\mathbb{Q}$, and $\mathbb{R}$, have a multiplicative identity 1. However, $2\mathbb{Z}$ does not have an identity element for multiplication. Note also that multiplication is not commutative in the matrix rings described in Example 2.

It is evident that $\{0\}$, with $0 + 0 = 0$ and $(0)(0) = 0$, gives a ring, the **zero ring**. Here 0 acts as multiplicative as well as additive identity. By Theorem 5.1, this is the only case in which 0 could act as a multiplicative identity, for if $0a = a$, we can deduce that $a = 0$. We shall exclude this trivial case when speaking of a multiplicative identity in a ring; that is, whenever we speak of a multiplicative identity, we will assume that it is nonzero.

DEFINITION 5.4 (Commutative Ring, Unity) A ring in which the multiplication is commutative is a **commutative ring**. A ring R with a multiplicative identity 1 such that $1x = x1 = x$ for all $x \in R$ is a **ring with unity**. A multiplicative identity in a ring is **unity**.

In a ring with unity 1 the distributive laws show that

$$\underbrace{(1 + 1 + \cdots + 1)}_{n \text{ summands}}\underbrace{(1 + 1 + \cdots + 1)}_{m \text{ summands}} = \underbrace{(1 + 1 + \cdots + 1)}_{nm \text{ summands}},$$

that is, $(n \cdot 1)(m \cdot 1) = (nm) \cdot 1$. The next example gives an application of this observation.

EXAMPLE 10 We claim that for integers r and s where $\gcd(r, s) = 1$, the rings $\mathbb{Z}_{rs}$ and $\mathbb{Z}_r \times \mathbb{Z}_s$ are isomorphic. Additively, they are both cyclic abelian groups of order rs with generators 1 and $(1, 1)$ respectively. Thus $\phi : \mathbb{Z}_{rs} \to \mathbb{Z}_r \times \mathbb{Z}_s$ defined by $\phi(n \cdot 1) = n \cdot (1, 1)$ is an additive group isomorphism. To check the multiplicative condition 2 of Definition 5.2, we use the observation preceding this example for the unity $(1, 1)$ in the ring $\mathbb{Z}_r \times \mathbb{Z}_s$, and compute

$$\phi(nm) = (nm) \cdot (1, 1) = [n \cdot (1, 1)][m \cdot (1, 1)] = \phi(n)\phi(m). \text{ ▲}$$

THEOREM 5.2 If R is a ring with unity, then this unity 1 is the only multiplicative identity.

PROOF We proceed exactly as we did for groups. Let 1 and $1'$ both be multiplicative identities in a ring R, and set up a competition. Regarding 1 as identity, we have.

$$(1)(1') = 1'.$$

Regarding $1'$ as identity, we have

$$(1)(1') = 1.$$

Thus $1 = 1'$. ◆

Note that a direct product $R_1 \times R_2 \times \cdots \times R_n$ of rings is commutative or has unity if and only if each R_i is commutative or has unity, respectively.

In a ring R with unity, the set R^* of nonzero elements, if closed under the ring multiplication, will be a multiplicative group if multiplicative inverses exist. A **multiplicative inverse** of an element a in a ring R with unity 1 is an element $a^{-1} \in R$ such that $aa^{-1} = a^{-1}a = 1$. Precisely as for groups, a multiplicative inverse for an element a in R is unique, if it exists at all (see Exercise 41). Theorem 5.1 shows that it would be hopeless to have a multiplicative inverse for 0 unless we wish to regard the set $\{0\}$, where $0 + 0 = 0$ and $(0)(0) = 0$, as a ring with 0 as both additive and multiplicative identity. We have agreed to exclude the zero ring when speaking of a ring with unity. We are thus led to discuss the existence of multiplicative inverses for nonzero elements in a ring with unity. There is unavoidably a lot of terminology to be defined in this introductory section on rings. We are almost done.

DEFINITION 5.5 (Unit, Field, Skew Field) Let R be a ring with unity. An element u in R is a **unit of** R if it has a multiplicative inverse in R. If every nonzero element of R is a unit, then R is a **division ring**. A **field** is a commutative division ring. A noncommutative division ring is a **skew field**.

EXAMPLE 11 Let us find the units in $\mathbb{Z}_{14}$. Of course, 1 and $-1 = 13$ are units. Since $(3)(5) = 1$ we see that 3 and 5 are units; therefore $-3 = 11$ and $-5 = 9$ are also units. None of the remaining elements of $\mathbb{Z}_{14}$ can be units, since no multiple of 2, 4, 6, 7, 8, or 10 can be one more than a multiple of 14; they all have a common factor, either 2 or 7, with 14. Section 5.3 will show that the units in $\mathbb{Z}_n$ are precisely those $m \in \mathbb{Z}_n$ such that $\gcd(m, n) = 1$. ▲

EXAMPLE 12 $\mathbb{Z}$ is not a field, since 2, for example, has no multiplicative inverse, so 2 is not a unit in $\mathbb{Z}$. The only units in $\mathbb{Z}$ are 1 and -1. However, $\mathbb{Q}$ and $\mathbb{R}$ are fields. An example of a skew-field is given in Section 5.7. ▲

We have the natural concepts of a subring of a ring and a subfield of a field. A **subring of a ring** is a subset of the ring that is a ring under induced operations from the whole ring; a **subfield** is defined similarly for a subset of a field. In fact, let us say here once and for all that if we have a set, together with a certain specified type of algebraic structure on the set, the resulting conglomeration being a **glob** (group, ring, field, integral domain, vector space, and so on), then any subset of this set, together with a natural induced algebraic structure *that yields an algebraic structure of the same type,* is a **subglob**. If K and L are globs, we shall let $K \leq L$ denote that K is a subglob of L and $K < L$ denote that $K \leq L$ but $K \neq L$.

Although fields were implicit in the early work on the solvability of equations by Abel and Galois, it was Leopold Kronecker (1823–1891) who in connection with his own work on this subject first published in 1881 a definition of what he called a "domain of rationality": "The domain of rationality $(R', R'', R''', \ldots)$ contains ... every one of those quantities which are rational functions of the quantities $R', R'', R''', \ldots$ with integral coefficients." Kronecker, however, who insisted that any mathematical subject must be constructible in finitely many steps, did not view the domain of rationality as a complete entity, but merely as a region in which took place various operations on its elements.

Richard Dedekind (1831–1916), the inventor of the Dedekind cut definition of a real number, considered a field as a completed entity. In 1871, he published the following definition in his supplement to the second edition of Dirichlet's text on number theory: "By a field we mean any system of infinitely many real or complex numbers, which in itself is so closed and complete, that the addition, subtraction, multiplication, and division of any two numbers always produces a number of the same system." Both Kronecker and Dedekind had, however, dealt with their varying ideas of this notion as early as the 1850s in their university lectures.

A more abstract definition of a field, similar to the one in the text, was given by Heinrich Weber (1842–1913) in a paper of 1893. Weber's definition, unlike that of Dedekind, specifically included fields with finitely many elements as well as other fields, such as function fields, which were not subfields of the field of complex numbers.

Finally, be careful not to confuse our use of the words *unit* and *unity*. Unity is the multiplicative identity, while a unit is any element having a multiplicative inverse. Thus the multiplicative identity or unity is a unit, but not every unit is unity. For example, -1 is a unit in $\mathbb{Z}$, but -1 is not unity, that is, $-1 \neq 1$.

Exercises 5.1

Computations

In Exercises 1 through 6, compute the product in the given ring.

1. $(12)(16)$ in $\mathbb{Z}_{24}$

2. $(16)(3)$ in $\mathbb{Z}_{32}$

3. $(11)(-4)$ in $\mathbb{Z}_{15}$

4. $(20)(-8)$ in $\mathbb{Z}_{26}$

5. $(2,3)(3,5)$ in $\mathbb{Z}_5 \times \mathbb{Z}_9$

6. $(-3,5)(2,-4)$ in $\mathbb{Z}_4 \times \mathbb{Z}_{11}$

In Exercises 7 through 13, decide whether the indicated operations of addition and multiplication are defined (closed) on the set, and give a ring structure. If a ring is not formed, tell why this is the case. If a ring is formed, state whether the ring is commutative, has unity, or is a field.

7. $n\mathbb{Z}$ with the usual addition and multiplication

8. $\mathbb{Z}^+$ with the usual addition and multiplication

9. $\mathbb{Z} \times \mathbb{Z}$ with addition and multiplication by components

10. $2\mathbb{Z} \times \mathbb{Z}$ with addition and multiplication by components

11. $\{a + b\sqrt{2} \mid a, b \in \mathbb{Z}\}$ with the usual addition and multiplication

12. $\{a + b\sqrt{2} \mid a, b \in \mathbb{Q}\}$ with the usual addition and multiplication

13. The set of all pure imaginary complex numbers ri for $r \in \mathbb{R}$ with the usual addition and multiplication

In Exercises 14 through 19, describe all units in the given ring.

14. $\mathbb{Z}$ **15.** $\mathbb{Z} \times \mathbb{Z}$ **16.** $\mathbb{Z}_5$

17. $\mathbb{Q}$ **18.** $\mathbb{Z} \times \mathbb{Q} \times \mathbb{Z}$ **19.** $\mathbb{Z}_4$

20. Consider the matrix ring $M_2(\mathbb{Z}_2)$.

   a. Find the **order** of the ring, that is, the number of elements in it.

   b. List all units in the ring.

21. If possible, give an example of a homomorphism $\phi: R \to R'$ where R and R' are rings with unity 1 and 1′, and where $\phi(1) \neq 0'$ but $\phi(1) \neq 1'$.

22. (Linear algebra) Consider the map det of $M_n(\mathbb{R})$ into $\mathbb{R}$ where $\det(A)$ is the determinant of the matrix A for $A \in M_n(\mathbb{R})$. Is det a ring homomorphism? Why or why not?

23. Describe all ring homomorphisms of $\mathbb{Z}$ into $\mathbb{Z}$.

24. Describe all ring homomorphisms of $\mathbb{Z}$ into $\mathbb{Z} \times \mathbb{Z}$.

25. Describe all ring homomorphisms of $\mathbb{Z} \times \mathbb{Z}$ into $\mathbb{Z}$.

26. How many homomorphisms are there of $\mathbb{Z} \times \mathbb{Z} \times \mathbb{Z}$ into $\mathbb{Z}$?

27. Consider this solution of the equation $X^2 = I_3$ in the ring $M_3(\mathbb{R})$.

$X^2 = I_3$ implies $X^2 - I_3 = 0$, the zero matrix, so factoring, we have $(X - I_3)(X + I_3) = 0$ whence either $X = I_3$ or $X = -I_3$.

Is this reasoning correct? If not, point out the error, and if possible, give a counterexample to the conclusion.

28. Find all solutions of the equation $x^2 + x - 6 = 0$ in the ring $\mathbb{Z}_{14}$ by factoring the quadratic polynomial. Compare with Exercise 27.

Concepts

29. Give an example of a ring having two elements a and b such that $ab = 0$ but neither a nor b is zero.

30. Give an example of a ring with unity 1 that has a subring with unity $1' \neq 1$. [*Hint:* Consider a direct product, or a subring of $\mathbb{Z}_6$.]

31. Mark each of the following true or false.
_____   a. Every field is also a ring.
_____   b. Every ring has a multiplicative identity.
_____   c. Every ring with unity has at least two units.
_____   d. Every ring with unity has at most two units.
_____   e. It is possible for a subset of some field to be a ring but not a subfield, under the induced operations.
_____   f. The distributive laws for a ring are not very important.
_____   g. Multiplication in a field is commutative.
_____   h. The nonzero elements of a field form a group under the multiplication in the field.
_____   i. Addition in every ring is commutative.
_____   j. Every element in a ring has an additive inverse.

Theory

32. Show that the multiplication defined on the set F of functions in Example 3 satisfies axioms $\mathscr{R}_2$ and $\mathscr{R}_3$ for a ring.

33. Show that the evaluation map ϕ_a of Example 7 satisfies the multiplicative requirement for a homomorphism.

34. Complete the argument outlined after Definition 5.3 to show that isomorphism gives an equivalence relation on a collection of rings.

35. Show that if U is the collection of all units in a ring $\langle R, +, \cdot \rangle$ with unity, then $\langle U, \cdot \rangle$ is a group. [*Warning:* Be sure to show that U is closed under multiplication.]

36. Show that $a^2 - b^2 = (a + b)(a - b)$ for all a and b in a ring R if and only if R is commutative.

37. Let $(R, +)$ be an abelian group. Show that $(R, +, \cdot)$ is a ring if we define $ab = 0$ for all $a, b \in R$.

38. Show that the rings $2\mathbb{Z}$ and $3\mathbb{Z}$ are not isomorphic. Show that the fields $\mathbb{R}$ and $\mathbb{C}$ are not isomorphic.

39. (Freshman exponentiation) Let p be a prime. Show that in the ring Z_p we have $(a + b)^p = a^p + b^p$ for all $a, b \in Z_p$ [*Hint:* Observe that the usual binomial expansion for $(a + b)^n$ is valid in a *commutative* ring.]

40. Show that the unity element in a subfield of a field must be the unity of the whole field, in contrast to Exercise 30 for rings.

41. Show that the multiplicative inverse of a unit in a ring with unity is unique.

42. An element a of a ring R is **idempotent** if $a^2 = a$.

a. Show that the set of all idempotent elements of a commutative ring is closed under multiplication.

b. Find all idempotents in the ring $\mathbb{Z}_6 \times \mathbb{Z}_{12}$.

43. (Linear algebra) Recall that for an $m \times n$ matrix A, the *transpose* A^T of A is the matrix whose jth column is the jth row of A. Show that if A is an $m \times n$ matrix such that $A^T A$ is invertible, then the *projection matrix* $P = A(A^T A)^{-1} A^T$ is idempotent.

44. An element a of a ring R is **nilpotent** if $a^n = 0$ for some $n \in \mathbb{Z}^+$. Show that if a and b are nilpotent elements of a *commutative* ring, then $a + b$ is also nilpotent.

45. Show that a ring R has no nonzero nilpotent element if and only if 0 is the only solution of $x^2 = 0$ in R.

46. Show that a subset S of a ring R gives a subring of R if and only if the following hold:

$0 \in S$;

$(a - b) \in S$ for all $a, b \in S$;

$ab \in S$ for all $a, b \in S$.

47. a. Show that an intersection of subrings of a ring R is again a subring of R.

b. Show that an intersection of subfields of a field F is again a subfield of F.

48. Let R be a ring, and let a be a fixed element of R. Let $I_a = \{x \in R \mid ax = 0\}$. Show that I_a is a subring of R.

49. Let R be a ring, and let a be a fixed element of R. Let R_a be the subring of R that is the intersection of all subrings of R containing a (see Exercise 47). The ring R_a is the **subring of R generated by** a. Show that the abelian group $\langle R_a, + \rangle$ is generated (in the sense of Section 1.4) by $\{a^n \mid n \in \mathbb{Z}^+\}$.

50. (Chinese Remainder Theorem for two congruences) Let r and s be positive integers such that $\gcd(r, s) = 1$. Use the isomorphism in Example 10 to show that for $m, n \in \mathbb{Z}$, there exists an integer x such that $x \equiv m \pmod{r}$ and $x \equiv n \pmod{s}$.

51. a. State and prove the generalization of Example 10 for a direct product with n factors,

b. Prove the Chinese Remainder Theorem: Let $a_i, b_i \in \mathbb{Z}^+$ for $i = 1, 2, \ldots, n$ and let $\gcd(b_i, b_j) = 1$ for $i \neq j$. Then there exists $x \in \mathbb{Z}^+$ such that $x \equiv a_i \pmod{b_i}$ for $i = 1, 2, \ldots, n$.

52. Consider $\langle S, +, \cdot \rangle$, where S is a set and $+$ and $\cdot$ are binary operations on S such that

$\langle S, + \rangle$ is a group,

$\langle S^*, \cdot \rangle$ is a group, where S^* consists of all elements of S except the additive identity,

$a(b + c) = (ab) + (ac)$ and $(a + b)c = (ac) + (bc)$ for all $a, b, c \in S$.

Show that $\langle S, +, \cdot \rangle$ is a division ring. [*Hint:* Apply the distributive laws to $(1 + 1)(a + b)$ to prove the commutativity of addition.]

53. A ring R is a **Boolean ring** if $a^2 = a$ for all $a \in R$, so that every element is idempotent. Show that every Boolean ring is commutative.

54. (For students having some knowledge of the laws of set theory) For a set S let $\mathcal{P}(S)$ be the collection of all subsets of S. Let binary operations $+$ and $\cdot$ on $\mathcal{P}(S)$ be defined by

$$A + B = (A \cup B) - (A \cap B) = \{x \mid x \in A \text{ or } x \in B \text{ but } x \notin (A \cap B)\}$$

and
$$A \cdot B = A \cap B$$
for $A, B \in \mathscr{P}(S)$.

a. Give the tables for $+$ and $\cdot$ for $\mathscr{P}(S)$, where $S = \{a, b\}$. [*Hint:* $\mathscr{P}(S)$ has four elements.]

b. Show that for *any* set S, $\langle \mathscr{P}(S), +, \cdot \rangle$ is a Boolean ring (see Exercise 53).

5.2

Integral Domains

While a careful treatment of polynomials is not given until Section 5.4, for purposes of motivation we shall make intuitive use of them in this section.

Divisors of Zero and Cancellation

One of the most important algebraic properties of our usual number system is that a product of two numbers can only be 0 if at least one of the factors is 0. We have used this fact many times in solving equations, perhaps without realizing that we were using it. Suppose, for example, we are asked to solve the equation

$$x^2 - 5x + 6 = 0.$$

The first thing we do is to factor the left side:

$$x^2 - 5x + 6 = (x - 2)(x - 3)$$

Then we conclude that the only possible values for x are 2 and 3. Why? The reason is that if x is replaced by any number a, the product $(a - 2)(a - 3)$ of the resulting numbers is 0 if and only if either $a - 2 = 0$ or $a - 3 = 0$.

EXAMPLE 1 Solve the equation $x^2 - 5x + 6 = 0$ in $\mathbb{Z}_{12}$.

Solution The factorization $x^2 - 5x + 6 = (x - 2)(x - 3)$ is still valid if we think of x as standing for any number in $\mathbb{Z}_{12}$. But in $\mathbb{Z}_{12}$, not only is $0a - a0 = 0$ for all $a \in \mathbb{Z}_{12}$, but also

$$(2)(6) = (6)(2) = (3)(4) = (4)(3) = (4)(3) = (3)(8) = (8)(3)$$
$$= (4)(6) = (6)(4) = (4)(9) = (9)(4) = (6)(6) = (6)(8)$$
$$= (8)(6) = (6)(10) = (10)(6) = (8)(9) = (9)(8) = 0.$$

Thus our equation has not only 2 and 3 as solutions, but also 6 and 11, for $(6 - 2)(6 - 3) = (4)(3) = 0$ and $(11 - 2)(11 - 3) = (9)(8) = 0$ in $\mathbb{Z}_{12}$. ▲

These ideas are of such importance that we formalize them in a definition.

DEFINITION 5.6 (Divisors of 0)   If a and b are two nonzero elements of a ring R such that $ab = 0$, then a and b are **divisors of 0** (or **0 divisors**).

Example 1 shows that in $\mathbb{Z}_{12}$ the elements 2, 3, 4, 6, 8, 9, and 10 are ~~not~~ divisors of 0. Note that these are exactly the numbers in $\mathbb{Z}_{12}$ that are not relatively prime to 12, and that is whose gcd with 12 is not 1. Our next theorem shows that this is an example of a general situation.

THEOREM 5.3   In the ring $\mathbb{Z}_n$, the divisors of 0 are precisely those elements that are not relatively prime to n.

PROOF   Let $m \in \mathbb{Z}_n$, where $m \neq 0$, and let the gcd of m and n be $d \neq 1$. Then

$$m\left(\frac{n}{d}\right) = \left(\frac{m}{d}\right)n,$$

and $(m/d)n$ gives 0 as a multiple of n. Thus $m(n/d) = 0$ in $\mathbb{Z}_n$, while neither m not n/d is 0, so m is a divisor of 0.

On the other hand, suppose $m \in \mathbb{Z}_n$ is relatively prime to n. If for $s \in \mathbb{Z}_n$ we have $ms = 0$, then n divides the product ms of m and s as elements in the ring $\mathbb{Z}$. Since n is relatively prime to m, Property 1 (set off by rules) in Section 1.4 shows that n divides s, so $s = 0$ in $\mathbb{Z}_n$. ◆

COROLLARY   If p is a prime, then $\mathbb{Z}_p$ has no divisors of 0.

PROOF   This corollary is immediate from Theorem 5.3. ◆

Another indication of the importance of the concept of 0 divisors is shown in the following theorem. Let R be a ring, and let $a, b, c \in R$. The **cancellation laws** hold in R if $ab = ac$ with $a \neq 0$ implies $b = c$, and $ba = ca$ with $a \neq 0$ implies $b = c$. These are multiplicative cancellation laws. Of course, the additive cancellation laws hold in R, since $\langle R, + \rangle$ is a group.

THEOREM 5.4   The cancellation laws hold in a ring R if and only if R has no divisors of 0.

PROOF   Let R be a ring in which the cancellation laws hold, and suppose $ab = 0$ for some $a, b \in R$. We must show that either a or b is 0. If $a \neq 0$, then $ab = a0$ implies that $b = 0$ by cancellation laws. Similarly, $b \neq 0$ implies that $a = 0$, so there can be no divisors of 0 if the cancellation laws hold.

Conversely, suppose that R has no divisors of 0, and suppose that $ab = ac$ with $a \neq 0$. Then

$$ab - ac = a(b - c) = 0.$$

Since $a \neq 0$, and since R has no divisors of 0, we must have $b - c = 0$, so $b = c$. A similar argument shows that $ba = ca$ with $a \neq 0$ implies $b = c$. ◆

Suppose that R is a ring with no divisors of 0. Then an equation $ax = b$, with $a \neq 0$, in R can have at most one solution x in R, for if $ax_1 = b$ and $ax_2 = b$, then $ax_1 = ax_2$, and by Theorem 5.4. $x_1 = x_2$, since R has no divisors of 0. If R has unity 1 and a is a unit in R with multiplicative inverse a^{-1}, then the solution x of $ax = b$ is $a^{-1}b$. In the case that R is commutative, in particular if R is a field, it is customary to denote $a^{-1}b$ and ba^{-1} (they are equal by commutativity) by the formal quotient b/a. This quotient notation must not be used in the event that R is not commutative, for then we do not know whether b/a denotes $a^{-1}b$ or ba^{-1}. In a field F it is usual to *define* a **quotient** b/a, where $a \neq 0$, as the solution x in F of the equation $ax = b$. This definition is consistent with our preceding remarks, and we shall be using this quotient notation when we work in a field. In particular, the multiplicative inverse of a nonzero element a of a field is $1/a$.

Integral Domains

The integers are really our most familiar number system. In terms of the algebraic properties we are discussing, $\mathbb{Z}$ is a commutative ring with unity and no divisors of 0. Surely this is responsible for the name that the next definition gives to such a structure.

DEFINITION 5.7 (Integral Domain) An **integral domain** D is a commutative ring with unity containing no divisors of 0.

Thus, if the coefficients of a polynomial are from an integral domain, one can solve a polynomial equation in which the polynomial can be factored into linear factors in the usual fashion by setting each factor equal to 0.

In our hierarchy of algebraic structures, an integral domain belongs between a commutative ring with unity and a field, as we shall show. Theorem 5.4 shows that the cancellation laws for multiplication hold in an integral domain.

EXAMPLE 2 We have seen that $\mathbb{Z}$ and $\mathbb{Z}_p$ for any prime p are integral domains, but $\mathbb{Z}_n$ is not an integral domain if n is not prime. A moment of thought shows that the direct product $R \times S$ of two nontrivial rings R and S is not an integral domain. Just observe that for $r \in R$ and $s \in S$ both nonzero, we have $(r, 0)(0, s) = (0, 0)$. ▲

EXAMPLE 3 Show that although $\mathbb{Z}_2$ is an integral domain, the matrix ring $M_2(\mathbb{Z}_2)$ has divisors of zero.

Solution We need only observe that

$$\begin{pmatrix} 1 & 0 \\ 0 & 0 \end{pmatrix}\begin{pmatrix} 0 & 0 \\ 1 & 0 \end{pmatrix} = \begin{pmatrix} 0 & 0 \\ 0 & 0 \end{pmatrix}. \quad ▲$$

Our next theorem shows that the structure of a field is still the most restrictive (that is, the richest) one we have defined.

THEOREM 5.5   Every field F is an integral domain.

PROOF   Let $a, b \in F$, and suppose that $a \neq 0$. Then if $ab = 0$, we have

$$\left(\frac{1}{a}\right)(ab) = \left(\frac{1}{a}\right)0 = 0.$$

But then

$$0 = \left(\frac{1}{a}\right)(ab) = \left[\left(\frac{1}{a}\right)a\right]b = 1b = b.$$

We have shown that $ab = 0$ with $a \neq 0$ implies that $b = 0$ in F, so there are no divisors of 0 in F. Of course, F is a commutative ring with unity, so our theorem is proved. ♦

Figure 5.1 gives a Venn diagram view of containment for the algebraic structures having two binary operations with which we will be chiefly concerned. In Exercise 18 we ask you to redraw this figure to include skew fields as well.

Thus far the only fields we know are $\mathbb{Q}$, $\mathbb{R}$, and $\mathbb{C}$. The corollary of the next theorem will exhibit some fields of finite order! The proof of this theorem is a personal favorite. It is done by counting. Counting is one of the most powerful techniques in mathematics.

THEOREM 5.6   Every finite integral domain is a field.

PROOF   Let

$$0, 1, a_1, \ldots, a_n$$

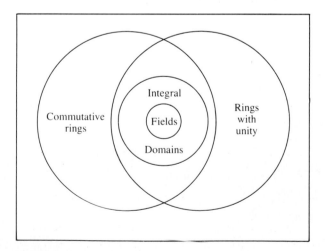

Figure 5.1   A collection of rings.

be all the elements of a finite integral domain D. We need to show that for $a \in D$, where $a \neq 0$, there exists $b \in D$ such that $ab = 1$. Now consider

$$a1, aa_1, \ldots, aa_n.$$

We claim that all these elements of D are distinct, for $aa_i = aa_j$ implies that $a_i = a_j$, by the cancellation laws that hold in an integral domain. Also, since D has no 0 divisors, none of these elements is 0. Hence by *counting*, we find that $a1, aa_1, \ldots, aa_n$ are the elements $1, a_1, \ldots, a_n$ in some order, so that either $a1 = 1$, that is, $a = 1$, or $aa_i = 1$ for some i. Thus a has a multiplicative inverse. ♦

COROLLARY   If p is a prime, then $\mathbb{Z}_p$ is a field. ⁄

PROOF   This corollary follows immediately from the fact that $\mathbb{Z}_p$ is an integral domain and from Theorem 5.6. ♦

The preceding corollary shows that when we consider the ring $M_n(\mathbb{Z}_p)$, we are talking about a ring of matrices over a *field*. In the typical undergraduate linear algebra course, only the arithmetic field properties are used in much of the work. Such notions as matrix reduction to solve linear systems, determinants, Cramer's rule, eigenvalues and eigenvectors, and similarity transformations to try to diagonalize a matrix are valid using matrices over any field; they depend only on the arithmetic properties of a field. Considerations of linear algebra involving notions of magnitude, such as least-squares approximate solutions or orthonormal bases, only make sense using fields where we have an idea of magnitude. The relation

$$p \cdot 1 = \underbrace{1 + 1 + \cdots + 1}_{p \text{ summands}} = 0$$

indicates that there can be no very natural notion of magnitude in the field $\mathbb{Z}_p$.

The Characteristic of a Ring

Let R be any ring. We might ask whether there is a positive integer n such that $n \cdot a = 0$ for all $a \in R$, where $n \cdot a$ means $a + a + \cdots + a$ for n summands, as explained in Section 5.1. For example, the integer m has this property for the ring $\mathbb{Z}_m$.

DEFINITION 5.8 (Characteristic of a Ring)   If for a ring R a positive integer n exists such that $n \cdot a = 0$ for all $a \in R$, then the least such positive integer is the **characteristic of the ring** R. If no such positive integer exists, then R is of **characteristic** 0.

We shall be using the concept of a characteristic chiefly for fields. Exercise 25 asks us to show that the characteristic of an integral domain is either 0 or a prime p.

EXAMPLE 4   The ring $\mathbb{Z}_n$ is of characteristic n, while $\mathbb{Z}$, $\mathbb{Q}$, $\mathbb{R}$, and $\mathbb{C}$ all have characteristic 0. ▲

At first glance, determination of the characteristic of a ring seems to be a tough job, unless the ring is obviously of characteristic 0. Do we have to examine *every* element a of the ring in accordance with Definition 5.8? Our final theorem of this section shows that if the ring has unity, it suffices to examine only $a = 1$.

THEOREM 5.7   If R is a ring with unity 1, then R has characteristic $n > 0$ if and only if n is the smallest positive integer such that $n \cdot 1 = 0$.

PROOF   By definition, if R has characteristic $n > 0$, then $n \cdot a = 0$ for all $a \in R$, so in particular $n \cdot 1 = 0$.

Conversely, suppose that n is a positive integer such that $n \cdot 1 = 0$. Then for any $a \in R$, we have

$$n \cdot a = a + a + \cdots + a = a(1 + 1 + \cdots + 1) = a(n \cdot 1) = a0 = 0.$$

Our theorem follows directly. ◆

Exercises 5.2

Computations

1. Find all solutions of the equation $x^3 - 2x^2 - 3x = 0$ in $\mathbb{Z}_{12}$.

2. Solve the equation $3x = 2$ in the field $\mathbb{Z}_7$; in the field $\mathbb{Z}_{23}$.

3. Find all solutions of the equation $x^2 + 2x + 2 = 0$ in $\mathbb{Z}_6$.

4. Find all solutions of $x^2 + 2x + 4 = 0$ in $\mathbb{Z}_6$.

In Exercises 5 through 10, find the characteristic of the given ring.

5. $2\mathbb{Z}$ **6.** $\mathbb{Z} \times \mathbb{Z}$ **7.** $\mathbb{Z}_3 \times 3\mathbb{Z}$

8. $\mathbb{Z}_3 \times \mathbb{Z}_3$ **9.** $\mathbb{Z}_3 \times \mathbb{Z}_4$ **10.** $\mathbb{Z}_6 \times \mathbb{Z}_{15}$

11. Let R be a commutative ring with unity of characteristic 4. Compute and simplify $(a + b)^4$ for $a, b \in R$.

12. Let R be a commutative ring with unity of characteristic 3. Compute and simplify $(a + b)^9$ for $a, b \in R$.

13. Let R be a commutative ring with unity of characteristic 3. Compute and simplify $(a + b)^6$ for $a, b \in R$.

14. Show that the matrix $\begin{bmatrix} 1 & 2 \\ 2 & 4 \end{bmatrix}$ is a divisor of zero in $M_2(\mathbb{Z})$.

Concepts

15. Mark each of the following true or false.
 _____ a. $n\mathbb{Z}$ has zero divisors if n is not prime.
 _____ b. Every field is an integral domain.
 _____ c. The characteristic of $n\mathbb{Z}$ is n.
 _____ d. As a ring, $\mathbb{Z}$ is isomorphic to $n\mathbb{Z}$ for all $n \geq 1$.
 _____ e. The cancellation law holds in any ring that is isomorphic to an integral domain.
 _____ f. Every integral domain of characteristic 0 is infinite.
 _____ g. The direct product of two integral domains is again an integral domain.
 _____ h. A divisor of zero in a commutative ring with unity can have no multiplicative inverse.
 _____ i. $n\mathbb{Z}$ is a subdomain of $\mathbb{Z}$.
 _____ j. $\mathbb{Z}$ is a subfield of $\mathbb{Q}$.

16. Each of the six regions in Fig. 5.1 corresponds to a certain type of a ring. Give an example of a ring in each of the six cells.

17. (For students who have had a semester of linear algebra) Let F be a field. Give five different characterizations of the elements A of $M_n(F)$ that are divisors of 0.

18. Redraw Fig. 5.1 to include a subset corresponding to skew fields.

Theory

19. An element a of a ring R is **idempotent** if $a^2 = a$. Show that a division ring contains exactly two idempotent elements.

20. Show that an intersection of subdomains of an integral domain D is again a subdomain of D.

21. Show that a finite ring R with unity and no divisors of 0 is a division ring. (It is actually a field, although commutativity is not easy to prove. See Theorem 5.27.) [*Note:* In your proof, to show that $a \neq 0$ is a unit, you must show that a "left multiplicative inverse" of $a \neq 0$ in R is also a "right multiplicative inverse."]

22. Let R be a ring that contains at least two elements. Suppose for each nonzero $a \in R$, there exists a unique $b \in R$ such that $aba = a$.

 a. Show that R has no divisors of 0.

 b. Show that $bab = b$.

 c. Show that R has unity.

 d. Show that R is a division ring.

23. Show that the characteristic of a subdomain of an integral domain D is equal to the characteristic of D.

24. Show that if D is an integral domain, then $\{n \cdot 1 \mid n \in \mathbb{Z}\}$ is a subdomain of D contained in every subdomain of D.

25. Show that the characteristic of an integral domain D must be either 0 or a prime p. [*Hint:* If the characteristic of D is mn, consider $(m \cdot 1)(n \cdot 1)$ in D.]

26. This exercise shows that every ring R can be enlarged (if necessary) to a ring S with unity, having the same characteristic as R. Let $S = R \times Z$ if R has characteristic 0, and $R \times Z_n$ if R has characteristic n. Let addition in S be the

usual addition by components, and let multiplication be defined by

$$(r_1, n_1)(r_2, n_2) = (r_1 r_2 + n_1 \cdot r_2 + n_2 \cdot r_1, n_1 n_2)$$

where $n \cdot r$ has the meaning explained in Section 5.1.

a. Show that S is a ring.

b. Show that S has unity.

c. Show that S and R have the same characteristic.

d. Show that the map $\phi : R \to S$ given by $\phi(r) = (r, 0)$ for $r \in R$ maps R isomorphically onto a subring of S.

Algebraic Coding

Binary linear codes were discussed in Section 2.5. Recall that an (n, k) binary code is a subgroup of $\mathbb{B}^n$. The binary alphabet $\mathbb{B} = \{0, 1\}$ consists of elements of the field $\mathbb{Z}_2$. Algebraic coding can be done using vectors with entries from any *finite* field. Thus ternary coding uses the alphabet $\{0, 1, 2\}$ from the field $\mathbb{Z}_3$, quintary code would use $\{0, 1, 2, 3, 4\}$ from the field $\mathbb{Z}_5$, etc. We will show in Chapter 8 that there exist finite fields not only of every prime order, but of every order that is a power of a prime. Thus there exist fields of orders 2, 3, 4, 5, 7, 8, 9, 11, 13, 16, 17, 19, ... In view of the use of computers, binary codes are currently the most important ones.

27. How many code words are there in a $(4, 2)$ ternary code? How many possible received words?

28. How many possible code words are there in a $(4, 2)$ code using as alphabet a finite field of 16 elements? How many possible received words?

5.3

Fermat's and Euler's Theorems

Fermat's Theorem

We know that as additive groups, $\mathbb{Z}_n$ and $\mathbb{Z}/n\mathbb{Z}$ are naturally isomorphic, with the coset $a + n\mathbb{Z}$ corresponding to a for each $a \in \mathbb{Z}_n$. Furthermore, addition of cosets in $\mathbb{Z}/n\mathbb{Z}$ may be performed by choosing any representatives, adding them in $\mathbb{Z}$, and finding the coset of $n\mathbb{Z}$ containing their sum. It is easy to see that $\mathbb{Z}/n\mathbb{Z}$ can be made into a ring by multiplying cosets in the same fashion, that is, by multiplying any chosen representatives. While we will be showing this in a more general situation in Chapter 6, we do this special case now. We need only show that such coset multiplication is well defined, because the associativity of multiplication and the distributive laws will follow immediately from those properties of the chosen representatives in $\mathbb{Z}$. To this end, choose representatives $a + rn$ and $b + sn$, rather than a and b, from the cosets $a + n\mathbb{Z}$ and $b + n\mathbb{Z}$. Then

$$(a + rn)(b + sn) = ab + (as + rb + rsn)n,$$

which is also an element of $ab + n\mathbb{Z}$. Thus the multiplication is well defined, and our cosets form a ring isomorphic to the ring $\mathbb{Z}_n$.

The following is a special case of Exercise 35 in Section 5.1.

For any field, the nonzero elements form a group under the field multiplication.

In particular, for $\mathbb{Z}_p$, the elements

$$1, 2, 3, \ldots, p - 1$$

form a group of order $p - 1$ under multiplication modulo p. Since the order of any element in a group divides the order of the group, we see that for $b \neq 0$ and $b \in \mathbb{Z}_p$, we have $b^{p-1} = 1$ in $\mathbb{Z}_p$. Using the fact that $\mathbb{Z}_p$ is isomorphic to the ring of cosets of the form $a + p\mathbb{Z}$ described above, we see at once that for any $a \in \mathbb{Z}$ not in the coset $0 + p\mathbb{Z}$, we must have

$$a^{p-1} \equiv 1 \,(\mathrm{mod}\, p).$$

This gives us at once the so-called Little Theorem of Fermat.

THEOREM 5.8 (Little Theorem of Fermat) If $a \in \mathbb{Z}$ and p is a prime not dividing a, then p divides $a^{p-1} - 1$, that is, $a^{p-1} \equiv 1 \,(\mathrm{mod}\, p)$ for $a \not\equiv 0 \,(\mathrm{mod}\, p)$.

◆ **H I S T O R I C A L   N O T E** ◆

The statement of Theorem 5.8 occurs in a letter from Pierre de Fermat (1601–1665) to Bernard Frenicle de Bessy, dated 18 October, 1640. Fermat's version of the theorem was that for any prime p and any geometric progression $a, a^2, \ldots, a^t, \ldots$, there is a least number a^T of the progression such that p divides $a^T - 1$. Furthermore, T divides $p - 1$ and p also divides all numbers of the form $a^{KT} - 1$. (It is curious that Fermat failed to note the condition that p not divide a; perhaps he felt that it was obvious that the result fails in that case.)

Fermat did not in the letter or elsewhere indicate a proof of the result and, in fact, never mentioned it again. But we can infer from other parts of this correspondence that Fermat's interest in this result came from his study of perfect numbers. (A perfect number is a positive integer m that is the sum of all of its divisors less than m; for example, $6 = 1 + 2 + 3$ is a perfect number.) Euclid had shown that $2^{n-1}(2^n - 1)$ is perfect if $2^n - 1$ is prime. The question then was to find methods for determining whether $2^n - 1$ was prime. Fermat noted that $2^n - 1$ was composite if n is composite and then derived from his theorem the result that if n is prime, the only possible divisors of $2^n - 1$ are those of the form $2kn + 1$. From this result he was able quickly to show, for example, that $2^{37} - 1$ was divisible by $223 = 2 \cdot 3 \cdot 37 + 1$.

COROLLARY If $a \in \mathbb{Z}$, then $a^p \equiv a \pmod{p}$ for any prime p.

PROOF The corollary follows from Theorem 5.8 if $a \not\equiv 0 \pmod{p}$. If $a \equiv 0 \pmod{p}$, then both sides reduce to 0 modulo p. ◆

This corollary will be of great importance to us later in our work with finite fields.

EXAMPLE 1 Let us compute the remainder of 8^{103} when divided by 13. Using Fermat's theorem, we have

$$8^{103} \equiv (8^{12})^8(8^7) \equiv (1^8)(8^7) \equiv 8^7 \equiv (-5)^7$$
$$\equiv (25)^3(-5) \equiv (-1)^3(-5) \equiv 5 \pmod{13}. \; ▲$$

EXAMPLE 2 Show that $2^{11,213} - 1$ is not divisible by 11.

Solution By Fermat's theorem, $2^{10} \equiv 1 \pmod{11}$, so

$$2^{11,213} - 1 \equiv \left[(2^{10})^{1,121} \cdot 2^3\right] - 1 \equiv \left[1^{1,121} \cdot 2^3\right] - 1$$
$$\equiv 2^3 - 1 \equiv 8 - 1 \equiv 7 \pmod{11}.$$

Thus the remainder of $2^{11,213} - 1$ when divided by 11 is 7, not 0. (The number 11,213 is prime, and it has been shown that $2^{11,213} - 1$ is a prime number. Primes of the form $2^p - 1$ where p is prime are known as **Mersenne primes**.) ▲

EXAMPLE 3 Show that for every integer n, the number $n^{33} - n$ is divisible by 15.

Solution This seems like an incredible result. It means that 15 divides $2^{33} - 2, 3^{33} - 3, 4^{33} - 4$, etc.

Now $15 = 3 \cdot 5$, and we shall use Fermat's theorem to show that $n^{33} - n$ is divisible by both 3 and 5 for every n. Note that $n^{33} - n = n(n^{32} - 1)$.

If 3 divides n, then surely 3 divides $n(n^{32} - 1)$. If 3 does not divide n, then by Fermat's theorem, $n^2 \equiv 1 \pmod{3}$ so

$$n^{32} - 1 \equiv (n^2)^{16} - 1 \equiv 1^{16} - 1 \equiv 0 \pmod{3},$$

and hence 3 divides $n^{32} - 1$.

If $n \equiv 0 \pmod{5}$, then $n^{33} - n \equiv 0 \pmod{5}$. If $n \not\equiv 0 \pmod{5}$, then by

Fermat's theorem, $n^4 \equiv 1 \pmod 5$, so

$$n^{32} - 1 \equiv (n^4)^8 - 1 \equiv 1^8 - 1 \equiv 0 \pmod 5.$$

Thus $n^{33} - n \equiv 0 \pmod 5$ for every n also. ▲

Euler's Generalization

Euler gave a generalization of Fermat's theorem. His generalization will follow at once from our next theorem, which is proved by *counting*, using essentially the same argument as in Theorem 5.6 of the preceding section.

THEOREM 5.9 The set G_n of nonzero elements of $\mathbb{Z}_n$ that are not 0 divisors forms a group under multiplication modulo n.

PROOF First we must show that G_n is closed under multiplication modulo n. Let $a, b \in G_n$. If $ab \notin G_n$, then there would exist $c \neq 0$ in $\mathbb{Z}_n$ such that $(ab)c = 0$. Now $(ab)c = 0$ implies that $a(bc) = 0$. Since $b \in G_n$ and $c \neq 0$, we have $bc \neq 0$ by definition of G_n. But then $a(bc) = 0$ would imply that $a \notin G_n$ contrary to assumption. *Note that we have shown that for any ring the set of elements that are not divisors of 0 is closed under multiplication.* No structure of $\mathbb{Z}_n$ other than ring structure has been involved so far.

We now show that G_n is a group. Of course, multiplication modulo n is associative, and $1 \in G_n$. It remains to show that for $a \in G_n$, there is $b \in G_n$ such that $ab = 1$. Let

$$1, a_1, \ldots, a_r$$

be the elements of G_n. The elements

$$a1, aa_1, \ldots, aa_r$$

are all different, for if $aa_i = aa_j$, then $a(a_i - a_j) = 0$, and since $a \in G_n$ and thus is not a divisor of 0, we must have $a_i - a_j = 0$ or $a_i = a_j$. Therefore by counting, we find that either $a1 = 1$, or some aa_i must be 1, so a has a multiplicative inverse. ◆

Note that the only property of $\mathbb{Z}_n$ used in this last theorem, other than the fact that it was a ring with unity, was that it was finite. In both Theorem 5.6 and Theorem 5.9 we have (in essentially the same construction) employed a counting argument. *Counting arguments are often simple, but they are among the most powerful tools of mathematics.*

Let n be a positive integer. Let $\varphi(n)$ be defined as the number of positive integers less than or equal to n and relatively prime to n.

EXAMPLE 4 Let $n = 12$. The positive integers less than or equal to 12 and relatively prime to 12 are $1, 5, 7,$ and 11, so $\varphi(12) = 4$. ▲

By Theorem 5.3, $\varphi(n)$ is the number of elements of $\mathbb{Z}_n$ that are not divisors of 0. This function $\varphi : \mathbb{Z}^+ \to \mathbb{Z}^+$ is the **Euler phi-function**. We can now describe Euler's generalization of Fermat's theorem.

THEOREM 5.10 (Euler's Theorem) If a is an integer relatively prime to n, then $a^{\varphi(n)} - 1$ is divisible by n, that is, $a^{\varphi(n)} \equiv 1 \pmod{n}$.

PROOF If a is relatively prime to n, then the coset $a + n\mathbb{Z}$ of $n\mathbb{Z}$ containing a contains an integer $b < n$ and relatively prime to n. Using the fact that multiplication of these cosets by multiplication modulo n of representatives is well defined, we have

$$a^{\varphi(n)} \equiv b^{\varphi(n)} \pmod{n}.$$

But by Theorems 5.3 and 5.9, b can be viewed as an element of the multiplicative group G_n of order $\varphi(n)$ consisting of the $\varphi(n)$ elements of $\mathbb{Z}_n$ relatively prime to n. Thus

$$b^{\varphi(n)} \equiv 1 \pmod{n},$$

and our theorem follows. ◆

EXAMPLE 5 Let $n = 12$. We saw in Example 4 that $\varphi(12) = 4$. Thus if we take any integer a relatively prime to 12, then $a^4 \equiv 1 \pmod{12}$. For example, with $a = 7$, we have $7^4 = (49)^2 = 2{,}401 = 12(200) + 1$, so $7^4 \equiv 1 \pmod{12}$. Of course, the easy way to compute $7^4 \pmod{12}$, without using Euler's theorem, is to compute it in $\mathbb{Z}_{12}$. In $\mathbb{Z}_{12}$, we have $7 = -5$ so

$$7^2 = (-5)^2 = (5)^2 = 1 \quad \text{and} \quad 7^4 = 1^2 = 1. \ ▲$$

Application to $ax \equiv b \pmod{m}$

Using Theorem 5.9, we can find all solutions of a linear congruence $ax \equiv b \pmod{m}$. We prefer to work with an equation in $\mathbb{Z}_m$ and interpret the results for congruences.

THEOREM 5.11 Let m be a positive integer and let $a \in \mathbb{Z}_m$ be relatively prime to m. For each $b \in \mathbb{Z}_m$, the equation $ax = b$ has a unique solution in $\mathbb{Z}_m$.

PROOF By Theorem 5.9, a is a unit in $\mathbb{Z}_m$ and $s = a^{-1}b$ is certainly a solution of the equation. Multiplying both sides of $ax = b$ on the left by a^{-1}, we see this is the only solution. ◆

Interpreting this theorem for congruences, we obtain at once the following corollary.

COROLLARY If a and m are relatively prime integers, then for any integer b, the congruence $ax \equiv b \pmod{m}$ has as solutions all integers in precisely one residue class modulo m.

Theorem 5.11 serves as a lemma for the general case.

THEOREM 5.12 Let m be a positive integer and let $a, b \in \mathbb{Z}_m$. Let d be the gcd of a and m. The equation $ax = b$ has a solution in $\mathbb{Z}_m$ if and only if d divides b. When d divides b, the equation has exactly d solutions in $\mathbb{Z}_m$.

PROOF First we show there is no solution of $ax = b$ in $\mathbb{Z}_m$ unless d divides b. Suppose $s \in \mathbb{Z}_m$ is a solution. Then $as - b = qm$ in $\mathbb{Z}$, so $b = as - qm$. Since d divides both a and m, we see that d divides the right-hand side of the equation $b = as - qm$, and hence divides b. Thus a solution s can exist only if d divides b.

Suppose now that d does divide b. Let

$$a = a_1 d, \qquad b = b_1 d, \qquad \text{and} \qquad m = m_1 d.$$

Then the equation $as - b = qm$ in $\mathbb{Z}$ can be rewritten as $d(a_1 s - b_1) = dqm_1$. We see that $as - b$ is a multiple of m if and only if $a_1 s - b_1$ is a multiple of m_1. Thus the solutions s of $ax = b$ in $\mathbb{Z}_m$ are precisely the elements that, read modulo m_1, yield solutions of $a_1 x = b_1$ in $\mathbb{Z}_{m_1}$. Now let $s \in \mathbb{Z}_{m_1}$ be the unique solution of $a_1 x = b_1$ in $\mathbb{Z}_{m_1}$ given by Theorem 5.11. The numbers in $\mathbb{Z}_m$ that reduce to s modulo m_1 are precisely those that can be computed in $\mathbb{Z}_m$ as

$$s, s + m_1, s + 2m_1, s + 3m_1, \ldots, s + (d-1)m_1.$$

Thus there are exactly d solutions of the equation in $\mathbb{Z}_m$. ◆

Theorem 5.12 gives us at once this classical result on the solutions of a linear congruence.

COROLLARY Let d be the gcd of positive integers a and m. The congruence $ax \equiv b \pmod{m}$ has a solution if and only if d divides b. When this is the case, the solutions are the integers in exactly d distinct residue classes modulo m.

Actually, our proof of Theorem 5.12 shows a bit more about the solutions of $ax \equiv b \pmod{m}$ than we stated in this corollary; namely, it shows that if any solution s is found, then the solutions are precisely all elements of the residue classes $(s + km_1) + (m\mathbb{Z})$ where $m_1 = m/d$ and k runs through the integers from 0 to $d - 1$. It also tells us that we can find such an s by finding $a_1 = a/d$ and $b_1 = b/d$, and solving $a_1 x \equiv b_1 \pmod{m_1}$. To solve this congruence, we may consider a_1 and b_1 to be replaced by their remainders modulo m_1 and solve the equation $a_1 x = b_1$ in $\mathbb{Z}_{m_1}$.

EXAMPLE 6   Find all solutions of the congruence $12x \equiv 27 \pmod{18}$.

Solution   The gcd of 12 and 18 is 6, and 6 is not a divisor of 27. Thus by the preceding corollary, there are no solutions. ▲

EXAMPLE 7   Find all solutions of the congruence $15x \equiv 27 \pmod{18}$.

Solution   The gcd of 15 and 18 is 3, and 3 does divide 27. Proceeding as explained before Example 6, we divide everything by 3 and consider the congruence $5x \equiv 9 \pmod{6}$, which amounts to solving the equation $5x = 3$ in $\mathbb{Z}_6$. Now the units in $\mathbb{Z}_6$ are 1 and 5, and 5 is clearly its own inverse in this group of units. Thus the solution in $\mathbb{Z}_6$ is $s = (5^{-1})(3) = (5)(3) = 3$. Consequently, the solutions of $15x \equiv 27 \pmod{18}$ are the integers in the residue classes

$$3 + 18\mathbb{Z} = \{\ldots, -33, -15, 3, 21, 39, \ldots\},$$
$$9 + 18\mathbb{Z} = \{\ldots, -27, -9, 9, 27, 45, \ldots\},$$
$$15 + 18\mathbb{Z} = \{\ldots, -21, -3, 15, 33, 51, \ldots\}. \;▲$$

Exercises 5.3

Computations

We will see later that the multiplicative group of nonzero elements of a finite field is cyclic. Illustrate this by finding a generator for this group for the given finite field.

1. $\mathbb{Z}_7$ **2.** $\mathbb{Z}_{11}$ **3.** $\mathbb{Z}_{17}$

4. Using Fermat's theorem, find the remainder of 3^{47} when it is divided by 23.

5. Use Fermat's theorem to find the remainder of 37^{49} when it is divided by 7.

6. Compute the remainder of $2^{(2^{17})} + 1$ when divided by 19. [*Hint:* You will need to compute the remainder of 2^{17} modulo 18.]

7. Make a table of values of $\varphi(n)$ for $n \leq 30$.

8. Compute $\varphi(p^2)$ where p is a prime.

9. Compute $\varphi(pq)$ where both p and q are primes.

10. Use Euler's generalization of Fermat's theorem to find the remainder of 7^{1000} when divided by 24.

In Exercises 11 through 18, describe all solutions of the given congruence, as we did in Examples 6 and 7.

11. $2x \equiv 6 \pmod{4}$ **12.** $22x \equiv 5 \pmod{15}$

13. $36x \equiv 15 \pmod{24}$ **14.** $45x \equiv 15 \pmod{24}$

15. $39x \equiv 125 \pmod{9}$ **16.** $41x \equiv 125 \pmod{9}$

17. $155x \equiv 75 \pmod{65}$ **18.** $39x \equiv 52 \pmod{130}$

19. Let p be a prime ≥ 3. Use Exercise 26 below to find the remainder of $(p - 2)!$ modulo p.

20. Using Exercise 26 below, find the remainder of 34! modulo 37.

21. Using Exercise 26 below, find the remainder of 49! modulo 53.

22. Using Exercise 26 below, find the remainder of 24! modulo 29.

Concepts

23. Mark each of the following true or false.

_____ a. $a^{p-1} \equiv 1 \pmod{p}$ for all integers a and primes p.

_____ b. $a^{p-1} \equiv 1 \pmod{p}$ for all integers a such that $a \neq 0 \pmod{p}$ for a prime p.

_____ c. $\varphi(n) \leq n$ for all $n \in \mathbb{Z}^+$.

_____ d. $\varphi(n) \leq n - 1$ for all $n \subset \mathbb{Z}^+$.

_____ e. The units in $\mathbb{Z}_n$ are the positive integers less than n and relatively prime to n.

_____ f. The product of two units in $\mathbb{Z}_n$ is always a unit.

_____ g. The product of two nonunits in $\mathbb{Z}_n$ may be a unit.

_____ h. The product of a unit and a nonunit in $\mathbb{Z}_n$ is never a unit.

_____ i. Every congruence $ax \equiv b \pmod{p}$, where p is a prime, has a solution.

_____ j. Let d be the gcd of positive integers a and m. If d divides b, then the congruence $ax \equiv b \pmod{m}$ has exactly d incongruent solutions.

24. Give the group multiplication table for the multiplicative group of units in $\mathbb{Z}_{12}$. To which group of order 4 is it isomorphic?

Theory

25. Show that 1 and $p - 1$ are the only elements of the field $\mathbb{Z}_p$ that are their own multiplicative inverse. [*Hint:* Consider the equation $x^2 - 1 = 0$.]

26. Using Exercise 25, deduce the half of *Wilson's theorem* that states that if p is a prime, then $(p - 1)! \equiv -1 \pmod{p}$. [The other half states that if n is an integer >1 such that $(n - 1)! \equiv -1 \pmod{n}$, then n is a prime. Just think what the remainder of $(n - 1)!$ would be modulo n if n is not a prime.]

27. Use Fermat's theorem to show that for any positive integer n, the integer $n^{37} - n$ is divisible by 383838. [*Hint:* $383838 = (37)(19)(13)(7)(3)(2)$.]

28. Referring to Exercise 27, find a number larger than 383838 that divides $n^{37} - n$ for all positive integers n.

5.4

The Field of Quotients of an Integral Domain

If an integral domain is such that every nonzero element has a multiplicative inverse, then it is a field. However, many integral domains, such as the integers $\mathbb{Z}$, do not form a field. This dilemma is not too serious. It is the purpose of this section to show that every integral domain can be regarded as being contained in a certain field, *a field of quotients of the integral domain.*

This field will be a minimal field containing the integral domain in a sense that we shall describe. For example, the integers are contained in the field $\mathbb{Q}$, whose elements can all be expressed as quotients of integers. Our construction of a field of quotients of an integral domain is exactly the same as the construction of the rational numbers from the integers, which often appears in a course in foundations of advanced calculus. To follow this construction through is such a good exercise in the use of definitions and the concept of isomorphism that we discuss it in some detail, although to write out, or to read, every last detail would be tedious. We can be motivated at every step by the way $\mathbb{Q}$ can be formed from $\mathbb{Z}$. Recall that the different representations of a rational number as a quotient of integers constituted our motivation for the discussion of equivalence relations in Section 0.2.

The Construction

Let D be an integral domain that we desire to enlarge to a field of quotients F. A coarse outline of the steps we take is as follows:

1. Define what the elements of F are to be.
2. Define the binary operations of addition and multiplication on F.
3. Check all the field axioms to show that F is a field under these operations.
4. Show that F can be viewed as containing D as an integral subdomain.

Steps 1, 2, and 4 are very interesting, and Step 3 is largely a mechanical chore. We proceed with the construction.

Step I Let D be a given integral domain, and form the Cartesian product

$$D \times D = \{(a, b) \mid a, b \in D\}$$

We are going to think of an ordered pair (a, b) as representing a *formal quotient* a/b, that is, if $D = \mathbb{Z}$, the pair $(2, 3)$ will eventually represent the number $\frac{2}{3}$ for us. The pair $(2, 0)$ represents no element of $\mathbb{Q}$ and suggests that we cut the set $D \times D$ down a bit. Let S be the subset of $D \times D$ given by

$$S = \{(a, b) \mid a, b \in D, b \neq 0\}.$$

Now S is still not going to be our field as is indicated by the fact that, with $D = \mathbb{Z}$, *different* pairs of integers such as $(2, 3)$ and $(4, 6)$ can represent the *same* rational number. We next define when two elements of S will eventually represent the same element of F, or, as we shall say, when two elements of S are *equivalent*.

DEFINITION 5.9 (Equivalent Pairs) Two elements (a, b) and (c, d) in S are **equivalent**, denoted by $(a, b) \sim (c, d)$, if and only if $ad = bc$.

Observe that this definition is reasonable, since the criterion for

$(a, b) \sim (c, d)$ is an equation $ad = bc$ involving elements in D and concerning the known multiplication in D. Note also that for $D = \mathbb{Z}$, the criterion gives us our usual definition of *equality*, for example, $\frac{2}{3} = \frac{4}{6}$, since $(2)(6) = (3)(4)$. The rational number that we usually denote by $\frac{2}{3}$ can be thought of as the collection of *all* quotients of integers that reduce to, or are equivalent to, $\frac{2}{3}$.

LEMMA 5.1   The relation $\sim$ between elements of the set S as just described is an equivalence relation.

PROOF   We must check the three properties of an equivalence relation.

Reflexive $(a, b) \sim (a, b)$ since $ab = ba$, for multiplication in D is commutative.

Symmetric   If $(a, b) \sim (c, d)$, then $ad = bc$. Since multiplication in D is commutative, we deduce that $cb = da$, and consequently $(c, d) \sim (a, b)$.

Transitive   If $(a, b) \sim (c, d)$ and $(c, d) \sim (r, s)$, then $ad = bc$ and $cs = dr$. Using these relations and the fact that multiplication in D is commutative, we have

$$asd = sad = sbc = bcs = bdr = brd.$$

Now $d \neq 0$, and D is an integral domain, so cancellation is valid; this is a crucial step in the argument. Hence from $asd = brd$ we obtain $as = br$, so that $(a, b) \sim (r, s)$. ◆

It is worth comparing the preceding proof with the demonstration in Example 4 of Section 0.2. The steps are identical.

We now know, in view of Theorem 0.1 in Section 0.2, that $\sim$ gives a partition of S into equivalence classes. To avoid long bars over extended expressions, we shall let $[(a, b)]$, rather than $\overline{(a, b)}$, be the equivalence class of (a, b) in S under the relation $\sim$. We now finish Step 1 by defining F to be the set of all equivalence classes $[(a, b)]$ for $(a, b) \in S$.

Step 2   The next lemma serves to define addition and multiplication in F. Observe that if $D = \mathbb{Z}$ and $[(a, b)]$ is viewed as $(a/b) \in \mathbb{Q}$, these definitions applied to $\mathbb{Q}$ give the usual operations.

LEMMA 5.2   For $[(a, b)]$ and $[(c, d)]$ in F, the equations

$$[(a, b)] + [(c, d)] = [(ad + bc, bd)]$$

and

$$[(a, b)][(c, d)] = [(ac, bd)]$$

give well-defined operations of addition and multiplication on F.

PROOF   Observe first that if $[(a, b)]$ and $[(c, d)]$ are in F, then (a, b) and (c, d) are in S, so $b \neq 0$ and $d \neq 0$. Since D is an integral domain, $bd \neq 0$, so both $(ad + bc, bd)$ and (ac, bd) are in S. (Note the crucial use here of the fact that D has no divisors of 0.) This shows that the right-hand sides of the defining equations are at least in F.

It remains for us to show that these operations of addition and multiplication are well defined. That is, they were defined by means of representatives in S of elements of F. We must show that if different representatives in S are chosen, the same element of F will result. To this end, suppose that $(a_1, b_1) \in [(a, b)]$ and $(c_1, d_1) \in [(c, d)]$. We must show that

$$(a_1d_1 + b_1c_1, b_1d_1) \in [(ad + bc, bd)]$$

and

$$(a_1c_1, b_1d_1) \in [(ac, bd)].$$

Now $(a_1, b_1) \in [(a, b)]$ means that $(a_1, b_1) \sim (a, b)$; that is,

$$a_1b = b_1a.$$

Similarly, $(c_1, d_1) \in [(c, d)]$ implies that

$$c_1d = d_1c.$$

Multiplying the first equation by d_1d, the second by b_1b, and adding the resulting equations, we obtain the following equation in D:

$$a_1bd_1d + c_1db_1b = b_1ad_1d + d_1cb_1b.$$

Using various axioms for an integral domain, we see that

$$(a_1d_1 + b_1c_1)bd = b_1d_1(ad + bc),$$

so

$$(a_1d_1 + b_1c_1, b_1d_1) \sim (ad + bc, bd),$$

giving $(a_1d_1 + b_1c_1, b_1d_1) \in [(ad + bc, bd)]$. This takes care of addition in F. For multiplication in F, on multiplying the equations $a_1b = b_1a$ and $c_1d = d_1c$, we obtain

$$a_1bc_1d = b_1ad_1c,$$

so, using axioms of D, we get

$$a_1c_1bd = b_1d_1ac,$$

which implies that

$$(a_1c_1, b_1d_1) \sim (ac, bd).$$

Thus $(a_1c_1, b_1d_1) \in [(ac, bd)]$, which completes the proof. ◆

It is important to *understand* the meaning of the last lemma and the necessity for proving it. This completes our Step 2.

Step 3 Step 3 is routine, but it is good for us to work through a few of these details. The reason for this is that we cannot work through them unless we *understand* what we have done. Thus working through them will contribute to our understanding of this construction. We list the things that must be proved and prove a couple of them. The rest are left to the exercises.

1. Addition in F is commutative.

> **PROOF**   Now $[(a, b)] + [(c, d)]$ is by definition $[(ad + bc, bd)]$. Also $[(c, d)] + [(a, b)]$ is by definition $[(cb + da, db)]$. We need to show that $(ad + bc, bd) \sim (cd + da, db)$. This is true, since $ad + bc = cb + da$ and $bd = db$, by the axioms of D. ◆

2. Addition is associative.
3. $[(0, 1)]$ is an identity for addition in F.
4. $[(-a, b)]$ is an additive inverse for $[(a, b)]$ in F.
5. Multiplication in F is associative.
6. Multiplication in F is commutative.
7. The distributive laws hold in F.
8. $[(1, 1)]$ is a multiplicative identity in F.
9. If $[(a, b)] \in F$ is not the additive identity, then $a \neq 0$ in D and $[(b, a)]$ is a multiplicative inverse for $[(a, b)]$.

PROOF   Let $[(a, b)] \in F$. If $a = 0$, then

$$a1 = b0 = 0,$$

so

$$(a, b) \sim (0, 1),$$

that is, $[(a, b)] = [(0, 1)]$. But $[(0, 1)]$ is the additive identity by Part 3. Thus if $[(a, b)]$ is not the additive identity in F, we have $a \neq 0$, so it makes sense to talk about $[(b, a)]$ in F. Now $[(a, b)][(b, a)] = [(ab, ba)]$. But in D we have $ab = ba$, so $(ab)1 = (ba)1$, and

$$(ab, ba) \sim (1, 1).$$

Thus

$$[(a, b)][(b, a)] = [(1, 1)],$$

and $[(1, 1)]$ is the multiplicative identity by Part 8. ◆

This completes Step 3.

Step 4 It remains for us to show that F can be regarded as containing D. To do this, we show that there is an isomorphism i of D with a subdomain of F. Then if we rename the image of D under i using the names of the elements of D, we will be done. The next lemma gives us this isomorphism. We use the letter i for this isomorphism to suggest *injection*; we will inject D into F.

LEMMA 5.3 The map $i : D \to F$ given by $i(a) = [(a, 1)]$ is an isomorphism of D with a subdomain of F.

PROOF For a and b in D, we have

$$i(a + b) = [(a + b, 1)].$$

Also,

$$i(a) + i(b) = [(a, 1)] + [(b, 1)] = [(a1 + 1b, 1)] = [(a + b, 1)].$$

so $i(a + b) = i(a) + i(b)$. Furthermore,

$$i(ab) = [(ab, 1)],$$

while

$$i(a)i(b) = [(a, 1)][(b, 1)] = [(ab, 1)],$$

so $i(ab) = i(a)i(b)$.

It remains for us to show only that i is one to one. If $i(a) = i(b)$, then

$$[(a, 1)] = [(b, 1)],$$

so $(a, 1) \sim (b, 1)$ giving $a1 = 1b$; that is,

$$a = b.$$

Thus i is an isomorphism of D with $i[D]$, and, of course, $i[D]$ is then a subdomain of F. ◆

Since $[(a, b)] = [(a, 1)][(1, b)] = [(a, 1)]/[(b, 1)] = i(a)/i(b)$ clearly holds in F, we have now proved the following theorem.

THEOREM 5.13 Any integral domain D can be enlarged to (or embedded in) a field F such that every element of F can be expressed as a quotient of two elements of D. (Such a field F is a **field of quotients of D**.)

Uniqueness

We said in the beginning that F could be regarded in some sense as a minimal field containing D. This is intuitively evident, since every field containing D must contain all elements a/b for every $a, b \in D$ with $b \neq 0$. The next theorem will show that every field containing D contains a subfield which is a

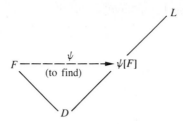

Figure 5.2

field of quotients of D, and that any two fields of quotients of D are isomorphic.

THEOREM 5.14 Let F be a field of quotients of D and let L be any field containing D. Then there exists a map $\psi: F \to L$ that gives an isomorphism of F with a subfield of L such that $\psi(a) = a$ for $a \in D$.

PROOF The lattice and mapping diagram in Fig. 5.2 may help you to visualize the situation for this theorem.

An element of F is of the form $a/_F b$ where $/_F$ denotes the quotient of $a \in D$ by $b \in D$ regarded as elements of F. We of course want to map $a/_F b$ onto $a/_L b$ where $/_L$ denotes the quotient of elements in L. The main job will be to show that such a map is well defined.

We must define $\psi: F \to L$, and we start by defining

$$\psi(a) = a \qquad \text{for} \qquad a \in D.$$

Every $x \in F$ is a quotient $a/_F b$ of some two elements a and b, $b \neq 0$, of D. Let us attempt to define ψ by

$$\psi(a/_F b) = \psi(a)/_L \psi(b).$$

We must first show that this map ψ is sensible and well defined. Since ψ is the identity on D, for $b \neq 0$ we have $\psi(b) \neq 0$, so our definition of $\psi(a/_F b)$ as $\psi(a)/_L \psi(b)$ makes sense. If $a/_F b = c/_F d$ in F, then $ad = bc$ in D, so $\psi(ad) = \psi(bc)$. But since ψ is the identity on D,

$$\psi(ad) = \psi(a)\psi(d) \qquad \text{and} \qquad \psi(bc) = \psi(b)\psi(c).$$

Thus

$$\psi(a)/_L \psi(b) = \psi(c)/_L \psi(d)$$

in L, so ψ is well defined.

The equations

$$\psi(xy) = \psi(x)\psi(y)$$

and

$$\psi(x + y) = \psi(x) + \psi(y)$$

follow easily from the definition of ψ on F and from the fact that ψ is the identity on D.

If $\psi(a/_F b) = \psi(c/_F d)$, we have

$$\psi(a)/_L \psi(b) = \psi(c)/_L \psi(d)$$

so

$$\psi(a)\psi(d) = \psi(b)\psi(c).$$

Since ψ is the identity on D, we then deduce that $ad = bc$, so $a/_F b = c/_F d$. Thus ψ is one to one.

By definition, $\psi(a) = a$ for $a \in D$. ◆

COROLLARY I Every field L containing an integral domain D contains a field of quotients of D.

PROOF In the proof of Theorem 5.14, every element of the subfield $\psi[F]$ of L is a quotient in L of elements of D. ◆

COROLLARY 2 Any two fields of quotients of an integral domain D are isomorphic.

PROOF Suppose in Theorem 5.14 that L is a field of quotients of D, so that every element x of L can be expressed in the form $a/_L b$ for $a, b \in D$. Then L is the field $\psi[F]$ of the proof of Theorem 5.14 and is thus isomorphic to F. ◆

Exercises 5.4

Computations

1. Describe the field F of quotients of the integral subdomain

$$D = \{n + mi \mid n, m \in \mathbb{Z}\}$$

of $\mathbb{C}$. "Describe" means give the elements of $\mathbb{C}$ that make up the field of quotients of D in $\mathbb{C}$.

2. Describe (in the sense of Exercise 1) the field F of quotients of the integral subdomain $D = \{n + m\sqrt{2} \mid n, m \in \mathbb{Z}\}$ of $\mathbb{R}$.

Concepts

3. Mark each of the following true or false.
   _____ a. $\mathbb{Q}$ is a field of quotients of $\mathbb{Z}$.
   _____ b. $\mathbb{R}$ is a field of quotients of $\mathbb{Z}$.
   _____ c. $\mathbb{R}$ is a field of quotients of $\mathbb{R}$.
   _____ d. $\mathbb{C}$ is a field of quotients of $\mathbb{R}$.

———— e. If D is a field, then any field of quotients of D is isomorphic to D.

———— f. The fact that D has no divisors of 0 was used strongly several times in the construction of a field F of quotients of the integral domain D.

———— g. Every element of an integral domain D is a unit in a field F of quotients of D.

———— h. Every nonzero element of an integral domain D is a unit in a field F of quotients of D.

———— i. A field of quotients F' of a subdomain D' of an integral domain D can be regarded as a subfield of some field of quotients of D.

———— j. Every field of quotients of $\mathbb{Z}$ is isomorphic to $\mathbb{Q}$.

4. Show by an example that a field F' of quotients of a proper subdomain D' of an integral domain D may also be a field of quotients for D.

Theory

5. Prove Part 2 of Step 3. You may assume any preceding part of Step 3.

6. Prove Part 3 of Step 3. You may assume any preceding part of Step 3.

7. Prove Part 4 of Step 3. You may assume any preceding part of Step 3.

8. Prove Part 5 of Step 3. You may assume any preceding part of Step 3.

9. Prove Part 6 of Step 3. You may assume any preceding part of Step 3.

10. Prove Part 7 of Step 3. You may assume any preceding part of Step 3.

11. Let R be a commutative ring, and let T be a nonempty subset of R closed under multiplication and containing neither 0 nor divisors of 0. Starting with $R \times T$ and otherwise exactly following the construction in this section, we can show that the ring R can be enlarged to a *partial ring of quotients $Q(R, T)$*. Think about this for fifteen minutes or so; look back over the construction and see why things still work. In particular, show the following:

a. $Q(R, T)$ has unity even if R does not.

b. In $Q(R, T)$, every nonzero element of T is a unit.

12. Prove from Exercise 11 that every commutative ring containing an element a that is not a divisor of 0 can be enlarged to a commutative ring with unity. Compare with Exercise 26 of Section 5.2.

13. With reference to Exercise 11, how many elements are there in the ring $Q(\mathbb{Z}_4, \{1, 3\})$?

14. With reference to Exercise 11, describe the ring $Q(\mathbb{Z}, \{2^n \mid n \in \mathbb{Z}^+\})$ by describing a subring of $\mathbb{R}$ to which it is isomorphic.

15. With reference to Exercise 11, describe the ring $Q(3\mathbb{Z}, \{6^n \mid n \in \mathbb{Z}^+\})$ by describing a subring of $\mathbb{R}$ to which it is isomorphic.

16. With reference to Exercise 11, suppose we drop the condition that T have no divisors of zero and just require that nonempty T not containing 0 be closed under multiplication. The attempt to enlarge R to a commutative ring with unity in which every nonzero element of T is a unit must fail if T contains an element a that is a divisor of 0 for a divisor of 0 cannot also be a unit. Try to discover where a construction parallel to that in the text but starting with $R \times T$ first runs into trouble. In particular, for $R = \mathbb{Z}_6$ and $T = \{1, 2, 4\}$, illustrate the first difficulty encountered. [*Hint:* It is in Step 1.]

5.5

Rings of Polynomials

Polynomials in an Indeterminate

We all have a pretty workable idea of what constitutes a *polynomial in x with coefficients in a ring R.* We can guess how to add and multiply such polynomials and know what is meant by the *degree* of a polynomial. We expect that the set $R[x]$ of all polynomials with coefficients in the ring R is itself a ring with the usual operations of polynomial addition and multiplication, and that R is a subring of $R[x]$. However, we will be working with polynomials from a slightly different viewpoint than the approach in high school algebra or calculus, and there are a few things that we want to say.

◆   **H I S T O R I C A L   N O T E**   ◆

The use of "*x*" and other letters near the end of the alphabet to represent an "indeterminate" is due to René Descartes (1596–1650). Earlier, Francois Viete (1540–1603) had used vowels for indeterminates and consonants for known quantities. Descartes is also responsible for the first publication of the factor theorem (Corollary 1, Theorem 5.18) in his work *The Geometry,* which appeared as an appendix to his *Discourse on Method* (1637). This work also contained the first publication of the basic concepts of analytic geometry; Descartes showed how geometric curves can be described algebraically.

Descartes was born to a wealthy family in La Haye, France; since he was always of delicate health, he formed the habit of spending his mornings in bed. It was at these times that he accomplished his most productive work. The *Discourse on Method* was Descartes's attempt to show the proper procedures for "searching for truth in the sciences." The first step in this process was to reject as absolutely false everything of which he had the least doubt; but, since it was necessary that he who was thinking was "something," he conceived his first principle of philosophy: "I think, therefore I am." The most enlightening parts of the *Discourse on Method,* however, are the three appendices: *The Optics, The Geometry,* and *The Meteorology.* It was here that Descartes provided examples of how he actually applied his method. Among the important ideas Descartes discovered and published in these works were the sine law of refraction of light, the basics of the theory of equations, and a geometrical explanation of the rainbow.

In 1649, Descartes was invited by Queen Christina of Sweden to come to Stockholm to tutor her. Unfortunately, the Queen required him, contrary to his long-established habits, to rise at an early hour. He soon contracted a lung disease and died in 1650.

In the first place, we will call x an **indeterminate** rather than a variable. Suppose, for example that our ring of coefficients is $\mathbb{Z}$. One of the polynomials in the ring $\mathbb{Z}[x]$ is $1x$, which we shall write simply as x. Now x is not 1 or 2 or any of the other elements of $\mathbb{Z}[x]$. Thus from now on we will never write such things as "$x = 1$" or "$x = 2$", as we have done before. We call x an indeterminate rather than a variable to emphasize this change. Also, we will never write an expression such as "$x^2 - 4 = 0$," simply because $x^2 - 4$ is not the additive identity in our ring $\mathbb{Z}[x]$. We are accustomed to speaking of "solving a polynomial equation," and will be spending a lot of time in the remainder of our text discussing this, but we will always refer to it as "finding a zero of a polynomial." In summary, we try to be careful in our discussion of algebraic structures not to say in one context that things are equal and in another context that they are not equal.

If a person knows nothing about polynomials, it is not an easy task to describe precisely the nature of a polynomial in x with coefficients in a ring R. If we just define such a polynomial to be a *finite formal sum*

$$\sum_{i=0}^{n} a_i x^i = a_0 + a_1 x + \cdots + a_n x^n,$$

where $a_i \in R$, we get ourselves into a bit of trouble. For surely $0 + a_1 x$ and $0 + a_1 x + 0x^2$ are different as formal sums, but we want to regard them as the same polynomial. A practical solution to this problem is to define a polynomial as an *infinite formal sum*

$$\sum_{i=0}^{\infty} a_i x^i = a_0 + a_1 x + \cdots + a_n x^n + \cdots,$$

where $a_i = 0$ for all but a finite number of values of i. Now there is no problem of having more than one formal sum represent what we wish to consider a single polynomial.

DEFINITION 5.10 (Polynomial) Let R be a ring. A **polynomial** $f(x)$ **with coefficients in** R is an infinite formal sum

$$\sum_{i=0}^{\infty} a_i x^i = a_0 + a_1 x + \cdots + a_n x^n + \cdots,$$

where $a_i \in R$ and $a_i = 0$ for all but a finite number of values of i. The a_i are **coefficients of** $f(x)$. If for some $i \geq 0$ it is true that $a_i \neq 0$, the largest such value of i is the **degree of** $f(x)$. If all $a_i = 0$, then the degree of $f(x)$ is undefined.

To simplify working with polynomials, let us agree that if $f(x) = a_0 + a_1 x + \cdots + a_n x^n + \cdots$ has $a_i = 0$ for $i > n$, then we may denote $f(x)$ by $a_0 + a_1 x + \cdots + a_n x^n$. Also, if R has unity, we will write a term $1x^k$ in such a sum as x^k. For example, in $\mathbb{Z}[x]$, we will write the polynomial $2 + 1x$ as $2 + x$. Finally, we shall agree that we may omit altogether from the formal sum any term $0x^i$, or a_0, if $a_0 = 0$ but not all $a_i = 0$. Thus $0, 2, x$, and $2 + x^2$ are all polynomials with coefficients in $\mathbb{Z}$. An element of R is a **constant polynomial**.

Addition and multiplication of polynomials with coefficients in a ring R are defined in a way familiar to us. If

$$f(x) = a_0 + a_1 x + \cdots + a_n x^n + \cdots$$

and

$$g(x) = b_0 + b_1 x + \cdots + b_n x^n + \cdots,$$

then for polynomial addition, we have

$$f(x) + g(x) = c_0 + c_1 x + \cdots + c_n x^n + \cdots,$$

where $c_n = a_n + b_n$, and for polynomial multiplication, we have

$$f(x)g(x) = d_0 + d_1 x + \cdots + d_n x^n + \cdots,$$

where $d_n = \sum_{i=0}^{n} a_i b_{n-i}$. Observe that both c_i and d_i are 0 for all but a finite number of values of i, so these definitions make sense. Note that $\sum_{i=0}^{n} a_i b_{n-i}$ need not equal $\sum_{i=0}^{n} b_i a_{n-i}$ if R is not commutative. With these definitions of addition and multiplication, we have the following theorem.

THEOREM 5.15 The set $R[x]$ of all polynomials in an indeterminate x with coefficients in a ring R is a ring under polynomial addition and multiplication. If R is commutative, then so is $R[x]$, and if R has unity 1, then 1 is also unity for $R[x]$.

PROOF That $\langle R[x], + \rangle$ is an abelian group is apparent. The associative law for multiplication and the distributive laws are straightforward, but slightly cumbersome, computations. We illustrate by proving the associative law.

Applying ring axioms to $a_i, b_j, c_k \in R$, we obtain

$$\left[\left(\sum_{i=0}^{\infty} a_i x^i\right)\left(\sum_{j=0}^{\infty} b_j x^j\right)\right]\left(\sum_{k=0}^{\infty} c_k x^k\right) = \left[\sum_{n=0}^{\infty}\left(\sum_{i=0}^{n} a_i b_{n-i}\right)x^n\right]\left(\sum_{k=0}^{\infty} c_k x^k\right)$$

$$= \sum_{s=0}^{\infty}\left[\sum_{n=0}^{s}\left(\sum_{i=0}^{n} a_i b_{n-i}\right)c_{s-n}\right]x^s$$

$$= \sum_{s=0}^{\infty}\left(\sum_{i+j+k=s} a_i b_j c_k\right)x^s$$

$$= \sum_{s=0}^{\infty}\left[\sum_{m=0}^{s} a_{s-m}\left(\sum_{j=0}^{m} b_j c_{m-j}\right)\right]x^s$$

$$= \left(\sum_{i=0}^{\infty} a_i x^i\right)\left[\sum_{m=0}^{\infty}\left(\sum_{j=0}^{m} b_j c_{m-j}\right)x^m\right]$$

$$= \left(\sum_{i=0}^{\infty} a_i x^i\right)\left[\left(\sum_{j=0}^{\infty} b_j x^j\right)\left(\sum_{k=0}^{\infty} c_k x^k\right)\right].$$

Whew!! The distributive laws are similarly proved.

The comments prior to the statement of the theorem show that $R[x]$ is a commutative ring if R is commutative, and a unity 1 in R is also unity for $R[x]$, view of the definition of multiplication in $R[x]$. ◆

Thus $\mathbb{Z}[x]$ is the ring of polynomials in the indeterminate x with integral coefficients, $\mathbb{Q}[x]$ the ring of polynomials in x with rational coefficients, and so on.

EXAMPLE I   In $\mathbb{Z}_2[x]$, we have

$$(x + 1)^2 = (x + 1)(x + 1) = x^2 + (1 + 1)x + 1 = x^2 + 1.$$

Still working in $\mathbb{Z}_2[x]$, we obtain

$$(x + 1) + (x + 1) = (1 + 1)x + (1 + 1) = 0x + 0 = 0. \; ▲$$

If R is a ring and x and y are two indeterminates, then we can form the ring $(R[x])[y]$, that is, the ring of polynomials in y with coefficients that are polynomials in x. Every polynomial in y with coefficients that are polynomials in x can be rewritten in a natural way as a polynomial in x with coefficients that are polynomials in y. This indicates that $(R[x])[y]$ is naturally isomorphic to $(R[y])[x]$, although a careful proof is tedious. We shall identify these rings by means of this natural isomorphism and shall consider this ring $R[x, y]$ the **ring of polynomials in two indeterminates x and y with coefficients in R**. The **ring $R[x_1, \ldots, x_n]$ of polynomials in the n indeterminates x_i with coefficients in R** is similarly defined.

We leave as Exercise 22 the proof that if D is an integral domain then so is $D[x]$. In particular, if F is a field, then $F[x]$ is an integral domain. Note that $F[x]$ is not a field, for x is not a unit in $F[x]$. That is, there is no polynomial $f(x) \in F[x]$ such that $xf(x) = 1$. By Theorem 5.13, one can construct the field of quotients $F(x)$ of $F[x]$. Any element in $F(x)$ can be represented as a quotient $f(x)/g(x)$ of two polynomials in $f[x]$ with $g(x) \neq 0$. We similarly define $F(x_1, \ldots, x_n)$ to be the field of quotients of $F[x_1, \ldots, x_n]$. This field $F(x_1, \ldots, x_n)$ is the **field of rational functions in n indeterminates over** F. These fields play a very important role in algebraic geometry.

The Evaluation Homomorphisms

We are now ready to proceed to show how homomorphisms can be used to study what we have always referred to as "solving a polynomial equation." Let E and F be fields, with F a subfield of E, that is, $F \leq E$. The next theorem asserts the existence of very important homomorphisms of $F[x]$ into E. *These homomorphisms will be the fundamental tools for most of the rest of our work.*

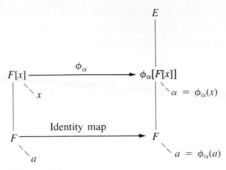

Figure 5.3

THEOREM 5.16 (The Evaluation Homomorphisms for Field Theory) Let F be a subfield of a field E, let α be any element of E, and let x be an indeterminate. The map $\phi_\alpha : F[x] \to E$ defined by

$$\phi_\alpha(a_0 + a_1 x + \cdots + a_n x^n) = a_0 + a_1 \alpha + \cdots + a_n \alpha^n$$

for $(a_0 + a_1 x + \cdots + a_n x^n) \in F[x]$ is a homomorphism of $F[x]$ into E. Also, $\phi_\alpha(x) = \alpha$, and ϕ_α maps F isomorphically by the identity map; that is, $\phi_\alpha(a) = a$ for $a \in F$. The homomorphism ϕ_α is **evaluation at** α.

PROOF The lattice and mapping diagram in Fig. 5.3 may help us to visualize this situation. The dashed lines indicate an element of the set. The theorem is really an immediate consequence of our definitions of addition and multiplication in $F[x]$. The map ϕ_α is well defined, that is, independent of our representation of $f(x) \in F[x]$ as a finite sum

$$a_0 + a_1 x + \cdots + a_n x^n,$$

since such a finite sum representing $f(x)$ can be changed only by insertion or deletion of terms $0x^i$, which does not affect the value of $\phi_\alpha(f(x))$.

If $f(x) = a_0 + a_1 x + \cdots + a_n x^n$, $g(x) = b_0 + b_1 x + \cdots + b_m x^m$, and $h(x) = f(x) + g(x) = c_0 + c_1 x + \cdots + c_r x^r$, then

$$\phi_\alpha(f(x) + g(x)) = \phi_\alpha(h(x)) = c_0 + c_1 \alpha + \cdots + c_r \alpha^r,$$

while

$$\phi_\alpha(f(x)) + \phi_\alpha(g(x)) = (a_0 + a_1 \alpha + \cdots + a_n \alpha^n)$$
$$+ (b_0 + b_1 \alpha + \cdots + b_m \alpha^m).$$

Since by definition of polynomial addition we have $c_i = a_i + b_i$, we see that

$$\phi_\alpha(f(x) + g(x)) = \phi_\alpha(f(x)) + \phi_\alpha(g(x)).$$

Turning to multiplication, we see that if

$$f(x)g(x) = d_0 + d_1 x + \cdots + d_s x^s,$$

then

$$\phi_\alpha(f(x)g(x)) = d_0 + d_1 \alpha + \cdots + d_s \alpha^s,$$

while

$$[\phi_\alpha(f(x))][\phi_\alpha(g(x))]$$
$$= (a_0 + a_1\alpha + \cdots + a_n\alpha^n)(b_0 + b_1\alpha + \cdots + b_m\alpha^m).$$

Since by definition of polynomial multiplication, $d_j = \sum_{i=0}^{j} a_i b_{j-i}$, we see that

$$\phi_\alpha(f(x)g(x)) = [\phi_\alpha(f(x))][\phi_\alpha(g(x))].$$

Thus ϕ_α is a homomorphism.

The very definition of ϕ_α applied to a constant polynomial $a \in F[x]$, where $a \in F$, gives $\phi_\alpha(a) = a$, so ϕ_α maps F isomorphically by the identity map. Again by definition of ϕ_α, we have $\phi_\alpha(x) = \phi_\alpha(1x) = 1\alpha = \alpha$. $\blacklozenge$

We point out that this theorem is valid with the identical proof if F and E are merely commutative rings with unity rather than fields. However, we shall be interested only in the case in which they are fields.

It is hard to overemphasize the importance of this simple theorem for us. It is the very foundation for all of our further work in field theory. It is so simple that it could justifiably be called an *observation* rather than a theorem. It was perhaps a little misleading to write out the proof because the polynomial notation makes it look so complicated that you may be fooled into thinking it is a difficult theorem.

EXAMPLE 2   Let F be $\mathbb{Q}$ and E be $\mathbb{R}$ in Theorem 5.16, and consider the evaluation homomorphism $\phi_0 : \mathbb{Q}[x] \to \mathbb{R}$. Here

$$\phi_0(a_0 + a_1x + \cdots + a_nx^n) = a_0 + a_10 + \cdots + a_n0^n = a_0.$$

Thus every polynomial is mapped onto its constant term. $\blacktriangle$

EXAMPLE 3   Let F be $\mathbb{Q}$ and E be $\mathbb{R}$ in Theorem 5.16 and consider the evaluation homomorphism $\phi_2 : \mathbb{Q}[x] \to \mathbb{R}$. Here

$$\phi_2(a_0 + a_1x + \cdots + a_nx^n) = a_0 + a_12 + \cdots + a_n2^n.$$

Note that

$$\phi_2(x^2 + x - 6) = 2^2 + 2 - 6 = 0,$$

Thus $x^2 + x - 6$ is in the kernel N of ϕ_2. Of course,

$$x^2 + x - 6 = (x - 2)(x + 3),$$

and the reason that $\phi_2(x^2 + x - 6) = 0$ is that $\phi_2(x - 2) = 2 - 2 = 0$. $\blacktriangle$

EXAMPLE 4   Let F be $\mathbb{Q}$ and E be $\mathbb{C}$ in Theorem 5.16 and consider the evaluation homomorphism $\phi_i : \mathbb{Q}[x] \to \mathbb{C}$. Here

$$\phi_i(a_0 + a_1 x + \cdots + a_n x^n) = a_0 + a_1 i + \cdots + a_n i^n$$

and $\phi_i(x) = i$. Note that

$$\phi_i(x^2 + 1) = i^2 + 1 = 0.$$

so $x^2 + 1$ is in the kernel N of ϕ_i. ▲

EXAMPLE 5   Let F be $\mathbb{Q}$ and let E be $\mathbb{R}$ in Theorem 5.16 and consider the evaluation homomorphism $\phi_\pi : \mathbb{Q}[x] \to \mathbb{R}$. Here

$$\phi_\pi(a_0 + a_1 x + \cdots + a_n x^n) = a_0 + a_1 \pi + \cdots + a_n \pi^n.$$

It can be proved that $a_0 + a_1 \pi + \cdots + a_n \pi^n = 0$ if and only if $a_i = 0$ for $i = 0, 1, \ldots, n$. Thus the kernel of ϕ_π is $\{0\}$, and ϕ_π is a one-to-one map. This shows that all *formal polynomials in π with rational coefficients* form a ring isomorphic to $\mathbb{Q}[x]$ in a natural way with $\phi_\pi(x) = \pi$. ▲

The New Approach

We now complete the connection between our new ideas and the classical concept of solving a polynomial equation. Rather than speak of *solving a polynomial equation*, we shall refer to *finding a zero of a polynomial.*

DEFINITION 5.11 (Zero of a Polynomial)   Let F be a subfield of a field E, and let α be an element of E. Let $f(x) = a_0 + a_1 x + \cdots + a_n x^n$ be in $F[x]$, and let $\phi_\alpha : F[x] \to E$ be the evaluation homomorphism of Theorem 5.16. Let $f(\alpha)$ denote

$$\phi_\alpha(f(x)) = a_0 + a_1 \alpha + \cdots + a_n \alpha^n.$$

If $f(\alpha) = 0$, then α is a **zero of** $f(x)$.

In terms of this definition, we can rephrase the classical problem of finding all real numbers r such that $r^2 + r - 6 = 0$ by letting $F = \mathbb{Q}$ and $E = \mathbb{R}$ and *finding all $\alpha \in \mathbb{R}$ such that*

$$\phi_\alpha(x^2 + x - 6) = 0,$$

that is, finding all zeros of $x^2 + x - 6$ in $\mathbb{R}$. Both problems have the same answer, since

$$\{\alpha \in \mathbb{R} \mid \phi_\alpha(x^2 + x - 6) = 0\} = \{r \in \mathbb{R} \mid r^2 + r - 6 = 0\} = \{2, -3\}.$$

It may seem that we have merely succeeded in making a simple problem seem quite complicated. In fact, *what we have done is to phrase the problem in the language of mappings, and we can now use all the mapping machinery that we have developed and will continue to develop for its solution.*

Our Basic Goal

We continue the attempt to put our future work in perspective. Chapter 6 is concerned with topics in ring theory that are analogous to the material on factor groups and homomorphisms for group theory. However, our aim in developing these analogous concepts for rings will be quite different from our aims in group theory. In group theory we used the concepts of factor groups and homomorphisms to study the structure of a given group and to determine the types of group structures of certain orders that could exist. We will be talking about homomorphisms and factor rings in Chapter 6 with an eye to finding zeros of polynomials, which is one of the oldest and most fundamental problems in algebra. Let us take a moment to talk about this aim in the light of mathematical history, using the language of "solving polynomial equations" to which we are accustomed.

We start with the Pythagorean school of mathematics of about 525 B.C. The Pythagoreans asserted with an almost fanatical fervor that all distances are **commensurable**; that is, given distances a and b, there should exist a unit of distance u and integers n and m such that $a = (n)(u)$ and $b = (m)(u)$. In terms of numbers, then, thinking of u as being one unit of distance, they maintained that all numbers are integers. This idea of commensurability can be rephrased according to our ideas as an assertion that all numbers are rational, for if a and b are rational numbers, then each is an integral multiple of the reciprocal of the least common multiple of their denominators. For example, if $a = \frac{7}{12}$ and $b = \frac{19}{15}$, then $a = (35)(\frac{1}{60})$ and $b = (76)(\frac{1}{60})$.

The Pythagoreans knew, of course, what is now called the *Pythagorean theorem*; that is, for a right triangle with legs of lengths a and b and a hypotenuse of length c,

$$a^2 + b^2 = c^2.$$

They also had to grant the existence of a hypotenuse of a right triangle having two legs of equal length, say one unit each. The hypotenuse of such a right triangle would, as we know, have to have a length of $\sqrt{2}$. Imagine then their consternation, dismay, and even fury when one of their society—according to some stories it was Pythagoras himself—came up with the embarrassing fact that is stated in our terminology in the following theorem.

THEOREM 5.17 The equation $x^2 = 2$ has no solutions in rational numbers. Thus $\sqrt{2}$ is not a rational number.

PROOF Suppose that m/n for $m, n \in \mathbb{Z}$ is a rational number such that $(m/n)^2 = 2$. We assume that we have cancelled any factors common to m and n, so that the fraction m/n is in lowest terms with $\gcd(m, n) = 1$. Then

$$m^2 = 2n^2,$$

where both m^2 and $2n^2$ are integers. Since m^2 and $2n^2$ are the same integer, and since 2 is a factor of $2n^2$, we see that 2 must be one of the factors of m^2. But as a square, m^2 has as factors the factors of m repeated twice. Thus m^2 must have two factors 2. Then $2n^2$ must have two factors 2, so n^2 must have 2 as a factor, and consequently n has 2 as a factor. We have deduced from $m^2 = 2n^2$ that both m and n must be divisible by 2, contradicting the fact that the fraction m/n is in lowest terms. Thus we have $2 \neq (m/n)^2$ for any $m, n \in \mathbb{Z}$. ◆

Thus the Pythagoreans ran right into the question of a solution of a polynomial equation, $x^2 - 2 = 0$. We refer the student to Shanks [36, Chapter 3], for a lively and totally delightful account of this Pythagorean dilemma and its significance in mathematics.

In our motivation of the definition of a group, we commented on the necessity of having negative numbers, so that equations such as $x + 2 = 0$ might have solutions. The introduction of negative numbers caused a certain amount of consternation in some philosophical circles. We can visualize 1 apple, 2 apples, and even $\frac{13}{11}$ apples, but how can we point to anything and say that it is -17 apples? Finally, consideration of the equation $x^2 + 1 = 0$ led

---

◆   **HISTORICAL   NOTE**   ◆

The solution of polynomial equations has been a goal of mathematics for nearly 4000 years. The Babylonians developed versions of the quadratic formula to solve quadratic equations. For example, to solve $x^2 - x = 870$, the Babylonian scribe instructed his students to take half of 1 ($\frac{1}{2}$), square it ($\frac{1}{4}$), and add that to 870. The square root of $870\frac{1}{4}$, namely $29\frac{1}{2}$, is then added to $\frac{1}{2}$ to give 30 as the answer. What the scribes did not discuss, however, was what to do if the square root in this process was not a rational number. Chinese mathematicians, however, from about 200 B.C., discovered a method similar to what is now called Horner's method to solve quadratic equations numerically; since they used a decimal system, they were able in principle to carry out the computation to as many places as necessary and could therefore ignore the distinction between rational and irrational solutions. The Chinese, in fact, extended their numerical techniques to polynomial equations of higher degree. In the Arab world the Persian poet-mathematician Omar Khayyam (1048–1131) developed methods for solving cubic equations geometrically by finding the point(s) of intersection of appropriately chosen conic sections, while Sharaf al-Din al-Tusi (died 1213) used, in effect, techniques of calculus to determine whether or not a cubic equation had a real positive root. It was the Italian Girolamo Cardano (1501–1576) who first published a procedure for solving cubic equations algebraically.

to the introduction of the number i. The very name of an "imaginary number" given to i shows how this number was regarded. Even today many students are led by this name to regard i with some degree of suspicion. The negative numbers were introduced to us at such an early stage in our mathematical development that we accepted them without question.

We first met polynomials in high school freshman algebra. The first problem there was to learn how to add, multiply, and factor polynomials. Then, in both freshman algebra and in the second course in algebra in high school, considerable emphasis was placed on solving polynomial equations. These topics are exactly those with which we shall be concerned. The difference is that while in high school only polynomials with real number coefficients were considered, *we shall be doing our work for polynomials with coefficients from any field.*

Once we have developed the machinery of homomorphisms and factor rings in Chapter 6, we will proceed with our **basic goal**: to show that given any polynomial of degree ≥ 1, where the coefficients of the polynomial may be from any field, we can find (or construct) a zero of this polynomial. After the machinery is developed in Chapter 6, the achievement of this goal will be very easy, and is really a very elegant piece of mathematics.

All this fuss may seem ridiculous, but just think back in history. This is the *culmination of more than* 2000 *years of mathematical endeavor in working with polynomial equations.* After achieving our *basic goal,* we shall spend the rest of our time studying the nature of these solutions of polynomial equations. We need have no fear in approaching this material. *We shall be dealing with familiar topics of high school algebra. This work should seem much more natural than group theory.*

In conclusion, we remark that the machinery of factor rings and ring homomorphisms is not really necessary in order for us to achieve our *basic goal.* For a direct demonstration, see Artin [27, p. 29]. However, factor rings and ring homomorphisms are fundamental ideas that we should grasp, and our *basic goal* will follow very easily once we have mastered them. We will further use these concepts effectively in the study of properties of solutions of polynomial equations.

Exercises 5.5

Computations

In Exercises 1 through 4, find the sum and the product of the given polynomials in the given polynomial ring.

1. $f(x) = 4x - 5$, $g(x) = 2x^2 - 4x + 2$ in $\mathbb{Z}_8[x]$.

2. $f(x) = x + 1$, $g(x) = x + 1$ in $\mathbb{Z}_2[x]$.

3. $f(x) = 2x^2 + 3x + 4$, $g(x) = 3x^2 + 2x + 3$ in $\mathbb{Z}_6[x]$.

4. $f(x) = 2x^3 + 4x^2 + 3x + 2$, $g(x) = 3x^4 + 2x + 4$ in $\mathbb{Z}_5[x]$.

5. How many polynomials are there of degree ≤ 3 in $\mathbb{Z}_2[x]$? (Include 0.)

6. How many polynomials are there of degree ≤ 2 in $\mathbb{Z}_5[x]$? (Include 0.)

In Exercises 7 through 11, $F = E = \mathbb{Z}_7$ in Theorem 5.16. Compute for the indicated evaluation homomorphism.

7. $\phi_2(x^2 + 3)$ **8.** $\phi_0(2x^3 - x^2 + 3x + 2)$

9. $\phi_3[(x^4 + 2x)(x^3 - 3x^2 + 3)]$

10. $\phi_5[(x^3 + 2)(4x^2 + 3)(x^7 + 3x^2 + 1)]$

11. $\phi_4(3x^{106} + 5x^{99} + 2x^{53})$
     [*Hint:* Use Fermat's theorem.]

In Exercises 12 through 15, find all zeros in the indicated finite field of the given polynomial with coefficients in that field. [*Hint:* One way is simply to try all candidates!]

12. $x^2 + 1$ in $\mathbb{Z}_2$ **13.** $x^3 + 2x + 2$ in $\mathbb{Z}_7$

14. $x^5 + 3x^3 + x^2 + 2x$ in $\mathbb{Z}_5$

15. $f(x)g(x)$ where $f(x) = x^3 + 2x^2 + 5$ and $g(x) = 3x^2 + 2x$ in $\mathbb{Z}_7$

16. Let $\phi_a : \mathbb{Z}_5[x] \to \mathbb{Z}_5$ be an evaluation homomorphism as in Theorem 5.16. Use Fermat's theorem to evaluate $\phi_3(x^{231} + 3x^{117} - 2x^{53} + 1)$.

17. Use Fermat's theorem to find all zeros in $\mathbb{Z}_5$ of
$$2x^{219} + 3x^{74} + 2x^{57} + 3x^{44}.$$

Concepts

18. Consider the element
$$f(x, y) = (3x^3 + 2x)y^3 + (x^2 - 6x + 1)y^2 + (x^4 - 2x)y + (x^4 - 3x^2 + 2)$$
of $(\mathbb{Q}[x])[y]$. Write $f(x, y)$ as it would appear if viewed as an element of $(\mathbb{Q}[y])[x]$.

19. Consider the evaluation homomorphism $\phi_5 : \mathbb{Q}[x] \to \mathbb{R}$. Find six elements in the kernel of ϕ_5.

20. Find a polynomial of degree > 0 in $\mathbb{Z}_4[x]$ that is a unit.

21. Mark each of the following true or false.

_____ a. The polynomial $(a_n x^n + \cdots + a_1 x + a_0) \in R[x]$ is 0 if and only if $a_i = 0$, for $i = 0, 1, \ldots, n$.

_____ b. If R is a commutative ring, then $R[x]$ is commutative.

_____ c. If D is an integral domain, then $D[x]$ is an integral domain.

_____ d. If R is a ring containing divisors of 0, then $R[x]$ has divisors of 0.

_____ e. If R is a ring and $f(x)$ and $g(x)$ in $R[x]$ are degrees 3 and 4, respectively, then $f(x)g(x)$ may be of degree 8 in $R[x]$.

_____ f. If R is any ring and $f(x)$ and $g(x)$ in $R[x]$ are of degrees 3 and 4, respectively, then $f(x)g(x)$ is always of degree 7.

_____ g. If F is a subfield E and $\alpha \in E$ is a zero of $f(x) \in F[x]$, then α is a zero of $h(x) = f(x)g(x)$ for all $g(x) \in F[x]$.

_____ h. If F is a field, then the units in $F[x]$ are precisely the units in F.

_____ i. If R is a ring, then x is never a divisor of 0 in $R[x]$.

_____ j. If R is a ring, then the zero divisors in $R[x]$ are precisely the zero divisors in R.

Theory

22. Prove that if D is an integral domain, then $D[x]$ is an integral domain.

23. Let D be an integral domain and x an indeterminate.

   a. Describe the units in $D[x]$.

   b. Find the units in $\mathbb{Z}[x]$.

   c. Find the units in $\mathbb{Z}_7[x]$.

24. Prove the left distributive law for $R[x]$, where R is a ring and x is an indeterminate.

25. Let F be a field of characteristic zero and let D be the formal polynomial differentiation map, so that

$$D(a_0 + a_1 x + a_2 x^2 + \cdots + a_n x^n) = a_1 + 2 \cdot a_2 x + \cdots + n \cdot a_n x^{n-1}.$$

   a. Show that $D : F[x] \to F[x]$ is a group homomorphism of $\langle F[x], + \rangle$ into itself. Is D a ring homomorphism?

   b. Find the kernel of D.

   c. Find the image of $F[x]$ under D.

26. Let F be a subfield of a field E.

   a. Define an *evaluation homomorphism*

$$\phi_{\alpha_1, \ldots, \alpha_n} : F[x_1, \ldots, x_n] \to E \qquad \text{for} \quad \alpha_i \in E,$$

stating the analog of Theorem 5.16.

   b. With $E = F = \mathbb{Q}$, compute $\phi_{-3,2}(x_1^2 x_2^3 + 3x_1^4 x_2)$.

   c. Define the concept of a *zero of a polynomial* $f(x_1, \ldots, x_n) \in F[x_1, \ldots, x_n]$ in a way analogouus to the definition in the text of a zero of $f(x)$.

27. Let R be a ring, and let R^R be the set of all functions mapping R into R. For $\phi, \psi \in R^R$, define the sum $\phi + \psi$ by

$$(\phi + \psi)(r) = \phi(r) + \gamma(r)$$

and the product $\phi \cdot \psi$ by

$$(\phi \cdot \psi)(r) = \phi(r)\psi(r)$$

for $r \in R$. Note that $\cdot$ is *not* function composition. Show that $\langle R^R, +, \cdot \rangle$ is a ring.

28. Referring to Exercise 27, let F be a field. An element ϕ of F^F is a **polynomial function on** F, if there exists $f(x) \in F[x]$ such that $\phi(a) = f(a)$ for all $u \in F$.

   a. Show that the set P_F is of all polynomial functions on F forms a subring of F^F.

b. Show that the ring P_F is not necessarily isomorphic to $F[x]$. [*Hint:* Show that if F is a finite field, P_F and $F[x]$ don't even have the same number of elements.]

29. Refer to Exercises 27 and 28 for the following questions.

a. How many elements are there in $Z_2^{Z_2}$? in $Z_3^{Z_3}$?

b. Classify $\langle Z_2^{Z_2}, + \rangle$ and $\langle Z_3^{Z_3}, + \rangle$ by Theorem 2.11, the fundamental theorem of finitely generated abelian groups.

c. Show that if F is a finite field, then $F^F = P_F$. [*Hint:* Of course, $P_F \subseteq F^F$. Let F have as elements $a_1, \ldots, a_n$. Note that if

$$f_i(x) = c(x - a_1) \cdots (x - a_{i-1})(x - a_{i+1}) \cdots (x - a_n),$$

then $f_i(a_j) = 0$ for $j \neq i$, and the value $f_i(a_i)$ can be controlled by the choice of $c \in F$. Use this to show that every function on F is a polynomial function.]

5.6

Factorization of Polynomials over a Field

Recall that we are concerned with finding zeros of polynomials. Let E and F be fields, with $F \le E$. Suppose that $f(x) \in F[x]$ factors in $F[x]$, so that $f(x) = g(x)h(x)$ for $g(x), h(x) \in F[x]$ and let $\alpha \in E$. Now for the evaluation homomorphism ϕ_α, we have

$$f(\alpha) = \phi_\alpha(f(x)) = \phi_\alpha(g(x)h(x)) = \phi_\alpha(g(x))\phi_\alpha(h(x)) = g(\alpha)h(\alpha).$$

Thus if $\alpha \in E$, then $f(\alpha) = 0$ if and only if either $g(\alpha) = 0$ or $h(\alpha) = 0$. The attempt to find a zero of $f(x)$ is reduced to the problem of finding a zero of a factor of $f(x)$. This is one reason why it is useful to study factorization of polynomials.

The Division Algorithm in F[x]

The following theorem is the basic tool for our work in this section. Note the similarity with the division algorithm for Z given in Section 1.4, the importance of which has been amply demonstrated. Division algorithms will be treated in a more general setting in Section 7.2.

THEOREM 5.18 (Division Algorithm for F[x])   Let

$$f(x) = a_n x^n + a_{n-1} x^{n-1} + \cdots + a_0$$

and

$$g(x) = b_m x^m + b_{m-1} x^{m-1} + \cdots + b_0$$

be two elements of $F[x]$, with a_n and b_m both nonzero elements of F and $m > 0$. Then there are unique polynomials $q(x)$ and $r(x)$ in $F[x]$ such that $f(x) = g(x)q(x) + r(x)$, with the degree of $r(x)$ less than the degree m of $g(x)$.

PROOF   Consider the set $S = \{f(x) - g(x)s(x) \mid s(x) \in F[x]\}$. Let $r(x)$ be an element of minimal degree in S. Then

$$f(x) = g(x)q(x) + r(x)$$

for some $q(x) \in F[x]$. We must show that the degree of $r(x)$ is less than m. Suppose that

$$r(x) = c_t x^t + c_{t-1} x^{t-1} + \cdots + c_0,$$

with $c_j \in F$ and $c_t \neq 0$ if $t \neq 0$. If $t \geq m$, then

$$f(x) - q(x)g(x) - (c_t/b_m)x^{t-m}g(x) = r(x) - (c_t/b_m)x^{t-m}g(x), \quad \text{(1)}$$

and the latter is of the form

$$r(x) - (c_t x^t + \text{terms of lower degree}),$$

which is a polynomial of degree lower than t, the degree of $r(x)$. However, the polynomial in Eq. (1) can be written in the form

$$f(x) - g(x)[q(x) + (c_t/b_m)x^{t-m}],$$

so it is in S, contradicting the fact that $r(x)$ was selected to have minimal degree in S. Thus the degree of $r(x)$ is less than the degree m of $g(x)$.

For uniqueness, if

$$f(x) = g(x)q_1(x) + r_1(x)$$

and

$$f(x) = g(x)q_2(x) + r_2(x),$$

then subtracting we have

$$g(x)[q_1(x) - q_2(x)] = r_2(x) - r_1(x).$$

Since the degree of $r_2(x) - r_1(x)$ is less than the degree of $g(x)$, this can only hold if $q_1(x) - q_2(x) = 0$ or $q_1(x) = q_2(x)$. Then we must have $r_2(x) - r_1(x) = 0$ or $r_1(x) = r_2(x)$. ◆

We can compute the polynomials $q(x)$ and $r(x)$ of Theorem 5.18 by long division just as we divided polynomials in $\mathbb{R}[x]$ in high school.

We give three important corollaries of Theorem 5.18. The first one appears in high school algebra for the special case $F[x] = \mathbb{R}[x]$. We phrase our proof in terms of the mapping (homomorphism) approach described in Section 5.5.

COROLLARY I (Factor Theorem) An element $a \in F$ is a zero of $f(x) \in F[x]$ if and only if $x - a$ is a factor of $f(x)$ in $F[x]$.

PROOF Suppose that for $a \in F$ we have $f(a) = 0$. By Theorem 5.18, there exist $q(x), r(x) \in F[x]$ such that

$$f(x) = (x - a)q(x) + r(x),$$

where the degree of $r(x)$ is less than 1. Then we must have $r(x) = c$ for

$c \in F$, so

$$f(x) = (x - a)q(x) + c.$$

Applying our evaluation homomorphism, $\phi_a : F[x] \to F$ of Theorem 5.16, we find

$$0 = f(a) = 0q(a) + c.$$

so it must be that $c = 0$. Then $f(x) = (x - a)q(x)$, so $x - a$ is a factor of $f(x)$.

Conversely, if $x - a$ is a factor of $f(x)$ in $F[x]$, where $a \in F$, then applying our evaluation homomorphism ϕ_a to $f(x) = (x - a)q(x)$, we have $f(a) = 0q(a) = 0$. ◆

The next corollary should also look familiar.

COROLLARY 2   A nonzero polynomial $f(x) \in F[x]$ of degree n can have at most n zeros in a field F.

PROOF   The preceding corollary shows that if $a_1 \in F$ is a zero of $f(x)$, then

$$f(x) = (x - a_1)q_1(x),$$

where of course the degree of $q_1(x)$ is $n - 1$. A zero $a_2 \in F$ of $q_1(x)$ then results in a factorization

$$f(x) = (x - a_1)(x - a_2)q_2(x).$$

Continuing this process, we arrive at

$$f(x) = (x - a_1) \cdots (x - a_r)q_r(x),$$

where $q_r(x)$ has no further zeros in F. Since the degree of $f(x)$ is r, at most r factors $(x - a_i)$ can appear on the right-hand side of the preceding equation, so $r \le n$. Also, if $b \ne a_i$ for $i = 1, \dots, r$ and $b \in F$, then

$$f(b) = (b - a_1) \cdots (b - a_r)q_r(b) \ne 0,$$

since F has no divisors of 0 and none of $b - a_i$ or $q_r(b)$ are 0 by construction. Hence the a_i for $i = 1, \dots, r \le n$ are all the zeros in F of $f(x)$. ◆

Our final corollary is concerned with the structure of the multiplicative group F^* of nonzero elements of a field F, rather than with factorization in $F[x]$. It may at first seem surprising that such a result follows from the division algorithm in $F[x]$, but recall that the result that a subgroup of a cyclic group is cyclic follows from the division algorithm in $\mathbb{Z}$.

COROLLARY 3   If G is a finite multiplicative subgroup of the multiplicative group $\langle F^*, \cdot \rangle$ of a field F, then G is cyclic. In particular, the multiplicative group of all nonzero elements of a finite field is cyclic.

PROOF By Theorem 2.11, as a finite abelian group, G is isomorphic to a direct product $\mathbb{Z}_{d_1} \times \mathbb{Z}_{d_2} \times \cdots \times \mathbb{Z}_{d_r}$, where each d_i is a power of a prime. Let us think of each of the $\mathbb{Z}_{d_i}$ as a cyclic group of order d_i in *multiplicative* notation. Let m be the least common multiple of all the d_i for $i = 1, 2, \ldots, r$; note that $m \le d_1 d_2 \cdots d_r$. If $a_i \in \mathbb{Z}_{d_i}$, then $a_i^{d_i} = 1$, so $a_i^m = 1$ since d_i divides m. Thus for all $\alpha \in G$, we have $\alpha^m = 1$, so every element of G is a zero of $x^m - 1$. But G has $d_1 d_2 \cdots d_r$ elements, while $x^m - 1$ can have at most m zeros in the field F by Corollary 2, so $m \ge d_1 d_2 \cdots d_r$. Hence $m = d_1 d_2 \cdots d_r$, so the primes involved in the prime powers $d_1, d_2, \ldots, d_r$ are distinct, and G is isomorphic to the cyclic group $\mathbb{Z}_m$. ◆

Exercises 5 through 8 ask us to find all generators of the cyclic groups of units for some finite fields. The fact that the multiplicative group of units of a finite field is cyclic has been applied in algebraic coding.

EXAMPLE I Let us work with polynomials in $\mathbb{Z}_5[x]$ and divide

$$f(x) = x^4 - 3x^3 + 2x^3 + 4x - 1$$

by $g(x) = x^2 - 2x + 3$ to find $q(x)$ and $r(x)$ of Theorem 5.18. The long division should be easy to follow, but remember that we are in $\mathbb{Z}_5[x]$, so, for example, $4x - (-3x) = 2x$.

$$
\begin{array}{r}
x^2 - x - 3 \\
x^2 - 2x + 3 \overline{\smash{\big)}\ x^4 - 3x^3 + 2x^2 + 4x - 1} \\
\underline{x^4 - 2x^3 + 3x^2} \\
-\ x^3 - x^2 + 4x \\
\underline{-\ x^3 + 2x^2 - 3x} \\
-\ 3x^2 + 2x - 1 \\
\underline{-\ 3x^2 + x - 4} \\
x + 3
\end{array}
$$

Thus

so $q(x) = x^2 - x - 3$, and $r(x) = x + 3$. ▲

EXAMPLE 2 Working again in $\mathbb{Z}_5[x]$, note that 1 is a zero of

$$(x^4 + 3x^3 + 2x + 4) \in \mathbb{Z}_5[x].$$

Thus by Corollary 1 of Theorem 5.18, we should be able to factor $x^4 + 3x^3 + 2x + 4$ into $(x - 1)q(x)$ in $\mathbb{Z}_5[x]$. Let us find the factorization by long division.

$$
\begin{array}{r}
x^3 + 4x^2 + 4x\ \ + 1 \\
x - 1\overline{\smash{\big)}\,x^4 + 3x^3 + \qquad\quad 2x + 4} \\
\underline{x^4 - \ x^3} \\
4x^3 \\
\underline{4x^3 - 4x^2} \\
4x^2 + 2x \\
\underline{4x^2 - 4x} \\
x + 4 \\
\underline{x - 1} \\
0
\end{array}
$$

Thus $x^4 + 3x^3 + 2x + 4 = (x - 1)(x^3 + 4x^2 + 4x + 1)$ in $\mathbb{Z}_5[x]$. Since 1 is seen to be a zero of $x^3 + 4x^2 + 4x + 1$ also, we can divide this polynomial by $x - 1$ and get

$$
\begin{array}{r}
x^2 + 4 \\
x - 1\overline{\smash{\big)}\,x^3 + 4x^2 + 4x + 1} \\
\underline{x^3 - \ x^2} \\
0 \ \ + 4x + 1 \\
\underline{4x - 4} \\
0
\end{array}
$$

Since $x^2 + 4$ still has 1 as a zero, we can divide again by $x - 1$ and get

$$
\begin{array}{r}
x\ \ + 1 \\
x - 1\overline{\smash{\big)}\,x^2 \qquad + 4} \\
\underline{x^2 - x} \\
x + 4 \\
\underline{x - 1} \\
0
\end{array}
$$

Thus $x^4 + 3x^3 + 2x + 4 = (x - 1)^3(x + 1)$ in $\mathbb{Z}_5[x]$. ▲

Irreducible Polynomials

Our next definition singles out a type of polynomial in $F[x]$ that will be of utmost importance to us. The concept is probably already familiar. We really *are* doing high school algebra in a more general setting.

DEFINITION 5.12 (Irreducible Polynomial) A nonconstant polynomial $f(x) \in F[x]$ is **irreducible over** F or is an **irreducible polynomial in** $F[x]$ if $f(x)$ cannot be expressed as a product $g(x)h(x)$ of two polynomials $g(x)$ and $h(x)$ in $F[x]$ both of lower degree than the degree of $f(x)$.

Note that the preceding definition concerns the concept *irreducible over F* and not just the concept *irreducible*. A polynomial $f(x)$ may be irreducible over F, but may not be irreducible if viewed over a larger field E containing F. We illustrate this.

EXAMPLE 3 Theorem 5.17 shows that $x^2 - 2$ viewed in $\mathbb{Q}[x]$ has no zeros in $\mathbb{Q}$. This shows that $x^2 - 2$ is irreducible over $\mathbb{Q}$, for a factorization $x^2 - 2 = (ax + b)(cx + d)$ for a, b, c, $d \in \mathbb{Q}$ would give rise to zeros of $x^2 - 2$ in $\mathbb{Q}$. However, $x^2 - 2$ viewed in $\mathbb{R}[x]$ is not irreducible over $\mathbb{R}$, because $x^2 - 2$ factors in $\mathbb{R}[x]$ into $(x - \sqrt{2})(x + \sqrt{2})$. ▲

It is worthwhile to remember that *the units in $F[x]$ are precisely the nonzero elements of F.* Thus we could have defined an irreducible polynomial $f(x)$ as a nonconstant polynomial such that in any factorization $f(x) = g(x)h(x)$ in $F[x]$, either $g(x)$ or $h(x)$ is a unit. This viewpoint will be elaborated in Chapter 7, which deals with factorization in rings more general than $F[x]$.

EXAMPLE 4 Let us show that $f(x) = x^3 + 3x + 2$ viewed in $\mathbb{Z}_5[x]$ is irreducible over $\mathbb{Z}_5$. If $x^3 + 3x + 2$ factored in $\mathbb{Z}_5[x]$ into polynomials of lower degree then there would exist at least one linear factor of $f(x)$ of the form $x - a$ for some $a \in \mathbb{Z}_5$. But then $f(a)$ would be 0, by Corollary 1 of Theorem 5.18. However, $f(0) = 2$, $f(1) = 1$, $f(-1) = -2$, $f(2) = 1$, and $f(-2) = -2$, showing that $f(x)$ has no zeros in $\mathbb{Z}_5$. Thus $f(x)$ is irreducible over $\mathbb{Z}_5$. This test for irreducibility by finding zeros works nicely for quadratic and cubic polynomials over a finite field with a small number of elements. ▲

Irreducible polynomials will play a very important role in our work from now on. The problem of determining whether a given $f(x) \in F[x]$ is irreducible over F may be difficult. We now give some criteria for irreducibility that are useful in certain cases. One technique for determining irreducibility of quadratic and cubic polynomials was illustrated in Examples 3 and 4. We formalize it in a theorem.

THEOREM 5.19 Let $f(x) \in F[x]$, and let $f(x)$ be of degree 2 or 3. Then $f(x)$ is reducible over F if and only if it has a zero in F.

PROOF If $f(x)$ is reducible so that $f(x) = g(x)h(x)$, where the degree of $g(x)$ and the degree of $h(x)$ are both less than the degree of $f(x)$, then since $f(x)$ is either quadratic or cubic, either $g(x)$ or $h(x)$ is of degree 1. If, say, $g(x)$ is of degree 1, then except for a possible factor in F, $g(x)$ is of the form $x - a$. Then $g(a) = 0$, which implies that $f(a) = 0$, so $f(x)$ has a zero in F.

Conversely, Corollary 1 of Theorem 5.18 shows that if $f(a) = 0$ for $a \in F$, then $x - a$ is a factor of $f(x)$, so $f(x)$ is reducible. ◆

We turn to some conditions for irreducibility over $\mathbb{Q}$ of polynomials in $\mathbb{Q}[x]$. The most important condition that we shall give is contained in the next theorem. We shall not prove this theorem here; it is proved in a more general situation in Section 7.1 (see the corollary of Lemma 7.6 there).

THEOREM 5.20 If $f(x) \in \mathbb{Z}[x]$, then $f(x)$ factors into a product of two polynomials of lower degrees r and s in $\mathbb{Q}[x]$ if and only if it has such a factorization with polynomials of the same degrees r and s in $\mathbb{Z}[x]$.

PROOF See the corollary of Lemma 7.6 in Section 7.1. ◆

COROLLARY If $f(x) = x^n + a_{n-1}x^{n-1} + \cdots + a_0$ is in $\mathbb{Z}[x]$ with $a_0 \neq 0$, and if $f(x)$ has a zero in $\mathbb{Q}$, then it has a zero m in $\mathbb{Z}$, and m must divide a_0.

PROOF If $f(x)$ has a zero a in $\mathbb{Q}$, then $f(x)$ has a linear factor $x - a$ in $\mathbb{Q}[x]$ by Corollary 1 of Theorem 5.18. But then by Theorem 5.20, $f(x)$ has a factorization with a linear factor in $\mathbb{Z}[x]$, so for some $m \in \mathbb{Z}$ we must have

$$f(x) = (x - m)(x^{n-1} + \cdots - a_0/m).$$

Thus a_0/m is in $\mathbb{Z}$, so m divides a_0. ◆

EXAMPLE 5 This corollary of Theorem 5.20 gives us another proof of the irreducibility of $x^2 - 2$ over $\mathbb{Q}$, for $x^2 - 2$ factors nontrivially in $\mathbb{Q}[x]$ if and only if it has a zero in $\mathbb{Q}$ by Theorem 5.19. By the corollary of Theorem 5.20, it has a zero in $\mathbb{Q}$ if and only if it has a zero in $\mathbb{Z}$, and moreover the only possibilities are the divisors ± 1 and ± 2 of 2. A check shows that none of these numbers is a zero of $x^2 - 2$. ▲

EXAMPLE 6 Let us use Theorem 5.20 to show that

$$f(x) = x^4 - 2x^2 + 8x + 1$$

viewed in $\mathbb{Q}[x]$ is irreducible over $\mathbb{Q}$. If $f(x)$ has a linear factor in $\mathbb{Q}[x]$, then it has a zero in $\mathbb{Z}$, and by the corollary of Theorem 5.20, this zero would have to be a divisor in $\mathbb{Z}$ of 1, that is, either ± 1. But $f(1) = 8$, and $f(-1) = -8$, so such a factorization is impossible.

If $f(x)$ factors into two quadratic factors in $\mathbb{Q}[x]$, then by Theorem 5.20, it has a factorization

$$(x^2 + ax + b)(x^2 + cx + d)$$

in $\mathbb{Z}[x]$. Equating coefficients of powers of x, we find that we must have

$$bd = 1, \qquad ad + bc = 8, \qquad ac + b + d = -2, \qquad \text{and} \qquad a + c = 0$$

for integers $a, b, c, d \in \mathbb{Z}$. From $bd = 1$, we see that either $b = d = 1$ or $b = d = -1$. In any case, $b = d$ and from $ad + bc = 8$, we deduce that $d(a + c) = 8$. But this is impossible since $a + c = 0$. Thus a factorization into two quadratic polynomials is also impossible and $f(x)$ is irreducible over $\mathbb{Q}$. ▲

We conclude our irreducibility criteria with the famous Eisenstein condition for irreducibility. An additional very useful criterion is given in Exercise 33.

THEOREM 5.21 (Eisenstein Polynomial) Let $p \in \mathbb{Z}$ be a prime. Suppose that $f(x) = a_n x^n + \cdots + a_0$ is in $\mathbb{Z}[x]$, and $a_n \not\equiv 0 \pmod{p}$, but $a_i \equiv 0 \pmod{p}$ for $i < n$, with $a_0 \not\equiv 0 \pmod{p^2}$. Then $f(x)$ is irreducible over $\mathbb{Q}$.

PROOF By Theorem 5.20 we need only show that $f(x)$ does not factor into polynomials of lower degree in $\mathbb{Z}[x]$. If

$$f(x) = (b_r x^r + \cdots + b_0)(c_s x^s + \cdots + c_0)$$

is a factorization in $\mathbb{Z}[x]$, with $b_r \neq 0$, $c_s \neq 0$ and $r, s < n$, then $a_0 \not\equiv 0$ $\pmod{p^2}$ implies that not both b_0 and c_0 are congruent to 0 modulo p. Suppose that $b_0 \not\equiv 0 \pmod{p}$ and $c_0 \equiv 0 \pmod{p}$. Now $a_n \not\equiv 0 \pmod{p}$ implies that $b_r, c_s \not\equiv 0 \pmod{p}$, since $a_n = b_r c_s$. Let m be the smallest value of k such that $c_k \not\equiv 0 \pmod{p}$. Then

$$a_m = b_0 c_m + b_1 c_{m-1} + \cdots + b_{m-i} c_i$$

for some $i, 0 \leq i < m$. Now b_0 and c_m neither congruent to 0 modulo p and $c_{m-1}, \ldots, c_i$ all congruent to 0 modulo p implies that $a_m \not\equiv 0$ modulo p, so $m = n$. Consequently, $s = n$, contradicting our assumption that $s < n$; that is, that our factorization was nontrivial. ◆

Note that if we take $p = 2$, the Eisenstein condition gives us still another proof of the irreducibility of $x^2 - 2$ over $\mathbb{Q}$.

EXAMPLE 7 Taking $p = 3$, we see by Theorem 5.21 that

$$25x^5 - 9x^4 + 3x^2 - 12$$

is irreducible over $\mathbb{Q}$. ▲

COROLLARY The cyclotomic polynomial

$$\Phi_p(x) = \frac{x^p - 1}{x - 1} = x^{p-1} + x^{p-2} + \cdots + x + 1$$

is irreducible over $\mathbb{Q}$ for any prime p.

PROOF Again by Theorem 5.20, we need only consider factorizations in $\mathbb{Z}[x]$. Let

$$g(x) = \Phi_p(x + 1) = \frac{(x + 1)^p - 1}{(x + 1) - 1} = \frac{x^p + \binom{p}{1} x^{p-1} + \cdots + px}{x}.$$

Then

$$g(x) = x^{p-1} + \binom{p}{1} x^{p-2} + \cdots + p$$

satisfies the Eisenstein condition for the prime p and is thus irreducible over $\mathbb{Q}$. But if $\Phi_p(x) = h(x)r(x)$ were a nontrivial factorization of $\Phi_p(x)$

in $\mathbb{Z}[x]$, then

$$\Phi_p(x + 1) = g(x) = h(x + 1)r(x + 1)$$

would give a nontrivial factorization of $g(x)$ in $\mathbb{Z}[x]$. Thus $\Phi_p(x)$ must also be irreducible over $\mathbb{Q}$. ◆

Uniqueness of Factorization in $F[x]$

Polynomials in $F[x]$ can be factored into a product of irreducible polynomials in $F[x]$ in an essentially unique way. We shall consider unique factorization in more general situations in Section 7.1. Since Chapter 7 might not be studied, we decided to include the special case for $F[x]$ here. This special case is all that is needed for Chapters 8 and 9.

For $f(x)$, $g(x) \in F[x]$ we say that $g(x)$ **divides** $f(x)$ **in** $F[x]$ if there exists $q(x) \in F[x]$ such that $f(x) = g(x)q(x)$. Note the similarity of the theorem that follows with Property (1) (set off by rules) for $\mathbb{Z}$ in Section 1.4.

THEOREM 5.22   Let $p(x)$ be an irreducible polynomial in $F[x]$. If $p(x)$ divides $r(x)s(x)$ for $r(x)$, $s(x) \in F[x]$, then either $p(x)$ divides $r(x)$ or $p(x)$ divides $s(x)$.

PROOF   We delay the proof of this theorem to Section 6.2. (See Theorem 6.15.) ◆

COROLLARY   If $p(x)$ is irreducible in $F[x]$ and $p(x)$ divides the product $r_1(x) \cdots r_n(x)$ for $r_i(x) \in F[x]$, then $p(x)$ divides $r_i(x)$ for at least one i.

PROOF   Using mathematical induction, we find that this is immediate from Theorem 5.22. ◆

THEOREM 5.23   If F is a field, then every nonconstant polynomial $f(x) \in F[x]$ can be factored in $F[x]$ into a product of irreducible polynomials, the irreducible polynomials being unique except for order and for unit (that is, nonzero constant) factors in F.

PROOF   Let $f(x) \in F[x]$ be a nonconstant polynomial. If $f(x)$ is not irreducible, then $f(x) = g(x)h(x)$, with the degree of $g(x)$ and the degree of $h(x)$ both less than the degree of $f(x)$. If $g(x)$ and $h(x)$ are both irreducible, we stop here. If not, at least one of them factors into polynomials of lower degree. Continuing this process (an induction argument really), we arrive at a factorization

$$f(x) = p_1(x)p_2(x) \cdots p_r(x),$$

where $p_i(x)$ is irreducible for $i = 1, 2, \ldots, r$.

It remains for us to show uniqueness. Suppose that

$$f(x) = p_1(x)p_2(x) \cdots p_r(x) = q_1(x)q_2(x) \cdots q_s(x)$$

are two factorizations of $f(x)$ into irreducible polynomials. Then by the corollary to Theorem 5.22, $p_1(x)$ divides some $q_j(x)$, let us assume $q_1(x)$. Since $q_1(x)$ is irreducible,

$$q_1(x) = u_1 p_1(x),$$

where $u_1 \neq 0$, but u_1 is in F and thus is a unit. Then substituting $u_1 p_1(x)$ for $q_1(x)$ and canceling, we get

$$p_2(x) \cdots p_r(x) = u_1 q_2(x) \cdots q_s(x).$$

By a similar argument, say $q_2(x) = u_2 p_2(x)$, so

$$p_3(x) \cdots p_r(x) = u_1 u_2 q_3(x) \cdots q_s(x).$$

Continuing in this manner, we eventually arrive at

$$1 = u_1 u_2 \cdots u_e q_{r+1}(x) \cdots q_s(x).$$

This is only possible if $s = r$, so that this equation is actually $1 = u_1 u_2 \cdots u_r$. Thus the irreducible factors $p_i(x)$ and $q_j(x)$ were the same except possibly for order and unit factors. ◆

EXAMPLE 8 Example 2 shows that the factorization of $x^4 + 3x^3 + 2x + 4$ in $\mathbb{Z}_5[x]$ is $(x - 1)^3(x + 1)$. These irreducible factors in $\mathbb{Z}_5[x]$ are only defined up to units in $\mathbb{Z}_5[x]$, that is, nonzero constants in $\mathbb{Z}_5$. For example, $(x - 1)^3(x + 1) = (x - 1)^2(2x - 2)(3x + 3)$. ▲

Exercises 5.6

Computations

In Exercises 1 through 4, find $q(x)$ and $r(x)$ as described by the division algorithm so that $f(x) = g(x)q(x) + r(x)$ with the degree of $r(x)$ less than the degree of $g(x)$.

1. $f(x) = x^6 + 3x^5 + 4x^2 - 3x + 2$ and $g(x) = x^2 + 2x - 3$ in $\mathbb{Z}_7[x]$.
2. $f(x) = x^6 + 3x^5 + 4x^2 - 3x + 2$ and $g(x) = 3x^2 + 2x - 3$ in $\mathbb{Z}_7[x]$.
3. $f(x) = x^5 - 2x^4 + 3x - 5$ and $g(x) = 2x + 1$ in $\mathbb{Z}_{11}[x]$.
4. $f(x) = x^4 + 5x^3 - 3x^2$ and $g(x) = 5x^2 - x + 2$ in $\mathbb{Z}_{11}[x]$.

In Exercises 5 through 8, find all generators of the cyclic multiplicative group of units of the given finite field. (Review the corollary of Theorem 1.9 in Section 1.4.)

5. $\mathbb{Z}_5$ 　　　　6. $\mathbb{Z}_7$ 　　　　7. $\mathbb{Z}_{17}$ 　　　　8. $\mathbb{Z}_{23}$

9. The polynomial $x^4 + 4$ can be factored into linear factors in $\mathbb{Z}_5[x]$. Find this factorization.

10. The polynomial $x^3 + 2x^2 + 2x + 1$ can be factored into linear factors in $\mathbb{Z}_7[x]$. Find this factorization.

11. The polynomial $2x^3 + 3x^2 - 7x - 5$ can be factored into linear factors in $\mathbb{Z}_{11}[x]$. Find this factorization.

12. Is $x^3 + 2x + 3$ an irreducible polynomial of $\mathbb{Z}_5[x]$? Why? Express it as a product of irreducible polynomials of $\mathbb{Z}_5[x]$.

13. Is $2x^3 + x^2 + 2x + 2$ an irreducible polynomial in $\mathbb{Z}_5[x]$? Why? Express it as a product of irreducible polynomials in $\mathbb{Z}_5[x]$.

14. Show that $f(x) = x^2 + 8x - 2$ is irreducible over $\mathbb{Q}$. Is $f(x)$ irreducible over $\mathbb{R}$? Over $\mathbb{C}$?

15. Repeat Exercise 14 with $g(x) = x^2 + 6x + 12$ in place of $f(x)$.

16. Demonstrate that $x^3 + 3x^2 - 8$ is irreducible over $\mathbb{Q}$.

17. Demonstrate that $x^4 - 22x^2 + 1$ is irreducible over $\mathbb{Q}$.

In Exercises 18 through 21, determine whether the polynomial in $\mathbb{Z}[x]$ satisfies an Eisenstein criterion for irreducibility over $\mathbb{Q}$.

18. $x^2 - 12$ 19. $8x^3 + 6x^2 - 9x + 24$

20. $4x^{10} - 9x^3 + 24x - 18$ 21. $2x^{10} - 25x^3 + 10x^2 - 30$

22. Find all zeros of $6x^4 + 17x^3 + 7x^2 + x - 10$ in $\mathbb{Q}$. (This is a tedious high school algebra problem. *You* might use a bit of analytic geometry and calculus and make a graph, or use Newton's method to see which are the best candidates for zeros.)

Concepts

23. Mark each of the following true or false.
    _____ a. $x - 2$ is irreducible over $\mathbb{Q}$.
    _____ b. $3x - 6$ is irreducible over $\mathbb{Q}$.
    _____ c. $x^2 - 3$ is irreducible over $\mathbb{Q}$.
    _____ d. $x^2 + 3$ is irreducible over $\mathbb{Z}_7$.
    _____ e. If F is a field, the units of $F[x]$ are precisely the nonzero elements of F.
    _____ f. If F is a field, the units of $F(x)$ are precisely the nonzero elements of F.
    _____ g. A polynomial $f(x)$ of degree n with coefficients in a field F can have at most n zeros in F.
    _____ h. A polynomial $f(x)$ of degree n with coefficients in a field F can have at most n zeros in any given field E such that $F \leq E$.
    _____ i. Every polynomial of degree 1 in $F[x]$ has at least one zero in the field F.
    _____ j. Each polynomial in $F[x]$ can have at most a finite number of zeros in the field F.

24. Find all odd prime numbers p such that $x + 2$ is a factor of $x^4 + x^3 + x^2 - x + 1$ in $\mathbb{Z}_p[x]$.

In Exercises 25 through 28, find all irreducible polynomials of the indicated degree in the given ring.

25. Degree 2 in $\mathbb{Z}_2[x]$ 26. Degree 3 in $\mathbb{Z}_2[x]$

27. Degree 2 in $\mathbb{Z}_3[x]$ **28.** Degree 3 in $\mathbb{Z}_3[x]$

29. Find the number of irreducible quadratic polynomials in $\mathbb{Z}_p[x]$, where p is a prime. [*Hint:* Find the number of reducible polynomials of the form $x^2 + ax + b$, then the number of reducible quadratics, and subtract this from the total number of quadratics.]

Theory

30. Show that the polynomial $x^p + a$ in $\mathbb{Z}_p[x]$ is not irreducible for any $a \in \mathbb{Z}_p$.

31. If F is a field and $a \neq 0$ is a zero of $f(x) = a_0 + a_1x + \cdots + a_nx^n$ in $F[x]$, show that $1/a$ is a zero of $a_n + a_{n-1}x + \cdots + a_0x^n$.

32. (Remainder Theorem) Let $f(x) \in F[x]$ where F is a field, and let $\alpha \in F$. Show that the remainder $r(x)$ when $f(x)$ is divided by $x - \alpha$, in accordance with the division algorithm, is $f(\alpha)$.

33. Let $\sigma_m : \mathbb{Z} \to \mathbb{Z}_m$ be the natural homomorphism given by $\sigma_m(a) =$ (the remainder of a when divided by m) for $a \in \mathbb{Z}$.

    a. Show that $\overline{\sigma_m} : \mathbb{Z}[x] \to \mathbb{Z}_m[x]$ given by

$$\overline{\sigma_m}(a_0 + a_1x + \cdots + a_nx^n) = \sigma_m(a_0) + \sigma_m(a_1)x + \cdots + \sigma_m(a_m)x^n$$

    is a homomorphism of $\mathbb{Z}[x]$ onto $\mathbb{Z}_m[x]$.

    b. Show that if $f(x) \in \mathbb{Z}[x]$ and $\overline{\sigma_m}(f(x))$ both have degree n and $\overline{\sigma_m}(f(x))$ does not factor in $\mathbb{Z}_m[x]$ into two polynomials of degree less than n, then $f(x)$ is irreducible in $\mathbb{Q}[x]$.

    c. Use part (b) to show that $x^3 + 17x + 36$ is irreducible in $\mathbb{Q}[x]$. [*Hint:* Try a prime value of m that simplifies the coefficients.]

5.7

Noncommutative Examples†

Thus far, the only example we have presented of a ring that is not commutative is the ring $M_n(F)$ of all $n \times n$ matrices with entries in a field F. We shall be doing almost nothing with noncommutative rings and skew fields. To show that there are other important noncommutative rings occurring very naturally in algebra, we give several examples of such rings.

Rings of Endomorphisms

Let A be any abelian group. A homomorphism of A into itself is an **endomorphism of** A. Let the set of all endomorphisms of A be Hom(A). Since the composition of two homomorphisms of A into itself is again such a homomorphism, we define multiplication on Hom(A) by function composition, and thus multiplication is associative.

---

† This section is not used in the remainder of the text.

To define addition, for $\phi, \psi \in \text{Hom}(A)$, we have to describe the value of $(\phi + \psi)$ on each $a \in A$. Define

$$(\phi + \psi)(a) = \phi(a) + \psi(a).$$

Since

$$
\begin{aligned}
(\phi + \psi)(a + b) &= \phi(a + b) + \psi(a + b) \\
&= [\phi(a) + \phi(b)] + [\psi(a) + \psi(b)] \\
&= [\phi(a) + \psi(a)] + [\phi(b) + \psi(b)] \\
&= (\phi + \psi)(a) + (\phi + \psi)(b)
\end{aligned}
$$

we see that $\phi + \psi$ is again in $\text{Hom}(A)$.

Since A is commutative, we have

$$(\phi + \psi)(a) = \phi(a) + \psi(a) = \psi(a) + \phi(a) = (\psi + \phi)(a)$$

for all $a \in A$, so $\phi + \psi = \psi + \phi$ and addition in $\text{Hom}(A)$ is commutative. The associativity of addition follows from

$$
\begin{aligned}
[\phi + (\psi + \theta)](a) &= \phi(a) + [(\psi + \theta)(a)] \\
&= \phi(a) + [\psi(a) + \theta(a)] \\
&= [\phi(a) + \psi(a)] + \theta(a) \\
&= (\phi + \psi)(a) + \theta(a) \\
&= [(\phi + \psi) + \theta](a).
\end{aligned}
$$

If e is the additive identity of A, then the homomorphism 0 defined by

$$0(a) = e$$

for $a \in A$ is an additive identity in $\text{Hom}(A)$. Finally, for

$$\phi \in \text{Hom}(A),$$

$-\phi$ defined by

$$(-\phi)(a) = -\phi(a)$$

is in $\text{Hom}(A)$, since

$$
\begin{aligned}
(-\phi)(a + b) &= -\phi(a + b) = -[\phi(a) + \phi(b)] \\
&= -\phi(a) - \phi(b) = (-\phi)(a) + (-\phi)(b),
\end{aligned}
$$

and $\phi + (-\phi) = 0$. Thus $\langle \text{Hom}(A), + \rangle$ is an abelian group.

Note that we have not yet used the fact that our functions are *homomorphisms* except to show that $\phi + \psi$ and $-\phi$ are again *homomorphisms*. Thus the set A^A of *all functions* from A into A is an abelian group under exactly the same definition of addition, and, of course, function composition again gives a nice associative multiplication in A^A. However, we do need the fact that these functions in $\text{Hom}(A)$ are homomorphisms now to prove the left distributive law in $\text{Hom}(A)$. Except for this left distributive law, $\langle A^A, +, \cdot \rangle$ satisfies all the axioms for a ring. Let ϕ, ψ, and θ be in $\text{Hom}(A)$, and let $a \in A$. Then

$$[\theta(\phi + \psi)](a) = \theta[(\phi + \psi)(a)] = \theta[\phi(a) + \psi(a)].$$

Since θ is a *homomorphism*,

$$\theta[\phi(a) + \psi(b)] = \theta[\phi(a)] + \theta[\psi(b)]$$
$$= (\theta\phi)(a) + (\theta\psi)(b)$$
$$= (\theta\phi + \theta\psi)(a).$$

Thus $\theta(\phi + \psi) = \theta\phi + \theta\psi$. The right distributive law causes no trouble, even in A^A, and follows from

$$[(\psi + \theta)\phi](a) = (\psi + \theta)[\phi(a)] = \psi[\phi(a)] + \theta[\phi(a)]$$
$$= (\psi\phi)(a) + (\theta\phi)(a) = (\psi\phi + \theta\phi)(a).$$

Thus we have proved the following theorem.

THEOREM 5.24   The set $\text{Hom}(A)$ of all endomorphisms of an abelian group A forms a ring under homomorphism addition and homomorphism multiplication (function composition).

Again, to show relevance to this chapter, we should give an example showing that $\text{Hom}(A)$ need not be commutative. Since function composition is in general not commutative, this seems reasonable to expect. However, $\text{Hom}(A)$ may be commutative in some cases. Indeed, Exercise 15 asks us to show that $\text{Hom}(\langle \mathbb{Z}, + \rangle)$ is commutative.

EXAMPLE 1   Consider the abelian group $\langle \mathbb{Z} \times \mathbb{Z}, + \rangle$ discussed in Section 2.4. We can specify an endomorphism of this group by giving its values on the generators $(1, 0)$ and $(0, 1)$ of the group. Define

$$\phi \in \text{Hom}(\langle \mathbb{Z} \times \mathbb{Z}, + \rangle)$$

by

$$\phi(1, 0) = (1, 0) \quad \text{and} \quad \phi(0, 1) = (1, 0).$$

Define ψ by

$$\psi(1, 0) = (0, 0) \quad \text{and} \quad \psi(0, 1) = (0, 1).$$

Note that ϕ maps everything onto the first factor of $\mathbb{Z} \times \mathbb{Z}$, and ψ collapses the first factor. Thus

$$(\psi\phi)(n, m) = \psi(n + m, 0) = (0, 0).$$

while

$$(\phi\psi)(n, m) = \phi(0, m) = (m, 0).$$

Hence $\phi\psi \neq \psi\phi$. ▲

EXAMPLE 2   Let F be a field, and let $\langle F(x), + \rangle$ be the additive group of the ring $F[x]$ of polynomials with coefficients in F. For this example, let us denote this additive group by $F[x]$, to simplify this notation. We can consider $\text{Hom}(F[x])$. One element of $\text{Hom}(F[x])$ acts on each polynomial in $F[x]$ by multiplying it by x. Let this endomorphism be X, so

$$X(a_0 + a_1x + a_2x^2 + \cdots + a_nx^n) = a_0x + a_1x^2 + a_2x^3 + \cdots + a_nx^{n+1}.$$

Another element of $\text{Hom}(F[x])$ is formal differentiation with respect to x. (The familiar formula "the derivative of a sum is the sum of the derivatives" guarantees that differentiation is an endomorphism of $F[x]$.) Let Y be this endomorphism, so

$$Y(a_0 + a_1x + a_2x^2 + \cdots + a_nx^n) = a_1 + 2a_2x + \cdots + na_nx^{n-1}.$$

Exercise 17 asks us to show that $YX - XY = 1$, where 1 is unity (the identity map) in $\text{Hom}(F[x])$. Thus $XY \neq YX$. Multiplication of polynomials in $F[x]$ by any element of F also gives an element of $\text{Hom}(F[x])$. The subring of $\text{Hom}(F[x])$ generated by X and Y and multiplications by elements of F is the Weyl Algebra and is important in quantum mechanics. ▲

Group Rings and Group Algebras

Let $G = \{g_i \mid i \in I\}$ be any multiplicative group, and let R be any commutative ring with unity. Let $R(G)$ be the set of all *formal sums*

$$\sum_{i \in I} a_i g_i$$

for $a_i \in R$ and $g_i \in G$, *where all but a finite number of the a_i are* 0. Define the sum of two elements of $R(G)$ by

$$\left(\sum_{i \in I} a_i g_i\right) + \left(\sum_{i \in I} b_i g_i\right) = \sum_{i \in I} (a_i + b_i)g_i.$$

Observe that $(a_i + b_i) = 0$ except for a finite number of indices i, so $\sum_{i \in I} (a_i + b_i)g_i$ is again in $R(G)$. It is immediate that $\langle R(G), + \rangle$ is an abelian group with additive identity $\sum_{i \in I} 0g_i$.

Multiplication of two elements of $R(G)$ is defined by the use of the multiplications in G and R as follows:

$$\left(\sum_{i \in I} a_i g_i\right)\left(\sum_{i \in I} b_i g_i\right) = \sum_{i \in I}\left(\sum_{g_j g_k = g_i} a_j b_k\right) g_i.$$

Naively, we formally distribute the sum $\sum_{i \in I} a_i g_i$ over the sum $\sum_{i \in I} b_i g_i$ and rename a term $a_j g_j b_k g_k$ by $a_j b_k g_k$ where $g_j g_k = g_i$ in G. Since a_i and b_i are 0 for all but a finite number of i, the sum $\sum_{g_j g_k = g_i} a_j b_k$ contains only a finite number of nonzero summands $a_j b_k \in R$ and may thus be viewed as an element of R. Again at most a finite number of such sums $\sum_{g_j g_k = g_i} a_j b_k$ are nonzero. Thus multiplication is closed on $R(G)$.

The distributive laws follow at once from the definition of addition and the formal way we used distributivity to define multiplication. For the associativity of multiplication

$$\left(\sum_{i \in I} a_i g_i\right)\left[\left(\sum_{i \in I} b_i g_i\right)\left(\sum_{i \in I} c_i g_i\right)\right] = \left(\sum_{i \in I} a_i g_i\right)\left[\sum_{i \in I}\left(\sum_{g_j g_k = g_i} b_j c_k\right) g_i\right]$$

$$= \sum_{i \in I}\left(\sum_{g_h g_j g_k = g_i} a_h b_j c_k\right) g_i$$

$$= \left[\sum_{i \in I}\left(\sum_{g_h g_j = g_i} a_h b_j\right) g_i\right]\left(\sum_{i \in I} c_i g_i\right)$$

$$= \left[\left(\sum_{i \in I} a_i g_i\right)\left(\sum_{i \in I} b_i g_i\right)\right]\left(\sum_{i \in I} c_i g_i\right).$$

Thus we have proved the following theorem.

THEOREM 5.25 If G is any multiplicative group, then $\langle R(G), +, \cdot \rangle$ is a ring.

If we rename the element $\sum_{i \in I} a_i g_i$ of $R(G)$, where $a_i = 0$ for $i \neq j$ and $a_j = 1$, by g_j, we see that $\langle R(G), \cdot \rangle$ can be considered to contain G naturally as a multiplicative subsystem. Thus, if G is not abelian, $R(G)$ is not a commutative ring.

DEFINITION 5.13 (Group Ring) The ring $R(G)$ defined above is the **group ring of G over R**. If F is a field, then $F(G)$ is the **group algebra of G over F**.

EXAMPLE 3 Let us give the addition and multiplication tables for the group algebra $\mathbb{Z}_2(G)$, where $G = \{e, a\}$ is cyclic of order 2. The elements of $\mathbb{Z}_2(G)$ are

$$0e + 0a, \quad 0e + 1a, \quad 1e + 0a, \quad \text{and} \quad 1e + 1a.$$

If we denote these elements in the obvious, natural way by

$$0, \quad a, \quad e, \quad \text{and} \quad e + a,$$

+	0	a	e	e + a
0	0	a	e	e + a
a	a	0	e + a	e
e	e	e + a	0	a
e + a	e + a	e	a	0

Table 5.1

	0	a	e	e + a
0	0	0	0	0
a	0	e	a	e + a
e	0	a	e	e + a
e + a	0	e + a	e + a	0

Table 5.2

respectively, we get Tables 5.1 and 5.2. For example, to see that $(e + a)(e + a) = 0$, we have

$$\left(1e + 1a\right)\left(1e + 1a\right) = (1 + 1)e + (1 + 1)a = 0e + 0a.$$

This example shows that a group algebra may have 0 divisors. Indeed, this is usually the case. ▲

The Quaternions

We have not yet given an example of a skew field. The *quaternions* of Hamilton are the standard example of a skew field; let us describe them.

Let the set Q be $\mathbb{R} \times \mathbb{R} \times \mathbb{R} \times \mathbb{R}$. Now $\langle \mathbb{R} \times \mathbb{R} \times \mathbb{R} \times \mathbb{R}, + \rangle$ is a group

---

◆ **HISTORICAL NOTE** ◆

Sir William Rowan Hamilton (1805–1865) discovered quaternions in 1843 while he was searching for a way to multiply number triplets (vectors in $\mathbb{R}^3$). Six years earlier he had developed the complex numbers abstractly as pairs (a, b) of real numbers with addition $(a, b) + (a', b') = (a + a', b + b')$ and multiplication $(a, b)(a', b') = (aa' - bb', ab' + a'b)$; he was then looking for an analogous multiplication for 3-vectors that was distributive and such that the length of the product vector was the product of the lengths of the factors. After many unsuccessful attempts to multiply vectors of the form $a + bi + cj$ (where $1, i, j$ are mutually perpendicular), he realized while walking along the Royal Canal in Dublin on October 16, 1843, that he needed a new "imaginary symbol" k to be perpendicular to the other three elements. He could not "resist the impulse ... to cut with a knife on a stone of Brougham Bridge" the fundamental defining formulas on page 337 for multiplying these quaternions.

The quaternions were the first known example of a strictly skew field. Though many others were subsequently discovered, it was eventually noted that none were finite. In 1909 Joseph Henry Maclagan Wedderburn (1882–1948), then a preceptor at Princeton University, gave the first proof of Theorem 5.27.

under addition by components, the direct product of $\mathbb{R}$ under addition with itself four times. This gives the operation of addition on Q. Let us rename certain elements of Q. We shall let

$$1 = (1, 0, 0, 0), \qquad i = (0, 1, 0, 0),$$
$$j = (0, 0, 1, 0), \qquad \text{and} \qquad k = (0, 0, 0, 1).$$

We furthermore agree to let

$$a_1 = (a_1, 0, 0, 0), \qquad a_2 i = (0, a_2, 0, 0),$$
$$a_3 j = (0, 0, a_3, 0) \qquad \text{and} \qquad a_4 k = (0, 0, 0, a_4).$$

In view of our definition of addition, we then have

$$(a_1, a_2, a_3, a_4) = a_1 + a_2 i + a_3 j + a_4 k.$$

Thus

$$(a_1 + a_2 i + a_3 j + a_4 k) + (b_1 + b_2 i + b_3 j + b_4 k)$$
$$= (a_1 + b_1) + (a_2 + b_2)i + (a_3 + b_3)j + (a_4 + b_4)k.$$

To define multiplication on Q, we start by defining

$$1a = a1 = a \qquad \text{for} \qquad a \in Q,$$
$$i^2 = j^2 = k^2 = -1,$$

and

$$ij = k, \qquad jk = i, \qquad ki = j, \qquad ji = -k, \qquad kj = -i, \qquad \text{and} \qquad ik = -j.$$

Note the similarity with the so-called cross product of vectors. These formulas are easy to remember if we think of the sequence

$$i, j, k, i, j, k.$$

The product from left to right of two adjacent elements is the next one to the right. The product from right to left of two adjacent elements is the negative of the next one to the left. We then define a product to be what it must be to make the distributive laws hold, namely,

$$(a_1 + a_2 i + a_3 j + a_4 k)(b_1 + b_2 i + b_3 j + b_4 k)$$
$$= (a_1 b_1 - a_2 b_2 - a_3 b_3 - a_4 b_4) + (a_1 b_2 + a_2 b_1 + a_3 b_4 - a_4 b_3)i$$
$$+ (a_1 b_3 - a_2 b_4 + a_3 b_1 + a_4 b_2)j$$
$$+ (a_1 b_4 + a_2 b_3 - a_3 b_2 + a_4 b_1)k.$$

Verification that Q is a skew field is now a tedious chore, some of which is assigned in an exercise. Since $ij = k$ and $ji = -k$, we see that multiplication is not commutative, so Q is definitely not a field. The only axiom that cannot be verified mechanically is the existence of a multiplicative inverse for

$a = a_1 + a_2i + a_3j + a_4k$, with not all $a_i = 0$. Computation shows that

$$(a_1 + a_2i + a_3j + a_4k)(a_1 - a_2i - a_3j - a_4k) = a_1^2 + a_2^2 + a_3^2 + a_4^2.$$

If we let

$$|a|^2 = a_1^2 + a_2^2 + a_3^2 + a_4^2 \quad \text{and} \quad \bar{a} = a_1 - a_2i - a_3j - a_4k,$$

we see that

$$\frac{\bar{a}}{|a|^2} = \frac{a_1}{|a|^2} - \left(\frac{a_2}{|a|^2}\right)i - \left(\frac{a_3}{|a|^2}\right)j - \left(\frac{a_4}{|a|^2}\right)k$$

is a multiplicative inverse for a. We consider that we have demonstrated the following theorem.

THEOREM 5.26   The quaternions Q form a skew field under addition and multiplication.

Note that $G = \{\pm 1, \pm i, \pm j, \pm k\}$ is a group of order 8 under quaternion multiplication. In terms of generators and relations, this group is generated by i and j, where

$$i^4 = 1, \quad j^2 = i^2 \quad \text{and} \quad ji = i^3j.$$

Since we saw in Example 5 of Section 4.8 that G_2, with presentation

$$(a, b : a^4 = 1, b^2 = a^2, ba = a^3b),$$

is a group of order 8, we must have $G_2 \simeq G$. This explains why the group G_2 of that example was called the *quaternion group*.

Algebra is not as rich in skew fields as it is in fields. For example, there are no finite skew fields. This is the content of a famous theorem of Wedderburn, which we state without proof.

THEOREM 5.27 (Wedderburn's Theorem)   A finite division ring is a field.

PROOF   See [24] for proof of Wedderburn's theorem. ◆

Exercises 5.7

Computations

In Exercises 1 through 3, let $G = \{e, a, b\}$ be a cyclic group of order 3 with identity element e. Write the element in the group algebra $\mathbb{Z}_5(G)$ in the form

$$re + sa + tb \quad \text{for} \quad r, s, t \in \mathbb{Z}_5.$$

1. $(2e + 3a + 0b) + (4e + 2a + 3b)$
2. $(2e + 3a + 0b)(4e + 2a + 3b)$
3. $(3e + 3a + 3b)^4$

In Exercises 4 through 7, write the element of Q in the form

$$a_1 + a_2 i + a_3 j + a_4 k \qquad \text{for} \qquad a_i \in \mathbb{R}.$$

4. $(i + 3j)(4 + 2j - k)$ **5.** $i^2 j^3 k j i^5$

6. $(i + j)^{-1}$ **7.** $[(1 + 3i)(4j + 3k)]^{-1}$

8. Referring to the group S_3 given in Example 4 of Section 2.1, compute the product
$$(0\rho_0 + 1\rho_1 + 0\rho_2 + 0\mu_1 + 1\mu_2 + 1\mu_3)(1\rho_0 + 1\rho_1 + 0\rho_2 + 1\mu_1 + 0\mu_2 + 1\mu_3)$$
in the group algebra $\mathbb{Z}_2(S_3)$.

9. Find the center of the group $\langle Q^*, \cdot \rangle$, where Q^* is the set of nonzero quaternions.

Concepts

10. Find two subsets of Q different from $\mathbb{C}$ and from each other, each of which is a field isomorphic to $\mathbb{C}$ under the induced addition and multiplication from Q.

11. Mark each of the following true or false.
_____ a. $M_n(F)$ has no divisors of 0 for any n.
_____ b. Every nonzero element of $M_2(\mathbb{Z}_2)$ is a unit.
_____ c. Hom(A) is always a ring with unity $\neq 0$ for every abelian group A.
_____ d. Hom(A) is never a ring with unity $\neq 0$ for any abelian group A.
_____ e. The subset Iso(A) of Hom(A), consisting of the isomorphisms of A onto A, forms a subring of Hom(A) for every abelian group A.
_____ f. $R(\langle \mathbb{Z}, + \rangle)$ is isomorphic to $\langle \mathbb{Z}, +, \cdot \rangle$ for every commutative ring R with unity.
_____ g. The group ring $R(G)$ of an abelian group G is a commutative ring for any commutative ring R with unity.
_____ h. The quaternions are a field.
_____ i. $\langle Q^*, \cdot \rangle$ is a group where Q^* is the set of nonzero quaternions.
_____ j. No subring of Q is a field.

12. Show each of the following by giving an example.

a. A polynomial of degree n with coefficients in a skew field may have more than n zeros in the skew field.

b. A finite multiplicative subgroup of a skew field need not be cyclic.

Theory

13. Let ϕ be the element of Hom$(\langle \mathbb{Z} \times \mathbb{Z}, + \rangle)$ given in Example 1, That example showed that ϕ is a right divisor of 0. Show that ϕ is also a left divisor of 0.

14. Show that $M_2(F)$ has at least six units for every field F. Exhibit these units. [*Hint:* F has at least two elements, 0 and 1.]

15. Show that Hom$(\langle \mathbb{Z}, + \rangle)$ is naturally isomorphic to $\langle \mathbb{Z}, +, \cdot \rangle$ and that Hom$(\langle \mathbb{Z}_n, + \rangle)$ is naturally isomorphic to $\langle \mathbb{Z}_n, +, \cdot \rangle$.

16. Show that Hom$(\langle \mathbb{Z}_2 \times \mathbb{Z}_2, + \rangle)$ is not isomorphic to $\langle \mathbb{Z}_2 \times \mathbb{Z}_2, +, \cdot \rangle$.

17. Referring to Example 2, show that $YX - XY = 1$.

18. It $G = \{e\}$, the group of one element, show that $R(G)$ is isomorphic to R for any ring R.

19. Prove the associative law for multiplication in Q. (This should cure you of wanting to verify any other skew field axioms for Q.)

CHAPTER SIX

◆

FACTOR RINGS AND IDEALS

◆

This chapter makes a study of rings analogous to the study of homomorphisms and factor groups in Chapter 3. Since $\langle R, + \rangle$ is a group for every ring R (actually an abelian group), the additive part of this theory is already done. We shall have to concern ourselves only with its multiplicative aspects. To make the analogy with the situation for groups as clear as possible, we shall develop this theory in the same way that we did for groups. This will give us another chance to master it.

Homomorphisms and Factor Rings

Homomorphisms

We defined the concepts of *homomorphism* and *isomorphism* for rings in Section 5.1, since we wished to talk about evaluation homomorphisms for polynomials and about isomorphic rings. We repeat some definitions here for easy reference. Recall that a homomorphism is a *structure-relating map*. A homomorphism for rings must relate both their additive structure and their multiplicative structure.

DEFINITION 6.1 (Ring Homomorphism; Analogue of Definition 3.1) A map ϕ of a ring R into a ring R' is a **homomorphism** if

$$\phi(a + b) = \phi(a) + \phi(b)$$

and

$$\phi(ab) = \phi(a)\phi(b)$$

for all elements a and b in R.

In Section 5.1, Example 7 defined evaluation homomorphisms, and Example 8 showed that the map $\gamma: \mathbb{Z} \to \mathbb{Z}_n$, where $\gamma(m)$ is the remainder of *m* when divided by *n*, is a homomorphism. We give another simple but very fundamental example of a homomorphism.

EXAMPLE I   **(Projection Homomorphisms)** Let $R_1, R_2, \ldots, R_n$ be rings. The map $\pi_i : R_1 \times R_2 \times \cdots \times R_n \to R_i$ defined by $\pi(r_1, r_2, \ldots, r_n) = r_i$ is a homomorphism, *projection onto the ith component*. The two required properties of a homomorphism hold for π_i since both addition and multiplication in the direct product are computed by addition and multiplication in each individual component. ▲

Properties of Homomorphisms

We work our way through the exposition of Section 3.1 but for ring homomorphisms.

THEOREM 6.1 (Analogue of Theorem 3.1)   Let ϕ be a homomorphism of a ring R into a ring R'. If 0 is the additive identity in R, then $\phi(0) = 0'$ is the additive identity in R', and if $a \in R$, then $\phi(-a) = -\phi(a)$. If S is a subring of R, then $\phi[S]$ is a subring of R'. Going the other way, if S' is a subring of R', then $\phi^{-1}[S']$ is a subring of R. Finally, if R has unity 1 and $\phi(1) \neq 0'$, then $\phi(1)$ is unity for $\phi[R]$. Loosely speaking, subrings correspond to subrings, and rings with unity to rings with unity under a ring homomorphism.

PROOF   Let ϕ be a homomorphism of a ring R into a ring R'. Since, in particular, ϕ can be viewed as a group homomorphism of $\langle R, + \rangle$ into $\langle R', +' \rangle$, Theorem 3.1 tells us that $\phi(0) = 0'$ is the additive identity of R' and that $\phi(-a) = -\phi(a)$.

Theorem 3.1 also tells us that if S is a subring of R, then, considering the additive group $\langle S, + \rangle$, the set $\langle \phi[S], +' \rangle$ gives a subgroup of $\langle R', +' \rangle$. If $\phi(s_1)$ and $\phi(s_2)$ are two elements of $\phi[S]$, then

$$\phi(s_1)\phi(s_2) = \phi(s_1 s_2)$$

and $\phi(s_1 s_2) \in \phi[S]$. Thus $\phi(s_1)\phi(s_2) \in \phi[S]$, so $\phi[S]$ is closed under multiplication. Consequently, $\phi[S]$ is a subring of R'.

Gong the other way, Theorem 3.1 also shows that if S' is a subring of R', then $\langle \phi^{-1}[S'], + \rangle$ is a subgroup of $\langle R, + \rangle$. Let $a, b \in \phi^{-1}[S']$, so that $\phi(a) \in S'$ and $\phi(b) \in S'$. Then

$$\phi(ab) = \phi(a)\phi(b).$$

Since $\phi(a)\phi(b) \in S'$, we see that $ab \in \phi^{-1}[S']$ so $\phi^{-1}[S']$ is closed under multiplication and thus is a subring of R.

Finally, if R, has unity 1, then for all $r \in R$,

$$\phi(r) = \phi(1r) = \phi(r1) = \phi(1)\phi(r) = \phi(r)\phi(1),$$

so $\phi(1)$ is a multiplicative identity for $\phi[R]$. If $\phi(1) \neq 0'$, then $\phi(1)$ is unity for $\phi[R]$. ◆

Note in Theorem 6.1 that $\phi(1)$ is unity for $\phi[R]$, but not necessarily for R' as we ask you to illustrate in Exercise 7.

DEFINITION 6.2 (Kernel; Analogue of Definition 3.3) Let a map $\phi: R \to R$ be a homomorphism of rings. The subring

$$\phi^{-1}[0'] = \{r \in R \mid \phi(r) = 0'\}$$

is the **kernel** of ϕ, denoted by $\mathrm{Ker}(\phi)$.

Now this $\mathrm{Ker}(\phi)$ is the same as the kernel of the group homomorphism of $\langle R, + \rangle$ into $\langle R', + \rangle$ given by ϕ. Theorem 3.2 and its corollary on group homomorphisms give us at once analogous results for ring homomorphisms.

THEOREM 6.2 (Analogue of Theorem 3.2) Let $\phi: R \to R'$ be a ring homomorphism, and let $H = \mathrm{Ker}(\phi)$. Let $a \in R$. Then $\phi^{-1}[\phi(a)] = a + H = H + a$, where $a + H = H + a$ is the coset containing a of the commutative additive group $\langle H, + \rangle$.

COROLLARY A ring homomorphism $\phi: R \to R'$ is a one-to-one map if and only if $\mathrm{Ker}(\phi) = \{0\}$.

Isomorphisms of Rings

Although we defined ring isomorphisms in Section 5.1, we repeat the definition here to continue our parallel to the exposition in Chapter 3. We are into topics in Section 3.2 already! We don't have to supply as much motivation, and we have done the additive part of the theory. We can just concentrate on the multiplication.

DEFINITION 6.3 (Isomorphism; Analogue of Definition 3.5) An **isomorphism** $\phi: R \to R'$ is a homomorphism that is one to one and onto R'.

We leave the proof of the next theorem to Exercises 16 through 18. Compare with Exercises 15 through 17 of Section 3.2.

THEOREM 6.3 (Analogue of Theorem 3.3) Let $\mathcal{R}$ be any collection of rings, and define $R \simeq R'$ for R and R' in $\mathcal{R}$ if there exists an isomorphism $\phi: R \to R'$. Then $\simeq$ is an equivalence relation.

In Section 3.2, we discussed how to show that groups are isomorphic and how to show they are not isomorphic. The same techniques hold for rings. To show that rings R and R' are isomorphic, we must actually exhibit a map $\phi : R \to R'$ that is an isomorphism. We can attempt to show that two rings are not isomorphic by finding an algebraic property of one not shared by the other.

EXAMPLE 2   Show that the fields $\mathbb{R}$ and $\mathbb{C}$ are not isomorphic.

Solution   An isomorphism $\phi : \mathbb{R} \to \mathbb{C}$ would have to map the multiplicative identity 1 of $\mathbb{R}$ into the multiplicative identity 1 of $\mathbb{C}$. Consequently, it must map the additive inverse -1 of 1 in $\mathbb{R}$ into -1 in $\mathbb{C}$. Since -1 is not the square of a real number in $\mathbb{R}$ but $-1 = i^2$ in $\mathbb{C}$, we see that the fields cannot be isomorphic. ▲

EXAMPLE 3   Show that the ring $M_2(\mathbb{R})$ of all 2×2 matrices with entries in $\mathbb{R}$ are not isomorphic to the field $\mathbb{C}$ of complex numbers.

Solution   The matrix ring $M_2(\mathbb{R})$ has divisors of zero, but $\mathbb{C}$ does not. ▲

Example 10 in Section 5.1 shows that if $\gcd(r, s) = 1$, then the rings $\mathbb{Z}_{rs}$ and $\mathbb{Z}_r \times \mathbb{Z}_s$ are isomorphic, and provides us with an illustration demonstrating a ring isomorphism. As another example, we give a fundamental isomorphism of linear algebra for those who have studied vector spaces. We do not prove here that the map we define is an isomorphism; that is part of the content of a course in linear algebra.

EXAMPLE 4   (**Linear Algebra**) Let V be an n-dimensional vector space with scalars in $\mathbb{R}$. The linear transformations $T : V \to V$ have a natural ring structure if we define addition by $(T_1 + T_2)(\mathbf{v}) = T_1(\mathbf{v}) + T_2(\mathbf{v})$ for $\mathbf{v} \in V$ and define multiplication to be function composition. One of the most useful facts in linear algebra is that this ring of transformations is isomorphic to the matrix ring $M_n(\mathbb{R})$. An isomorphism ϕ that can be used to enable the study of a transformation T by studying the matrix $\phi(T)$ is defined as follows. Let $(\mathbf{b}_1, \mathbf{b}_2, \ldots, \mathbf{b}_n)$ be an ordered basis for V, and for each $\mathbf{b}_j$, find scalars c_{ij} such that $T(\mathbf{b}_j) = c_{1j}\mathbf{b}_1 + c_{2j}\mathbf{b}_2 + \cdots + c_{nj}\mathbf{b}_n$. Let $\phi(T)$ be the matrix C whose entry in the ith row and jth column is c_{ij}. Then ϕ is an isomorphism; addition of transformations corresponds to addition of matrices and composition of transformations corresponds to multiplication of matrices. Indeed, multiplication of matrices is defined precisely so that this is the case. We mention that the field $\mathbb{R}$ of scalars can be replaced by $\mathbb{Q}$, $\mathbb{C}$, or any other field, and the result will still be true. ▲

Factor (Quotient) Rings

We are now ready to describe the analogue for rings of Section 3.3. We start with the analogue of Theorem 3.6.

THEOREM 6.4 (Analogue of Theorem 3.6) Let $\phi: R \to R'$ be a ring homomorphism with kernel H. Then the additive cosets of H form a ring R/H whose binary operations are defined by choosing representatives. That is, the sum of two cosets is defined by

$$(a + H) + (b + H) = (a + b) + H,$$

and the product of the cosets is defined by

$$(a + H)(b + H) = (ab) + H.$$

Also, the map $\mu: R/H \to \phi[R]$ defined by $\mu(a + H) = \phi(a)$ is an isomorphism.

PROOF Once again, the additive part of the theory is done for us in Theorem 3.6. We proceed to check the multiplicative aspects.

We must first show that multiplication of cosets by choosing representatives is well defined. To this end, let $h_1, h_2 \in H$ and consider the representatives $a + h_1$ of $a + H$ and $b + h_2$ of $b + H$. Let

$$c = (a + h_1)(b + h_2) = ab + ah_2 + h_1b + h_1h_2.$$

We must show that this element c lies in the coset $ab + H$. Since $ab + H = \phi^{-1}[\phi(ab)]$, we need only show that $\phi(c) = \phi(ab)$. Since ϕ is a homomorphism and $\phi(h) = 0'$ for $h \in H$, we obtain

$$\begin{aligned}
\phi(c) &= \phi(ab + ah_2 + h_1b + h_1h_2) \\
&= \phi(ab) + \phi(ah_2) + \phi(h_1b) + \phi(h_1h_2) \\
&= \phi(ab) + \phi(a)0' + 0'\phi(b) + 0'0' \\
&= \phi(ab) + 0' + 0' + 0' = \phi(ab).
\end{aligned} \tag{1}$$

Thus multiplication by choosing representatives is well defined.

To show that R/H is a ring, it remains to show that the associative property for multiplication and the distributive laws hold in R/H. Since addition and multiplication are computed by choosing representatives, these properties follow at once from corresponding properties in $\mathbb{R}$.

Theorem 3.6 shows that the map μ defined in the statement of Theorem 6.4 is well defined, one to one, onto $\phi[R]$, and satisfies the additive property for a homomorphism. Multiplicatively, we have

$$\begin{aligned}
\mu[(a + H)(b + H)] &= \mu(ab + H) = \phi(ab) \\
&= \phi(a)\phi(b) = \mu(a + H)\mu(b + H).
\end{aligned}$$

This completes the demonstration that μ is an isomorphism. ◆

EXAMPLE 5   Example 8 of Section 5.1 shows that the map $\phi:\mathbb{Z} \to \mathbb{Z}_n$ defined by $\phi(m) = r$, where r is the remainder of m when divided by r, is an isomorphism. Since $\text{Ker}(\phi) = n\mathbb{Z}$, Theorem 6.4 shows that $\mathbb{Z}/n\mathbb{Z}$ is a ring where operations on residue classes can be computed by choosing representatives and performing the corresponding operation in $\mathbb{Z}$. The theorem also shows that this ring $\mathbb{Z}/n\mathbb{Z}$ is isomorphic to $\mathbb{Z}_n$. We used these facts in Chapter 5 without a detailed proof. ▲

It remains only to characterize those subrings H of a ring R such that multiplication of additive cosets of H by choosing representatives is well defined. The coset multiplication in Theorem 6.4 was shown to be well defined in Eq. (1). The success of Eq. (1) is due to the fact that $\phi(ah_2) = \phi(h_1b) = \phi(h_1h_2) = 0'$. That is, if $h \in H$ where $H = \text{Ker}(\phi)$, then for every $a, b \in R$ we have $ah \in H$ and $hb \in H$. This suggests Theorem 6.5 below, which is the analogue of Theorem 3.7.

THEOREM 6.5 (Analogue of Theorem 3.7)   Let H be a subring of the ring R. Multiplication of additive cosets of H is well defined by the equation

$$(a + H)(b + H) = ab + H$$

if and only if $ah \in H$ and $hb \in H$ for all $a, b \in R$ and $h \in H$.

PROOF   Suppose first that $ah \in H$ and $hb \in H$ for all $a, b \in R$ and all $h \in H$. Let $h_1, h_2 \in H$ so that $a + h_1$ and $b + h_2$ are also representatives of the cosets $a + H$ and $b + H$ containing a and b. Then

$$(a + h_1)(b + h_1) = ab + ah_2 + h_1b + h_1h_2.$$

Since ah_2 and h_1b and h_1h_2 are all in H by hypothesis, we see that $(a + h_1)(b + h_2) \in ab + H$.

Conversely, suppose that multiplication of additive cosets by representatives is well defined. Let $a \in R$ and consider the coset product $(a + H)H$. Choosing representatives $a \in (a + H)$ and $0 \in H$, we see that $(a + H)H = a0 + H = 0 + H = H$. Since we can also compute $(a + H)H$ by choosing $a \in (a + H)$ and any $h \in H$, we see that $ah \in H$ for any $h \in H$. A similar argument starting with the product $H(b + H)$ shows that $hb \in H$ for any $h \in H$. ◆

In group theory, normal subgroups are precisely the type of substructure of groups required to form a factor group with a well-defined operation on cosets given by operating with chosen representatives. Theorem 6.5 shows that in ring theory, the analogous substructure must be a subring H of a ring R such that $aH \subseteq H$ and $Hb \subseteq H$ for all $a, b \in R$, where $aH = \{ah \mid h \in H\}$ and $Hb = \{hb \mid h \in H\}$. From now on we will usually denote such a substructure by N rather than H. Recall that we started using N to mean a normal subgroup in Section 3.4.

DEFINITION 6.4 (Ideal; Analogue of Definition 3.4)   A subring N of a ring R satisfying the properties

$$aN \subseteq N \quad \text{and} \quad Nb \subseteq N \qquad \text{for all } a, b \in R$$

is an **ideal**.

EXAMPLE 6   We see that $n\mathbb{Z}$ is an ideal in the ring $\mathbb{Z}$ since we know it is a subring, and $s(nm) = (nm)s = n(ms) \in n\mathbb{Z}$ for all $s \in \mathbb{Z}$. ▲

EXAMPLE 7   Let F be the ring of all functions mapping $\mathbb{R}$ into $\mathbb{R}$, and let C be the subring of F consisting of all the constant functions in F. Is C an ideal in F? Why?

◆   **H I S T O R I C A L   N O T E**   ◆

It was Ernst Eduard Kummer (1810–1893) who introduced the concept of an "ideal complex number" in 1847 in order to preserve the notion of unique factorization in certain rings of algebraic integers. In particular, Kummer wanted to be able to factor into primes numbers of the form $a_0 + a_1\alpha + a_2\alpha^2 + \cdots + a_{p-1}\alpha^{p-1}$, where α is a complex root of $x^p = 1$ (p prime) and the a_i are ordinary integers. Kummer had noticed that the naive definition of primes as "unfactorable numbers" does not lead to the expected results; the product of two such "unfactorable" numbers may well be divisible by other "unfactorable" numbers. Kummer defined "ideal prime factors" and "ideal numbers" in terms of certain congruence relationships; these "ideal factors" were then used as the divisors necessary to preserve unique factorization. By use of these, Kummer was in fact able to prove certain cases of Fermat's Last Theorem, which states that $x^n + y^n = z^n$ has no solutions $x, y, z \in \mathbb{Z}^+$ if $n > 2$.

It turned out that an "ideal number," which was in general not a "number" at all, was uniquely determined by the set of integers it "divided." Richard Dedekind took advantage of this fact to identify the ideal factor with this set; he therefore called the set itself an ideal and proceeded to show that it satisfied the definition given in the text. Dedekind was then able to define the notions of prime ideal and product of two ideals and show that any ideal in the ring of integers of any algebraic number field could be written uniquely as a product of prime ideals.

Solution   It is not true that the product of a constant function with every function is again a constant function. For example, the product of $\sin x$ and 2 is the function $2 \sin x$. Thus C is not an ideal of F. ▲

EXAMPLE 8   Let F be as in the preceding example, and let N be the subring of all functions f such that $f(2) = 0$. Is N an ideal in F? Why?

Solution   Let $f \in N$ and let $g \in F$. Then $(fg)(2) = f(2)g(2) = 0g(2) = 0,$ so $fg \in N$. Similarly, we find that $gf \in N$. Therefore N is an ideal of F. We could also have proved this by just observing that N is the kernel of the evaluation homomorphism $\phi_2 : F \to \mathbb{R}$. ▲

Once we know that multiplication by choosing representatives is well defined on additive cosets of a subring N of R, the associative law for multiplication and the distributives laws for these cosets follow at once from the same properties in R. We have at once this corollary of Theorem 6.5.

> **COROLLARY (Analogue of the Corollary of Theorem 3.7)**   Let N be an ideal of a ring R. Then the additive cosets of N form a ring R/N with the binary operations defined by
>
> $$(a + N) + (b + N) = (a + b) + N$$
>
> and
>
> $$(a + N)(b + N) = ab + N.$$

> **DEFINITION 6.5 (Factor or Quotient Ring; Analogue of Definition 3.7)**   The ring R/N in the preceding corollary is the **factor ring** (or **quotient ring**) **of R modulo N.**

If we use the term *quotient ring*, be sure not to confuse it with the notion of the *field of quotients* of an integral domain, discussed in Section 5.4.

Fundamental Homomorphism Theorem

To complete our analogy with Chapter 3, we give the analogues of Theorems 3.8 and 3.9 of Section 3.3.

> **THEOREM 6.6 (Analogue of Theorem 3.8)**   Let N be an ideal of a ring R. Then $\gamma : R \to R/N$ given by $\gamma(x) = x + N$ is a ring homomorphism with kernel N.

> **PROOF**   The additive part is done in Theorem 3.8. Turning to the multiplicative question, we see that
>
> $$\gamma(xy) = (xy) + N = (x + N)(y + N) = \gamma(x)\gamma(y). \quad \blacklozenge$$

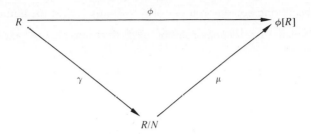

Figure 6.1

THEOREM 6.7 (Fundamental Homomorphism Theorem; Analogue of Theorem 3.9) Let $\phi:R \to R'$ be a ring homomorphism with kernel N. Then $\phi[R]$ is a ring, and the map $\mu:R/N \to \phi[R]$ given by $\mu(x + N) = \phi(x)$ is an isomorphism. If $\gamma:R \to R/N$ is the homomorphism given by $\gamma(x) = x + N$, then for each $x \in R$, we have $\phi(x) = \mu\gamma(x)$.

PROOF This follows at once from Theorems 6.4 and 6.6. Figure 6.1 is the analogue of Fig. 3.6 in Section 3.3. ◆

EXAMPLE 9 Example 6 shows that $n\mathbb{Z}$ is an ideal of $\mathbb{Z}$, so we can form the factor ring $\mathbb{Z}/n\mathbb{Z}$. Example 8 of Section 5.1 shows that $\phi:\mathbb{Z} \to \mathbb{Z}_n$ where $\phi(m)$ is the remainder of m modulo n is a homomorphism, and we see that $\text{Ker}(\phi) = n\mathbb{Z}$. Theorem 6.7 then shows that the map $\mu:\mathbb{Z}/n\mathbb{Z} \to \mathbb{Z}_n$ where $\mu(m + n\mathbb{Z})$ is the remainder of m modulo n is well defined and is an isomorphism. ▲

In summary, every ring homomorphism with domain R gives rise to a factor ring R/N, and every factor ring R/N gives rise to a homomorphism R into R/N. An *ideal* in ring theory is analogous to a *normal subgroup* in the group theory. Both are the type of substructure needed to form a factor structure.

We should now add an addendum to Theorem 6.1 on properties of homomorphisms. Let $\phi:R \to R'$ be a homomorphism, and let N be an ideal of R. Then $\phi[N]$ is an ideal of $\phi[R]$, although it need not be an ideal of R'. Also, if N' is an ideal of either $\phi[R]$ or of R', then $\phi^{-1}[N']$ is an ideal of R. We leave the proof of this to Exercise 23.

Exercises 6.1

Computations

1. Describe all ring homomorphisms of $\mathbb{Z}$ into $\mathbb{Z}$. [*Hint:* By group theory, a homomorphism of a ring R is determined by its values on an additive generating set of the group $\langle R, + \rangle$. Describe the homomorphisms by giving their values on the generator 1 of $\langle \mathbb{Z}, + \rangle$.]

2. a. Describe all ring homomorphisms of $\mathbb{Z} \times \mathbb{Z}$ into $\mathbb{Z}$. [See the hint of Exercise 1.]

   b. Describe all ring homomorphisms of $\mathbb{Z} \times \mathbb{Z}$ into $\mathbb{Z} \times \mathbb{Z}$. [See the hint of Exercise 1.]

3. Find all positive integers n such that $\mathbb{Z}_n$ contains a subring isomorphic to $\mathbb{Z}_2$.

4. Find all ideals N of $\mathbb{Z}_{12}$. In each case compute $\mathbb{Z}_{12}/N$; that is, find a known ring to which the quotient ring is isomorphic.

5. Give addition and multiplication tables for $2\mathbb{Z}/8\mathbb{Z}$. Are $2\mathbb{Z}/8\mathbb{Z}$ and $\mathbb{Z}_4$ isomorphic rings?

Concepts

6. Let F be the ring of all functions mapping $\mathbb{R}$ into $\mathbb{R}$ and having derivatives of all orders. Differentiation gives a map $\delta:F \to F$ where $\delta(f(x)) = f'(x)$. Is δ a homomorphism? Why? Give the connection between this exercise and Example 7.

7. Give an example of a ring homomorphism $\phi:R \to R'$ when R has unity 1 and $\phi(1) \neq 0'$, but $\phi(1)$ is not unity for R'.

8. Mark each of the following true or false.
   _____ a. The concept of a ring homomorphism is closely connected with the idea of a factor ring.
   _____ b. A homomorphism is to a ring as an isomorphism is to a group.
   _____ c. A ring homomorphism is one to one if and only if the kernel is $\{0\}$.
   _____ d. $\mathbb{Q}$ is an ideal in $\mathbb{R}$.
   _____ e. Every ideal in a ring is a subring of the ring.
   _____ f. Every subring of every ring is an ideal of the ring.
   _____ g. Every quotient ring of every commutative ring is again a commutative ring.
   _____ h. The rings $\mathbb{Z}/4\mathbb{Z}$ and $\mathbb{Z}_4$ are isomorphic.
   _____ i. An ideal N in a ring with unity R is all of R if and only if $1 \in N$.
   _____ j. The concept of an ideal is to the concept of a ring as the concept of a normal subgroup is to the concept of a group.

9. Let R be a ring. Observe that $\{0\}$ and R are both ideals of R. Are the factor rings R/R and $R/\{0\}$ of real interest? Why?

10. Give an example to show that a factor ring of an integral domain may be a field.

11. Give an example to show that a factor ring of an integral domain may have divisors of 0.

12. Give an example to show that a factor ring of a ring with divisors of 0 may be an integral domain.

13. Find a subring of the ring $\mathbb{Z} \times \mathbb{Z}$ that is not an ideal of $\mathbb{Z} \times \mathbb{Z}$.

14. A student is asked to prove that a quotient ring of a ring R modulo an ideal N is commutative if and only if $(rs - sr) \in N$ for all $r, s \in R$. The student starts out:

    Assume R/N is commutative. Then $rs = sr$ for all $r, s \in R/N$.

    a. Why does the instructor reading this expect nonsense from there on?

b. What should the student have written?

c. Prove the assertion. (Note the "if and only if.")

Theory

15. Let $R = \{a + b\sqrt{2} \mid a, b \in \mathbb{Z}\}$ and let R' consist of all 2×2 matrices of the form $\begin{bmatrix} a & 2b \\ b & a \end{bmatrix}$ for $a, b \in \mathbb{Z}$. Show that R and R' are isomorphic rings. (Be sure to show first that they are rings.)

16. With reference to Theorem 6.3, show that the relation $\simeq$ is reflexive.

17. With reference to Theorem 6.3, show that the relation $\simeq$ is symmetric.

18. With reference to Theorem 6.3, show that the relation $\simeq$ is transitive.

19. Show that each homomorphism of a field is either one to one or maps everything onto 0.

20. Show that if R, R', and R'' are rings, and if $\phi : R \to R'$ and $\psi : R' \to R''$ are homomorphisms, then the composite function $\psi\phi : R \to R''$ is a homomorphism. (Use Exercise 40 of Section 3.1.)

21. Let R be a commutative ring with unity of prime characteristic p. Show that the map $\phi_p : R \to R$ given by $\phi(a) = a^p$ is a homomorphism (the **Frobenius homomorphism**).

22. Let R and R' be rings and let $\phi : R \to R'$ be a ring homomorphism such that $\phi[R] \neq \{0'\}$. Show that if R has unity 1 and R' has no 0 divisors, then $\phi(1)$ is unity for R'.

23. Let $\phi : R \to R'$ be a homomorphism, and let N be an ideal of R. Show that $\phi[N]$ is an ideal of $\phi[R]$, but give an example to show that it need not be an ideal of R'. Let N' be an ideal of either $\phi[R]$ or of R'. Show that $\phi^{-1}[N']$ is an ideal of R.

24. Let F be a field, and let S be any subset of $F \times F \times \cdots \times F$ for n factors. Show that the set N_S of all $f(x_1, \ldots, x_n) \in F[x_1, \ldots, x_n]$ that have every element $(a_1, \ldots, a_n)$ of S as a zero (see Exercise 26 of Section 5.5) is an ideal in $F[x_1, \ldots, x_n]$. This is of importance in algebraic geometry.

25. Show that a factor ring of a field is either the trivial ring of one element or is isomorphic to the field.

26. Show that if R is a ring with unity and N is an ideal of R such that $N \neq R$, then R/N is a ring with unity.

27. Let R be a commutative ring and let $a \in R$. Show that $I_a = \{x \in R \mid ax = 0\}$ is an ideal of R.

28. Show that an intersection of ideals of a ring R is again an ideal of R.

29. Let R and R' be rings and let N and N' be ideals of R and R', respectively. Let ϕ be a homomorphism of R into R'. Show that ϕ induces a natural homomorphism $\phi_* : R/N \to R'/N'$ if $\phi[N] \subseteq N'$. (Use Exercise 34 of Section 3.3.)

30. Let ϕ be a homomorphism of a ring R with unity onto a ring R'. Let u be a unit in R. Show that $\phi(u)$ is a unit in R' if and only if no unit of R is in the kernel of ϕ.

31. An element a of a ring R is **nilpotent** if $a^n = 0$ for some $n \in \mathbb{Z}^+$. Show that the collection of all nilpotent elements in a commutative ring R is an ideal, the **radical of R**.

32. Referring to the definition given in Exercise 31, find the radical of the ring $\mathbb{Z}_{12}$ and observe that it is one of the ideals of $\mathbb{Z}_{12}$ found in Exercise 4. What is the radical of $\mathbb{Z}$? of $\mathbb{Z}_{32}$?

33. Referring to Exercise 31, show that if N is the radical of a commutative ring R, then R/N has as radical the trivial ideal $\{0 + N\}$.

34. Let R be a commutative ring and N an ideal of R. Referring to Exercise 31, show that if every element of N is nilpotent and the radical of R/N is R/N, then the radical of R is R.

35. Let R be a commutative ring and N an ideal of R. Show that the set $\sqrt{N}$ of all $a \in R$, such that $a^n \in N$ for some $n \in \mathbb{Z}^+$, is an ideal of R, the **radical of N**. Is this terminology consistent with that in Exercise 31?

36. Referring to Exercise 35, show by examples that for proper ideals N of a commutative ring R,

a. $\sqrt{N}$ need not equal N,   b. $\sqrt{N}$ may equal N.

37. What is the relationship of the ideal $\sqrt{N}$ of Exercise 35 to the radical of R/N (see Exercise 31)? Word your answer carefully.

38. (Second isomorphism theorem for rings) Let M and N be ideals of a ring R and let

$$M + N = \{m + n \mid m \in M, n \in N\}.$$

Show that $M + N$ is an ideal of R and that $(M + N)/N$ is naturally isomorphic to $M/(M \cap N)$. (Use Theorem 4.2.)

39. (Third isomorphism theorem for rings) Let M and N be ideals of a ring R such that $M \le N$. Show that there is a natural isomorphism mapping R/N onto $(R/M)/(N/M)$. (Use Theorem 4.3.)

40. Show that $\phi : \mathbb{C} \to M_2(\mathbb{R})$ given by

$$\phi(a + bi) = \begin{pmatrix} a & b \\ -b & a \end{pmatrix}$$

for $a, b \in \mathbb{R}$ gives an isomorphism of $\mathbb{C}$ with the subring $\phi[\mathbb{C}]$ of $M_2(\mathbb{R})$.

41. Let R be a ring with unity and let $\text{Hom}(\langle R, + \rangle)$ be the ring of endomorphisms of $\langle R, + \rangle$ as described in Section 5.7. Let $a \in R$, and let $\lambda_a : R \to R$ be given by

$$\lambda_a(x) = ax$$

for $x \in R$.

a. Show that λ_a is an endomorphism of $\langle R, + \rangle$.

b. Show that $R' = \{\lambda_a \mid a \in R\}$ is a subring of $\text{Hom}(\langle R, + \rangle)$.

c. Prove the analogue of Cayley's theorem for R by showing that R' of (b) is isomorphic to R.

6.2

Prime and Maximal Ideals

Exercises 10 through 12 of the preceding section asked us to provide examples of factor rings R/N where R and R/N have very different structural properties. We start with some examples of this situation, and in the process, provide solutions to those exercises.

EXAMPLE I As was shown in the corollary of Theorem 5.6 of Section 5.2 $\mathbb{Z}_p$, which is isomorphic to $\mathbb{Z}/p\mathbb{Z}$, is a field for p a prime. *Thus a factor ring of an integral domain may be a field.* ▲

EXAMPLE 2 The subset $N = \{0, 3\}$ of $\mathbb{Z}_6$ is easily seen to be an ideal of $\mathbb{Z}_6$, and $\mathbb{Z}_6/N$ has three elements, $0 + N$, $1 + N$, and $2 + N$. These add and multiply in such a fashion as to show that $\mathbb{Z}_6/N \simeq \mathbb{Z}_3$ under the correspondence

$$(0 + N) \leftrightarrow 0, \qquad (1 + N) \leftrightarrow 1, \qquad (2 + N) \leftrightarrow 2.$$

This example shows that *if R is not even an integral domain, that is, if R has zero divisors, it is still possible for R/N to be a field.* ▲

EXAMPLE 3 The ring $\mathbb{Z} \times \mathbb{Z}$ is not an integral domain, for

$$(0, 1)(1, 0) = (0, 0),$$

showing that $(0, 1)$ and $(1, 0)$ are 0 divisors. Let $N = \{(0, n) \mid n \in \mathbb{Z}\}$. Now N is an ideal of $\mathbb{Z} \times \mathbb{Z}$, and $(\mathbb{Z} \times \mathbb{Z})/N$ is isomorphic to $\mathbb{Z}$ under the correspondence $[(m, 0) + N] \leftrightarrow m$, where $m \in \mathbb{Z}$. Thus a *factor* ring of a ring may be an integral domain, even though the original ring is not. ▲

EXAMPLE 4 Note that $\mathbb{Z}$ is an integral domain, but $\mathbb{Z}/6\mathbb{Z} \simeq \mathbb{Z}_6$ is not. The preceding examples showed that a factor ring may have a structure that seems *better* than the original ring. This example indicates that the structure of a factor ring may seem *worse* than that of the original ring. ▲

Every ring R has two ideals, the **improper ideal** R and the **trivial ideal** $\{0\}$. For these ideals, the factor rings are R/R, which has only one element, and $R/\{0\}$, which is isomorphic to R. These are uninteresting cases. Just as for a subgroup of a group, a **proper nontrivial ideal** of a ring R is an ideal N of R such that $N \neq R$ and $N \neq \{0\}$.

While factor rings of rings and integral domains may be of great interest, as the above examples indicate, the corollary of Theorem 6.8, which follows, shows that a factor ring of a field is really not useful to us.

THEOREM 6.8 If R is a ring with unity, and N is an ideal of R containing a unit, then $N = R$.

PROOF Let N be an ideal of R, and suppose that $u \in N$ for some unit u in R. Then the condition $rN \subseteq N$ for all $r \in R$ implies, if we take $r = u^{-1}$ and $u \in N$, that $1 = u^{-1}u$ is in N. But then $rN \subseteq N$ for all $r \in R$ implies that $r1 = r$ is in N for all $r \in R$, so $N = R$. ◆

COROLLARY A field contains no proper nontrivial ideals.

PROOF Since every nonzero element of a field is a unit, it follows at once from Theorem 6.8 that an ideal of a field F is either $\{0\}$ or all of F. ◆

Maximal and Prime Ideals

We now take up the question of when a factor ring of a ring is a field and when it is an integral domain. The analogy with groups in Chapter 3 can be stretched a bit further to cover the case in which the factor ring is a field.

DEFINITION 6.6 (Maximal Ideal; Analogue of Definition 3.10, Section 3.4) A **maximal ideal of a ring** R is an ideal M different from R such that there is no proper ideal N of R properly containing M.

EXAMPLE 5 Let p be a prime positive integer. We know that $\mathbb{Z}/p\mathbb{Z}$ is isomorphic to $\mathbb{Z}_p$. Forgetting about multiplication for the moment and regarding $\mathbb{Z}/p\mathbb{Z}$ and $\mathbb{Z}_p$ as additive groups, we know that $\mathbb{Z}_p$ is a simple group, and consequently $p\mathbb{Z}$ must be a maximal normal subgroup of $\mathbb{Z}$ by Theorem 3.14. Since $\mathbb{Z}$ is an abelian group and every subgroup is a normal subgroup, we see that $p\mathbb{Z}$ is a maximal proper subgroup of $\mathbb{Z}$. Since $p\mathbb{Z}$ is an ideal of the ring $\mathbb{Z}$, it follows that $p\mathbb{Z}$ is a maximal ideal of $\mathbb{Z}$. We know that $\mathbb{Z}/p\mathbb{Z}$ is isomorphic to the ring $\mathbb{Z}_p$, and that $\mathbb{Z}_p$ is actually a field. Thus $\mathbb{Z}/p\mathbb{Z}$ is a field. This illustrates the next theorem. ▲

THEOREM 6.9 (Analogue of Theorem 3.14) Let R be a commutative ring with unity. Then M is a maximal ideal of R if and only if R/M is a field.

PROOF Suppose M is a maximal ideal in R. Observe that if R is a commutative ring with unity, then R/M is also a commutative ring with unity if $M \neq R$, which is the case if M is maximal. Let $(a + M) \in R/M$, with $a \notin M$, so that $a + M$ is not the additive identity of R/M. We must show that $a + M$ has a multiplicative inverse in R/M. Let

$$N = \{ra + m/r \in R, m \in M\}.$$

Then $\langle N, + \rangle$ is a group, for

$$(r_1a + m_1) + (r_2a + m_2) = (r_1 + r_2)a + (m_1 + m_2),$$

and the latter is clearly in N, as are

$$0 = 0a + 0 \quad \text{and} \quad -(ra + m) = (-r)a + (-m).$$

Now

$$r_1(ra + m) = (r_1r)a + r_1m$$

shows that $r_1(ra + m) \in N$ for $r_1 \in R$, and since R is a commutative ring, $(ra + m)r_1 \in N$ also. Thus N is an ideal. But

$$a = 1a + 0.$$

shows that $a \in N$, and for $m \in M$,

$$m = 0a + m$$

shows that $M \subseteq N$. Hence N is an ideal of R properly containing M, since $a \in N$ and $a \notin M$. Since M is maximal, we must have $N = R$. In particular, $1 \in N$. Then by definition of N, there is $b \in R$ and $m \in M$ such that $1 = ba + m$. Therefore,

$$1 + M = ba + M = (b + M)(a + M),$$

so $b + M$ is a multiplicative inverse of $a + M$.

Conversely, suppose that R/M is a field. By the final paragraph of Section 6.1, if N is any ideal of R such that $M \subset N \subset R$ and γ is the canonical homomorphism of R onto R/M, then $\gamma[N]$ is an ideal of R/M with $\{(0 + M)\} \subset \gamma[N] \subset R/M$. But this is contrary to the corollary of Theorem 6.8, which states that the field R/M contains no proper nontrivial ideals. Hence if R/M is a field, M is maximal. ◆

EXAMPLE 6 Since $\mathbb{Z}/n\mathbb{Z}$ is isomorphic to $\mathbb{Z}_n$ and $\mathbb{Z}_n$ is a field if and only if n is a prime, we see that the maximal ideals of $\mathbb{Z}$ are precisely the ideals $p\mathbb{Z}$ for prime positive integers p. ▲

COROLLARY A commutative ring with unity is a field if and only if it has no proper nontrivial ideals.

PROOF The corollary of Theorem 6.8 shows that a field has no proper nontrivial ideals.

Conversely, if a commutative ring R with unity has no proper nontrivial ideals, then $\{0\}$ is a maximal ideal and $R/\{0\}$, which is isomorphic to R, is a field by Theorem 6.9. ◆

We now turn to the question of characterizing, for a commutative ring R with unity, the ideals $N \neq R$ such that R/N is an integral domain. The answer here is rather obvious. The factor ring R/N will be an integral domain if and only if $(a + N)(b + N) = N$ implies that either

$$a + N = N \quad \text{or} \quad b + N = N.$$

This is exactly the statement that R/N has no divisors of 0, since the coset N plays the role of 0 in R/N. Looking at representatives, we see that this condition amounts to saying that $ab \in N$ implies that either $a \in N$ or $b \in N$.

EXAMPLE 7   The ideals of $\mathbb{Z}$ are of the form $n\mathbb{Z}$. We have seen that $\mathbb{Z}/n\mathbb{Z} \simeq \mathbb{Z}_n$ and that $\mathbb{Z}_n$ is an integral domain if and only if n is a prime. Thus the ideals $n\mathbb{Z}$ such that $\mathbb{Z}/n\mathbb{Z}$ is an integral domain are of the form $p\mathbb{Z}$, where p is a prime. Of course, $\mathbb{Z}/p\mathbb{Z}$ is actually a field, so that $p\mathbb{Z}$ is a maximal ideal of $\mathbb{Z}$. Note that for a product rs of integers to be in $p\mathbb{Z}$, the prime p must divide either r or s. The role of prime integers in this example makes the use of the word *prime* in the next definition more reasonable. ▲

> **DEFINITION 6.7 (Prime Ideal)**   An ideal $N \neq R$ in a commutative ring R is a **prime ideal** if $ab \in N$ implies that either $a \in N$ or $b \in N$ for $a, b \in R$.

EXAMPLE 8   Note that $\mathbb{Z} \times \{0\}$ is a prime ideal of $\mathbb{Z} \times \mathbb{Z}$, for if $(a, b)(c, d) \in \mathbb{Z} \times \{0\}$ then we must have $bd = 0$ in $\mathbb{Z}$. This implies that either $b = 0$ so $(a, b) \in \mathbb{Z} \times \{0\}$ or $d = 0$ so $(c, d) \in \mathbb{Z} \times \{0\}$. Observe that $(\mathbb{Z} \times \mathbb{Z})/(\mathbb{Z} \times \{0\})$ is isomorphic to $\mathbb{Z}$, which is an integral domain. ▲

Our remarks preceding Example 7 constitute a proof of the following theorem, which is illustrated by Exanple 8.

> **THEOREM 6.10**   Let R be a commutative ring with unity, and let $N \neq R$ be an ideal in R. Then R/N is an integral domain if and only if N is a prime ideal in R.

> **COROLLARY**   Every maximal ideal in a commutative ring R with unity is a prime ideal.

> **PROOF**   If M is maximal in R, then R/M is a field, hence an integral domain, and therefore M is a prime ideal by Theorem 6.10. ◆

The material that has just been presented regarding maximal and prime ideals is very important and we shall be using it quite a lot. We should keep the main ideas well in mind. We must know and understand the definitions of maximal and prime ideals and must remember the following facts that we have demonstrated.

For a commutative ring R with unity:

1. An ideal M of R is maximal if and only if R/M is a field.

2. An ideal N of R is prime if and only if R/N is an integral domain.
3. Every maximal ideal of R is a prime ideal.

Prime Fields

We now proceed to show that the rings $\mathbb{Z}$ and $\mathbb{Z}_n$ form foundations upon which all rings with unity rest, and that $\mathbb{Q}$ and $\mathbb{Z}_p$ perform a similar service for all fields. Let R be any ring with unity 1. Recall that by $n \cdot 1$ we mean $1 + 1 + \cdots + 1$ for n summands for $n > 0$, and $(-1) + (-1) + \cdots + (-1)$ for $|n|$ summands for $n < 0$, while $n \cdot 1 = 0$ for $n = 0$.

THEOREM 6.11 If R is a ring with unity 1, then the map $\phi : \mathbb{Z} \to R$ given by

$$\phi(n) = n \cdot 1$$

for $n \in \mathbb{Z}$ is a homomorphism of $\mathbb{Z}$ into R.

PROOF Observe that

$$\phi(n + m) = (n + m) \cdot 1 = (n \cdot 1) + (m \cdot 1) = \phi(n) + \phi(m).$$

The distributive laws in R show that

$$\underbrace{(1 + 1 + \cdots + 1)}_{n \text{ summands}} \underbrace{(1 + 1 + \cdots + 1)}_{m \text{ summands}} = \underbrace{(1 + 1 + \cdots + 1)}_{nm \text{ summands}}.$$

Thus $(n \cdot 1)(m \cdot 1) = (nm) \cdot 1$ for $n, m > 0$. Similar arguments with the distributive laws show that for all $n, m \in \mathbb{Z}$, we have

$$(n \cdot 1)(m \cdot 1) = (nm) \cdot 1.$$

Thus

$$\phi(nm) = (nm) \cdot 1 = (n \cdot 1)(m \cdot 1) = \phi(n)\phi(m). \quad \blacklozenge$$

COROLLARY If R is a ring with unity and characteristic $n > 1$, then R contains a subring isomorphic to $\mathbb{Z}_n$. If R has characteristic 0, then R contains a subring isomorphic to $\mathbb{Z}$.

PROOF The map $\phi : \mathbb{Z} \to R$ given by $\phi(m) = m \cdot 1$ for $m \in \mathbb{Z}$ is a homomorphism by Theorem 6.11. The kernel must be an ideal in $\mathbb{Z}$. All ideals in $\mathbb{Z}$ are of the form $s\mathbb{Z}$ for some $s \in \mathbb{Z}$. By Theorem 5.7 of Section 5.2, we see that if R has characteristic $n > 0$, then the kernel of ϕ is $n\mathbb{Z}$. Then the image $\phi[\mathbb{Z}] \leq R$ is isomorphic to $\mathbb{Z}/n\mathbb{Z} \simeq \mathbb{Z}_n$. If the characteristic of R is 0, then $m \cdot 1 \neq 0$ for all $m \neq 0$, so the kernel of ϕ is $\{0\}$. Thus, the image $\phi[\mathbb{Z}] \leq R$ is isomorphic to $\mathbb{Z}$. $\quad \blacklozenge$

THEOREM 6.12 A field F is either of prime characteristic p and contains a subfield isomorphic to $\mathbb{Z}_p$ or of characteristic 0 and contains a subfield isomorphic to $\mathbb{Q}$.

PROOF If the characteristic of F is not 0, the above corollary shows that F contains a subring isomorphic to $\mathbb{Z}_n$. Then n must be a prime p, or F would have 0 divisors. If F is of characteristic 0, then F must contain a subring isomorphic to $\mathbb{Z}$. In this case the corollaries of Theorem 5.14 of Section 5.4 show that F must contain a field of quotients of this subring and that this field of quotients must be isomorphic to $\mathbb{Q}$. ◆

Thus every field contains either a subfield isomorphic to $\mathbb{Z}_p$ for some prime p or a subfield isomorphic to $\mathbb{Q}$. These fields $\mathbb{Z}_p$ and $\mathbb{Q}$ are the fundamental building blocks on which all fields rest.

DEFINITION 6.8 (Prime Field) The fields $\mathbb{Z}_p$ and $\mathbb{Q}$ are **prime fields**.

Ideal Structure in F[x]

Throughout the rest of this section, we assume that F is a field. We give the next definition for a general commutative ring R with unity, although we are only interested in the case $R = F[x]$. Note that for a commutative ring R with unity and $a \in R$, the set $\{ra \mid r \in R\}$ is an ideal in R that contains the element a.

DEFINITION 6.9 (Principal Ideal) If R is a commutative ring with unity and $a \in R$, the ideal $\{ra \mid r \in R\}$ of all multiples of a is the **principal ideal generated by** a and is denoted by $\langle a \rangle$. An ideal N of R is a **principal ideal** if $N = \langle a \rangle$ for some $a \in R$.

EXAMPLE 9 The ideal $\langle x \rangle$ in $F[x]$ consists of all polynomials in $F[x]$ having zero constant term. ▲

The next theorem is another simple but very important application of the division algorithm for $F[x]$. (See Theorem 5.18 of Section 5.6.) The proof of this theorem is to the division algorithm in $F[x]$ as the proof that a subgroup of a cyclic group is cyclic is to the division algorithm in $\mathbb{Z}$.

THEOREM 6.13 If F is a field, every ideal in $F[x]$ is principal.

PROOF Let N be an ideal of $F[x]$. If $N = \{0\}$, then $N = \langle 0 \rangle$. Suppose that $N \neq \{0\}$, and let $g(x)$ be a nonzero element of N of minimal degree. If the degree of $g(x)$ is 0, then $g(x) \in F$ and is a unit, so $N = F[x] = \langle 1 \rangle$ by Theorem 6.8, so N is principal. If the degree of $g(x)$ is ≥ 1, let $f(x)$ be any element of N. Then by Theorem 5.18, $f(x) = g(x)q(x) - r(x)$, where (degree $r(x)$) < (degree $g(x)$). Now $f(x) \in N$ and $g(x) \in N$ imply that $f(x) - g(x)q(x) = r(x)$ is in N by definition of an ideal. Since $g(x)$ is a nonzero element of minimal degree in N, we must have $r(x) = 0$. Thus $f(x) = g(x)q(x)$ and $N = \langle g(x) \rangle$. ◆

We can now characterize the maximal ideals of $F[x]$. This is a crucial step in achieving our **basic goal**: to show that any nonconstant polynomial $f(x)$ in $F[x]$ has a zero in some field E containing F.

THEOREM 6.14 An ideal $\langle p(x) \rangle \neq \{0\}$ of $F[x]$ is maximal if and only if $p(x)$ is irreducible over F.

PROOF Suppose that $\langle p(x) \rangle \neq \{0\}$ is a maximal ideal of $F[x]$. Then $\langle p(x) \rangle \neq F[x]$, so $p(x) \notin F$. Let $p(x) = f(x)g(x)$ be a factorization of $p(x)$ in $F[x]$. Since $\langle p(x) \rangle$ is a maximal ideal and hence also a prime ideal, $(f(x)g(x)) \in \langle p(x) \rangle$ implies that $f(x) \in \langle p(x) \rangle$ or $g(x) \in \langle p(x) \rangle$; that is, either $f(x)$ or $g(x)$ has $p(x)$ as a factor. But then we can't have the degrees of both $f(x)$ and $g(x)$ less than the degree of $p(x)$. This shows that $p(x)$ is irreducible over F.

Conversely, if $p(x)$ is irreducible over F, suppose that N is an ideal such that $\langle p(x) \rangle \subseteq N \subseteq F[x]$. Now N is a principal ideal by Theorem 6.13, so $N = \langle g(x) \rangle$ for some $g(x) \in N$. Then $p(x) \in N$ implies that $p(x) = g(x)q(x)$ for some $q(x) \in F[x]$. But $p(x)$ is irreducible, which implies that either $g(x)$ or $q(x)$ is of degree 0. If $g(x)$ is of degree 0, that is, a non-zero constant in F, then $g(x)$ is a unit in $F[x]$, so $\langle g(x) \rangle = N = F[x]$. If $q(x)$ is of degree 0, then $q(x) = c$, where $c \in F$, and $g(x) = (1/c)p(x)$ is in $\langle p(x) \rangle$, so $N = \langle p(x) \rangle$. Thus $\langle p(x) \rangle \subset N \subset F[x]$ is impossible, so $\langle p(x) \rangle$ is maximal. ◆

EXAMPLE 10 Example 4 of Section 5.6 shows that $x^3 + 3x + 2$ is irreducible in $\mathbb{Z}_5[x]$. Thus $\mathbb{Z}_5[x]/\langle x^3 + 3x + 2 \rangle$ is a field. Similarly, Theorem 5.17 of Section 5.5 shows that $x^2 - 2$ is irreducible in $\mathbb{Q}[x]$, so $\mathbb{Q}[x]/\langle x^2 - 2 \rangle$ is a field. We shall examine such fields in more detail later. ▲

Application to Unique Factorization in F[x]

In Section 5.6, we stated without proof Theorem 6.15, which follows. (See Theorem 5.22 of Section 5.6.) Assuming this theorem, we proved in Section 5.6 that factorization of polynomials in $F[x]$ into irreducible polynomials is unique, except for order of factors and units in F. We delayed the proof of Theorem 6.15 until now since the machinery we have developed enables us to give such a simple, five-line proof. This proof fills the gap in our proof of unique factorization in $F[x]$.

THEOREM 6.15 Let $p(x)$ be an irreducible polynomial in $F[x]$. If $p(x)$ divides $r(x)s(x)$ for $r(x)$, $s(x) \in F[x]$, then either $p(x)$ divides $r(x)$ or $p(x)$ divides $s(x)$.

PROOF Suppose $p(x)$ divides $r(x)s(x)$. Then $r(x)s(x) \in \langle p(x) \rangle$, which is maximal by Theorem 6.14. Therefore, $\langle p(x) \rangle$ is a prime ideal by the corollary to Theorem 6.10. Hence $r(x)s(x) \in p(x)$ implies that either

$r(x) \in \langle p(x) \rangle$, giving $p(x)$ divides $r(x)$, or that $s(x) \in \langle p(x) \rangle$, giving $p(x)$ divides $s(x)$. ◆

A Preview of Our Basic Goal

We close this section with an outline of the demonstration in Section 8.1 of our basic goal. We have all the ideas for the proof at hand now; perhaps you can fill in the details from this outline.

Basic goal: Let F be a field and let $f(x)$ be a nonconstant polynomial in $F[x]$. Show that there exists a field E containing F and containing a zero α of $f(x)$.

Outline of the Proof

1. Let $p(x)$ be an irreducible factor of $f(x)$ in $F[x]$.
2. Let E be the *field* $F[x]/\langle p(x) \rangle$. (See Theorems 6.14 and 6.9.)
3. Show that no two different elements of F are in the same coset of $F[x]/\langle p(x) \rangle$, and deduce that we may consider F to be (isomorphic to) a subfield of E.
4. Let α be the coset $x + \langle p(x) \rangle$ in E. Show that for the evaluation homomorphism $\phi_\alpha : F[x] \to E$, we have $\phi_\alpha(f(x)) = 0$. That is, α is a zero of $f(x)$ in E.

An example of a field constructed according to this outline is given in Section 8.1. There, we give addition and multiplication tables for the field $\mathbb{Z}_2[x]/\langle x^2 + x + 1 \rangle$. We show there that this field has just four elements, the cosets

$$0 + \langle x^2 + x + 1 \rangle, \qquad 1 + \langle x^2 + x + 1 \rangle, \qquad x + \langle x^2 + x + 1 \rangle,$$

and

$$x + 1 + \langle x^2 + x + 1 \rangle.$$

We rename these four cosets 0, 1, α, and $\alpha + 1$ respectively, and obtain Tables 8.1 and 8.2 for addition and multiplication in this 4-element field. To see how these tables are constructed, remember that we are in a field of characteristic 2, so that $\alpha + \alpha = \alpha(1 + 1) = \alpha 0 = 0$. Remember also that α is a zero of $x^2 + x + 1$, so that $\alpha^2 + \alpha + 1 = 0$ and consequently $\alpha^2 = -\alpha - 1 = \alpha + 1$.

Exercises 6.2

Computations

1. Find all prime ideals and all maximal ideals of $\mathbb{Z}_6$.
2. Find all prime ideals and all maximal ideals of $\mathbb{Z}_{12}$.
3. Find all prime ideals and all maximal ideals of $\mathbb{Z}_2 \times \mathbb{Z}_2$.
4. Find all prime ideals and all maximal ideals of $\mathbb{Z}_2 \times \mathbb{Z}_4$.

5. Find all $c \in \mathbb{Z}_3$ such that $\mathbb{Z}_3[x]/\langle x^2 + c \rangle$ is a field.

6. Find all $c \in \mathbb{Z}_3$ such that $\mathbb{Z}_3[x]/\langle x^3 + x^2 + c \rangle$ is a field.

7. Find all $c \in \mathbb{Z}_3$ such that $\mathbb{Z}_3[x]/\langle x^3 + cx^2 + 1 \rangle$ is a field.

8. Find all $c \in \mathbb{Z}_5$ such that $\mathbb{Z}_5[x]/\langle x^2 + x + c \rangle$ is a field.

9. Find all $c \in \mathbb{Z}_5$ such that $\mathbb{Z}_5[x]/\langle x^2 + cx + 1 \rangle$ is a field.

Concepts

10. Mark each of the following true or false.
    _____ a. Every prime ideal of every commutative ring with unity is a maximal ideal.
    _____ b. Every maximal ideal of every commutative ring with unity is a prime ideal.
    _____ c. $\mathbb{Q}$ is its own prime subfield.
    _____ d. The prime subfield of $\mathbb{C}$ is $\mathbb{R}$.
    _____ e. Every field contains a subfield isomorphic to a prime field.
    _____ f. A ring with zero divisors may contain one of the prime fields as a subring.
    _____ g. Every field of characteristic zero contains a subfield isomorphic to $\mathbb{Q}$.
    _____ h. Let F be a field. Since $F[x]$ has no divisors of 0, every ideal of $F[x]$ is a prime ideal.
    _____ i. Let F be a field. Every ideal of $F[x]$ is a principal ideal.
    _____ j. Let F be a field. Every principal ideal of $F[x]$ is a maximal ideal.

11. Find a maximal ideal of $\mathbb{Z} \times \mathbb{Z}$.

12. Find a prime ideal of $\mathbb{Z} \times \mathbb{Z}$ that is not maximal.

13. Find a nontrivial proper ideal of $\mathbb{Z} \times \mathbb{Z}$ that is not prime.

14. Is $\mathbb{Q}[x]/\langle x^2 - 5x + 6 \rangle$ a field? Why?

15. Is $\mathbb{Q}[x]/\langle x^2 - 6x + 6 \rangle$ a field? Why?

Theory

16. Let R be a finite commutative ring with unity. Show that every prime ideal in R is a maximal ideal.

17. The corollary of Theorem 6.11 tells us that every ring with unity contains a subring isomorphic to either $\mathbb{Z}$ or some $\mathbb{Z}_n$. Is it possible that a ring with unity may simultaneously contain two subrings isomorphic to $\mathbb{Z}_n$ and $\mathbb{Z}_m$ for $n \neq m$? If it is possible, give an example. If it is impossible, prove it.

18. Continuing Exercise 17, is it possible that a ring with unity may simultaneously contain two subrings isomorphic to the fields $\mathbb{Z}_p$ and $\mathbb{Z}_q$ for two different primes p and q? Give an example or prove it is impossible.

19. Following the idea of Exercise 18, is it possible for an integral domain to contain two subrings isomorphic to $\mathbb{Z}_p$ and $\mathbb{Z}_q$ for $p \neq q$ and p and q both prime? Give reasons or an illustration.

20. Prove directly from the definitions of maximal and prime ideals that every maximal ideal of a commutative ring R with unity is a prime ideal. [*Hint:* Suppose M is maximal in R, $ab \in M$, and $a \notin M$. Consider the ideal of R generated by a and M.]

21. Show that N is a maximal ideal in a ring R if and only if R/N is a **simple ring**, that is, it has no proper nontrivial ideals. (Compare with Theorem 3.14 of Section 3.4.)

22. Prove that if F is a field, every proper nontrivial prime ideal of $F[x]$ is maximal.

23. Let F be a field and $f(x)$, $g(x) \in F[x]$. Show that $f(x)$ divides $g(x)$ if and only if $g(x) \in \langle f(x) \rangle$.

24. Let F be a field and let $f(x)$, $g(x) \in F[x]$. Show that
$$N = \{r(x)f(x) + s(x)g(x) \mid r(x), s(x) \in F[x]\}$$
is an ideal of $F[x]$. Show that if $f(x)$ and $g(x)$ have different degrees and $N \neq F[x]$, then $f(x)$ and $g(x)$ cannot both be irreducible over F.

25. Use Theorem 6.13 to prove the *equivalence* of these two theorems.

Fundamental Theorem of Algebra: Every nonconstant polynomial in $\mathbb{C}[x]$ has a zero in $\mathbb{C}$.

Nullstellensatz for $\mathbb{C}[x]$: Let $f_1(x), \ldots, f_r(x) \in \mathbb{C}[x]$ and suppose that every $\alpha \in \mathbb{C}$ that is a zero of all r of these polynomials is also a zero of a polynomial $g(x)$ in $\mathbb{C}[x]$. Then some power of $g(x)$ is in the smallest ideal of $\mathbb{C}[x]$ that contains the r polynomials $f_1(x), \ldots, f_r(x)$.

There is a sort of arithmetic of ideals in a ring. The next three exercises define sum, product, and quotient of ideals.

26. If A and B are ideals of a ring R, the **sum** $A + B$ **of A and B** is defined by
$$A + B = \{a + b \mid a \in A, b \in B\}.$$
a. Show that $A + B$ is an ideal.   b. Show that $A \subseteq A + B$ and $B \subseteq A + B$.

27. Let A and B be ideals of a ring R. The **product** AB **of A and B** is defined by
$$AB = \left\{ \sum_{i=1}^{n} a_i b_i \mid a_i \in A, b_i \in B, n \in \mathbb{Z}^+ \right\}.$$
a. Show that AB is an ideal in R.   b. Show that $AB \subseteq (A \cap B)$.

28. Let A and B be ideals of a *commutative* ring R. The **quotient** $A : B$ **of A by B** is defined by
$$A : B = \{r \in R \mid rb \in A \text{ for all } b \in B\}.$$
Show that $A : B$ is an ideal of R.

29. Show that for a field F, the set S of all matrices of the form
$$\begin{pmatrix} a & b \\ 0 & 0 \end{pmatrix}$$
for $a, b \in F$ is a **right ideal** but not a **left ideal** of $M_2(F)$. That is show that S is a subring closed under multiplication on the *right* by any element of $M_2(F)$, but is not closed under *left* multiplication.

30. Show that the matrix ring $M_2(\mathbb{Z}_2)$ is a simple ring; that is, $M_2(\mathbb{Z}_2)$ has no proper nontrivial ideals.

◆

CHAPTER SEVEN

FACTORIZATION†

◆

We know that every positive integer can be factored into a product of prime positive integers in a unique way, up to the order of the factors. We have shown that a polynomial in $F[x]$ can be factored into a product of irreducible polynomials and that the factorization is unique except for the order of factors and elements of F. In Section 7.1, we study factorization in a general integral domain D, defining what we mean by an irreducible, by a prime, and by unique factorization. We then show that unique factorization holds in every integral domain with the property that every ideal is a principal ideal. Section 7.2 describes the notion of a Euclidean valuation on an integral domain, and develops the Euclidean algorithm for finding greatest common divisors. Section 7.3 introduces the Gaussian integers and multiplicative norms. The section concludes with a proof of Fermat's theorem characterizing the prime integers in $\mathbb{Z}^+$ that can be expressed as a sum of two squares.

7.1

Unique Factorization Domains

The integral domain $\mathbb{Z}$ is our standard example of an integral domain in which there is unique factorization into primes (irreducibles). Section 5.6 showed that for a field F, $F[x]$ is also such an integral domain with unique factorization. In order to discuss analogous ideas in an arbitrary integral domain, we shall give several definitions, some of which are repetitions of earlier ones. It is nice to have them all in one place for reference.

DEFINITION 7.1 (Factor) Let D be an integral domain and $a, b \in D$. if there exists $c \in D$ such that $b = ac$, then a **divides** b (or a **is a factor of** b), denoted by $a \mid b$.

DEFINITION 7.2 (Unit, Associates) An element u of an integral domain D is a **unit of** D if u divides 1, that is, if u has a multiplicative inverse in D. Two elements $a, b \in D$ are **associates in** D if $a = bu$, where u is a unit in D.

---

† This chapter is not used in the remainder of the text.

◆ **H I S T O R I C A L   N O T E** ◆

The question of unique factorization in an integral domain was first
raised in public in connection with the attempted proof by Gabriel
Lamé (1795–1870) of Fermat's Last Theorem,* the conjecture that
$x^n + y^n = z^n$ has no nontrivial integral solutions for $n > 2$. It is not
hard to show that the conjecture is true if it can be proved when n is an
odd prime p. At a meeting of the Paris Academy on March 1, 1847,
Lamé announced that he had proved the theorem and presented a
sketch of the proof. Lamé's idea was first to factor $x^p + y^p$ over the
complex numbers as:

$$x^p + y^p = (x + y)(x + \alpha y)(x + \alpha^2 y) \cdots (x + \alpha^{p-1} y)$$

where α is a primitive pth root of unity. He next proposed to show that
if the factors in this expression are relatively prime and if $x^p + y^p = z^p$,
then each of the p factors must be a pth power. He could then
demonstrate that this Fermat equation would be true for a triple x', y',
z' each smaller than the original triple. This would lead to an infinite
descending sequence of positive integers, an impossibility that would
prove the theorem.

   After Lamé finished his announcement, however, Joseph Liouville
(1809–1882) cast serious doubts on the proof, noting that the conclusion
that each of the relatively prime factors was a pth power because their
product was a pth power depended on the result that any integer can be
uniquely factored into a product of primes. It was by no means clear
that "integers" of the form $x + \alpha^k y$ had this unique factorization
property. Although Lamé attempted to overcome Liouville's objections,
the matter was settled on May 24 when Liouville produced a letter from
Ernst Kummer noting that in 1844 he had already proved that unique
factorization failed in the domain $\mathbb{Z}[\alpha]$, where α is a 23rd root of unity.

---

* As this book went to press it was learned that this theorem had apparently been solved.
See the Note on page 392.

Exercise 21, asks us to show that this criterion for a and b to be associates
is an equivalence relation on D.

EXAMPLE I   The only units in $\mathbb{Z}$ are 1 and -1. Thus the only associates
of 26 in $\mathbb{Z}$ are 26 and -26. ▲

DEFINITION 7.3 (Irreducible)   A nonzero element p that is not a unit
of an integral domain D is an **irreducible of** D if in any factorization
$p = ab$ in D either a or b is a unit.

Note that an associate of an irreducible p is again an irreducible, for if $p = uc$ for a unit u, then any factorization of c provides a factorization of p.

DEFINITION 7.4 (UFD) An integral domain D is a **unique factorization domain** (abbreviated UFD) if the following conditions are satisfied:

1. Every element of D that is neither 0 nor a unit can be factored into a product of a finite number of irreducibles.
2. If $p_1 \cdots p_r$ and $q_1 \cdots q_s$ are two factorizations of the same element of D into irreducibles, then $r = s$ and the q_j can be renumbered so that p_i and q_i are associates.

EXAMPLE 2 Theorem 5.23, Section 5.6, shows that for a field F, $F[x]$ is a UFD. Also we know that $\mathbb{Z}$ is a UFD; we have made frequent use of this fact, although we have never proved it. For example, in $\mathbb{Z}$ we have

$$24 = (2)(2)(3)(2) = (-2)(-3)(2)(2).$$

Here 2 and -2 are associates, as are 3 and -3. Thus except for order and associates, the irreducible factors in these two factorizations of 24 are the same. ▲

Recall that the *principal ideal* $\langle a \rangle$ of D consists of all multiples of the element a. After just one more definition we can describe what we wish to achieve in this section.

DEFINITION 7.5 (PID) An integral domain D is a **principal ideal domain** (abbreviated PID) if every ideal in D is a principal ideal.

Our purpose in this section is to prove two exceedingly important theorems:

1. Every PID is a UFD. (Theorem 7.2)
2. If D is a UFD, then $D[x]$ is a UFD. (Theorem 7.3)

The fact that $F[x]$ is a UFD, where F is a field (by Theorem 5.23), illustrates both theorems. For by Theorem 6.13, $F[x]$ is a PID. Also, since F has no nonzero elements that are not units, F satisfies our definition for a UFD. Thus Theorem 7.3 would give another proof that $F[x]$ is a UFD, except for the fact that we shall actually use Theorem 5.23 quite often in proving Theorem 7.3. In the following section we shall study properties of a certain special class of UFDs, the *Euclidean domains.*

Let us proceed to prove the two theorems.

Every PID Is a UFD

The steps leading up to Theorem 5.23 and its proof indicate the way for our proof of Theorem 7.2. Much of the material will be repetitive. We inefficiently handled the special case of $F[x]$ separately in Theorem 5.23, since it was easy and was the only case we shall need for our field theory in general.

To prove that an integral domain D is a UFD, it is necessary to show that both Conditions 1 and 2 of the definition of a UFD are satisfied. For our special case of $F[x]$ in Theorem 5.23, Condition 1 was very easy and resulted from an argument that in a factorization of a polynomial of degree > 0 into a product of two nonconstant polynomials, the degree of each factor was less than the degree of the original polynomial. Thus we couldn't keep on factoring indefinitely without running into unit factors, that is, polynomials of degree 0. For the general case of a PID, it is harder to show that this is so. We now turn to this problem. We shall need one more set-theoretic concept.

DEFINITION 7.6 (Union)   If $\{A_i \,|\, i \in I\}$ is a collection of sets, then the **union $\bigcup_{i \in I} A_i$ of the sets** A_i is the set of all x such that $x \in A_i$ for at least one $i \in I$.

LEMMA 7.1 (Ascending Chain Condition for a PID)   Let D be a PID. If $N_1 \subseteq N_2 \subseteq \cdots$ is a monotonic ascending chain of ideals N_i, then there exists a positive integer r such that $N_r = N_s$ for all $s \geq r$. Equivalently, every strictly ascending chain of ideals (all inclusions proper) in a PID is of finite length. We express this by saying that the **ascending chain condition** (ACC) holds for ideals in a PID.

PROOF   Let $N_1 \subseteq N_2 \subseteq \cdots$ be a monotonic ascending chain of ideals N_i in D. Let $N = \bigcup_i N_i$. We claim that N is an ideal in D. Let $a, b \in N$. Then there are ideals N_{i_1} and N_{i_2} in the chain, with $a \in N_{i_1}$ and $b \in N_{i_2}$. Now either $N_{i_1} \subseteq N_{i_2}$ or $N_{i_2} \subseteq N_{i_1}$; let us assume that $N_{i_1} \subseteq N_{i_2}$, so both a and b are in N_{i_2}. This implies that $a \pm b$ and ab are in N_{i_2}, so $a \pm b$ and ab are in N. Taking $a = 0$, we see that $b \in N$ implies $-b \in N$, and $0 \in N$ since $0 \in N_1$. Thus N is a subring of D. For $a \in N$ and $d \in D$, we must have $a \in N_{i_1}$ for some N_{i_1}. Then since N_{i_1} is an ideal, $da = ad$ is in N_{i_1}. Therefore, $da \in \bigcup_i N_i$, that is, $da \in N$. Hence N is an ideal.

Now as an ideal in D that is a PID, $N = \langle c \rangle$ for some $c \in D$. Since $N = \bigcup_i N_i$, we must have $c \in N_r$ for some $r \in \mathbb{Z}^+$. For $s \geq r$, we have

$$\langle c \rangle \subseteq N_r \subseteq N_s \subseteq N = \langle c \rangle.$$

Thus $N_r = N_s$ for $s \geq r$.

The equivalence with the ACC is immediate. ◆

In what follows, it will be useful to remember that for $a, b \in D$,

$$\langle a \rangle \subseteq \langle b \rangle \text{ if and only if } b \text{ divides } a, \text{ and}$$

$$\langle a \rangle = \langle b \rangle \text{ if and only if } a \text{ and } b \text{ are associates.}$$

For the first property, note that $\langle a \rangle \subseteq \langle b \rangle$ if and only if $a \in \langle b \rangle$, which is true if and only if $a = bd$ for some $d \in D$, so that b divides a. Using this first property, we see that $\langle a \rangle = \langle b \rangle$ if and only if $a = bc$ and $b = ad$ for some $c, d \in D$. But then $a = adc$ and by canceling, we obtain $1 = dc$. Thus d and c are units so a and b are associates.

We can now prove Condition 1 of the definition of a UFD for an integral domain that is a PID.

THEOREM 7.1 Let D be a PID. Every element that is neither 0 nor a unit in D is a product of irreducibles.

PROOF Let $a \in D$, where a is neither 0 nor a unit. We first show that a has at least one irreducible factor. If a is an irreducible, we are done. If a is not an irreducible, then $a = a_1 b_1$, where *neither* a_1 nor b_1 is a unit. Now

$$\langle a \rangle \subset \langle a_1 \rangle,$$

for $\langle a \rangle \subseteq \langle a_1 \rangle$ follows from $a = a_1 b_1$, and if $\langle a \rangle = \langle a_1 \rangle$, then a and a_1 would be associates and b_1 would be a unit, contrary to construction. Continuing this procedure then, starting now with a_1, we arrive at a strictly ascending chain of ideals

$$\langle a \rangle \subset \langle a_1 \rangle \subset \langle a_2 \rangle \subset \cdots.$$

By the ACC in Lemma 7.1, this chain terminates with some $\langle a_r \rangle$, and a_r must then be irreducible. Thus a has an irreducible factor a_r.

By what we have just proved, for an element a that is neither 0 nor a unit in D, either a is irreducible or $a = p_1 c_1$ for p_1 an irreducible and c_1 not a unit. By an argument similar to the one just made, in the latter case we can conclude that $\langle a \rangle \subset \langle c_1 \rangle$. If c_1 is not irreducible, then $c_1 = p_2 c_2$ for an irreducible p_2 with c_2 not a unit. Continuing, we get a strictly ascending chain of ideals

$$\langle a \rangle \subset \langle c_1 \rangle \subset \langle c_2 \rangle \subset \cdots.$$

This chain must terminate, by the ACC in Lemma 7.1, with some $c_r = q_r$ that is an irreducible. Then $a = p_1 p_2 \cdots p_r q_r$. ◆

This completes our demonstration of Condition 1 of the definition of a UFD. Let us turn to Condition 2. Our arguments here are parallel to those leading to Theorem 5.23. The results we encounter along the way are of some interest in themselves.

LEMMA 7.2 (Generalization of Theorem 6.14) An ideal $\langle p \rangle$ in a PID is maximal if and only if p is an irreducible.

PROOF Let $\langle p \rangle$ be a maximal ideal of D, a PID. Suppose that $p = ab$ in D. Then $\langle p \rangle \subseteq \langle a \rangle$. Suppose that $\langle a \rangle = \langle p \rangle$. Then a and p would be associates, so b must be a unit. If $\langle a \rangle \neq \langle p \rangle$, then we must have $\langle a \rangle = \langle 1 \rangle = D$, since $\langle p \rangle$ is maximal. But then a and 1 are associates, so a is a unit. Thus, if $p = ab$, either a or b must be a unit. Hence p is an irreducible of D.

Conversely, suppose that p is an irreducible in D. Then if $\langle p \rangle \subseteq \langle a \rangle$, we must have $p = ab$. Now if a is a unit, then $\langle a \rangle = \langle 1 \rangle = D$. If a is not a unit, then b must be a unit, so there exists $u \in D$ such that $bu = 1$. Then

$pu = abu = a$, so $\langle a \rangle \subseteq \langle p \rangle$, and we have $\langle a \rangle = \langle p \rangle$. Thus $\langle p \rangle \subseteq \langle a \rangle$ implies that either $\langle a \rangle = D$ or $\langle a \rangle = \langle p \rangle$, and $\langle p \rangle \neq D$ or p would be a unit. Hence $\langle p \rangle$ is a maximal ideal. ◆

LEMMA 7.3 (Generalization of Theorem 6.15) In a PID, if an irreducible p divides ab, then either $p \mid a$ or $p \mid b$.

PROOF Let D be a PID and suppose that for an irreducible p in D we have $p \mid ab$. Then $(ab) \in \langle p \rangle$. Since every maximal ideal in D is a prime ideal by the corollary to Theorem 6.10, $(ab) \in \langle p \rangle$ implies that either $a \in \langle p \rangle$ or $b \in \langle p \rangle$, giving either $p \mid a$ or $p \mid b$. ◆

COROLLARY If p is an irreducible in a PID and p divides the product $a_1 a_2 \cdots a_n$ for $a_i \in D$, then $p \mid a_i$ for at least one i.

PROOF Proof of this corollary is immediate from Lemma 7.3 if we use mathematical induction. ◆

DEFINITION 7.7 (Prime) A nonzero nonunit element p of an integral domain D with the property that $p \mid ab$ implies either $p \mid a$ or $p \mid b$ is a **prime**.

Lemma 7.3 focused our attention on the defining property of a prime. In Exercises 19 and 20, we ask you to show that a prime in an integral domain is always an irreducible and that in a UFD an irreducible is also a prime. Thus the concepts of prime and irreducible coincide in a UFD. Example 3 will exhibit an integral domain containing some irreducibles that are not primes, so the concepts do not coincide in every domain.

EXAMPLE 3 Let F be a field. Let D be the subdomain of $F[x, y]$ generated by F, x^3, xy, and y^3. Then x^3, xy, and y^3 are irreducibles in D, but

$$(x^3)(y^3) = (xy)(xy)(xy).$$

Since xy divides $x^3 y^3$ but not x^3 or y^3, we see that xy is not a prime. Similar arguments show that neither x^3 nor y^3 is a prime. ▲

The defining property of a prime is precisely what is needed to establish uniqueness of factorization, Condition 2 in the definition of a UFD. We now complete the proof of Theorem 7.2 by demonstrating the uniqueness of factorization in a PID.

THEOREM 7.2 (Generalization of Theorem 5.23) Every PID is a UFD.

PROOF Theorem 7.1 shows that if D is a PID, then each $a \in D$, where a is neither 0 not a unit, has a factorization

$$a = p_1 p_2 \cdots p_r$$

into irreducibles. It remains for us to show uniqueness. Let

$$a = q_1 q_2 \cdots q_s$$

be another such factorization into irreducibles. Then we have $p_1 \mid (q_1 q_2 \cdots q_s)$, which implies that $p_1 \mid q_{j_1}$ for some j_1 by the corollary of Lemma 7.3. By changing the order of the q_j if necessary, we can assume that $j_1 = 1$ so $p_1 \mid q_1$. Then $q_1 = p_1 u_1$, and since p_1 is an irreducible, u_1 is a unit, so p_1 and q_1 are associates. We have then

$$p_1 p_2 \cdots p_r = p_1 u_1 q_2 \cdots q_s,$$

so by the cancellation law in D,

$$p_2 \cdots p_r = u_1 q_2 \cdots q_s.$$

Continuing this process, starting with p_2 and so on, we finally arrive at

$$1 = u_1 u_2 \cdots u_r q_{r+1} \cdots q_s.$$

Since the q_j are irreducibles, we must have $r = s$.   ◆

Example 6 at the end of this section will show that the converse to Theorem 7.2 is false. That is, a UFD need not be a PID.

Many algebra texts start by proving the following corollary of Theorem 7.2. We have assumed that you were familiar with this corollary and used it freely in our other work.

COROLLARY (Fundamental Theorem of Arithmetic)   The integral domain $\mathbb{Z}$ is a UFD.

PROOF   We have seen that all ideals in $\mathbb{Z}$ are of the form $n\mathbb{Z} = \langle n \rangle$ for $n \in \mathbb{Z}$. Thus $\mathbb{Z}$ is a PID, and Theorem 7.2 applies.   ◆

It is worth noting that the proof that $\mathbb{Z}$ is a PID was really way back in the corollary of Theorem 1.7, Section 1.4. We proved Theorem 1.7 by using the division algorithm for $\mathbb{Z}$ exactly as we proved, in Theorem 6.13, that $F[x]$ is a PID by using the division algorithm for $F[x]$. In Section 7.2, we shall examine this parallel more closely.

If D Is a UFD, then D[x] Is a UFD

We now start the proof of Theorem 7.3, our second main result for this section. The idea of the argument is as follows. Let D be a UFD. We can form a field of quotients F of D. Then $F[x]$ is a UFD by Theorem 5.23, and we shall show that we can recover a factorization for $f(x) \in D[x]$ from its factorization in $F[x]$. It will be necessary to compare the irreducibles in $F[x]$ with those in $D[x]$, of course. This approach, which we prefer as more intuitive than some more efficient modern ones, is essentially due to Gauss.

DEFINITION 7.8 (Primitive Polynomial)　Let D be a UFD. A nonconstant polynomial

$$f(x) = a_0 + a_1x + \cdots + a_nx^n$$

in $D[x]$ is **primitive** if the only common divisors of all the a_i are units of D.

EXAMPLE 4　In $\mathbb{Z}[x]$, $4x^2 + 3x + 2$ is primitive, but $4x^2 + 6x + 2$ is not, since 2, a nonunit in $\mathbb{Z}$, is a common divisor of 4, 6, and 2. ▲

Observe that every nonconstant irreducible in $D[x]$ must be a primitive polynomial.

LEMMA 7.4　If D is a UFD, then for every nonconstant $f(x) \in D[x]$ we have $f(x) = (c)g(x)$, where $c \in D$, $g(x) \in D[x]$, and $g(x)$ is primitive. The element c is unique up to a unit factor in D and is the **content of** $f(x)$. Also $g(x)$ is unique up to a unit factor in D.

PROOF　Let $f(x) \in D[x]$ be given, where $f(x)$ is a nonconstant polynomial. Since D is a UFD, each coefficient of $f(x)$ can be factored into a finite product of irreducibles in D, uniquely up to order and associates. Imagine each coefficient of $f(x)$ to be so factored.

The number c to be constructed is really a greatest common divisor in D of the coefficients of $f(x)$. To form c, we proceed as follows. If p is a particular irreducible dividing every coefficient in $f(x)$, replace every occurrence of every associate of p in a factorization of a coefficient by pu for some unit u. Continuing this procedure for another irreducible q that is not an associate of p but appears in the factorization of every coefficient of $f(x)$, and so on, we arrive eventually at a factorization of the coefficients of $f(x)$ in which each irreducible p_i appearing in the factorization of one coefficient and dividing all coefficients actually appears in the factorization of all coefficients, but no other associate of p_i appears in the factorization of any coefficient. Let $c = \prod_i p_i^{v_i}$, where the product is taken over all irreducibles p_i appearing in the factorizations of all the coefficients in this adjusted factorization, and were v_i is the greatest integer such that $p_i^{v_i}$ divides all the coefficients. Then we have $f(x) = (c)g(x)$, where $c \in D$, $g(x) \in D[x]$, and $g(x)$ is primitive by construction.

For uniqueness, if also $f(x) = (d)h(x)$ for $d \in D$, $h(x) \in D[x]$, and $h(x)$ primitive, then each irreducible factor of c must divide d and conversely. By setting $(c)g(x) = (d)h(x)$ and canceling irreducible factors of c into d, we arrive at $(u)g(x) = (v)h(x)$ for a unit $u \in D$. But then v must be a unit of D or we would be able to cancel irreducible factors of v into u. Thus u and v are both units, so c is unique up to a unit factor. From $f(x) = (c)g(x)$, we see that the primitive polynomial $g(x)$ is also unique up to a unit factor. ▲

EXAMPLE 5 In $\mathbb{Z}[x]$,

$$4x^2 + 6x - 8 = (2)(2x^2 + 3x - 4),$$

where $2x^3 + 3x - 4$ is primitive. ▲

LEMMA 7.5 (Gauss's Lemma) If D is a UFD, then a product of two primitive polynomials in $D[x]$ is again primitive.

PROOF Let

$$f(x) = a_0 + a_1x + \cdots + a_nx^n$$

and

$$g(x) = b_0 + b_1x + \cdots + b_mx^m$$

be primitive in $D[x]$, and let $h(x) = f(x)g(x)$. Let p be an irreducible in D. Then p does not divide all a_i and p does not divide all b_j, since $f(x)$ and $g(x)$ are primitive. Let a_r be the first coefficient of $f(x)$ not divisible by p; that is, $p \mid a_i$ for $i < r$, but $p \nmid a_r$ (that is, p does not divide a_r). Similarly, let $p \mid b_j$ for $j < s$, but $p \nmid b_s$. The coefficient of x^{r+s} in $h(x) = f(x)g(x)$ is

$$c_{r+s} = (a_0b_{r+s} + \cdots + a_{r-1}b_{s+1}) + a_rb_s + (a_{r+1}b_{s-1} + \cdots + a_{r+s}b_0).$$

Now $p \mid a_i$ for $i < r$ implies that

$$p \mid (a_0b_{r+s} + \cdots + a_{r-1}b_{s+1}),$$

and also $p \mid b_j$ for $j < s$ implies that

$$p \mid (a_{r+1}b_{s-1} \cdots + a_{r+s}b_0).$$

But p does not divide a_r or b_s, so p does not divide a_rb_s, and consequently p does not divide c_{r+s}. This shows that given any irreducible $p \in D$, there is some coefficient of $f(x)g(x)$ not divisible by p. Thus $f(x)g(x)$ is primitive. ◆

COROLLARY If D is a UFD, then a finite product of primitive polynomials in $D[x]$ is again primitive.

PROOF This corollary follows from Lemma 7.5 by induction. ◆

Now let D be a UFD and let F be a field of quotients of D. By Theorem 5.23, $F[x]$ is a UFD. As we said earlier, we shall show that $D[x]$ is a UFD by carrying a factorization in $F[x]$ of $f(x) \in D[x]$ back into one in $D[x]$. The next lemma relates the nonconstant irreducibles of $D[x]$ to those of $F[x]$. This is the last important step.

LEMMA 7.6 Let D be a UFD and let F be a field of quotients of D. Let $f(x) \in D[x]$, where (degree $f(x)$) > 0. If $f(x)$ is an irreducible in $D[x]$, then $f(x)$ is also an irreducible in $F[x]$. Also, if $f(x)$ is primitive in $D[x]$ and irreducible in $F[x]$, then $f(x)$ is irreducible in $D[x]$.

PROOF   Suppose that a nonconstant $f(x) \in D[x]$ factors into polynomials of lower degree in $F[x]$, that is,

$$f(x) = r(x)s(x)$$

for $r(x)$, $s(x) \in F[x]$. Then since F is a field of quotients of D, each coefficient in $r(x)$ and $s(x)$ is of the form a/b for some $a, b \in D$. By clearing denominators, we can get

$$(d)f(x) = r_1(x)s_1(x)$$

for $d \in D$, and $r_1(x)$, $s_1(x) \in D[x]$, where the degrees of $r_1(x)$ and $s_1(x)$ are the degrees of $r(x)$ and $s(x)$, respectively. By Lemma 7.4, $f(x) = (c)g(x)$, $r_1(x) = (c_1)r_2(x)$, and $s_1(x) = (c_2)s_2(x)$ for primitive polynomials $g(x)$, $r_2(x)$, and $s_2(x)$, and $c, c_1, c_2 \in D$. Then

$$(dc)g(x) = (c_1 c_2)r_2(x)s_2(x),$$

and by Lemma 7.5, $r_2(x)s_2(x)$ is primitive. By the uniqueness part of Lemma 7.4, $c_1 c_2 = dcu$ for some unit u in D. But then

$$(dc)g(x) = (dcu)r_2(x)s_2(x),$$

so

$$f(x) = (c)g(x) = (cu)r_2(x)s_2(x).$$

We have shown that if $f(x)$ factors nontrivially in $F[x]$, then $f(x)$ factors nontrivially into polynomials of the same degrees in $D[x]$. Thus if $f(x) \in D[x]$ is irreducible in $D[x]$, it must be irreducible in $F[x]$.
     A nonconstant $f(x) \in D[x]$ that is primitive in $D[x]$ and irreducible in $F[x]$ is also irreducible in $D[x]$, since $D[x] \subseteq F[x]$. ◆

Lemma 7.6 shows that if D is a UFD, the irreducibles in $D[x]$ are precisely the irreducibles in D, together with the nonconstant primitive polynomials that are irreducible in $F[x]$, where F is a field of quotients of $D[x]$.
     The preceding lemma is very important in its own right. This is indicated by the following corollary, a special case of which was our Theorem 5.20, Section 5.6. (We admit that it does not seem very sensible to call a special case of a corollary of a lemma a theorem. The label assigned to a result depends somewhat on the context in which it appears.)

COROLLARY   If D is a UFD and F is a field of quotients of D, then a nonconstant $f(x) \in D[x]$ factors into a product of two polynomials of lower degrees r and s in $F[x]$ if and only if it has a factorization into polynomials of the same degrees r and s in $D[x]$.

PROOF   It was shown in the proof of Lemma 7.6 that if $f(x)$ factors into a product of two polynomials of lower degree in $F[x]$, then it has a factorization into polynomials of the same degrees in $D[x]$ (see the next to last sentence of the first paragraph of the proof).

The converse holds since $D[x] \subseteq F[x]$. ◆

We are now prepared to prove our main theorem. We shall repeat the construction of the proof of Lemma 7.6 again; it is the core of this whole argument.

THEOREM 7.3 If D is a UFD, then $D[x]$ is a UFD.

PROOF Let $f(x) \in D[x]$, where $f(x)$ is neither 0 nor a unit. If $f(x)$ is of degree 0, we are done, since D is a UFD. Suppose that (degree $f(x)$) > 0, and let us view $f(x)$ as an element in $F[x]$, where F is a field of quotients of D. By Theorem 5.23, $f(x) = p_1(x) \cdots p_r(x)$ in $F[x]$, where $p_i(x)$ is irreducible in $F[x]$. Since F is a field of quotients of D, each coefficient in each $p_i(x)$ is of the form a/b for some $a, b \in D$. Clearing all denominators in the usual fashion we arrive at

$$(d)f(x) = q_1(x) \cdots q_r(x)$$

for $d, q_i(x) \in D[x]$. Since each $p_i(x)$ was irreducible in $F[x]$, we see that $q_i(x)$, which is $p_i(x)$ multiplied by a *unit in F*, is also irreducible in $F[x]$. By Lemma 7.4, $f(x) = (c)g(x)$ and $q_i(x) = (c_i)q_i'(x)$ in $D[x]$ for $g(x)$ and $q_i'(x)$ primitive. Then

$$(dc)g(x) = (c_1 \cdots c_r)q_1'(x) \cdots q_r'(x),$$

where by Lemma 7.5, the product $q_1'(x) \cdots q_r'(x)$ is primitive. By the uniqueness part of Lemma 7.4, we see that

$$c_1 \cdots c_r = dcu$$

for some unit u in D. Then

$$(dc)g(x) = (dcu)q_1'(x) \cdots q_r'(x),$$

or

$$f(x) = (c)g(x) = (cu)q_1'(x) \cdots q_r'(x).$$

Now cu can be factored into irreducibles in D. Also $q_1'(x), \ldots, q_r'(x)$ are irreducible in $D[x]$, since they are primitive and irreducible in $F[x]$ by construction. Thus we have shown that we can factor $f(x)$ into a product of irreducibles in $D[x]$.

The factorization of $f(x) \in D[x]$, where $f(x)$ has degree 0, is unique since D is a UFD; see the comment following Lemma 7.6. If $f(x)$ has degree greater than 0, we can view any factorization of $f(x)$ into irreducibles in $D[x]$ as a factorization in $F[x]$ into units (that is, the factors in D) and irreducible polynomials in $F[x]$ by Lemma 7.6. By Theorem 5.23, these polynomials are unique, except for possible constant factors in F. But as an irreducible in $D[x]$, each polynomial of degree > 0 appearing in the factorization of $f(x)$ in $D[x]$ is primitive. By the uniqueness part of

Lemma 7.4, this shows that these polynomials are unique in $D[x]$ up to unit factors, that is, associates. The product of the irreducibles in D in the factorization of $f(x)$ is the content of $f(x)$, which is again unique up to a unit factor by Lemma 7.4. Thus all irreducibles in $D[x]$ appearing in the factorization are unique up to order and associates. ◆

COROLLARY If F is a field and $x_1, \ldots, x_n$ are indeterminates, then $F[x_1, \ldots, x_n]$ is a UFD.

PROOF By Theorem 5.23, $F[x_1]$ is a UFD. By Theorem 7.3, so is $(F[x_1])[x_2] = F[x_1, x_2]$. Continuing in this procedure, we see (by induction) that $F[x_1, \ldots, x_n]$ is a UFD. ◆

We have seen that a PID is a UFD. The corollary of Theorem 7.3 makes it easy for us to give an example that shows that *not every* UFD *is a PID*.

EXAMPLE 6 Let F be a field and let x and y be indeterminates. Then $F[x, y]$ is a UFD by the corollary of Theorem 7.3. Consider the set N of all polynomials in x and y in $F[x, y]$ having constant term 0. Then N is an ideal, but not a principal ideal. Thus $F[x, y]$ is not a PID. ▲

Another example of a UFD that is not a PID is $\mathbb{Z}[x]$, as shown in Exercise 10, Section 7.2.

Exercises 7.1

Computations

In Exercises 1 through 8, determine whether the element is an irreducible of the indicated domain.

1. 5 in $\mathbb{Z}$ **2.** -17 in $\mathbb{Z}$

3. 14 in $\mathbb{Z}$ **4.** $2x - 3$ in $\mathbb{Z}[x]$

5. $2x - 10$ in $\mathbb{Z}[x]$ **6.** $2x - 3$ in $\mathbb{Q}[x]$

7. $2x - 10$ in $\mathbb{Q}[x]$ **8.** $2x - 10$ in $\mathbb{Z}_{11}[x]$

9. If possible, give four different associates of $2x - 7$ viewed as an element of $\mathbb{Z}[x]$; of $\mathbb{Q}[x]$; of $\mathbb{Z}_{11}[x]$.

10. Factor the polynomial $4x^2 - 4x + 8$ into a product of irreducibles viewing it as an element of the integral domain $\mathbb{Z}[x]$; of the integral domain $\mathbb{Q}[x]$; of the integral domain $\mathbb{Z}_{11}[x]$.

In Exercises 11 through 14, express the given polynomial as the product of its content with a primitive polynomial in the indicated UFD.

11. $18x^2 - 12x + 48$ in $\mathbb{Z}[x]$ **12.** $18x^2 - 12x + 48$ in $\mathbb{Q}[x]$

13. $2x^2 - 3x + 6$ in $\mathbb{Z}[x]$ **14.** $2x^2 - 3x + 6$ in $\mathbb{Z}_7[x]$

Concepts

15. Mark each of the following true or false.
    _____ a. Every field is a UFD.
    _____ b. Every field is a PID.
    _____ c. Every PID is a UFD.
    _____ d. Every UFD is a PID.
    _____ e. $\mathbb{Z}[x]$ is a UFD.
    _____ f. Any two irreducibles in any UFD are associates.
    _____ g. If D is a PID, then $D[x]$ is a PID.
    _____ h. If D is a UFD, then $D[x]$ is a UFD.
    _____ i. In any UFD, if $p \mid a$ for an irreducible p, then p itself appears in every factorization of a.
    _____ j. A UFD has no divisors of 0.

16. Let D be a UFD. Describe the irreducibles in $D[x]$ in terms of the irreducibles in D and the irreducibles in $F[x]$, where F is a field of quotients of D.

17. Lemma 7.6 states that if D is a UFD with a field of quotients F, then a nonconstant irreducible $f(x)$ of $D[x]$ is also an irreducible of $F[x]$. Show by an example that a $g(x) \in D[x]$ that is an irreducible of $F[x]$ need not be an irreducible of $D[x]$.

18. All our work in this section was restricted to integral domains. Taking the same definitions in this section but for a commutative ring with unity, consider factorizations into irreducibles in $\mathbb{Z} \times \mathbb{Z}$. What can happen? Consider in particular $(1, 0)$.

Theory

19. Prove that if p is a prime in an integral domain D, then p is an irreducible.

20. Prove that if p is an irreducible in a UFD, then q is a prime.

21. For an integral domain D, show that the relation $a \sim b$ if a is an associate of b (that is, if $a = bu$ for u a unit in D) is an equivalence relation on D.

22. Let D be an integral domain. Exercise 35, Section 5.1 showed that $\langle U, \cdot \rangle$ is a group where U is the set of units of D. Show that the set $D^* - U$ of nonunits of D excluding 0 is closed under multiplication. Is this set a group under the multiplication of D?

23. Let D be a UFD. Show that a nonconstant divisor of a primitive polynomial in $D[x]$ is again a primitive polynomial.

24. Show that in a PID, every ideal is contained in a maximal ideal. [*Hint:* Use Lemma 7.1.]

25. Factor $x^3 - y^3$ into irreducibles in $\mathbb{Q}[x, y]$ and prove that each of the factors is irreducible.

There are several other concepts often considered that are similar in character to the ascending chain condition on ideals in a ring. The following three exercises concern some of these concepts.

26. Let R be any ring. The **ascending chain condition** (ACC) **for ideals** holds in R if every strictly increasing sequence $N_1 \subset N_2 \subset N_3 \subset \cdots$ of ideals in R is of finite length. The **maximum condition** (MC) **for ideals** holds in R if every nonempty set S of ideals in R contains an ideal not properly contained in any other ideal of the set S. The **finite basis condition** (FBC) **for ideals** holds in R if for each ideal N in R, there is a finite set $B_N = \{b_1, \ldots, b_n\} \subseteq N$ such that N is the intersection of all ideals of R containing B_N. The set B_N is a **finite basis for** N.

Show that for every ring R, the conditions ACC, MC, and FBC are equivalent.

27. Let R be any ring. The **descending chain condition** (DCC) **for ideals** holds in R if every strictly decreasing sequence $N_1 \supset N_2 \supset N_3 \supset \cdots$ of ideals in R is of finite length. The **minimum condition** (mC) **for ideals** holds in R if given any set S of ideals of R, there is an ideal of S that does not properly contain any other ideal in the set S.

Show that for every ring, the conditions DCC and mC are equivalent.

28. Give an example of a ring in which ACC holds but DCC does not hold. (See Exercises 26 and 27.)

7.2

Euclidean Domains

We have remarked several times on the importance of division algorithms. Our first contact with them was the *division algorithm for* $\mathbb{Z}$ in Section 1.4. This algorithm was immediately used to prove the important theorem that a subgroup of a cyclic group is cyclic, that is, has a single generator. The *division algorithm for* $F[x]$ appeared in Theorem 5.18 and was used in a completely analogous way to show that $F[x]$ is a PID, that is, that every ideal in $F[x]$ has a single generator. Now a modern technique of mathematics is to take some clearly related situations and to try to bring them under one roof by abstracting the important ideas common to them. The following definition is an illustration of this technique. Let us see what we can develop by starting with the existence of a fairly general division algorithm in an integral domain.

DEFINITION 7.9 (Valuation) A **Euclidean valuation on an integral domain** D is a function v mapping the nonzero elements of D into the nonnegative integers such that the following conditions are satisfied:

1. For all $a, b \in D$ with $b \neq 0$, there exist q and r in D such that $a = bq + r$, where either $r = 0$ or $v(r) < v(b)$.
2. For all $a, b \in D$, where neither a nor b is 0, $v(a) \leq v(ab)$.

An integral domain D is a **Euclidean domain** if there exists a Euclidean valuation on D.

The importance of Condition 1 is clear from our discussion. The importance of Condition 2 is that it will enable us to characterize the units of a Euclidean domain D.

EXAMPLE 1   The integral domain $\mathbb{Z}$ is a Euclidean domain, for the function v defined by $v(n) = |n|$ for $n \neq 0$ in $\mathbb{Z}$ is a Euclidean valuation on $\mathbb{Z}$. Condition 1 holds by the division algorithm for $\mathbb{Z}$. Condition 2 follows from $|ab| = |a|\,|b|$ and $|a| \geq 1$ for $a \neq 0$ in $\mathbb{Z}$. ▲

EXAMPLE 2   If F is a field, then $F[x]$ is a Euclidean domain, for the function v defined by $v(f(x)) = (\text{degree } f(x))$ for $f(x) \in F[x]$, and $f(x) \neq 0$ is a Euclidean valuation. Condition 1 holds by Theorem 5.18, and Condition 2 holds since the degree of the product of two polynomials is the sum of their degrees. ▲

Of course, we should give some examples of Euclidean domains other than these familiar ones that motivated the definition. We shall do this in the next section. In view of the opening remarks, we anticipate the following theorem.

THEOREM 7.4   Every Euclidean domain is a PID.

PROOF   Let D be a Euclidean domain with a Euclidean valuation v, and let N be an ideal in D. If $N = \{0\}$, then $N = \langle 0 \rangle$ and N is principal. Suppose that $N \neq \{0\}$. Then there exists $b \neq 0$ in N. Let us choose b such that $v(b)$ is minimal among all $v(n)$ for $n \in N$. We claim that $N = \langle b \rangle$. Let $a \in N$. Then by Condition 1 for a Euclidean domain, there exist q and r in D such that

$$a = bq + r,$$

where either $r = 0$ or $v(r) < v(b)$. Now $r = a - bq$ and $a, b \in N$, so that $r \in N$, since N is an ideal. Thus $v(r) < v(b)$ is impossible by our choice of b. Hence $r = 0$, so $a = bq$. Since a was any element of N, we see that $N = \langle b \rangle$. ◆

COROLLARY   A Euclidean domain is a UFD.

PROOF   By Theorem 7.4, a Euclidean domain is a PID and by Theorem 7.2, a PID is a UFD. ◆

Finally, we should mention that while a Euclidean domain is a PID by Theorem 7.4, not every PID is a Euclidean domain. Examples of PIDs that are not Euclidean are not easily found, however.

Arithmetic in Euclidean Domains

We shall now investigate some properties of Euclidean domains related to their multiplicative structure. We emphasize that the arithmetic structure of a Euclidean domain is *intrinsic to the domain* and is not affected in any way by a Euclidean valuation v on the domain. A Euclidean valuation is merely a useful tool for possibly throwing some light on this arithmetic structure of the domain. The arithmetic structure of a domain D is completely determined by the set D and the two binary operations $+$ and $\cdot$ on D.

Let D be a Euclidean domain with a Euclidean valuation v. We can use Condition 2 of a Euclidean valuation to characterize the units of D.

THEOREM 7.5 For a Euclidean domain with a Euclidean valuation v, $v(1)$ is minimal among all $v(a)$ for nonzero $a \in D$, and $u \in D$ is a unit if and only if $v(u) = v(1)$.

PROOF Condition 2 for v tells us at once that for $a \neq 0$,

$$v(1) \leq v(1a) = v(a).$$

On the other hand, if u is a unit in D, then

$$v(u) \leq v(uu^{-1}) = v(1).$$

Thus

$$v(u) = v(1)$$

for a unit u in D.

Conversely, suppose that a nonzero $u \in D$ is such that $v(u) = v(1)$. Then by the division algorithm, there exist q and r in D such that

$$1 = uq + r,$$

where either $r = 0$ or $v(r) < v(u)$. But since $v(u) = v(1)$ is minimal over all $v(d)$ for nonzero $d \in D$, $v(r) < v(u)$ is impossible. Hence $r = 0$ and $1 = uq$, so u is a unit. ◆

EXAMPLE 3 For $\mathbb{Z}$ with $v(n) = |n|$, the minimum of $v(n)$ for nonzero $n \in \mathbb{Z}$ is 1, and 1 and -1 are the only elements of $\mathbb{Z}$ with $v(n) = 1$. Of course, 1 and -1 are exactly the units of $\mathbb{Z}$. ▲

EXAMPLE 4 For $F[x]$ with $v(f(x)) = $ (degree $f(x)$) for $f(x) \neq 0$, the minimum value of $v(f(x))$ for all nonzero $f(x) \in F[x]$ is 0. The nonzero polynomials of degree 0 are exactly the nonzero elements of F, and these are precisely the units of $F[x]$. ▲

We emphasize that everything we prove here holds in *every* Euclidean domain, in particular in $\mathbb{Z}$ and $F[x]$. We are going to prove some nice, classical results about greatest common divisors in a Euclidean domain. We surely have a naive idea of what a greatest common divisor (gcd) of two elements a and b in a UFD should be. Simply take a and b and factor them both, adjusting the factorizations by units, so that if any irreducible divides both a and b, it either appears in both factorizations, and no other of its associates appears, or it does not appear in either factorization; that is, one of its associates appears instead. We then obtain a gcd of a and b by multiplying together all irreducibles appearing in both factorizations, taking each irreducible to the highest power for which it divides both a and b. We have used the concept freely for $\mathbb{Z}$ in group theory already. Since an irreducible appearing in a factorization is only defined up to a unit factor, we see that a gcd must also be defined only up to a unit factor in a UFD. It is for this reason that we say "a" gcd, rather than "the" gcd, of a and b. It is customary to give the following more elegant definition of a gcd in a UFD, describing the same concept we have just discussed.

DEFINITION 7.10 (gcd) Let D be a UFD. An element $d \in D$ is a **greatest common divisor** (abbreviated gcd) **of elements a and b in** D if $d \mid a$, $d \mid b$, and also $c \mid d$ for all c dividing both a and b.

EXAMPLE 5 In $\mathbb{Z}$, a gcd of 18 and 48 is 6. Another one is -6. In $\mathbb{Q}[x]$, a gcd of $x^2 - 2x + 1$ and $x^2 + x - 2$ is $x - 1$. Another one is $2(x - 1)$, since 2 is a unit in $\mathbb{Q}[x]$. Still another one is $(15/13)(x - 1)$. However, in $\mathbb{Z}[x]$, the only gcd's of $x^2 - 2x + 1$ and $x^2 + x - 2$ are $x - 1$ and $-(x - 1)$, for 1 and -1 are the only units in $\mathbb{Z}[x]$. ▲

As indicated after Example 4, we can show that any a and b in a UFD have a gcd by factoring a and b into irreducibles. (Actually, a gcd exists for any number of elements from a UFD. The proof of Lemma 7.4 shows in more detail how such a gcd can be constructed from factorizations into irreducibles.) Every PID is a UFD, so of course gcd's also exist in a PID. There is a nice proof for PIDs, using Definition 7.10 rather than factorization, that we give now.

THEOREM 7.6 If D is a PID and a and b are nonzero elements of D, then there exists a gcd of a and b. Furthermore, each gcd of a and b can be expressed in the form $\lambda a + \mu b$ for some $\lambda, \mu \in D$.

PROOF Consider the set

$$N = \{ra + sb \mid r, s \in D\}.$$

Since

$$\left(r_1 a + s_1 b\right) \pm \left(r_2 a + s_2 b\right) = \left(r_1 \pm r_2\right)a + \left(s_1 \pm s_2\right)b$$

and

$$t(ra + sb) = (tr)a + (ts)b$$

for $t \in D$, it follows that N is an ideal of D. Now $N = \langle d \rangle$ for some $d \in D$. Then $d \mid (ra + sb)$ for all $r, s \in D$, and taking first $s = 0$ with $r = 1$ and then $r = 0$ with $s = 1$, we see that $d \mid a$ and $d \mid b$. Also if $c \mid a$ and $c \mid b$, then $c \mid (ra + sb)$ for all $ra + sb$, that is, $c \mid n$ for all $n \in N$. Hence $c \mid d$. Thus d is a gcd of a and b.

For d as just constructed, $d \in N$ implies that there exist $\lambda, \mu \in D$ such that $d = \lambda a + \mu b$. But the definition of a gcd shows that if d_1 is also a gcd of a and b, then $d \mid d_1$ and $d_1 \mid d$. Thus

$$d_1 = vd = (v\lambda)a + (v\mu)b = \lambda_1 a + \mu_1 b. \quad \blacklozenge$$

The preceding proof was extremely elegant, but not at all constructive. Of course we can find a gcd of a and b if we can factor them into irreducibles, but such factorizations can be very tough to do. However, if a UFD is actually Euclidean, and we know a Euclidean valuation, there is an easy constructive way to find gcd's, as the next theorem shows.

THEOREM 7.7 (Euclidean Algorithm)   Let D be a Euclidean domain with a Euclidean valuation v, and let a and b be nonzero elements of D. Let r_1 be as in Condition 1 for a Euclidean valuation, that is,

$$a = bq_1 + r_1,$$

where either $r_1 = 0$ or $v(r_1) < v(b)$. If $r_1 \neq 0$, let r_2 be such that

$$b = r_1 q_2 + r_2,$$

where either $r_2 = 0$ or $v(r_2) < v(r_1)$. In general, let r_{i+1} be such that

$$r_{i-1} = r_i q_{i+1} + r_{i+1},$$

◆ H I S T O R I C A L   N O T E ◆

The Euclidean algorithm appears in Euclid's *Elements* as propositions 1 and 2 of Book VII, where it is used as here to find the greatest common divisor of two integers. Euclid uses it again in Book X (propositions 2 and 3) to find the greatest common measure of two magnitudes (if it exists) and to determine whether two magnitudes are incommensurable.

The algorithm appears again in the *Brahmesphutasiddhanta* (Correct Astronomical System of Brahma) (628) of the seventh-century Indian mathematician and astronomer Brahmagupta. To solve the indeterminate equation $rx + c = sy$ in integers, Brahmagupta uses Euclid's procedure to "reciprocally divide" r by s until he reaches the final nonzero remainder. By then using, in effect, a substitution procedure based on the various quotients and remainders, he produces a straightforward algorithm for finding the smallest positive solution to his equation.

The thirteenth-century Chinese algebraist Ch'in Chiu-Shao also used the Euclidean algorithm in his solution of the so-called Chinese Remainder problem published in the *Shu-Shu Chiu Chang* (Mathematical Treatise in Nine Sections) (1247). Ch'in's goal was to display a method for solving the system of congruences $N \equiv r_i$ (mod m_i). As part of that method he needed to solve congruences of the form $Nx \equiv 1$ (mod m), where N and m are relatively prime. The solution to a congruence of this form is again found by a substitution procedure, different from the Indian one, using the quotients and remainders from the Euclidean algorithm applied to N and m. It is not known whether the common element in the Indian and Chinese algorithms, the Euclidean algorithm itself, was discovered independently in these cultures or was learned from Greek sources.

where either $r_{i+1} = 0$ or $v(r_{i+1}) < v(r_i)$. Then the sequence $r_1, r_2, \ldots$ must terminate with some $r_s = 0$. If $r_1 = 0$, then b is a gcd of a and b. If $r_1 \neq 0$ and r_s is the first $r_i = 0$, then a gcd of a and b is r_{s-1}.

PROOF Since $v(r_i) < v(r_{i-1})$ and $v(r_i)$ is a nonnegative integer, it follows that after some finite number of steps we must arrive at some $r_s = 0$.

If $r_1 = 0$, then $a = bq_1$, and b is a gcd of a and b. Suppose $r_1 \neq 0$. Then if $d \mid a$ and $d \mid b$, we have

$$d \mid (a - bq_1),$$

so $d \mid r_1$. However, if $d_1 \mid r_1$ and $d_1 \mid b$, then

$$d_1 \mid (bq_1 + r_1),$$

so $d_1 \mid a$. Thus the set of common divisors of a and b is the same set as the set of common divisors of b and r_1. By a similar argument, if $r_2 \neq 0$, the set of common divisors of b and r_1 is the same set as the set of common divisors of r_1 and r_2. Continuing this process, we see finally that the set of common divisors of a and b is the same set as the set of common divisors of r_{s-2} and r_{s-1}, where r_s is the first r_i equal to 0. Thus a gcd of r_{s-2} and r_{s-1} is also a gcd of a and b. But the equation

$$r_{s-2} = q_s r_{s-1} + r_s = q_s r_{s-1}$$

shows that a gcd of r_{s-2} and r_{s-1} is r_{s-1}. ◆

EXAMPLE 6 Let us illustrate the Euclidean algorithm for the Euclidean valuation $| \ |$ on $\mathbb{Z}$ by computing a gcd of 22,471 and 3,266. We just apply the division algorithm over and over again, and the last nonzero remainder is a gcd. We label the numbers obtained as in Theorem 7.7 to further illustrate the statement and proof of the theorem. The computations are easily checked.

$$a = 22{,}471$$
$$b = 3{,}266$$

$$22{,}471 = (3{,}266)6 + 2{,}875 \qquad r_1 = 2{,}875$$
$$3{,}266 = (2{,}875)1 + 391 \qquad r_2 = 391$$
$$2{,}875 = (391)7 + 138 \qquad r_3 = 138$$
$$391 = (138)2 + 115 \qquad r_4 = 115$$
$$138 = (115)1 + 23 \qquad r_5 = 23$$
$$115 = (23)5 + 0 \qquad r_6 = 0$$

Thus $r_5 = 23$ is a gcd of 22,471 and 3,266. We found a gcd without factoring! This is important, for sometimes it is very difficult to find a factorization of an integer into primes. ▲

EXAMPLE 7 Note that the division algorithm Condition 1 in the definition of a Euclidean valuation says nothing about r being "positive." In computing a gcd in $\mathbb{Z}$ by the Euclidean algorithm for $| \ |$, as in Example 6, it is surely to our interest to make $|r_i|$ as small as possible in each division. Thus, repeating Example 6, it would be more efficient to write

$$a = 22{,}471$$
$$b = 3{,}266$$

$$
\begin{aligned}
22{,}471 &= (3{,}266)7 - 391 & r_1 &= -391 \\
3{,}266 &= (391)8 + 138 & r_2 &= 138 \\
391 &= (138)3 - 23 & r_3 &= -23 \\
138 &= (23)6 + 0 & r_4 &= 0
\end{aligned}
$$

We can change the sign of r_i from negative to positive when we wish since the divisors of r_i and $-r_i$ are the same. ▲

Exercises 7.2

Computations

In Exercises 1 through 5, state whether the given function v is a Euclidean valuation for the given integral domain.

1. The function v for $\mathbb{Z}$ given by $v(n) = n^2$ for nonzero $n \in \mathbb{Z}$

2. The function v for $\mathbb{Z}[x]$ given by $v(f(x)) = (\text{degree of } f(x))$ for $f(x) \in \mathbb{Z}[x]$, $f(x) \neq 0$

3. The function v for $\mathbb{Z}[x]$ given by $v(f(x)) = (\text{the absolute value of the coefficient of the highest degree nonzero term of } f(x))$ for nonzero $f(x) \in \mathbb{Z}[x]$

4. The function v for $\mathbb{Q}$ given by $v(a) = a^2$ for nonzero $a \in \mathbb{Q}$

5. The function v for $\mathbb{Q}$ given by $v(a) = 50$ for nonzero $a \in \mathbb{Q}$

6. By referring to Example 7, actually express the gcd 23 in the form $\lambda(22{,}471) + \mu(3{,}266)$ for $\lambda, \mu \in \mathbb{Z}$. [*Hint:* From the next to the last line of the computation in Example 7, $23 = (138)3 - 391$. From the line before that, $138 = 3{,}266 - (391)8$, so substituting, you get $23 = [3{,}266 - (391)8]3 - 391$, and so on. That is, work your way back up to actually find values for λ and μ.]

7. Find a gcd of 49,349 and 15,555 in $\mathbb{Z}$.

8. Following the idea of Exercise 6 and referring to Exercise 7, express the positive gcd of 49,349 and 15,555 in $\mathbb{Z}$ in the form $\lambda(49{,}349) + \mu(15{,}555)$ for $\lambda, \mu \in \mathbb{Z}$.

9. Find a gcd of

$$x^{10} - 3x^9 + 3x^8 - 11x^7 + 11x^6 - 11x^5 + 19x^4 - 13x^3 + 8x^2 - 9x + 3$$

and

$$x^6 - 3x^5 + 3x^4 - 9x^3 + 5x^2 - 5x + 2$$

in $\mathbb{Q}[x]$.

Concepts

10. Let us consider $\mathbb{Z}[x]$.
    a. Is $\mathbb{Z}[x]$ a UFD? Why?
    b. Show that $\{a + xf(x) \mid a \in 2\mathbb{Z}, f(x) \in \mathbb{Z}[x]\}$ is an ideal in $\mathbb{Z}[x]$.
    c. Is $\mathbb{Z}[x]$ a PID? (Consider part (b).)
    d. Is $\mathbb{Z}[x]$ a Euclidean domain? Why?

11. Mark each of the following true or false.

_____ a. Every Euclidean domain is a PID.

_____ b. Every PID is a Euclidean domain.

_____ c. Every Euclidean domain is a UFD.

_____ d. Every UFD is a Euclidean domain.

_____ e. A gcd of 2 and 3 in $\mathbb{Q}$ is $\frac{1}{2}$.

_____ f. The Euclidean algorithm gives a constructive method for finding a gcd of two integers.

_____ g. If v is a Euclidean valuation on a Euclidean domain D, then $v(1) \leq v(a)$ for all nonzero $a \in D$.

_____ h. If v is a Euclidean valuation on a Euclidean domain D, then $v(1) < v(a)$ for all nonzero $a \in D$, $a \neq 1$.

_____ i. If v is a Euclidean valuation on a Euclidean domain D, then $v(1) < v(a)$ for all nonzero nonunits $a \in D$.

_____ j. For any field F, $F[x]$ is a Euclidean domain.

12. Does the choice of a particular Euclidean valuation v on a Euclidean domain D influence the arithmetic structure of D in any way? Explain.

Theory

13. Let D be a Euclidean domain and let v be a Euclidean valuation on D. Show that if a and b are associates in D, then $v(a) = v(b)$.

14. Let D be a Euclidean domain and let v be a Euclidean valuation on D. Show that for nonzero $a, b \in D$, one has $v(a) < v(ab)$ if and only if b is not a unit of D. [*Hint:* Argue from Exercise 13 that $v(a) < v(ab)$ implies that b is not a unit of D. Using the Euclidean algorithm, show that $v(a) = v(ab)$ implies $\langle a \rangle = \langle ab \rangle$. Conclude that if b is not a unit, then $v(a) < v(ab)$.]

15. Prove or disprove the following statement: If v is a Euclidean valuation on Euclidean domain D, then $\{a \in D \mid v(a) > v(1)\} \cup \{0\}$ is an ideal of D.

16. Show that every field is a Euclidean domain.

17. Let v be a Euclidean valuation on a Euclidean domain D.

a. Show that if $s \in \mathbb{Z}$ such that $s + v(1) > 0$, then $\eta : D^* \to \mathbb{Z}$ defined by $\eta(a) = v(a) + s$ for nonzero $a \in D$ is a Euclidean valuation on D. As usual, D^* is the set of nonzero elements of D.

b. Show that for $r \in \mathbb{Z}^+$, $\lambda : D^* \to \mathbb{Z}$ given by $\lambda(a) = r \cdot v(a)$ for nonzero $a \in D$ is a Euclidean valuation on D.

c. Show that there exists a Euclidean valuation μ on D such that $\mu(1) = 1$ and $\mu(a) > 100$ for all nonzero nonunits $a \in D$.

18. Let D be a UFD. An element c in D is a **least common multiple** (abbreviated lcm) **of two elements a and b in D** if $a \mid c$, $b \mid c$ and if c divides every element of D that is divisible by both a and b. Show that every two nonzero elements a and b of a Euclidean domain D have an lcm in D. [*Hint:* Show that all common multiples, in the obvious sense, of both a and b form an ideal of D.]

19. Use the last statement in Theorem 7.6 to show that two nonzero elements $r, s \in \mathbb{Z}$ generate the group $\langle \mathbb{Z}, + \rangle$ if and only if r and s, viewed as integers in the domain $\mathbb{Z}$, are **relatively prime,** that is, have a gcd of 1.

20. Using the last statement in Theorem 7.6, show that for nonzero $a, b, n \in \mathbb{Z}$, the congruence $ax = b \pmod{n}$ has a solution in $\mathbb{Z}$ if a and n are relatively prime.

21. Generalize Exercise 20 by showing that for nonzero $a, b, n \in \mathbb{Z}$, the congruence $ax \equiv b \pmod{n}$ has a solution in $\mathbb{Z}$ if and only if the positive gcd of a and n in $\mathbb{Z}$ divides b. Interpret this result in the ring $\mathbb{Z}_n$.

22. Following the idea of Exercises 6 and 21, outline a constructive method for finding a solution in $\mathbb{Z}$ of the congruence $ax \equiv b \pmod{n}$ for nonzero $a, b, n \in \mathbb{Z}$, if the congruence does have a solution. Use this method to find a solution of the congruence $22x \equiv 18 \pmod{42}$.

7.3

Gaussian Integers and Norms

Gaussian Integers

We should give an example of a Euclidean domain different from $\mathbb{Z}$ and $F[x]$.

DEFINITION 7.11 (Gaussian Integer)   A **Gaussian integer** is a complex number $a + bi$, where $a, b \in \mathbb{Z}$. For a Gaussian integer $\alpha = a + bi$, the **norm** $N(\alpha)$ **of** α is $a^2 + b^2$.

◆   **HISTORICAL NOTE**   ◆

In his *Disquisitiones Arithmeticae*, Gauss studied in detail the theory of quadratic residues, that is, the theory of solutions to the congruence $x^2 \equiv p \pmod{q}$ and proved the famous quadratic reciprocity theorem showing the relationship between the solutions of the congruences $x^2 \equiv p \pmod{q}$ and $x^2 \equiv q \pmod{p}$ where p and q are primes. In attempting to generalize his results to theories of quartic residues, however, Gauss realized that it was much more natural to consider the Gaussian integers rather than the ordinary integers.

Gauss's investigations of the Gaussian integers are contained in a long paper published in 1832 in which he proved various analogies between them and the ordinary integers. For example, after noting that there are four units (invertible elements) among the Gaussian integers, namely 1, -1, i and $-i$, and defining the norm as in Definition 7.11, he generalized the notion of a prime integer by defining a prime Gaussian integer to be one that cannot be expressed as the product of two other integers, neither of them units, He was then able to determine which Gaussian integers are prime: A Gaussian integer that is not real is prime if and only if its norm is a real prime, which can only be 2 or of the form $4n + 1$. The real prime $2 = (1 + i)(1 - i)$ and real primes congruent to 1 modulo 4 like $13 = (2 + 3i)(2 - 3i)$ factor as the product of two Gaussian primes. Real primes of the form $4n + 3$ like 7 and 11 are still prime in the domain of Gaussian integers. See Exercise 10.

We shall let $\mathbb{Z}[i]$ be the set of all Gaussian integers. The following lemma gives some basic properties of the norm function N on $\mathbb{Z}[i]$ and leads to a demonstration that the function v defined by $v(\alpha) = N(\alpha)$ for nonzero $\alpha \in \mathbb{Z}[i]$ is a Euclidean valuation on $\mathbb{Z}[i]$. Note that the Gaussian integers include all the **rational integers**, that is, all the elements of $\mathbb{Z}$.

LEMMA 7.7   In $\mathbb{Z}[i]$, the following properties of the norm function N hold for all $\alpha, \beta \in \mathbb{Z}[i]$:

1. $N(\alpha) \geq 0$.
2. $N(\alpha) = 0$ if and only if $\alpha = 0$.
3. $N(\alpha\beta) = N(\alpha)N(\beta)$.

PROOF   If we let $\alpha = a_1 + a_2i$ and $\beta = b_1 + b_2i$, these results are all straightforward computations. We leave the proof of these properties as an exercise (see Exercise 11). ♦

LEMMA 7.8 $\mathbb{Z}[i]$ is an integral domain.

PROOF   It is obvious that $\mathbb{Z}[i]$ is a commutative ring with unity. We show that there are no divisors of 0. Let $\alpha, \beta \in \mathbb{Z}[i]$. Using Lemma 7.7, if $\alpha\beta = 0$, then

$$N(\alpha)N(\beta) = N(\alpha\beta) = N(0) = 0.$$

Thus $\alpha\beta = 0$ implies that $N(\alpha) = 0$ or $N(\beta) = 0$. By Lemma 7.7 again, this implies that either $\alpha = 0$ or $\beta = 0$. Thus $\mathbb{Z}[i]$ has no divisors of 0, so $\mathbb{Z}[i]$ is an integral domain. ♦

Of course, since $\mathbb{Z}[i]$ is a subring of $\mathbb{C}$, where $\mathbb{C}$ is the field of complex numbers, it is really obvious that $\mathbb{Z}[i]$ has no 0 divisors. We gave the argument of Lemma 7.8 to illustrate the use of the multiplicative property 3 of the norm function N and to avoid going outside of $\mathbb{Z}[i]$ in our argument.

THEOREM 7.8   The function v given by $v(\alpha) = N(\alpha)$ for nonzero $\alpha \in \mathbb{Z}[i]$ is a Euclidean valuation on $\mathbb{Z}[i]$. Thus $\mathbb{Z}[i]$ is a Euclidean domain.

PROOF   Note that for $\beta = b_1 + b_2i \neq 0$, $N(b_1 + b_2i) = b_1^2 + b_2^2$, so $N(\beta) \geq 1$. Then for all $\alpha, \beta \neq 0$ in $\mathbb{Z}[i]$, $N(\alpha) \leq N(\alpha)N(\beta) = N(\alpha\beta)$. This proves Condition 2 for a Euclidean valuation.

It remains to prove the division algorithm, Condition 1, for N. Let $\alpha, \beta \in \mathbb{Z}[i]$, with $\alpha = a_1 + a_2i$ and $\beta = b_1 + b_2i$, where $\beta \neq 0$. We must find σ and ρ in $\mathbb{Z}[i]$ such that $\alpha = \beta\sigma + \rho$, where either $\rho = 0$ or $N(\rho) < N(\beta) = b_1^2 + b_2^2$. Let us put $\sigma = q_1 + q_2i$, where q_1 and q_2 are *rational integers* in $\mathbb{Z}$ to be determined. Then ρ will have to have the form

$$\rho = (a_1 + a_2 i) - (b_1 + b_2 i)(q_1 + q_2 i)$$
$$= (a_1 - b_1 q_1 + b_2 q_2) + (a_2 - b_1 q_2 - b_2 q_1)i.$$

We have to try to find *rational integers* q_1 and q_2 such that

$$N(\rho) = (a_1 - b_1 q_1 + b_2 q_2)^2 + (a_2 - b_1 q_2 - b_2 q_1)^2 < b_1^2 + b_2^2,$$

that is, such that

$$\frac{(a_1 - b_1 q_1 + b_2 q_2)^2}{b_1^2 + b_2^2} + \frac{(a_2 - b_1 q_2 - b_2 q_1)^2}{b_1^2 + b_2^2} < 1.$$

Recall that

$$\frac{(a_1 - b_1 q_1 + b_2 q_2)^2}{b_1^2 + b_2^2}$$

is exactly the square of the distance d in the Euclidean plane from a point (q_1, q_2) to the line l with equation $a_1 - b_1 X + b_2 Y = 0$. Similarly,

$$\frac{(a_2 - b_1 q_2 - b_2 q_1)^2}{b_1^2 + b_2^2}$$

is the square of the distance d' from (q_1, q_2) to the line l' with equation $a_2 - b_2 X - b_1 Y = 0$. Note that l is perpendicular to l'. Let P be the point of intersection of these two lines as shown in Fig. 7.1. From this figure, we see that $d^2 + (d')^2$ is the square of the distance from (q_1, q_2) to P. Thus we must show that there is a point (q_1, q_2) *with integral coordinates* and with a distance from P whose square is less than 1, Since P is contained within or on the boundary of some square of unit side such that both coordinates of each vertex are integers, we see that if (q_1, q_2) is chosen to be a point with integer coordinates *as close as possible* to P, its distance from P can be at most half the length of a diagonal of square, that is, at most $\sqrt{2}/2$ (see Fig. 7.2). Thus the square of this distance from P is at most $\frac{1}{2}$, which is less than 1. ◆

We could have proved the division algorithm for the function N purely algebraically; the algebraic proof is easier, shorter to describe, and more

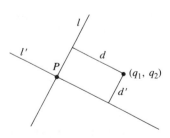

Figure 7.1

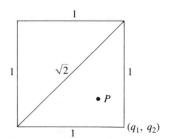

Figure 7.2

useful in applying the Euclidean algorithm for N (see Exercises 7 and 15). But we confess a certain fondness for our geometric argument. Besides, it is nice to have variety in the proofs in a text. We leave an algebraic proof to the exercises (see Exercise 14).

EXAMPLE I   We can now apply all our results of Section 7.2 to $\mathbb{Z}[i]$. In particular, since $N(1) = 1$, the units of $\mathbb{Z}[i]$ are exactly the $\alpha = a_1 + a_2i$ with $N(\alpha) = a_1^2 + a_2^2 = 1$. From the fact that a_1 and a_2 are integers, it follows that the only possibilities are $a_1 = \pm1$ with $a_2 = 0$, or $a_1 = 0$ with $a_2 = \pm1$. Thus the units of $\mathbb{Z}[i]$ are ±1 and $\pm i$. One can also use the Euclidean algorithm to compute a gcd of two nonzero elements. We leave such computations to the exercises. Finally, note that while 5 is an irreducible in $\mathbb{Z}$, 5 is no longer an irreducible in $\mathbb{Z}[i]$, for $5 = (1 + 2i)(1 - 2i)$, and neither $1 + 2i$ nor $1 - 2i$ is a unit. ▲

Multiplicative Norms

Let us point out again that for an integral domain D, *the arithmetic concepts of irreducibles and units are intrinsic to the integral domain itself* and are not affected in any way by a valuation or norm that may be defined on the domain. However, as the preceding section and our work thus far in this section show, a suitably defined valuation or norm may be of help in determining the arithmetic structure of D. This is strikingly illustrated in *algebraic number theory*, where for a domain of *algebraic integers* we consider many different valuations of the domain, each doing its part in helping to determine the arithmetic structure of the domain. In a domain of algebraic integers, we have essentially one valuation for each irreducible (up to associates), and each such valuation gives information concerning the behavior in the integral domain of the irreducible to which it corresponds. This is an example of the importance of studying properties of elements in an algebraic structure by means of mappings associated with them. We shall be doing this for zeros of polynomials in Chapter 8.

Let us study integral domains that have a multiplicative norm satisfying the properties of N on $\mathbb{Z}[i]$ given in Lemma 7.7.

DEFINITION 7.12 (Norm)   Let D be an integral domain. A **multiplicative norm N on D** is a function mapping D into the integers $\mathbb{Z}$ such that the following conditions are satisfied:

1. $N(\alpha) \geq 0$ for all $\alpha \in D$.
2. $N(\alpha) = 0$ if and only if $\alpha = 0$.
3. $N(\alpha\beta) = N(\alpha)N(\beta)$ for all $\alpha, \beta \in D$.

THEOREM 7.9   If D is an integral domain with a multiplicative norm N, then $N(1) = 1$ and $N(u) = 1$ for every unit u in D. If, furthermore, every

α such that $N(\alpha) = 1$ is a unit in D, then an element π in D, with $N(\pi) = p$ for a prime $p \in \mathbb{Z}$, is an irreducible of D.

PROOF   Let D be an integral domain with a multiplicative norm N. Then

$$N(1) = N((1)(1)) = N(1)N(1)$$

shows that $N(1) = 1$. Also, if u is a unit in D, then

$$1 = N(1) = (uu^{-1}) = N(u)N(u^{-1}).$$

Since $N(u)$ is a nonnegative integer, this implies that $N(u) = 1$.

Now suppose that the units of D are *exactly* the elements of norm 1. Let $\pi \in D$ be such that $N(\pi) = p$, where p is a prime in $\mathbb{Z}$. Then if $\pi = \alpha\beta$, we have

$$p = N(\pi) = N(\alpha)N(\beta),$$

so either $N(\alpha) = 1$ or $N(\beta) = 1$. By assumption, this means that either α or β is a unit of D. Thus π is an irreducible of D. ◆

EXAMPLE 2   On $\mathbb{Z}[i]$, the function N defined by $N(a + bi) = a^2 + b^2$ gives a multiplicative norm in the sense of our definition. We saw that the function v given by $v(\alpha) = N(\alpha)$ for nonzero $\alpha \in \mathbb{Z}[i]$ is a Euclidean valuation on $\mathbb{Z}[i]$, so the units are precisely the elements α of $\mathbb{Z}[i]$ with $N(\alpha) = N(1) = 1$. Thus the second part of Theorem 7.9 applies in $\mathbb{Z}[i]$. We saw in Example 1 that 5 is not an irreducible in $\mathbb{Z}[i]$, for $5 = (1 + 2i)(1 - 2i)$. Since $N(1 + 2i) = N(1 - 2i) = 1^2 + 2^2 = 5$ and 5 is a prime in $\mathbb{Z}$, we see from Theorem 7.9 that $1 + 2i$ and $1 - 2i$ are both irreducibles in $\mathbb{Z}[i]$. ▲

As an application of multiplicative norms, we shall now give another example of an integral domain that is *not* a UFD. We saw one example in Example 3, Section 7.1. The following is the standard illustration.

EXAMPLE 3   Let $\mathbb{Z}[\sqrt{-5}] = \{a + ib\sqrt{5} \mid a, b \in \mathbb{Z}\}$. As a subset of the complex numbers closed under addition, subtraction, and multiplication, and containing 0 and 1, $\mathbb{Z}[\sqrt{-5}]$ is an integral domain. Define N on $\mathbb{Z}[\sqrt{-5}]$ by

$$N(a + b\sqrt{-5}) = a^2 + 5b^2.$$

(Here $\sqrt{-5} = i\sqrt{5}$.) Clearly, $N(\alpha) \geq 0$ and $N(\alpha) = 0$ if and only if $\alpha = a + b\sqrt{-5} = 0$. That $N(\alpha\beta) = N(\alpha)N(\beta)$ is a straightforward computation that we leave to the exercises (see Exercise 12). Let us find all candidates for units in $\mathbb{Z}[\sqrt{-5}]$ by finding all elements α in $\mathbb{Z}[\sqrt{-5}]$ with $N(\alpha) = 1$. If $\alpha = a + b\sqrt{-5}$, and $N(\alpha) = 1$, we must have $a^2 + 5b^2 = 1$ for *integers a and b*. This is only possible if $b = 0$ and $a = \pm 1$. Hence ± 1 are the only candidates for units. Since ± 1 are units, they are then precisely the units in $\mathbb{Z}[\sqrt{-5}]$.

Now in $\mathbb{Z}[\sqrt{-5}]$, we have $21 = (3)(7)$ and also

$$21 = (1 + 2\sqrt{-5})(1 - 2\sqrt{-5}).$$

If we can show that $3, 7, 1 + 2\sqrt{-5}$, and $1 - 2\sqrt{-5}$ are all irreducibles in $\mathbb{Z}[\sqrt{-5}]$, we will then know that $\mathbb{Z}[\sqrt{-5}]$ cannot be a UFD, since neither 3 nor 7 is $\pm(1 + 2\sqrt{-5})$.

Suppose that $3 = \alpha\beta$. Then

$$9 = N(3) = N(\alpha)N(\beta)$$

shows that we must have $N(\alpha) = 1, 3$, or 9. If $N(\alpha) = 1$, then α is a unit. If $\alpha = a + b\sqrt{-5}$, then $N(\alpha) = a^2 + 5b^2$, and for no choice of integers a and b is $N(\alpha) = 3$. If $N(\alpha) = 9$, then $N(\beta) = 1$, so β is a unit. Thus from $3 = \alpha\beta$, we can conclude that either α or β is a unit. Therefore, 3 is an irreducible in $\mathbb{Z}[\sqrt{-5}]$. A similar argument shows that 7 is also an irreducible in $7[\sqrt{-5}]$.

If $1 + 2\sqrt{-5} = \gamma\delta$, we have

$$21 = N(1 + 2\sqrt{-5}) = N(\gamma)N(\delta).$$

so $N(\gamma) = 1, 3, 7$, or 21. We have seen that there is no element of $\mathbb{Z}[\sqrt{-5}]$ of norm 3 or 7. This either $N(\gamma) = 1$, and γ is a unit, or $N(\gamma) = 21$, so $N(\delta) = 1$, and δ is a unit. Therefore, $1 + 2\sqrt{-5}$ is an irreducible in $\mathbb{Z}[\sqrt{-5}]$. A parallel argument shows that $1 - 2\sqrt{-5}$ is also an irreducible in $\mathbb{Z}[\sqrt{-5}]$.

In summary, we have shown that

$$\mathbb{Z}[\sqrt{-5}] = \{a + ib\sqrt{5} \mid a, b \in \mathbb{Z}\}$$

is an integral domain but not a UFD. In particular, there are two *different* factorizations

$$21 = 3 \cdot 7 = (1 + 2\sqrt{-5})(1 - 2\sqrt{-5})$$

of 21 into irreducibles. These irreducibles cannot be primes, for the property of a prime enables us to prove uniqueness of factorization (see the proof of Theorem 7.2). ▲

We conclude with a classical application, determining which primes p in $\mathbb{Z}$ are equal to a sum of squares of two integers in $\mathbb{Z}$. For example, $2 = 1^2 + 1^2$, $5 = 1^2 + 2^2$, and $13 = 2^2 + 3^2$ are sums of squares. Since we have now answered this question for the only even prime number, 2, we can restrict ourselves to odd primes.

THEOREM 7.10 (Fermat's $p = a^2 + b^2$ Theorem)   Let p be an odd prime in $\mathbb{Z}$. Then $p = a^2 + b^2$ for integers a and b in $\mathbb{Z}$ if and only $p \equiv 1$ (mod 4).

PROOF   First, suppose that $p = a^2 + b^2$. Now a and b cannot both be even or both be odd since p is an odd number. If $a = 2r$ and $b = 2s + 1$, then $a^2 + b^2 = 4r^2 + 4(s^2 + s) + 1$, so $p \equiv 1$ (mod 4). This takes care of one direction for this "if and only if" theorem.

For the other direction, we assume that $p \equiv 1 \pmod 4$. Now the multiplicative group of nonzero elements of the finite field $\mathbb{Z}_p$ is cyclic, and has order $p - 1$. Since 4 is a divisor of $p - 1$, we see that $\mathbb{Z}_p$ contains an element n of multiplicative order 4. It follows that n^2 has multiplicative order 2, so $n^2 = -1$ in $\mathbb{Z}_p$. Thus in $\mathbb{Z}$, we have $n^2 \equiv -1 \pmod p$, so p divides $n^2 + 1$ in $\mathbb{Z}$.

Viewing p and $n^2 + 1$ in $\mathbb{Z}[i]$, we see that p divides $n^2 + 1 = (n + i)(n - i)$. Suppose that p is irreducible in $\mathbb{Z}[i]$; then p would have to divide $n + i$ or $n - i$, or both. If p divides both $n + i$ and $n - i$, then p divides $(n + i) - (n - i) = 2i$, contradicting the fact that p is an odd prime so that $N(p) > 4$. (See Example 2 for the norm in $\mathbb{Z}[i]$.) Thus p divides just one of these factors, say $n + i$. It follows that $N(p) = p^2$ is a divisor of $N(n + i) = n^2 + 1 = (n + i)(n - i)$. Since p does not divide $n - i$, we see that p^2 divides $n + i$ so $N(p^2) = p^4$ divides $N(n + i) = n^2 + 1$. Continuing in this fashion, we see that $n + i$ would have to be divisible by arbitrarily high powers of p, which is impossible. A symmetric argument can be made if p divides $n - i$ but not $n + i$. Thus our assumption that p is irreducible in $\mathbb{Z}[i]$ must be false.

Since p is not irreducible in $\mathbb{Z}[i]$, we have $p = (a + bi)(c + di)$ where neither $a + bi$ nor $c + di$ is a unit. Taking norms, we have $p^2 = (a^2 + b^2)(c^2 + d^2)$ where neither $a^2 + b^2 = 1$ nor $c^2 + d^2 = 1$. Consequently, we have $p = a^2 + b^2$, which completes our proof. [Since $a^2 + b^2 = (a + bi)(a - bi)$, we see that this is the factorization of p, that is, $c + di = a - bi$.] ◆

Exercise 10 asks you to determine which primes p in $\mathbb{Z}$ remain irreducible in $\mathbb{Z}[i]$.

Exercises 7.3

Computations

In Exercises 1 through 4, factor the Gaussian integer into a product of irreducibles in $\mathbb{Z}[i]$. [*Hint:* Since an irreducible factor of $\alpha \in \mathbb{Z}[i]$ must have norm >1 and dividing $N(\alpha)$, there are only a finite number of Gaussian integers $a + bi$ to consider as possible irreducible factors of a given α. Divide α by each of them in $\mathbb{C}$, and see for which ones the quotient is again in $\mathbb{Z}[i]$.]

1. 5   **2.** 7   **3.** $4 + 3i$ **4.** $6 - 7i$

5. Show that 6 does not factor uniquely (up to associates) into irreducibles in $\mathbb{Z}[\sqrt{-5}]$. Exhibit two different factorizations.

6. Consider $\alpha = 7 + 2i$ and $\beta = 3 - 4i$ in $\mathbb{Z}[i]$. Find σ and ρ in $\mathbb{Z}[i]$ such that

$$\alpha = \beta\sigma + \rho \qquad \text{with} \qquad N(\rho) < N(\beta).$$

[*Hint:* Use the construction in the hint of Exercise 14.]

7. Use a Euclidean algorithm in $\mathbb{Z}[i]$ to find a gcd of $8 + 6i$ and $5 - 15i$ in $\mathbb{Z}[i]$. [*Hint:* Use the construction in the hint of Exercise 14.]

Concepts

8. Mark each of the following true or false.

_____ a. $\mathbb{Z}[i]$ is a PID.

_____ b. $\mathbb{Z}[i]$ is a Euclidean domain.

_____ c. Every integer in $\mathbb{Z}$ is a Gaussian integer.

_____ d. Every complex number is a Gaussian integer.

_____ e. A Euclidean algorithm holds in $\mathbb{Z}[i]$.

_____ f. A multiplicative norm on an integral domain is sometimes an aid in finding irreducibles of the domain.

_____ g. If N is a multiplicative norm on an integral domain D, then $N(u) = 1$ for every unit u of D.

_____ h. If F is a field, then the function N defined by $N(f(x)) = $ (degree of $f(x)$) is a multiplicative norm on $F[x]$.

_____ i. If F is a field, then the function defined by $N(f(x)) = 2^{(\text{degree of } f(x))}$ for $f(x) \neq 0$ and $N(0) = 0$ is a multiplicative norm on $F[x]$ according to our definition.

_____ j. $\mathbb{Z}[\sqrt{-5}]$ as an integral domain but not a UFD.

9. Let D be an integral domain with a multiplicative norm N such that $N(\alpha) = 1$ for $\alpha \in D$ if and only if α is a unit of D. Let π be such that $N(\pi)$ is minimal among all $N(\beta) > 1$ for $\beta \in D$. Show that π is an irreducible of D.

10. a. Show that 2 is equal to the product of a unit and the square of an irreducible in $\mathbb{Z}[i]$.

b. Show that an odd prime p in $\mathbb{Z}$ is irreducible in $\mathbb{Z}[i]$ if and only if $p \equiv 3 \pmod 4$. (Use Theorem 7.10.)

11. Prove Lemma 7.7.

12. Prove that N of Example 3 is multiplicative, that is, that $N(\alpha\beta) = N(\alpha)N(\beta)$ for $\alpha, \beta \in \mathbb{Z}[\sqrt{-5}]$.

13. Let D be an integral domain with a multiplicative norm N such that $N(\alpha) = 1$ for $\alpha \in D$ if and only if α is a unit of D. Show that every nonzero nonunit of D has a factorization into irreducibles in D.

14. Prove algebraically that the division algorithm holds in $\mathbb{Z}[i]$ for v given by $v(\alpha) = N(\alpha)$ for nonzero $\alpha \in \mathbb{Z}[i]$. [*Hint:* For α and β in $\mathbb{Z}[i]$ with $\beta \neq 0$, $\alpha/\beta = r + si$ in $\mathbb{C}$ for $r, s \in \mathbb{Q}$. Let q_1 and q_2 be rational integers in $\mathbb{Z}$ as close as possible to the rational numbers r and s respectively. Show that for $\sigma = q_1 + q_2 i$ and $\rho = \alpha - \beta\sigma$, we have $N(\rho) < N(\beta)$, by showing that

$$N(\rho)/N(\beta) = |(\alpha/\beta) - \sigma|^2 < 1.$$

Here $|\ |$ is the usual absolute value for elements of $\mathbb{C}$.]

15. Use a Euclidean algorithm in $\mathbb{Z}[i]$ to find a gcd of $16 + 7i$ and $10 - 5i$ in $\mathbb{Z}[i]$. [*Hint:* Use the construction in the hint of Exercise 14.]

16. Let $\langle \alpha \rangle$ be a nonzero principal ideal in $\mathbb{Z}[i]$.

a. Show that $\mathbb{Z}[i]/\langle \alpha \rangle$ is a finite ring. [*Hint:* Use the division algorithm.]

b. Show that if π is an irreducible of $\mathbb{Z}[i]$, then $\mathbb{Z}[i]/\langle \pi \rangle$ is a field.

c. Referring to (b), find the order and characteristic of each of the following fields.

   i) $\mathbb{Z}[i]/\langle 3 \rangle$ ii) $\mathbb{Z}[i]/\langle 1 + i \rangle$ iii) $\mathbb{Z}[i]/\langle 1 + 2i \rangle$

17. Let $n \in \mathbb{Z}^+$ be square free, that is, not divisible by the square of any prime integer. Let $\mathbb{Z}[\sqrt{-n}] = \{a + ib\sqrt{n} \mid a, b \in \mathbb{Z}\}$.

   a. Show that the norm N, defined by $N(\alpha) = a^2 + nb^2$ for $\alpha = a + ib\sqrt{n}$, is a multiplicative norm on $\mathbb{Z}[\sqrt{-n}]$.

   b. Show that $N(\alpha) = 1$ for $\alpha \in \mathbb{Z}[\sqrt{-n}]$ if and only if α is a unit of $\mathbb{Z}[\sqrt{-n}]$.

   c. Show that every nonzero $\alpha \in \mathbb{Z}[\sqrt{-n}]$ that is not a unit has a factorization into irreducibles in $\mathbb{Z}[\sqrt{-n}]$. [*Hint:* Use (b).]

18. Repeat Exercise 17 for $\mathbb{Z}[\sqrt{n}] = \{a + b\sqrt{n} \mid a, b \in \mathbb{Z}\}$, with N defined by $N(\alpha) = |a^2 - nb^2|$ for $\alpha = a + b\sqrt{n}$ in $\mathbb{Z}[\sqrt{n}]$.

19. Show by a construction analogous to that given in the hint for Exercise 14 that the division algorithm holds in the integral domains $\mathbb{Z}[\sqrt{-2}]$, $\mathbb{Z}[\sqrt{2}]$, and $\mathbb{Z}[\sqrt{3}]$ for $v(\alpha) = N(\alpha)$ for nonzero α in one of these domains (see Exercises 17 and 18). (Thus these domains are Euclidean. See Hardy and Wright [29] for a discussion of which domains $\mathbb{Z}[\sqrt{n}]$ and $\mathbb{Z}[\sqrt{-n}]$ are Euclidean.)

Note: Fermat's last theorem dates to 1637. Fermat noted in the margin of a book that he had proved this result, but did not leave any record of the proof. It is called his "last theorem" since all the other theorems he stated without recording proofs have subsequently been proved. Many have tried to find a proof of this last theorem without success; many competent mathematicians have announced proofs only to have errors found in their work, as illustrated in the historical note.

As this text is in production, a front-page article in the June 24, 1993, edition of the *New York Times* reports the apparent success of Andrew Wiles at Princeton University in proving the theorem. A follow-up article appeared June 29. Becoming fascinated by the theorem at age 10, Andrew Wiles spent much time thinking about the theorem over the next 30 years, making a very concentrated effort at proof for the last 7 of those years. In June 1993, in a series of three lectures at Cambridge University in England to mathematicians capable of understanding his work, he presented his proposed proof of the theorem. No one present saw any flaws in his work. The proof takes about 200 manuscript pages. If it holds up after publication and much scrutiny, it will stand as a major mathematical achievement in this century. One wonders, with the pace of science today, whether any mathematician could now make a mathematical conjecture whose status (true, false, or undecidable) could not be established, despite intense effort by the best mathematicians, for another 350 years.

◆

CHAPTER EIGHT

EXTENSION FIELDS

◆

With the foundation of mappings and factor structures that we now have, we can gain real insight into the structure of fields. This chapter has to be the high point of our text. With the machinery we have developed, information now comes pouring out at an almost alarming rate. The results are elegant, and readily achieved.

In Section 8.1, we achieve our *basic goal,* showing that every nonconstant polynomial has a zero in some field. In the process we introduce the concepts of an extension field, and of algebraic elements and transcendental elements. We will find finite fields that are new to us.

Section 8.2 presents some notions concerning vector spaces, which will be review for many students. However, this section should not be skipped, since the examples indicate how we will be using these ideas.

Section 8.3 continues the discussion of field extensions, using the vector space concepts and demonstrates the existence of fields F with the property that every polynomial in $F[x]$ has a zero in F. The complex number field $\mathbb{C}$ has this property. Such richness in structure makes $\mathbb{C}$ a much better field to work with than the real number field $\mathbb{R}$. Unfortunately, pencil and paper arithmetic computations in $\mathbb{C}$ are more tedious than in $\mathbb{R}$, but the advent of the computer has largely erased that difficulty.

Section 8.4 is devoted to an application to classical plane geometry, and discusses what geometric constructions can be achieved using a straightedge and compass. In particular, we shall see that it is impossible to trisect an angle or to square a circle.

Section 8.5 gives a description of all finite fields.

Section 8.6, which can be skipped, catalogues some further types of algebraic structures, just for the sake of reference and information.

8.1

Introduction to Extension Fields

Our Basic Goal Achieved

We are now in a position to achieve our **basic goal**, which, loosely stated, is to show that every nonconstant polynomial has a zero. This will be stated more

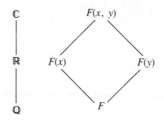

Figure 8.1

precisely and proved in Theorem 8.1, We first introduce some new terminology for some old ideas.

DEFINITION 8.1 (Extension Field) A field E is an **extension field of a field** F if $F \leq E$.

Thus $\mathbb{R}$ is an extension field of $\mathbb{Q}$, and $\mathbb{C}$ is an extension field of both $\mathbb{R}$ and $\mathbb{Q}$. As in the study of groups, it will often be convenient to use lattice diagrams to picture extension fields, the largest field being on top. We illustrate this in Fig. 8.1. A configuration where there is just one single column of fields, as at the left-hand side of Fig. 8.1, is often referred to, without any precise definition, as a **tower of fields.**

Now for our *basic goal*! This great and important result follows quickly and elegantly from the techniques we now have at our disposal.

THEOREM 8.1 (Kronecker's Theorem) (Basic Goal) Let F be a field and let $f(x)$ be a nonconstant polynomial in $F[x]$. Then there exists an extension field E of F and an $\alpha \in E$ such that $f(\alpha) = 0$.

PROOF By Theorem 5.23, $f(x)$ has a factorization in $F[x]$ into polynomials that are irreducible over F. Let $p(x)$ be an irreducible polynomial in such a factorization. It is clearly sufficient to find an extension field E of F containing an element α such that $p(\alpha) = 0$.

By Theorem 6.14, $\langle p(x) \rangle$ is a maximal ideal in $F[x]$, so $F[x]/\langle p(x) \rangle$ is a field. We claim that F can be identified with a subfield of $F[x]/\langle p(x) \rangle$ in a natural way by use of the map $\psi : F \to F[x]/\langle p(x) \rangle$ given by

$$\psi(a) = a + \langle p(x) \rangle$$

for $a \in F$. This map is one to one, for if $\psi(a) = \psi(b)$, that is, if $a + \langle p(x) \rangle = b + \langle p(x) \rangle$ for some $a, b \in F$, then $(a - b) \in \langle p(x) \rangle$, so

◆   **H I S T O R I C A L   N O T E**   ◆

Leopold Kronecker is known for his insistence on constructability of mathematical objects. As he noted, "God made the integers; all else is the work of man." Thus, he wanted to be able to construct new "domains of rationality" (fields) by using only the existence of integers and indeterminates and *not* by considering them as subfields of an already existing field of complex numbers. Hence in an 1881 paper Kronecker created an extension field by simply adjoining to a given field a root α of an irreducible nth degree polynomial $p(x)$; that is, his new field consisted of expressions rational in the original field elements and his new root α with the condition that $p(\alpha) = 0$. The proof of the theorem presented in the text (Theorem 8.1) dates from the twentieth century.

Kronecker completed his dissertation in 1845 at the University of Berlin. For many years thereafter he managed the family business, ultimately becoming financially independent. He then returned to Berlin, where he was elected to the Academy of Sciences and thus permitted to lecture at the university. On the retirement of Kummer, he became a professor at Berlin and with Karl Weierstrass (1815–1897) directed the influential mathematics seminar.

$a - b$ must be a multiple of the polynomial $p(x)$, which is of degree ≥ 1. Now $a, b \in F$ implies that $a - b$ is in F. Thus we must have $a - b = 0$, so $a = b$. We defined addition and multiplication in $F[x]/\langle p(x) \rangle$ by choosing any representatives, so we may choose $a \in (a + \langle p(x) \rangle)$. Thus ψ is a homomorphism that maps F one-to-one onto a subfield of $F[x]/\langle p(x) \rangle$. We identify F with $\{a + \langle p(x) \rangle \mid a \in F\}$ by means of this map ψ. Thus we shall view $E = F[x]/\langle p(x) \rangle$ as an extension field of F. We have now manufactured our desired extension field E of F. It remains for us to show that E contains a zero of $p(x)$.

Let us set

$$\alpha = x + \langle p(x) \rangle,$$

so $\alpha \in E$. Consider the evaluation homomorphism $\phi_\alpha : F[x] \to E$, given by Theorem 5.16. If $p(x) = a_0 + a_1 x + \cdots + a_n x^n$, where $a_i \in F$, then we have

$$\phi_\alpha(p(x)) = a_0 + a_1(x + \langle p(x) \rangle) + \cdots + a_n(x + \langle p(x) \rangle)^n$$

in $E = F[x]/\langle p(x) \rangle$. *But we can compute in* $F[x]/\langle p(x) \rangle$ *by choosing representatives, and x is a representative of the coset* $\alpha = x + \langle p(x) \rangle$. Therefore,

$$p(\alpha) = (a_0 + a_1 x + \cdots + a_n x^n) + \langle p(x) \rangle$$
$$= p(x) + \langle p(x) \rangle = \langle p(x) \rangle = 0$$

in $F[x]/\langle p(x) \rangle$. We have found an element α in $E = F[x]/\langle p(x) \rangle$ such that $p(\alpha) = 0$, and therefore $f(\alpha) = 0$. ◆

We illustrate the construction involved in the proof of Theorem 8.1 by two examples.

EXAMPLE I   Let $F = \mathbb{R}$, and let $f(x) = x^2 + 1$, which is well known to have no zeros in $\mathbb{R}$ and thus is irreducible over $\mathbb{R}$ by Theorem 5.19. Then $\langle x^2 + 1 \rangle$ is a maximal ideal in $\mathbb{R}[x]$, so $\mathbb{R}[x]/\langle x^2 + 1 \rangle$ is a field. Identifying $r \in \mathbb{R}$ with $r + \langle x^2 + 1 \rangle$ in $\mathbb{R}[x]/\langle x^2 + 1 \rangle$, we can view $\mathbb{R}$ as a subfield of $E = \mathbb{R}[x]/\langle x^2 + 1 \rangle$. Let

$$\alpha = x + \langle x^2 + 1 \rangle.$$

Computing in $\mathbb{R}[x]/\langle x^2 + 1 \rangle$, we find

$$\alpha^2 + 1 = (x + \langle x^2 + 1 \rangle)^2 + (1 + \langle x^2 + 1 \rangle)$$
$$= (x^2 + 1) + \langle x^2 + 1 \rangle = 0.$$

Thus α is a zero of $x^2 + 1$. We shall identify $\mathbb{R}[x]/\langle x^2 + 1 \rangle$ with $\mathbb{C}$ at the close of this section. ▲

EXAMPLE 2   Let $F = \mathbb{Q}$, and let $f(x) = x^4 - 5x^2 + 6$. This time $f(x)$ factors in $\mathbb{Q}[x]$ into $(x^2 - 2)(x^2 - 3)$, both factors being irreducible over $\mathbb{Q}$, as we have seen. We can start with $x^2 - 2$ and construct an extension field E of $\mathbb{Q}$ containing α such that $\alpha^2 - 2 = 0$, or we can construct an extension field K of $\mathbb{Q}$ containing an element β such that $\beta^2 - 3 = 0$. The construction in either case is just as in Example 1. ▲

Algebraic and Transcendental Elements

As we said before, most of the rest of this text is devoted to the study of zeros of polynomials. We commence this study by putting an element of an extension field E of a field F into one of two categories.

DEFINITION 8.2 (Algebraic, Transcendental)   An element α of an extension field E of a field F is **algebraic over** F if $f(\alpha) = 0$ for some nonzero $f(x) \in F[x]$. If α is not algebraic over F, then α is **transcendental over** F.

EXAMPLE 3 $\mathbb{C}$ is an extension field of $\mathbb{Q}$. Since $\sqrt{2}$ is a zero of $x^2 - 2$, we see that $\sqrt{2}$ is an algebraic element over $\mathbb{Q}$. Also, i is an algebraic element over $\mathbb{Q}$, being a zero of $x^2 + 1$. ▲

EXAMPLE 4   It is well known (but not easy to prove) that the real numbers π and e are transcendental over $\mathbb{Q}$. Here e is the base for the natural logarithms. ▲

Just as we do not speak simply of an *irreducible polynomial*, but rather of an *irreducible polynomial over F*, similarly we don't speak simply of an *algebraic element*, but rather of an *element algebraic over F*. The following illustration shows the reason for this.

EXAMPLE 5   The real number π is transcendental over $\mathbb{Q}$, as we stated in Example 4. However, π is algebraic over $\mathbb{R}$, for it is a zero of $(x - \pi) \in \mathbb{R}[x]$. ▲

EXAMPLE 6   It is easy to see that the real number $\sqrt{1 + \sqrt{3}}$ is algebraic over $\mathbb{Q}$. For if $\alpha = \sqrt{1 + \sqrt{3}}$, then $\alpha^2 = 1 + \sqrt{3}$, so $\alpha^2 - 1 = \sqrt{3}$ and $(\alpha^2 - 1)^2 = 3$. Therefore $\alpha^4 - 2\alpha^2 - 2 = 0$, so α is a zero of $x^4 - 2x^2 - 2$, which is in $\mathbb{Q}[x]$. ▲

To connect these ideas with those of number theory, we give the following definition.

DEFINITION 8.3 (Algebraic and Transcendental Numbers)   An element of $\mathbb{C}$ that is algebraic over $\mathbb{Q}$ is an **algebraic number**. A **transcendental number** is an element of $\mathbb{C}$ that is transcendental over $\mathbb{Q}$.

There is an extensive and elegant theory of algebraic numbers.

The next theorem gives a useful characterization of algebraic and transcendental elements over F in an extension field E of F. It also illustrates the importance of our evaluation homomorphisms ϕ_α. *Note that once more we are describing our concepts in terms of mappings.*

THEOREM 8.2   (Let E be an extension field of a field F and let $\alpha \in E$. Let $\phi_\alpha : F[x] \to E$ be the evaluation homomorphism of $F[x]$ into E such that $\phi_\alpha(a) = a$ for $a \in F$ and $\phi_\alpha(x) = \alpha$. Then α is transcendental over F if and only if ϕ_α gives an isomorphism of $F[x]$ with a subdomain of E, that is, if and only if ϕ_α is a one-to-one map.

PROOF   Now α is transcendental over F if and only if $f(\alpha) \neq 0$ for all nonconstant $f(x) \in F[x]$, which is true (by definition) if and only if $\phi_\alpha(f(x)) \neq 0$ for all nonconstant $f(x) \in F[x]$, which is true if and only if the kernel of ϕ_α is $\{0\}$, that is, if and only if ϕ_α is a one-to-one map.   ♦

The Irreducible Polynomial for α over F

Consider the extension field $\mathbb{R}$ of $\mathbb{Q}$. We know that $\sqrt{2}$ is algebraic over $\mathbb{Q}$, being a zero of $x^2 - 2$. Of course, $\sqrt{2}$ is also a zero of $x^3 - 2x$ and of $x^4 - 3x^2 + 2 = (x^2 - 2)(x^2 - 1)$. All these other polynomials having $\sqrt{2}$ as a zero were multiples of $x^2 - 2$. The next theorem shows that this is an illustration of a general situation. This theorem plays a central role in our later work.

THEOREM 8.3   Let E be an extension field of F, and let $\alpha \in E$, where α is algebraic over F. Then there is an irreducible polynomial $p(x) \in F[x]$ such that $p(\alpha) = 0$. This irreducible polynomial $p(x)$ is uniquely determined up to a constant factor in F and is a polynomial of minimal degree ≥ 1 in $F[x]$ having α as a zero. If $f(\alpha) = 0$ for $f(x) \in F[x]$, with $f(x) \neq 0$, then $p(x)$ divides $f(x)$.

PROOF   Let ϕ_α be the evaluation homomorphism of $F[x]$ into E, given by Theorem 5.16. The kernel of ϕ_α is an ideal and by Theorem 6.13 it must be a principal ideal generated by some $p(x) \in F[x]$. Now $\langle p(x) \rangle$ consists precisely of those elements of $F[x]$ having α as a zero. Thus, if $f(\alpha) = 0$ for $f(x) \neq 0$, then $f(x) \in \langle p(x) \rangle$, so $p(x)$ divides $f(x)$. Thus $p(x)$ is a polynomial of minimal degree ≥ 1 having α as a zero, and any other such polynomial of the same degree as $p(x)$ must be of the form $(a)p(x)$ for some $a \in F$.

It only remains for us to show that $p(x)$ is irreducible. If $p(x) = r(x)s(x)$ were a factorization of $p(x)$ into polynomials of lower degree, then $p(\alpha) = 0$ would imply that $r(\alpha)s(\alpha) = 0$, so either $r(\alpha) = 0$ or $s(\alpha) = 0$, since E is a field. This would contradict the fact that $p(x)$ is of minimal degree ≥ 1 such that $p(\alpha) = 0$. Thus $p(x)$ is irreducible.   ♦

By multiplying by a suitable constant in F, we can assume that the coefficient of the highest power of x appearing in $p(x)$ of Theorem 8.3 is 1. Such a polynomial having 1 as the coefficient of the highest power of x appearing is a **monic polynomial.**

DEFINITION 8.4 (irr(α, F))   Let E be an extension field of a field F, and let $\alpha \in E$ be algebraic over F. The unique monic polynomial $p(x)$ of Theorem 8.3 is the **irreducible polynomial for α over F** and will be denoted by irr(α, F). The degree of irr(α, F) is the **degree of α over F**, denoted by deg(α, F).

EXAMPLE 7   We know that $\mathrm{irr}(\sqrt{2}, \mathbb{Q}) = x^2 - 2$. Referring to Example 6, we see that for $\alpha = \sqrt{1 + \sqrt{3}}$ in $\mathbb{R}$, α is a zero of $x^4 - 2x^2 - 2$, which is in $\mathbb{Q}[x]$. Since $x^4 - 2x^2 - 2$ is irreducible over $\mathbb{Q}$ (by Eisenstein with $p = 2$, or by application of the technique of Example 6 of Section 5.6), we see that

$$\mathrm{irr}(\sqrt{1 + \sqrt{3}}, \mathbb{Q}) = x^4 - 2x^2 - 2.$$

Thus $\sqrt{1 + \sqrt{3}}$ is algebraic of degree 4 over $\mathbb{Q}$. ▲

Just as we must speak of an element α as *algebraic over F* rather than simply as *algebraic*, we must speak of the *degree of α over F* rather than the *degree of α.* To take a trivial illustration, $\sqrt{2} \in \mathbb{R}$ is algebraic of degree 2 over $\mathbb{Q}$ but algebraic of degree 1 over $\mathbb{R}$, for $\mathrm{irr}(\sqrt{2}, \mathbb{R}) = x - \sqrt{2}$.

The quick development of the theory here is due to the machinery of homomorphisms and ideal theory that we now have at our disposal. Note especially our constant use of the evaluation homomorphisms ϕ_α.

Simple Extensions

Let E be an extension field of a field F, and let $\alpha \in E$. Let ϕ_α be the evaluation homomorphism of $F[x]$ into E with $\phi_\alpha(a) = a$ for $a \in F$ and $\phi_\alpha(x) = \alpha$, as in Theorem 5.16. We consider two cases.

Case I *Suppose α is algebraic over F.* Then as in Theorem 8.3, the kernel of ϕ_α is $\langle \mathrm{irr}(\alpha, F)\rangle$ and by Theorem 6.14, $\langle \mathrm{irr}(\alpha, F)\rangle$ is a maximal ideal of $F[x]$. Therefore, $F[x]/\langle \mathrm{irr}(\alpha, F)\rangle$ is a field and is isomorphic to the image $\phi_\alpha[F[x]]$ in E. This subfield $\phi_\alpha[F[x]]$ of E is then the smallest subfield of E containing F and α. We shall denote this field by $F(\alpha)$.

Case II *Suppose α is transcendental over F.* Then by Theorem 8.2, ϕ_α gives an isomorphism of $F[x]$ with a subdomain of E. Thus in this case $\phi_\alpha[F[x]]$ is *not* a field but an integral domain that we shall denote by $F[\alpha]$. By Corollary 1 of Theorem 5.14, E contains a field of quotients of $F[\alpha]$, which is thus the smallest subfield of E containing F and α. As in Case I, we denote this field by $F(\alpha)$.

EXAMPLE 8 Since π is transcendental over $\mathbb{Q}$, the field $\mathbb{Q}(\pi)$ is isomorphic to the field $\mathbb{Q}(x)$ of rational functions over $\mathbb{Q}$ in the indeterminate x. Thus from a structural viewpoint, an element that is transcendental over a field F behaves as though it were an indeterminate over F. ▲

DEFINITION 8.5 (Simple Extension) An extension field E of a field F is a **simple extension of** F if $E = F(\alpha)$ for some $\alpha \in E$.

Many important results appear throughout this section. We have now developed so much machinery that results are starting to pour out of our efficient plant at an alarming rate. The next theorem gives us insight into the nature of the field $F(\alpha)$ in the case where α is algebraic over F.

THEOREM 8.4 Let E be a simple extension $F(\alpha)$ of a field F, and let α be algebraic over F. Let the degree of $\mathrm{irr}(\alpha, F)$ be $n \geq 1$. Then every element β of $E = F(\alpha)$ can be uniquely expressed in the form

$$\beta = b_0 + b_1\alpha + \cdots + b_{n-1}\alpha^{n-1},$$

where the b_i are in F.

PROOF For the usual evaluation homomorphism ϕ_α, every element of

$$F(\alpha) = \phi_\alpha[F[x]]$$

is of the form $\phi_\alpha(f(x)) = f(\alpha)$, for formal polynomial in α with coefficients in F. Let

$$\mathrm{irr}(\alpha, F) = p(x) = x^n + a_{n-1}x^{n-1} + \cdots + a_0.$$

Then $p(\alpha) = 0$, so

$$\alpha^n = -a_{n-1}\alpha^{n-1} - \cdots - a_0.$$

This equation in $F(\alpha)$ can be used to express every monomial α^m for $m \geq n$ in terms of powers of α that are less than n. For example,

$$\alpha^{n+1} = \alpha\alpha^n = -a_{n-1}\alpha^n - a_{n-2}\alpha^{n-1} - \cdots - a_0\alpha$$

$$= -a_{n-1}(-a_{n-1}\alpha^{n-1} - \cdots - a_0) - a_{n-2}\alpha^{n-1} - \cdots - a_0\alpha.$$

Thus, if $\beta \in F(\alpha)$, β can be expressed in the required form

$$\beta = b_0 + b_1\alpha + \cdots + b_{n-1}\alpha^{n-1}.$$

For uniqueness, if

$$b_0 + b_1\alpha + \cdots + b_{n-1}\alpha^{n-1} = b_0' + b_1'\alpha + \cdots + b_{n-1}'\alpha^{n-1}$$

+	0	1	α	$1 + \alpha$
0	0	1	α	$1 + \alpha$
1	1	0	$1 + \alpha$	α
α	α	$1 + \alpha$	0	1
$1 + \alpha$	$1 + \alpha$	α	1	0

Table 8.1

for $b_i' \in F$, then

$$(b_0 - b_0') + (b_1 - b_1')x + \cdots + (b_{n-1} - b_{n-1}')x^{n-1} = g(x)$$

is in $F[x]$ and $g(\alpha) = 0$. Also, the degree of $g(x)$ is less than the degree of $\mathrm{irr}(\alpha, F)$. Since $\mathrm{irr}(\alpha, F)$ is a nonzero polynomial of minimal degree in $F[x]$ having α as a zero, we must have $g(x) = 0$. Therefore, $b_i - b_i' = 0$, so

$$b_i = b_i',$$

and the uniqueness of the b_i is established. ◆

We give an impressive example illustrating Theorem 8.4.

EXAMPLE 9   The polynomial $p(x) = x^2 + x + 1$ in $\mathbb{Z}_2[x]$ is irreducible over $\mathbb{Z}_2$ by Theorem 5.19, since neither element 0 nor element 1 of $\mathbb{Z}_2$ is a zero of $p(x)$. By Theorem 8.1, we know that there is an extension field E of $\mathbb{Z}_2$ containing a zero α of $x^2 + x + 1$. By Theorem 8.4, $\mathbb{Z}_2(\alpha)$ has as elements $0 + 0\alpha$, $1 + 0\alpha$, $0 + 1\alpha$, and $1 + 1\alpha$, that is, 0, 1, α, and $1 + \alpha$. *This gives us a new finite field of four elements!* The addition and multiplication tables for this field are shown in Tables 8.1 and 8.2. For example, to compute $(1 + \alpha)(1 + \alpha)$ in $\mathbb{Z}_2(\alpha)$, we observe that since $p(\alpha) = \alpha^2 + \alpha + 1 = 0$, then

$$\alpha^2 = -\alpha - 1 = \alpha + 1.$$

Therefore,

$$(1 + \alpha)(1 + \alpha) = 1 + \alpha + \alpha + \alpha^2 = 1 + \alpha^2 = 1 + \alpha + 1 = \alpha. \;\blacktriangle$$

Finally, we can use Theorem 8.4 to fulfill our promise of Example 1 and

	0	1	α	$1 + \alpha$
0	0	0	0	0
1	0	1	α	$1 + \alpha$
α	0	α	$1 + \alpha$	1
$1 + \alpha$	0	$1 + \alpha$	1	α

Table 8.2

show that $\mathbb{R}[x]/\langle x^2 + 1\rangle$ is isomorphic to the field $\mathbb{C}$ of complex numbers. We saw in Example 1 that we can view $\mathbb{R}[x]/\langle x^2 + 1\rangle$ as an extension field of $\mathbb{R}$. Let

$$\alpha = x + \langle x^2 + 1\rangle.$$

Then $\mathbb{R}(\alpha) = \mathbb{R}[x]/\langle x^2 + 1\rangle$ and consists of all elements of the form $a + b\alpha$ for $a, b \in \mathbb{R}$, by Theorem 8.4. But since $\alpha^2 + 1 = 0$, we see that α plays the role of $i \in \mathbb{C}$, and $a + b\alpha$ plays the role of $(a + bi) \in \mathbb{C}$. Thus $\mathbb{R}(\alpha) \simeq \mathbb{C}$. *This is the elegant algebraic way to construct $\mathbb{C}$ from $\mathbb{R}$.*

Exercises 8.1

Computations

In Exercises 1 through 5, show that the given number $\alpha \in \mathbb{C}$ is algebraic over $\mathbb{Q}$ by finding $f(x) \in \mathbb{Q}[x]$ such that $f(\alpha) = 0$.

1. $1 + \sqrt{2}$ **2.** $\sqrt{2} + \sqrt{3}$ **3.** $1 + i$

4. $\sqrt{1 + \sqrt[3]{2}}$ **5.** $\sqrt{\sqrt[3]{2} - i}$

In Exercises 6 through 8, find irr$(\alpha, \mathbb{Q})$ and deg$(\alpha, \mathbb{Q})$ for the given algebraic number $\alpha \in \mathbb{C}$. Be prepared to prove that your polynomials are irreducible over $\mathbb{Q}$ if challenged to do so.

6. $\sqrt{3 - \sqrt{6}}$ **7.** $\sqrt{(\frac{1}{3}) + \sqrt{7}}$ **8.** $\sqrt{2} + i$

In Exercises 9 through 16, classify the given $\alpha \in \mathbb{C}$ as algebraic or transcendental over the given field F. If α is algebraic over F, find deg(α, F).

9. $\alpha = i, F = \mathbb{Q}$ **10.** $\alpha = 1 + i, F = \mathbb{R}$

11. $\alpha = \sqrt{\pi}, F = \mathbb{Q}$ **12.** $\alpha = \sqrt{\pi}, F = \mathbb{R}$

13. $\alpha = \sqrt{\pi}, F = \mathbb{Q}(\pi)$ **14.** $\alpha = \pi^2, F = \mathbb{Q}$

15. $\alpha = \pi^2, F = \mathbb{Q}(\pi)$ **16.** $\alpha = \pi^2, F = \mathbb{Q}(\pi^3)$

17. Refer to Example 9 of the text. The polynomial $x^2 + x + 1$ has a zero α in $\mathbb{Z}_2(\alpha)$ and thus must factor into a product of linear factors in $(\mathbb{Z}_2(\alpha))[x]$. Find this factorization. [*Hint:* Divide $x^2 + x + 1$ by $x - \alpha$ by long division, using the fact that $\alpha^2 = \alpha + 1$.]

18. a. Show that the polynomial $x^2 + 1$ is irreducible in $\mathbb{Z}_3[x]$.

  b. Let α be a zero of $x^2 + 1$ in an extension field of $\mathbb{Z}_3$. As in Example 9, give the multiplication and addition tables for the nine elements of $\mathbb{Z}_3(\alpha)$, written in the order $0, 1, 2, \alpha, 2\alpha, 1 + \alpha, 1 + 2\alpha, 2 + \alpha$, and $2 + 2\alpha$.

Concepts

19. Mark each of the following true or false.
    _____ a. The number π is transcendental over $\mathbb{Q}$.
    _____ b. $\mathbb{C}$ is a simple extension of $\mathbb{R}$.
    _____ c. Every element of a field F is algebraic over F.
    _____ d. $\mathbb{R}$ is an extension field of $\mathbb{Q}$.
    _____ e. $\mathbb{Q}$ is an extension field of $\mathbb{Z}_2$.
    _____ f. Let $\alpha \in \mathbb{C}$ be algebraic over $\mathbb{Q}$ of degree n. If $f(\alpha) = 0$ for nonzero $f(x) \in \mathbb{Q}[x]$, then (degree $f(x)) \geq n$.
    _____ g. Let $\alpha \in \mathbb{C}$ be algebraic over $\mathbb{Q}$ of degree n. If $f(\alpha) = 0$ for nonzero $f(x) \in \mathbb{R}[x]$, then (degree $f(x)) \geq n$.
    _____ h. Every nonconstant polynomial in $F[x]$ has a zero in some extension field of F.
    _____ i. Every nonconstant polynomial in $F[x]$ has a zero in every extension field of F.
    _____ j. If x is an indeterminate, $\mathbb{Q}[\pi] \simeq \mathbb{Q}[x]$.

20. We have stated without proof that π and e are transcendental over $\mathbb{Q}$.

    a. Find a subfield F of $\mathbb{R}$ such that π is algebraic of degree 3 over F.

    b. Find a subfield E of $\mathbb{R}$ such that e^2 is algebraic of degree 5 over E.

21. a. Show that $x^3 + x^2 + 1$ is irreducible over $\mathbb{Z}_2$.

    b. Let α be a zero of $x^3 + x^2 + 1$ in an extension field of $\mathbb{Z}_2$. Show that $x^3 + x^2 + 1$ factors into three linear factors in $(\mathbb{Z}_2(\alpha))[x]$ by actually finding this factorization. [*Hint:* Every element of $\mathbb{Z}_2(\alpha)$ is of the form
    $$a_0 + a_1\alpha + a_2\alpha^2 \qquad \text{for} \quad a_i = 0, 1.$$
    Divide $x^3 + x^2 + 1$ by $x - \alpha$ by long division. Show that the quotient also has a zero in $\mathbb{Z}_2(\alpha)$ by simply trying the eight possible elements. Then complete the factorization.]

22. Let E be an extension field of $\mathbb{Z}_2$ and let $\alpha \in E$ be algebraic of degree 3 over $\mathbb{Z}_2$. Classify the groups $\langle \mathbb{Z}_2(\alpha), + \rangle$ and $\langle (\mathbb{Z}_2(\alpha))^*, \cdot \rangle$ according to the fundamental theorem of finitely generated abelian groups. As usual, $(\mathbb{Z}_2(\alpha))^*$ is the set of nonzero elements of $\mathbb{Z}_2(\alpha)$.

23. Let E be an extension field of a field F and let $\alpha \in E$ be algebraic over F. The polynomial irr(α, F) is sometimes referred to as the **minimal polynomial for α over F**. Why is this designation appropriate?

Theory

24. Let E be an extension field of F, and let $\alpha, \beta \in E$. Suppose α is transcendental over F but algebraic over $F(\beta)$. Show that β is algebraic over $F(\alpha)$.

25. Let E be an extension field of a finite field F, where F has q elements. Let $\alpha \in E$ be algebraic over F of degree n. Prove that $F(\alpha)$ has q^n elements.

26. a. Show that there exists an irreducible polynomial of degree 3 in $\mathbb{Z}_3[x]$.

    b. Show from part (a) that there exists a finite field of 27 elements. [*Hint:* Use Exercise 25.]

27. Consider the prime field $\mathbb{Z}_p$ of characteristic $p \neq 0$.

    a. Show that, for $p \neq 2$, not every element in $\mathbb{Z}_p$ is a square of an element of $\mathbb{Z}_p$. [*Hint:* $1^2 = (p - 1)^2 = 1$ in $\mathbb{Z}_p$. Deduce the desired conclusion *by counting.*]

    b. Using part (a), show that there exist finite fields of p^2 elements for every prime p in $\mathbb{Z}^+$.

28. Let E be an extension field of a field F and let $\alpha \in E$ be transcendental over F. Show that every element of $F(\alpha)$ that is not in F is also transcendental over F.

29. Show that $\{a + b(\sqrt[3]{2}) + c(\sqrt[3]{2})^2 \mid a, b, c \in \mathbb{Q}\}$ is a subfield of $\mathbb{R}$ by using the ideas of this section, rather than by a formal verification of the field axioms. [*Hint:* Use Theorem 8.4.]

30. Following the idea of Exercise 26, show that there exists a field of 8 elements; of 16 elements; of 25 elements.

31. Let F be a finite field of characteristic p. Show that every element of F is algebraic over the prime field $\mathbb{Z}_p \leq F$. [*Hint:* Let F^* be the set of nonzero elements of F. Apply group theory to the group $\langle F^*, \cdot \rangle$ to show that every $\alpha \in F^*$ is a zero of some polynomial in $\mathbb{Z}_p[x]$ of the form $x^n - 1$.]

32. Use Exercises 25 and 31 to show that every finite field is of prime-power order, that is, it has a prime-power number of elements.

8.2

Vector Spaces

The notions of a vector space, scalars, independent vectors, and bases may be familiar. In this section, we present these ideas where the scalars may be elements of any field. We use Greek letters like α and β for vectors since, in our application, the vectors will be elements of an extension field E of a field F. The proofs are all identical with those often given in a first course in linear algebra. If these ideas are familiar, we suggest studying Examples 3, 5, 7, 9, and 10, and then reading Theorem 8.8 and its proof. If the examples and the theorem are understood, then do some exercises and proceed to the next section.

Definition and Elementary Properties

The topic of vector spaces is the cornerstone of linear algebra. Since linear algebra is not the subject for study in this text, our treatment of vector spaces will be brief, designed to develop only the concepts of linear independence and dimension that we need for our field theory.

    The terms *vector* and *scalar* are probably familiar from calculus. Here we allow scalars to be elements of any field, not just the real numbers, and develop the theory by axioms just as for the other algebraic structures we have studied.

DEFINITION 8.6 (Vector Space)   Let F be a field. A **vector space over** F (or F-**vector space**) consists of an abelian group V under addition together with an operation of scalar multiplication of each element of V by each element of F on the left, such that for all $a, b \in F$ and $\alpha, \beta \in V$ the following conditions are satisfied:

$\mathscr{V}_1.$ $a\alpha \in V.$

$\mathscr{V}_2.$ $a(b\alpha) = (ab)\alpha.$

$\mathscr{V}_3.$ $(a + b)\alpha = (a\alpha) + (b\alpha).$

$\mathscr{V}_4.$ $a(\alpha + \beta) = (a\alpha) + (a\beta).$

$\mathscr{V}_5.$ $1\alpha = \alpha.$

The elements of V are **vectors** and the elements of F are **scalars**. When only one field F is under discussion, we drop the reference to F and refer to a *vector space*.

Note that multiplication for a vector space is not a binary operation on one set in the sense we defined it in Section 1.1. It is rather a rule that

---

◆   **HISTORICAL NOTE**   ◆

The ideas behind the abstract notion of a vector space occurred in many concrete examples during the nineteenth century and earlier. For example, William Rowan Hamilton dealt with complex numbers explicitly as pairs of real numbers and, as noted in Section 5.7, also dealt with triples and eventually quadruples of real numbers in his invention of the quaternions. In these cases, the "vectors" turned out to be objects which could both be added and multiplied by scalars, using "reasonable" rules for both of these operations. Other examples of such objects included differential forms (things under integral signs) and algebraic integers (see Section 9.7).

Although Hermann Grassmann (1809–1877) succeeded in working out a detailed theory of n-dimensional spaces in his *Die lineale Ausdehnungslehre* of 1844 and 1862, the first mathematician to give an abstract definition of a vector space equivalent to Definition 8.6 was Giuseppe Peano (1858–1932) in his *Calcolo geometrico* of 1888. Peano's aim in the book, as the title indicates, was to develop a geometric calculus. According to Peano, such a calculus "consists of a system of operations analogous to those of algebraic calculus, but in which the objects with which the calculations are performed are, instead of numbers, geometrical objects," Curiously, Peano's work had no immediate effect on the mathematical scene. His definition did not enter the mathematical mainstream until Hermann Weyl (1885–1955) essentially repeated it in his definition of an "affine geometry" in his *Space-Time-Matter* of 1918.

associates an element $a\alpha$ of V with each ordered pair (a, α), consisting of an element a of F and an element α of V. This can be viewed as a *function* mapping $F \times V$ into V. The *nice* way to define a binary operation on a set S is similarly to say that it is a function from $S \times S$ into S, but we wished to be more naive in Section 1.1. Both the additive identity for V, the 0-vector, and the additive identity for F, the 0-scalar, will be denoted by 0.

EXAMPLE 1 Consider the abelian group $\langle \mathbb{R}^n, + \rangle = \mathbb{R} \times \mathbb{R} \times \cdots \times \mathbb{R}$ for n factors, which consists of ordered n-tuples under addition by components. Define scalar multiplication for scalars in $\mathbb{R}$ by

$$r\alpha = (ra_1, \ldots, ra_n)$$

for $r \in \mathbb{R}$ and $\alpha = (a_1, \ldots, a_n) \in \mathbb{R}^n$. With these operations, $\mathbb{R}^n$ becomes a vector space over $\mathbb{R}$. The axioms for a vector space are readily checked. In particular, $\mathbb{R}^2 = \mathbb{R} \times \mathbb{R}$ as a vector space over $\mathbb{R}$ can be viewed as all "vectors whose starting points are the origin of the Euclidean plane" in the sense often studied in calculus courses. ▲

EXAMPLE 2 For any field F, $F[x]$ can be viewed as a vector space over F, where addition of vectors is ordinary addition of polynomials in $F[x]$ and scalar multiplication $a\alpha$ of an element of $F[x]$ by an element of F is ordinary multiplication in $F[x]$. The axioms $\mathcal{V}_1$ through $\mathcal{V}_5$ for a vector space then follow immediately from the axioms for the integral domain $F[x]$. ▲

EXAMPLE 3 Let E be an extension field of a field F. Then E can be regarded as a vector space over F, where addition of vectors is the usual addition in E and scalar multipliction $a\alpha$ is the usual field multiplication in E with $a \in F$ and $\alpha \in E$. The axioms follow at once from the field axioms for E. Here our field of scalars is actually a subset of our space of vectors. *It is this example that is the important one for us.* ▲

We are assuming nothing about vector spaces from previous work and shall prove everything we need from the definition, even though the results may be familiar from calculus.

THEOREM 8.5 If V is a vector space over F, then $0\alpha = 0$, $a0 = 0$ and $(-a)\alpha = a(-\alpha) = -(a\alpha)$ for all $a \in F$ and $\alpha \in V$.

PROOF The equation $0\alpha = 0$ is to be read "(0-scalar)α = 0-vector," and likewise, $a0 = 0$ is to be read "a(0-vector) = 0-vector." The proofs here are very similar to those in Theorem 5.1 for a ring and again depend heavily on the distributive laws $\mathcal{V}_3$ and $\mathcal{V}_4$. Now

$$(0\alpha) = (0 + 0)\alpha = (0\alpha) + (0\alpha)$$

is an equation in the abelian group $\langle V, + \rangle$, so by the group cancellation law, $0 = 0\alpha$. Likewise, from

$$a0 = a(0 + 0) = a0 + a0,$$

we conclude that $a0 = 0$. Then

$$0 = 0\alpha = (a + (-a))\alpha = a\alpha + (-a)\alpha,$$

so $(-a)\alpha = -(a\alpha)$. Likewise, from

$$0 = a0 = a(\alpha + (-\alpha)) = a\alpha + a(-\alpha),$$

we conclude that $a(-\alpha) = -(a\alpha)$ also. ◆

Linear Independence and Bases

DEFINITION 8.7 (Span, Linear Combination)   Let V be a vector space over F. The vectors in a subset $S = \{\alpha_i \mid i \in I\}$ of V **span** (or **generate**) V if for every $\beta \in V$, we have

$$\beta = a_1\alpha_{i_1} + a_2\alpha_{i_2} + \cdots + a_n\alpha_{i_n}$$

for some $a_j \in F$ and $\alpha_{i_j} \in S$, $j = 1, \ldots, n$. A vector $\sum_{j=1}^{n} a_j\alpha_{i_j}$ is a **linear combination of the** α_{i_j}.

EXAMPLE 4   In the vector space $\mathbb{R}^n$ over $\mathbb{R}$ of Example 1, the vectors

$$(1, 0, \ldots, 0), (0, 1, \ldots, 0), \ldots, (0, 0, \ldots, 1)$$

clearly span $\mathbb{R}^n$, for

$$(a_1, a_2, \ldots, a_n) = a_1(1, 0, \ldots, 0) + a_2(0, 1, \ldots, 0) + \cdots$$
$$+ a_n(0, 0, \ldots, 1).$$

Also, the monomials x^m for $m \geq 0$ span $F[x]$ over F, the vector space of Example 2. ▲

EXAMPLE 5   Let F be a field and E an extension field of F. Let $\alpha \in E$ be algebraic over F. Then $F(\alpha)$ is a vector space over F and by Theorem 8.4, it is spanned by the vectors in $\{1, \alpha, \ldots, \alpha^{n-1}\}$, where $n = \deg(\alpha, F)$. *This is the important example for us.* ▲

DEFINITION 8.8 (Finite Dimensional)   A vector space V over a field F is **finite dimensional** if there is a finite subset of V whose vectors span V.

EXAMPLE 6   Example 4 shows that $\mathbb{R}^n$ is finite dimensional. The vector space $F[x]$ over F is *not* finite dimensional, since polynomials of arbitrarily large degree could not be linear combinations of elements of any *finite* set of polynomials. ▲

EXAMPLE 7   If $F \leq E$ and $\alpha \in E$ is algebraic over the field F, Example 5 shows that $F(\alpha)$ is a finite-dimensional vector space over F. *This is the most important example for us.* ▲

The next definition contains the most important idea in this section.

DEFINITION 8.9 (Independence)   The vectors in a subset $S = \{\alpha_i \mid i \in I\}$ of a vector space V over a field F are **linearly independent over** F if $\sum_{j=1}^n a_j\alpha_{i_j} = 0$ implies that $a_j = 0$ for $j = 1, \ldots, n$. If the vectors are not linearly independent over F, they are **linearly dependent over** F.

Thus the vectors in $\{\alpha_i \mid i \in I\}$ are linearly independent over F if the only way the 0-vector can be expressed as a linear combination of the vectors α_i is to have all scalar coefficients equal to 0. If the vectors are linearly dependent over F, then there exist $a_j \in F$ for $j = 1, \ldots, n$ such that $\sum_{j=1}^n a_j\alpha_{i_j} = 0$, where not all $a_j = 0$.

EXAMPLE 8   Observe that the vectors spanning the space $\mathbb{R}^n$ that are given in Example 4 are linearly independent over $\mathbb{R}$. Likewise, the vectors in $\{x^m \mid m \geq 0\}$ are linearly independent vectors of $F[x]$ over F. Note that $(1, -1), (2, 1)$, and $(-3, 2)$ are linearly dependent in $\mathbb{R}^2$ over $\mathbb{R}$, since

$$7(1, -1) + (2, 1) + 3(-3, 2) = (0, 0) = 0. \quad ▲$$

EXAMPLE 9   Let E be an extension field of a field F, and let $\alpha \in E$ be algebraic over F. If $\deg(\alpha, F) = n$, then by Theorem 8.4, every element of $F(\alpha)$ can be *uniquely* expressed in the form

$$b_0 + b_1\alpha + \cdots + b_{n-1}\alpha^{n-1}$$

for $b_i \in F$. In particular, $0 = 0 + 0\alpha + \cdots + 0\alpha^{n-1}$ must be a *unique* such expression for 0. Thus the elements $1, \alpha, \ldots, \alpha^{n-1}$ are linearly independent vectors in $F(\alpha)$ over the field F. They also span $F(\alpha)$, so by the next definition, $1, \alpha, \ldots, \alpha^{n-1}$ form a *basis* for $F(\alpha)$ over F. *This is the important example for us.* In fact, this is the reason we are doing this material on vector spaces. ▲

DEFINITION 8.10 (Basis)   If V is a vector space over a field f, the vectors in a subset $B = \{\beta_i \mid i \in I\}$ of V form a **basis for** V **over** F if they span V and are linearly independent.

Dimension

The only other results we wish to prove about vector spaces are that every finite-dimensional vector space has a basis, and that any two bases of a finite-dimensional vector space have the same number of elements. Both these facts are true without the assumption that the vector space is finite dimensional, but the proofs require more knowledge of set theory than we are assuming, and the finite-dimensional case is all we need. First we give an easy lemma.

LEMMA 8.1   Let V be a vector space over a field F, and let $\alpha \in V$. If α is a linear combination of vectors β_i for $i = 1, \ldots, m$ and each β_i is a linear combination of vectors γ_j for $j = 1, \ldots, n$, then α is a linear combination of the γ_j.

PROOF   Let $\alpha = \sum_{i=1}^{m} a_i \beta_i$, and let $\beta_i = \sum_{j=1}^{n} b_{ij} \gamma_j$, where a_i and b_{ij} are in F. Then

$$\alpha = \sum_{i=1}^{m} a_i \left(\sum_{j=1}^{n} b_{ij} \gamma_j \right) = \sum_{j=1}^{n} \left(\sum_{i=1}^{m} a_i b_{ij} \right) \gamma_j,$$

and $(\sum_{i=1}^{m} a_i b_{ij}) \in F$. ◆

THEOREM 8.6   In a finite-dimensional vector space, every finite set of vectors spanning the space contains a subset that is a basis.

PROOF   Let V be finite dimensional over F, and let vectors $\alpha_1, \ldots, \alpha_n$ in V span V. Let us list the α_i in a row. Examine each α_i in succession, starting at the left with $i = 1$, and discard the first α_j that is some linear combination of the preceding α_i for $i < j$. Then continue, starting with the following α_{j+1}, and discard the next α_k that is some linear combination of its remaining predecessors, and so on. When we reach α_n after a finite number of steps, those α_i remaining in our list are such that none is a linear combination of the preceding α_i in this reduced list. Lemma 8.1 shows that any vector that is a linear combination of the original collection of α_i is still a linear combination of our reduced, and possibly smaller, set in which no α_i is a linear combination of its predecessors. Thus the vectors in the reduced set of α_i again span V.

For the reduced set, suppose that

$$a_1 \alpha_{i_1} + \cdots + a_r \alpha_{i_r} = 0$$

for $i_1 < i_2 < \cdots < i_r$ and that some $a_j \neq 0$. We may assume from Theorem 8.5 that $a_r \neq 0$, or we could drop $a_r \alpha_{i_r}$ from the left side of the equation. Then, using Theorem 8.5 again, we obtain

$$\alpha_{i_r} = \left(-\frac{a_1}{a_r} \right) \alpha_{i_1} + \cdots + \left(-\frac{a_{r-1}}{a_r} \right) \alpha_{i_{r-1}},$$

which shows that α_{i_r} is a linear combination of its predecessors, contradicting our construction. Thus the vectors α_i in the reduced set both span V and are linearly independent, so they form a basis for V over F. ◆

COROLLARY A finite-dimensional vector space has a finite basis.

PROOF By definition, a finite-dimensional vector space has a finite set of vectors that span the space. Theorem 8.6 completes the proof. ◆

The next theorem is the culmination of our work on vector spaces.

THEOREM 8.7 Let $S = \{\alpha_1, \ldots, \alpha_r\}$ be a finite set of linearly independent vectors of a finite-dimensional vector space V over a field F. Then S can be enlarged to a basis for V over F, Furthermore, if $B = \{\beta_1, \ldots, \beta_n\}$ is any basis for V over F, then $r \leq n$.

PROOF By the corollary of Theorem 8.8, there is a basis $B = \{\beta_1, \ldots, \beta_n\}$ for V over F. Consider the finite sequence of vectors

$$\alpha_1, \ldots, \alpha_r, \beta_1, \ldots, \beta_n.$$

These vectors span V, since B is a basis. Following the technique, used in Theorem 8.6, of discarding in turn each vector that is a linear combination of its remaining predecessors, working from left to right, we arrive at a basis for V. Observe that no α_i is cast out, since the α_i are linearly independent. Thus S can be enlarged to a basis for V over F.

For the second part of the conclusion, consider the sequence

$$\alpha_1, \beta_1, \ldots, \beta_r.$$

These vectors are not linearly independent over F, because α_1 is a linear combination

$$\alpha_1 = b_1\beta_1 + \cdots + b_n\beta_n,$$

since the β_i form a basis. Thus

$$\alpha_1 + (-b_1)\beta_1 + \cdots + (-b_n)\beta_n = 0.$$

The vectors in the sequence do span V, and if we form a basis by the technique of working from left to right and casting out in turn each vector that is a linear combination of its remaining predecessors, at least one β_i must be cast out, giving a basis

$$\{\alpha_1, \beta_1^{(1)}, \ldots, \beta_m^{(1)}\},$$

where $m \leq n - 1$. Applying the same technique to the sequence of vectors

$$\alpha_1, \alpha_2, \beta_1^{(1)}, \ldots, \beta_m^{(1)},$$

we arrive at a new basis

$$\{\alpha_1, \alpha_2, \beta_1^{(2)}, \ldots, \beta_s^{(2)}\},$$

with $s \leq n - 2$. Continuing, we arrive finally at a basis

$$\{\alpha_1, \ldots, \alpha_r, \beta_1^{(r)}, \ldots, \beta_t^{(r)}\},$$

where $0 \leq t \leq n - r$. Thus $r \leq n$. ◆

COROLLARY Any two bases of a finite-dimensional vector space V over F have the same number of elements.

PROOF Let $B = \{\beta_1, \ldots, \beta_n\}$ and $B' = \{\beta_1', \ldots, \beta_m'\}$ be two bases. Then by Theorem 8.7, regarding B as an independent set of vectors and B' as a basis, we see that $n \leq m$. A symmetric argument gives $m \leq n$, so $m = n$. ◆

DEFINITION 8.11 (Dimension) If V is a finite-dimensional vector space over a field F, the number of elements in a basis (independent of the choice of basis, as just shown) is the **dimension of V over F**.

EXAMPLE 10 Let E be an extension field of a field F, and let $\alpha \in E$. Example 9 shows that if α is algebraic over F and $\deg(\alpha, F) = n$, then the dimension of $F(\alpha)$ as a vector space over F is n. *This is the important example for us.* ▲

An Application to Field Theory

We collect the results of field theory contained in Examples 3, 5, 7, 9, and 10, and incorporate them into one theorem. The last sentence of this theorem gives an additional elegant application of these vector space ideas to field theory.

THEOREM 8.8 Let E be an extension field of F, and let $\alpha \in E$ be algebraic over F. If $\deg(\alpha, F) = n$, then $F(\alpha)$ is an n-dimensional vector space over F with basis $\{1, \alpha, \ldots, \alpha^{n-1}\}$. Furthermore, every element β of $F(\alpha)$ is algebraic over F, and $\deg(\beta, F) \leq \deg(\alpha, F)$.

PROOF We have shown everything in the preceding examples except the very important result stated in the last sentence of the above theorem. Let $\beta \in F(\alpha)$, where α is algebraic over F of degree n. Consider the elements

$$1, \beta, \beta^2, \ldots, \beta^n.$$

These can't be $n + 1$ distinct elements of $F(\alpha)$ that are linearly independent over F, for by Theorem 8.7, any basis of $F(\alpha)$ over F would have to contain at least as many elements as are in any set of linearly independent vectors over F. However, the basis $\{1, \alpha, \ldots, \alpha^{n-1}\}$ has just n elements. If $\beta^i = \beta^j$, then $\beta^i - \beta^j = 0$, so in any case there exist $b_i \in F$ such that

$$b_0 + b_1\beta + b_2\beta^2 + \cdots + b_n\beta^n = 0,$$

where not all $b_i = 0$. Then $f(x) = b_n x^n + \cdots + b_1 x + b_0$ is a nonzero element of $F[x]$ such that $f(\beta) = 0$. Therefore, β is algebraic over F and $\deg(\beta, F)$ is at most n. ◆

Exercises 8.2

Computations

1. Find three bases for $\mathbb{R}^2$ over $\mathbb{R}$, no two of which have a vector in common.

In Exercises 2 and 3, determine whether the given set of vectors is a basis for $\mathbb{R}^3$ over $\mathbb{R}$.

2. $\{(1, 1, 0), (1, 0, 1), (0, 1, 1)\}$ **3.** $\{(-1, 1, 2), (2, -3, 1), (10, -14, 0)\}$

In Exercises 4 through 9, give a basis for the indicated vector space over the field.

4. $\mathbb{Q}(\sqrt{2})$ over $\mathbb{Q}$ **5.** $\mathbb{R}(\sqrt{2})$ over $\mathbb{R}$

6. $\mathbb{Q}(\sqrt[3]{2})$ over $\mathbb{Q}$ **7.** $\mathbb{C}$ over $\mathbb{R}$

8. $\mathbb{Q}(i)$ over $\mathbb{Q}$ **9.** $\mathbb{Q}(\sqrt[4]{2})$ over $\mathbb{Q}$

10. According to Theorem 8.8, the element $1 + \alpha$ of $\mathbb{Z}_2(\alpha)$ of Example 9 of Section 8.1 is algebraic over $\mathbb{Z}_2$. Find the irreducible polynomial for $1 + \alpha$ in $\mathbb{Z}_2[x]$.

Concepts

11. Mark each of the following true or false.

    _____ a. The sum of two vectors is a vector.

    _____ b. The sum of two scalars is a vector.

    _____ c. The product of two scalars is a scalar.

    _____ d. The product of a scalar and a vector is a vector.

    _____ e. Every vector space has a finite basis.

    _____ f. The vectors in a basis are linearly dependent.

    _____ g. The 0-vector may be part of a basis.

    _____ h. If $F \leq E$ and $\alpha \in E$ is algebraic over the field F, then α^2 is algebraic over F.

    _____ i. If $F \leq E$ and $\alpha \in E$ is algebraic over the field F, then $\alpha + \alpha^2$ is algebraic over F.

    _____ j. Every vector space has a basis.

The exercises that follow deal with the further study of vector spaces. In many cases, we are asked to define for vector spaces some concept that is analogous to one we have studied for other algebraic structures. These exercises should improve our ability to recognize parallel and related situations in algebra. Any of these exercises may assume knowledge of concepts defined in the preceding exercises.

12. Let V be a vector space over a field F.

    a. Define a *subspace of the vector space V over F*.

    b. Prove that an intersection of subspaces of V is again a subspace of V over F.

13. Let V be a vector space over a field F, and let $S = \{\alpha_i \mid i \in I\}$ be a nonempty collection of vectors in V.

    a. Using Exercise 12(b), define the *subspace of V generated by S*.

    b. Prove that the vectors in the subspace of V generated by S are precisely the (finite) linear combinations of vectors in S. (Compare with Theorem 1.11.)

14. Let $V_1, \ldots, V_n$ be vector spaces over the same field F. Define the *direct sum* $V_1 \oplus \cdots \oplus V_n$ *of the vector spaces* V_i for $i = 1, \ldots, n$, and show that the direct sum is again a vector space over F.

15. Generalize Example 1 to obtain the vector space F^n of ordered n-tuples of elements of F over the field F, for any field F. What is a basis for F^n?

16. Define an *isomorphism of a vector space V over a field F with a vector space V'* *over the same field F*.

Theory

17. Prove that if V is a finite-dimensional vector space over a field F, then a subset $\{\beta_i, \beta_2, \ldots, \beta_n\}$ of V is a basis for V over F if and only if every vector in V can be expressed *uniquely* as a linear combination of the β_i.

18. Let F be any field. Consider the "system of m simultaneous linear equations in n unknowns"

$$a_{11}X_1 + a_{12}X_2 + \cdots + a_{1n}X_n = b_1,$$

$$a_{21}X_1 + a_{22}X_2 + \cdots + a_{2n}X_n = b_2,$$

$$\vdots$$

$$a_{m1}X_1 + a_{m2}X_2 + \cdots + a_{mn}X_n = b_m,$$

where $a_{ij}, b_j \in F$.

    a. Show that the "system has a solution" if and only if the vector $\beta = (b_1, \ldots, b_m)$ of F^m lies in the subspace of F^m generated by the vectors $\alpha_j = (a_{ij}, \ldots, a_{mj})$. (This result is straightforward to prove, being practically the definition of a solution, but should really be regarded as the *fundamental existence theorem for a simultaneous solution of a system of linear equations*.)

    b. From part (a), show that if $n = m$ and $\{\alpha_j \mid j = 1, \ldots, n\}$ is a basis for F^n, then the system always has a unique solution.

19. Prove that every finite-dimensional vector space V of dimension n over a field F is isomorphic to the vector space F^n of Exercise 15.

20. Let V and V' be vector spaces over the same field F. A function $\phi : V \rightarrow V'$ is a **linear transformation of V into V'** if the following conditions are satisfied for all $\alpha, \beta \in V$ and $a \in F$:

$$\phi(\alpha + \beta) = \phi(\alpha) + \phi(\beta).$$

$$\phi(a\alpha) = a(\phi(\alpha)).$$

    a. If $\{\beta_i \mid i \in I\}$ is a basis for V over F, show that a linear transformation $\phi : V \rightarrow V'$ is completely determined by the vectors $\phi(\beta_i) \in V'$.

b. Let $\{\beta_i \mid i \in I\}$ be a basis for V, and let $\{\beta_i' \mid i \in I\}$ be any set of vectors, not necessarily distinct, of V'. Show that there exists exactly one linear transformation $\phi : V \to V'$ such that $\phi(\beta_i) = \beta_i'$.

21. Let V and V' be vector spaces over the same field F, and let $\phi : V \to V'$ be a linear transformation.

   a. To what concept that we have studied for the algebraic structures of groups and rings does the concept of a *linear transformation* correspond?

   b. Define the *kernel* (or *nullspace*) of ϕ, and show that it is a subspace of V.

   c. Describe when ϕ is an isomorphism of V with V'.

22. Let V be a vector space over a field F, and let S be a subspace of V. Define the *quotient space* V/S, and show that it is a vector space over F.

23. Let V and V' be vector spaces over the same field F, and let V be finite dimensional over F. Let $\dim(V)$ be the dimension of the vector space V over F. Let $\phi : V \to V'$ be a linear transformation.

   a. Show that $\phi[V]$ is a subspace of V'.

   b. Show that $\dim(\phi[V]) = \dim[V] + \dim(\text{kernel } \phi)$. [*Hint:* Choose a convenient basis for V, using Theorem 8.7. For example, enlarge a basis for (kernel ϕ) to a basis for V.]

8.3

Algebraic Extensions

Finite Extensions

In Theorem 8.8 we saw that if E is an extension field of a field F and $\alpha \in E$ is algebraic over F, then every element of $F(\alpha)$ is algebraic over F. In studying zeros of polynomials in $F[x]$, we shall be interested almost exclusively in extensions of F containing only elements algebraic over F.

DEFINITION 8.12 (Algebraic Extension) An extension field E of a field F is an **algebraic extension of** F if every element in E is algebraic over F.

DEFINITION 8.13 (Finite Extension) If an extension field E of a field F is of finite dimension n as a vector space over F, then E is a **finite extension of degree** n **over** F. We shall let $[E:F]$ be the degree n of E over F.

We shall often use the fact that if E is a finite extension of F, then $[E:F] = 1$ if and only if $E = F$. We need only observe that by Theorem 8.7, $\{1\}$ can always be enlarged to a basis for E over F. Thus $[E:F] = 1$ if and only if $E = F(1) = F$.

Let us repeat the argument of Theorem 8.8 to show that a finite extension E of a field F must be an algebraic extension of F.

THEOREM 8.9   A finite extension field E of a field F is an algebraic extension of F.

PROOF   We must show that for $\alpha \in E$, α is algebraic over F. By Theorem 8.7 if $[E:F] = n$, then

$$1, \alpha, \ldots, \alpha^n$$

cannot be linearly independent elements, so there exist $a_i \in F$ such that

$$a_n \alpha^n + \cdots + a_1 \alpha + a_0 = 0,$$

and not all $a_i = 0$. Then $f(x) = a_n x^n + \cdots + a_1 x + a_0$ is a nonzero polynomial in $F[x]$, and $f(\alpha) = 0$. Therefore, α is algebraic over F. ◆

We cannot overemphasize the importance of our next theorem. It plays a role in field theory analogous to the role of the theorem of Lagrange in group theory. While its proof follows easily from our brief work with vector spaces, it is a tool of incredible power. We shall later be using the theorem constantly in our Galois theory arguments. Also, an elegant application of it in the section that follows shows the impossibility of performing certain geometric constructions with a straightedge and a compass. *Never underestimate a theorem that counts something.*

THEOREM 8.10   If E is a finite extension field of a field F, and K is a finite extension field of E, then K is a finite extension of F, and

$$[K:F] = [K:E][E:F].$$

PROOF   Let $\{a_i \mid i = 1, \ldots, n\}$ be a basis for E as a vector space over F, and let $\{\beta_j \mid j = 1, \ldots, m\}$ be a basis for K as a vector space over E. The theorem will be proved if we can show that the mn elements $\alpha_i \beta_j$ form a basis for K, viewed as a vector space over F.

Let γ be any element of K. Since the β_j form a basis for K over E, we have

$$\gamma = \sum_{j=1}^{m} b_j \beta_j$$

for $b_j \in E$. Since the α_i form a basis for E over F, we have

$$b_j = \sum_{i=1}^{n} a_{ij} \alpha_i$$

for $a_{ij} \in F$. Then

$$\gamma = \sum_{j=1}^{m} \left(\sum_{i=1}^{n} a_{ij} \alpha_i \right) \beta_j = \sum_{i,j} a_{ij} (\alpha_i \beta_j),$$

so the mn vectors $\alpha_i \beta_j$ span K over F.

It remains for us to show that the mn elements $\alpha_i \beta_j$ are independent over F. Suppose that $\sum_{i,j} c_{ij}(\alpha_i \beta_j) = 0$, with $c_{ij} \in F$. Then

$$\sum_{j=1}^{m} \left(\sum_{i=1}^{n} c_{ij}\alpha_i \right) \beta_j = 0,$$

and $(\sum_{i=1}^{n} c_{ij}\alpha_i) \in E$. Since the elements β_j are independent over E, we must have

$$\sum_{i=1}^{n} c_{ij}\alpha_i = 0$$

for all j. But now the α_i are independent over F, so $\sum_{i=1}^{n} c_{ij}\alpha_i = 0$ implies that $c_{ij} = 0$ for all i and j. Thus the $\alpha_i\beta_j$ not only span K over F but also are independent over F. Thus they form a basis for K over F. ◆

Note that we proved this theorem by actually exhibiting a basis. It is worth remembering that if $\{\alpha_i \,|\, i = 1, \ldots, n\}$ is a basis for E over F and $\{\beta_j \,|\, j = 1, \ldots, m\}$ is a basis for K over E, for fields $F \le E \le K$, then the set $\{\alpha_i\beta_j\}$ of mn products is a basis for K over F. Figure 8.2 gives a diagram for this situation. We shall illustrate this further in a moment.

COROLLARY I   If F_i is a field for $i = 1, \ldots, r$ and F_{i+1} is a finite extension of F_i, then F_r is a finite extension of F_1, and

$$\left[F_r : F_1 \right] = \left[F_r : F_{r-1} \right]\left[F_{r-1} : F_{r-2} \right] \cdots \left[F_2 : F_1 \right].$$

PROOF   The proof is a straightforward extension of Theorem 8.10 by induction. ◆

COROLLARY 2   If E is an extension field of F, $\alpha \in E$ is algebraic over F, and $\beta \in F(\alpha)$, then $\deg(\beta, F)$ divides $\deg(\alpha, F)$.

PROOF   By Theorem 8.8, $\deg(\alpha, F) = [F(\alpha):F]$ and $\deg(\beta, F) = [F(\beta):F]$. We have $F \le F(\beta) \le F(\alpha)$, so by Theorem 8.10 $[F(\beta):F]$ divides $[F(\alpha):F]$. ◆

The following example illustrates a type of argument one often makes using Theorem 8.10 or its corollaries.

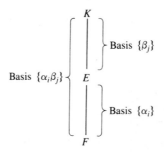

Figure 8.2

EXAMPLE 1 By Corollary 2 of Theorem 8.10, there is no element of $\mathbb{Q}(\sqrt{2})$ that is a zero of $x^3 - 2$. Note that $\deg(\sqrt{2}, \mathbb{Q}) = 2$, while a zero of $x^3 - 2$ is of degree 3 over $\mathbb{Q}$, but 3 does not divide 2. ▲

Let E be an extension field of a field F, and let α_1, α_2 be elements of E, not necessarily algebraic over F. By definition, $F(\alpha_1)$ is the smallest extension field of F in E that contains α_1. Similarly, $(F(\alpha_1))(\alpha_2)$ can be characterized as the smallest extension field of F in E containing both α_1 and α_2. We could equally well have started with α_2, so $(F(\alpha_1))(\alpha_2) = (F(\alpha_2))(\alpha_1)$. We denote this field by $F(\alpha_1, \alpha_2)$. Similarly, for $\alpha_i \in E$, $F(\alpha_1, \ldots, \alpha_n)$ is the smallest extension field of F in E containing all the α_i for $i = 1, \ldots, n$. We obtain the field $F(\alpha_1, \ldots, \alpha_n)$ from the field F by **adjoining to F the elements** α_i in E. Exercise 47 of Section 5.1 shows that, analogous to an intersection of subgroups of a group, an intersection of subfields of a field E is again a subfield of E. Thus $F(\alpha_1, \ldots, \alpha_n)$ can be characterized as the intersection of all subfields of E containing F and all the α_i for $i = 1, \ldots, n$.

EXAMPLE 2 Consider $\mathbb{Q}(\sqrt{2})$. Theorem 8.8 shows that $\{1, \sqrt{2}\}$ is a basis for $\mathbb{Q}(\sqrt{2})$ over $\mathbb{Q}$. By computing, we find $\mathrm{irr}(\sqrt{2} + \sqrt{3}, \mathbb{Q}) = x^4 - 10x^2 + 1$, so $[\mathbb{Q}(\sqrt{2} + \sqrt{3}):\mathbb{Q}] = 4$. Thus $(\sqrt{2} + \sqrt{3}) \notin \mathbb{Q}(\sqrt{2})$, so $\sqrt{3} \notin \mathbb{Q}(\sqrt{2})$. Consequently, $\{1, \sqrt{3}\}$ is a basis for $\mathbb{Q}(\sqrt{2}, \sqrt{3}) = (\mathbb{Q}(\sqrt{2}))(\sqrt{3})$ over $\mathbb{Q}(\sqrt{2})$. The proof of Theorem 8.10 (see the comment following the theorem) then shows that $\{1, \sqrt{2}, \sqrt{3}, \sqrt{6}\}$ is a basis for $\mathbb{Q}(\sqrt{2}, \sqrt{3})$ over $\mathbb{Q}$. ▲

EXAMPLE 3 Let $2^{1/3}$ be the real cube root of 2 and $2^{1/2}$ be the positive square root of 2. Then, as we saw in Example 1, $2^{1/3} \notin \mathbb{Q}(2^{1/2})$. Thus $[\mathbb{Q}(2^{1/2}, 2^{1/3}):\mathbb{Q}(2^{1/2})] = 3$. Then $\{1, 2^{1/2}\}$ is a basis for $\mathbb{Q}(2^{1/2})$ over $\mathbb{Q}$, and $\{1, 2^{1/3}, 2^{2/3}\}$ is a basis for $\mathbb{Q}(2^{1/2}, 2^{1/3})$ over $\mathbb{Q}(2^{1/2})$. Furthermore, by Theorem 8.10 (see the comment following the theorem),

$$\{1, 2^{1/2}, 2^{1/3}, 2^{5/6}, 2^{2/3}, 2^{7/6}\}$$

is a basis for $\mathbb{Q}(2^{1/2}, 2^{1/3})$ over $\mathbb{Q}$. Since $2^{7/6} = (2)2^{1/6}$, we have $2^{1/6} \in \mathbb{Q}(2^{1/2}, 2^{1/3})$. Now $2^{1/6}$ is a zero of $x^6 - 2$, which is irreducible over $\mathbb{Q}$, by Eisenstein's criterion, with $p = 2$. Thus

$$\mathbb{Q} \le \mathbb{Q}(2^{1/6}) \le \mathbb{Q}(2^{1/2}, 2^{1/3})$$

and by Theorem 8.10,

$$6 = [\mathbb{Q}(2^{1/2}, 2^{1/3}):\mathbb{Q}] = [\mathbb{Q}(2^{1/2}, 2^{1/3}):\mathbb{Q}(2^{1/6})][\mathbb{Q}(2^{1/6}):\mathbb{Q}]$$
$$= [\mathbb{Q}(2^{1/2}, 2^{1/3}):\mathbb{Q}(2^{1/6})](6).$$

Therefore, we must have

$$[\mathbb{Q}(2^{1/2}, 2^{1/3}):\mathbb{Q}(2^{1/6})] = 1,$$

so $\mathbb{Q}(2^{1/2}, 2^{1/3}) = \mathbb{Q}(2^{1/6})$, by the comment preceding Theorem 8.9. ▲

Example 3 shows that it is possible for an extension $F(\alpha_1, \ldots, \alpha_n)$ of a field F to be actually a simple extension, even though $n > 1$.

Let us characterize extensions of F of the form $F(\alpha_1, \ldots, \alpha_n)$ in the case that all the α_i are algebraic over F.

THEOREM 8.11 Let E be an algebraic extension of a field F. Then there exist a finite number of elements $\alpha_1, \ldots, \alpha_n$ in E such that $E = F(\alpha_1, \ldots, \alpha_n)$ if and only if E is a finite-dimensional vector space over F, that is, if and only if E is a finite extension of F.

PROOF Suppose that $E = F(\alpha_1, \ldots, \alpha_n)$. Since E is an algebraic extension of F, each α_i is algebraic over F, so each α_i is algebraic over every extension field of F in E. Thus $F(\alpha_1)$ is algebraic over F, and in general, $F(\alpha_1, \ldots, \alpha_j)$ is algebraic over $F(\alpha_1, \ldots, \alpha_{j-1})$ for $j = 2, \ldots, n$. Corollary 1 of Theorem 8.10 applied to the sequence of finite extensions

$$F, F(\alpha_1), F(\alpha_1, \alpha_2), \ldots, F(\alpha_1, \ldots, \alpha_n) = E$$

then shows that E is a finite extension of F.

Conversely, suppose that E is a finite algebraic extension of F. If $[E:F] = 1$, then $E = F(1) = F$, and we are done. If $E \neq F$, let $\alpha_1 \in E$, where $\alpha_1 \notin F$. Then $[F(\alpha_1):F] > 1$. If $F(\alpha_1) = E$, we are done; if not, let $\alpha_2 \in E$, where $\alpha_2 \notin F(\alpha_1)$. Continuing this process, we see from Theorem 8.10 that since $[E:F]$ is finite, we must arrive at α_n such that

$$F(\alpha_1, \ldots, \alpha_n) = E. \quad \blacklozenge$$

Algebraically Closed Fields and Algebraic Closures

We have not yet observed that if E is an extension of a field F and $\alpha, \beta \in E$ are algebraic over F, then so are $\alpha + \beta$, $\alpha\beta$, $\alpha - \beta$, and α/β, if $\beta \neq 0$. This follows from Theorem 8.11 and is also included in the following theorem.

THEOREM 8.12 Let E be an extension field of F. Then

$$\bar{F}_E = \{\alpha \in E \mid \alpha \text{ is algebraic over } F\}$$

is a subfield of E, the **algebraic closure of F in E.**

PROOF Let $\alpha, \beta \in \bar{F}_E$. Then Theorem 8.11 shows that $F(\alpha, \beta)$ is a finite extension of F, and by Theorem 8.9 every element of $F(\alpha, \beta)$ is algebraic over F, that is, $F(\alpha, \beta) \subseteq \bar{F}_E$. Thus $\bar{F}_E$ contains $\alpha + \beta$, $\alpha\beta$, $\alpha - \beta$, and also contains α/β for $\beta \neq 0$, so $\bar{F}_E$ is a subfield of E. $\quad \blacklozenge$

COROLLARY The set of all algebraic numbers forms a field.

PROOF Proof of this corollary is immediate from Theorem 8.12, because the set of all algebraic numbers is the algebraic closure of $\mathbb{Q}$ in $\mathbb{C}$. ♦

It is well known that the complex numbers have the property that every nonconstant polynomial in $\mathbb{C}[x]$ has a zero in $\mathbb{C}$. This is known as the *Fundamental Theorem of Algebra.* An analytic proof of this theorem is given later in this chapter. We now give a definition generalizing this important concept to other fields.

DEFINITION 8.14 (Algebraically Closed) A field F is **algebraically closed** if every nonconstant polynomial in $F[x]$ has a zero in F.

The next theorem shows that the concept of a field being algebraically closed can also be defined in terms of factorization of polynomials over the field.

THEOREM 8.13 A field F is algebraically closed if and only if every nonconstant polynomial in $F[x]$ factors in $F[x]$ into linear factors.

PROOF Let F be algebraically closed, and let $f(x)$ be a nonconstant polynomial in $F[x]$. Then $f(x)$ has a zero $a \in F$. By Corollary 1 of Theorem 5.18, $x - a$ is a factor of $f(x)$, so $f(x) = (x - a)g(x)$. Then if $g(x)$ is nonconstant, it has a zero $b \in F$, and $f(x) = (x - a)(x - b)h(x)$. Continuing, we get a factorization of $f(x)$ in $F[x]$ into linear factors.

Conversely, suppose that every nonconstant polynomial of $F[x]$ has a factorization into linear factors. If $ax - b$ is a linear factor of $f(x)$, then b/a is a zero of $f(x)$. Thus F is algebraically closed. ♦

COROLLARY An algebraically closed field F has no proper algebraic extensions, that is, no algebraic extensions E with $F < E$.

PROOF Let E be an algebraic extension of F, so $F \le E$. Then if $\alpha \in E$, we have $\text{irr}(\alpha, F) = x - \alpha$, by Theorem 8.13, since F is algebraically closed. Thus $\alpha \in F$, and we must have $F = E$. ♦

In a moment we shall show that just as there exists an algebraically closed extension $\mathbb{C}$ of the real numbers $\mathbb{R}$, for any field F there exists similarly an algebraic extension $\bar{F}$ of F, with the property that $\bar{F}$ is algebraically closed. Section 9.2 will show that such an extension $\bar{F}$ is unique, up to isomorphism, of course. Naively, to find $\bar{F}$ we proceed as follows. If not every polynomial $f(x)$ in $F[x]$ has a zero, then adjoin a zero α of such an $f(x)$ to F, thus obtaining the field $F(\alpha)$. *Theorem 8.1, Kronecker's theorem, is strongly used here, of course.* If $F(\alpha)$ is still not algebraically closed, then continue the process further. The trouble is that, contrary to the situation for the algebraic closure $\mathbb{C}$ of $\mathbb{R}$, we may have to do this a (possibly large) infinite number of times. It can be shown (see Exercises 27 and 30) that $\bar{\mathbb{Q}}$ is isomorphic to the field of all algebraic numbers, and that we cannot obtain $\bar{\mathbb{Q}}$ from $\mathbb{Q}$ by adjoining a finite number of algebraic numbers. We shall have to first discuss some set-theoretic machinery, *Zorn's lemma,* in order to be able to handle such a situation. This machinery is a bit complex, so we are putting the proof under a separate heading. The existence theorem for $\bar{F}$ is very important, and we state it here so that we will know this fact, even if we do not study the proof.

THEOREM 8.14   Every field F has an **algebraic closure**, that is, an algebraic extension $\bar{F}$ that is algebraically closed.

It is well known that $\mathbb{C}$ is an algebraically closed field. We recall an analytic proof for the student who has had a course in functions of a complex variable. There are algebraic proofs, but they are much longer.

THEOREM 8.15 (Fundamental Theorem of Algebra)   The field $\mathbb{C}$ of complex numbers is an algebraically closed field.

PROOF   Let the polynomial $f(z) \in \mathbb{C}[z]$ have no zero in $\mathbb{C}$. Then $1/f(z)$ gives an entire function; that is, $1/f$ is analytic everywhere. Also if $f \notin \mathbb{C}$, $\lim_{|c|\to\infty} |f(c)| = \infty$, so $\lim_{|c|\to\infty} |1/f(c)| = 0$. Thus $1/f$ must be bounded in the plane. Hence by Liouville's theorem of complex function theory, $1/f$ is constant, and thus f is constant. Therefore, a nonconstant polynomial in $\mathbb{C}[z]$ must have a zero in $\mathbb{C}$, so $\mathbb{C}$ is algebraically closed. ◆

Proof of the Existence of an Algebraic Closure

We shall prove that every field has an algebraic extension that is algebraically closed. Mathematics students should have the opportunity to see some proof involving the *Axiom of Choice* by the time they finish college. This is a natural place for such a proof. We shall use an equivalent form, *Zorn's lemma,* of the Axiom of Choice. To state Zorn's lemma, we have to give a set-theoretic definition.

DEFINITION 8.15 (Partial Ordering)   A **partial ordering of a set** S is given by a relation $\leq$ defined for certain ordered pairs of elements of S such that the following conditions are satisfied:

1. $a \leq a$ for all $a \in S$ (**reflexive law**).
2. If $a \leq b$ and $b \leq a$, then $a = b$ (**antisymmetric law**).
3. If $a \leq b$ and $b \leq c$, then $a \leq c$ (**transitive law**).

In a *partially* ordered set, not every two elements need by **comparable**; that is, for $a, b \in S$, we need not have either $a \leq b$ or $b \leq a$. As usual, $a < b$ denotes $a \leq b$ but $a \neq b$.

A subset T of a partially ordered set S is a **chain** if every two elements a and b in T are comparable, that is, either $a \leq b$ or $b \leq a$ (or both). An element $u \in S$ is an **upper bound for a subset** A of partially ordered set S if $a \leq u$ for all $a \in A$. Finally, an element m of a partially ordered set S is **maximal** if there is no $s \in S$ such that $m < s$.

EXAMPLE 4   The collection of all subsets of a set forms a partially ordered set under the relation $\leq$ given by $\subseteq$. For example, if the whole set is $\mathbb{R}$, we have $\mathbb{Z} \subseteq \mathbb{Q}$. Note, however, that for $\mathbb{Z}$ and $\mathbb{Q}^+$, neither $\mathbb{Z} \subseteq \mathbb{Q}^+$ nor $\mathbb{Q}^+ \subseteq \mathbb{Z}$. ▲

ZORN'S LEMMA   If S is a partially ordered set such that every chain in S has an upper bound in S, then S has at least one maximal element.

There is no question of *proving* Zorn's lemma. The lemma is equivalent to the Axiom of Choice. Thus we are really taking Zorn's lemma here as an *axiom* for our set theory. Refer to the literature for a statement of the Axiom of Choice and a proof of its equivalence to Zorn's lemma.

◆   **H I S T O R I C A L   N O T E**   ◆

The Axiom of Choice, although used implicitly in the 1870s and 1880s, was first stated explicitly by Ernst Zermelo in 1904 in connection with his proof of the well-ordering theorem, the result that for any set A, there exists an order-relation $<$ such that every nonempty subset B of A contains a least element. Zermelo's Axiom of Choice asserted that, given any set M and the set S of all subsets of M, there always exists a "choice" function, a function $f : S \rightarrow M$ such that $f(M') \in M$ for every M' in S. Zermelo noted, in fact, that "this logical principle cannot . . . be reduced to a still simpler one, but it is applied without hesitation everywhere in mathematical deduction." A few years later he included this axiom in his collection of axioms for set theory, a collection which was slightly modified in 1930 into what is now called Zermelo–Fraenkel set theory, the axiom system generally used today as a basis of that theory.

Zorn's lemma was introduced by Max Zorn in 1935. Although he realized that it was equivalent to the well-ordering theorem (itself equivalent to the Axiom of Choice), he claimed that his lemma was more natural to use in algebra because the well-ordering theorem was somehow a "transcendental" principle. Other mathematicians soon agreed with his reasoning. The lemma appeared in 1939 in the first volume of Nicolas Bourbaki's *Eléments de mathématique: Les structures fondamentales de l'analyse*. It was used consistently in that work and quickly became an essential part of the mathematician's toolbox.

Zorn's lemma is often useful when we want to show the existence of a largest or maximal structure of some kind. If a field F has an algebraic extension $\bar{F}$ that is algebraically closed, then $\bar{F}$ will certainly be a maximal algebraic extension of F, for since $\bar{F}$ is algebraically closed, it can have no proper algebraic extensions.

The idea of our proof of Theorem 8.14 is very simple. Given a field F, we shall first describe a class of algebraic extensions of F that is so large that it must contain (up to isomorphism) any conceivable algebraic extension of F. We then define a partial ordering, the ordinary subfield ordering, on this class, and show that the hypotheses of Zorn's lemma are satisfied. By Zorn's lemma, there will exist a maximal algebraic extension $\bar{F}$ of F in this class. We shall then argue that, as a maximal element, this extension $\bar{F}$ can have no proper algebraic extensions, so it must be algebraically closed.

Our proof differs a bit from the one found in many texts. We like it because it uses no algebra other than that derived from Theorems 8.1 and 8.10. Thus it throws into sharp relief the tremendous strength of both Kronecker's theorem and Zorn's lemma. The proof looks long, but only because we are writing out every little step. To the professional mathematician, the construction of the proof from the information in the preceding paragraph is a routine matter. This proof was suggested to the author during his graduate student days by a fellow graduate student, Norman Shapiro, who also had a strong preference for it.

We are now ready to carry out our proof of Theorem 8.14, which we restate here.

THEOREM Every field F has an algebraic closure $\bar{F}$.

PROOF   It can be shown in set theory that given any set, there exists a set with *strictly more* elements. Suppose we form a set

$$A = \{\omega_{f_i} \mid f \in F[x]; i = 0, \ldots, (\text{degree } f)\}$$

that has an element for every possible zero of any $f(x) \in F[x]$. Let Ω be a set with strictly more elements than A. By forming $\Omega \cup F$ if necessary, we can assume $F \subset \Omega$. Consider all possible fields that are algebraic extensions of F and that, as sets, consist of elements of Ω. One such algebraic extension is F itself. If E is any extension field of F, and if $\gamma \in E$ is a zero of $f(x) \in F[x]$ for $\gamma \notin F$ and $\deg(\gamma, F) = n$, then renaming γ by ω for $\omega \in \Omega$ and $\omega \notin F$, and renaming elements $a_0 + a_1\gamma + \cdots + a_{n-1}\gamma^{n-1}$ of $F(\gamma)$ by distinct elements of Ω as the a_i range over F, we can consider our renamed $F(\gamma)$ to be an algebraic extension field $F(\omega)$ of F, with $F(\omega) \subset \Omega$ and $f(\omega) = 0$. The set Ω has enough elements to form $F(\omega)$, since Ω has more than enough elements to provide n different zeros for each element of each degree n in any subset of $F[x]$.

All algebraic extension fields E_j of F, with $E_j \subseteq \Omega$, form a set

$$S = \{E_j \mid j \in J\}$$

that is partially ordered under our usual subfield inclusion $\leq$. One element of S is F itself. The preceding paragraph shows that if F is far away from being algebraically closed, there will be many fields E_j in S.

Let $T = \{E_{j_k}\}$ be a chain in S, and let $W = \bigcup_k E_{j_k}$. We now make W into a field. Let $\alpha, \beta \in W$. Then there exist $E_{j_1}, E_{j_2} \in S$, with $\alpha \in E_{j_1}$ and $\beta \in E_{j_2}$. Since T is a chain, one of the fields E_{j_1} and E_{j_2} is a subfield of the other, say $E_{j_1} \leq E_{j_2}$. Then $\alpha, \beta \in E_{j_2}$, and we use the field operations of E_{j_2} to *define* the sum of α and β in W as $(\alpha + \beta) \in E_{j_2}$ and, likewise, the product as $(\alpha\beta) \in E_{j_2}$. These operations are well defined in W; they are independent of our choice of E_{j_2}, since if $\alpha, \beta \in E_{j_3}$ also, for E_{j_3} in T, then one of the fields E_{j_2} and E_{j_3} is a subfield of the other, since T is a chain. Thus we have operations of addition and multiplication defined on W.

All the field axioms for W under these operations now follow from the fact that these operations were defined in terms of addition and multiplication in fields. Thus, for example, $1 \in F$ serves as multiplicative identity in W, since for $\alpha \in W$, if $1, \alpha \in E_{j_1}$, then we have $1\alpha = \alpha$ in E_{j_1}, so $1\alpha = \alpha$ in W, by definition of multiplication in W. Also, as further illustration, to check the distributive laws, let $\alpha, \beta, \gamma \in W$. Since T is a chain, we can find one field in T containing all three elements α, β, and γ, and in this field the distributive laws for α, β, and γ hold. Thus they hold in W. Therefore, we can view W as a field, and by construction, $E_{j_k} \leq W$ for every $E_{j_k} \in T$.

If we can show that W is algebraic over F, then $W \in S$ will be an upper bound for T. But if $\alpha \in W$, then $\alpha \in E_{j_1}$ for some E_{j_1} in T, so α is algebraic over F. Hence W is an algebraic extension of F and is an upper bound for T.

The hypotheses of Zorn's lemma are thus fulfilled, so there is a maximal element $\bar{F}$ of S. We claim that $\bar{F}$ is algebraically closed. Let $f(x) \in \bar{F}[x]$, where $f(x) \notin \bar{F}$. Suppose that $f(x)$ has no zero in $\bar{F}$. Since Ω has many more elements than $\bar{F}$ has, we can take $\omega \in \Omega$, where $\omega \notin \bar{F}$, and form a field $\bar{F}(\omega) \subseteq \Omega$, with ω a zero of $f(x)$, as we saw in the first paragraph of this proof. Let β be in $\bar{F}(\omega)$. Then by Theorem 8.8, β is a zero of a polynomial

$$g(x) = \alpha_0 + \alpha_1 x + \cdots + \alpha_n x^n$$

in $\bar{F}[x]$, with $\alpha_i \in \bar{F}$, and hence α_i algebraic over F. Then by Theorem 8.11, $F(\alpha_0, \ldots, \alpha_n)$ is a finite extension of F, and since β is algebraic over $F(\alpha_0, \ldots, \alpha_n)$, we also see that $F(\alpha_0, \ldots, \alpha_n, \beta)$ is a finite extension over $F(\alpha_0, \ldots, \alpha_n)$. Theorem 8.10 then shows that $F(\alpha_0, \ldots, \alpha_n, \beta)$ is a finite extension of F, so by Theorem 8.9, β is algebraic over F. Hence $\bar{F}(\omega) \in S$ and $\bar{F} < \bar{F}(\omega)$, which contradicts the choice of $\bar{F}$ as maximal in S. Thus $f(x)$ must have had a zero in $\bar{F}$, so $\bar{F}$ is algebraically closed. ◆

The mechanics of the preceding proof are routine to the professional mathematician. Since it may be the first proof that we have ever seen using Zorn's lemma, we wrote the proof out in detail.

Exercises 8.3

Computations

In Exercises 1 through 13, find the degree and a basis for the given field extension. Be prepared to justify your answers.

1. $\mathbb{Q}(\sqrt{2})$ over $\mathbb{Q}$

2. $\mathbb{Q}(\sqrt{2}, \sqrt{3})$ over $\mathbb{Q}$

3. $\mathbb{Q}(\sqrt{2}, \sqrt{3}, \sqrt{5})$ over $\mathbb{Q}$

4. $\mathbb{Q}(\sqrt[3]{2}, \sqrt{3})$ over $\mathbb{Q}$

5. $\mathbb{Q}(\sqrt{2}, \sqrt[3]{2})$ over $\mathbb{Q}$

6. $\mathbb{Q}(\sqrt{2} + \sqrt{3})$ over $\mathbb{Q}$

7. $\mathbb{Q}(\sqrt{2}\sqrt{3})$ over $\mathbb{Q}$

8. $\mathbb{Q}(\sqrt{2}, \sqrt[3]{5})$ over $\mathbb{Q}$

9. $\mathbb{Q}(\sqrt[3]{2}, \sqrt[3]{6}, \sqrt[3]{24})$ over $\mathbb{Q}$

10. $\mathbb{Q}(\sqrt{2}, \sqrt{6})$ over $\mathbb{Q}(\sqrt{3})$

11. $\mathbb{Q}(\sqrt{2} + \sqrt{3})$ over $\mathbb{Q}(\sqrt{3})$

12. $\mathbb{Q}(\sqrt{2}, \sqrt{3})$ over $\mathbb{Q}(\sqrt{2} + \sqrt{3})$

13. $\mathbb{Q}(\sqrt{2}, \sqrt{6} + \sqrt{10})$ over $\mathbb{Q}(\sqrt{3} + \sqrt{5})$

Concepts

14. Show by an example that for a proper extension field E of a field F, the algebraic closure of F in E need not be algebraically closed.

15. Mark each of the following true or false.

_____ a. Every finite extension of a field is an algebraic extension.

_____ b. Every algebraic extension of a field is a finite extension.

_____ c. The top field of a finite tower of finite extensions of fields is a finite extension of the bottom field.

_____ d. $\mathbb{R}$ is algebraically closed.

_____ e. $\mathbb{Q}$ is its own algebraic closure in $\mathbb{R}$, that is $\mathbb{Q}$ is **algebraically closed in** $\mathbb{R}$.

_____ f. $\mathbb{C}$ is algebraically closed in $\mathbb{C}(x)$, where x is an indeterminate.

_____ g. $\mathbb{C}(x)$ is algebraically closed, where x is an indeterminate.

_____ h. The field $\mathbb{C}(x)$ has no algebraic closure, since $\mathbb{C}$ already contains all algebraic numbers.

_____ i. An algebraically closed field must be of characteristic 0.

_____ j. If E is an algebraically closed extension field of F, then E is an algebraic extension of F.

Theory

16. Let $(a + bi) \in \mathbb{C}$ for $a, b \in \mathbb{R}$ with $b \neq 0$. Show that $\mathbb{C} = \mathbb{R}(a + bi)$.

17. Show that if E is a finite extension of a field F and $[E:F]$ is a prime number, then E is a simple extension of F and, indeed, $E = F(\alpha)$ for every $\alpha \in E$ not in F.

18. Prove that $x^2 - 3$ is irreducible over $\mathbb{Q}(\sqrt[3]{2})$.

19. What degree field extensions can we obtain by successively adjoining to a field F a square root of an element of F not a square in F, then square root of some nonsquare in this new field, and so on? Argue from this that a zero of

$x^{14} - 3x^2 + 12$ over $\mathbb{Q}$ can never be expressed as a rational function of square roots of rational functions of square roots, and so on of elements of $\mathbb{Q}$.

20. Let E be a finite extension field of F. Let D be an integral domain such that $F \subseteq D \subseteq E$. Show that D is a field.

21. Prove in detail that $\mathbb{Q}(\sqrt{3} + \sqrt{7}) = \mathbb{Q}(\sqrt{3}, \sqrt{7})$.

22. Generalizing Exercise 21, show that if $\sqrt{a} + \sqrt{b} \neq 0$, then $\mathbb{Q}(\sqrt{a} + \sqrt{b}) = \mathbb{Q}(\sqrt{a}, \sqrt{b})$ for all a and b in $\mathbb{Q}$. [*Hint:* Compute $(\sqrt{a} + \sqrt{b})^2$.]

23. Let E be a finite extension of a field F, and let $p(x) \in F[x]$ be irreducible over F and have degree that is not a divisor of $[E:F]$. Show that $p(x)$ has no zeros in E.

24. Let E be an extension field of F. Let $\alpha \in E$ be algebraic of odd degree over F. Show that α^2 is algebraic of odd degree over F, and $F(\alpha) = F(\alpha^2)$.

25. Show that if F, E, and K are fields with $F \leq E \leq K$, then K is algebraic over F if and only if E is algebraic over F, and K is algebraic over E. (You must *not* assume the extensions are finite.)

26. Let E be an extension field of a field F. Prove that every $\alpha \in E$ that is not in the algebraic closure $\bar{F}_E$ of F in E is transcendental over $\bar{F}_E$.

27. Let E be an algebraically closed extension field of a field F. Show that the algebraic closure $\bar{F}_E$ of F in E is algebraically closed. (Applying this exercise to $\mathbb{C}$ and $\mathbb{Q}$, we see that the field of all algebraic numbers is an algebraically closed field.)

28. Show that if E is an algebraic extension of a field F and contains all zeros in $\bar{F}$ of every $f(x) \in F[x]$, then E is an algebraically closed field.

29. Show that no finite field of odd characteristic is algebraically closed. (Actually, no finite field of characteristic 2 is algebraically closed either.) [*Hint:* By counting, show that for such a finite field F, some polynomial $x^2 - a$, for some $a \in F$, has no zero in F. See Exercise 27, Section 8.1.]

30. Prove that, as asserted in the text, the algebraic closure of $\mathbb{Q}$ in $\mathbb{C}$ is not a finite extension of $\mathbb{Q}$.

31. Argue that every finite extension field of $\mathbb{R}$ is either $\mathbb{R}$ itself or is isomorphic to $\mathbb{C}$.

32. Use Zorn's lemma to show that every proper ideal of a ring R with unity is contained in some maximal ideal.

8.4

Geometric Constructions

In this section we digress briefly to give an application demonstrating the power of Theorem 8.13. For a more detailed study of geometric constructions, you are referred to Courant and Robbins [44, Chapter III].

We are interested in what types of figures can be constructed with a compass and a straightedge in the sense of classical Euclidean plane geometry. We shall discuss the impossibility of trisecting certain angles and other classical questions.

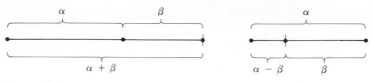

Figure 8.3

Constructible Numbers

Let us imagine that we are given only a single line segment that we shall define to be *one unit* in length. A real number α is **constructible** if we can construct a line segment of length $|\alpha|$ in a finite number of steps from this given segment of unit length by using a straightedge and a compass. Recall that using a straightedge and a compass it is possible, among other things, to erect a perpendicular to a given line at a known point on the line and to find a line passing through a given point and parallel to a given line. Our first result is the following theorem.

THEOREM 8.16   If α and β are constructible real numbers, then so are $\alpha + \beta$, $\alpha - \beta$, $\alpha\beta$, and α/β, if $\beta \neq 0$.

PROOF   We are given that α and β are constructible, so there are line segments of lengths $|\alpha|$ and $|\beta|$ available to us. For $\alpha, \beta > 0$, extend a line segment of length α with the straightedge. Start at one end of the original segment of length α, and lay off on the extension the length β with the compass. This constructs a line segment of length $\alpha + \beta$; $\alpha - \beta$ is similarly constructible (see Fig. 8.3). If α and β are not both positive, an obvious breakdown into cases according to their signs shows that $\alpha + \beta$ and $\alpha - \beta$ are still constructible.

The construction of $\alpha\beta$ is indicated in Fig. 8.4. We shall let $\overline{OA}$ be the line segment from the point O to the point A, and shall let $|OA|$ be the length of this line segment. If $\overline{OA}$ is of length $|\alpha|$, find a line l through O not containing $\overline{OA}$. Then find the points P and B on l such that $\overline{OP}$ is of length 1 and $\overline{OB}$ is of length $|\beta|$. Draw $\overline{PA}$ and construct l' through B, parallel to $\overline{PA}$ and intersecting $\overline{OA}$ extended at Q. By similar triangles, we

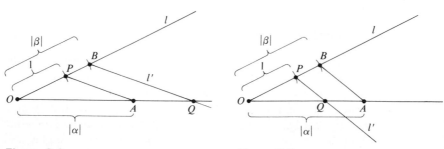

Figure 8.4   **Figure 8.5**

have

$$\frac{1}{|\alpha|} = \frac{|\beta|}{|\overline{OQ}|},$$

so $\overline{OQ}$ is of length $|\alpha\beta|$.

Finally, Fig. 8.5 shows that α/β is constructible if $\beta \neq 0$. Let $\overline{OA}$ be of length $|\alpha|$, and find l through O not containing $\overline{OA}$. Then find B and P on l such that $\overline{OB}$ is of length $|\beta|$ and $\overline{OP}$ is of length 1. Draw $\overline{BA}$ and construct l' through P, parallel to $\overline{BA}$, and intersecting $\overline{OA}$ at Q. Again by similar triangles, we have

$$\frac{|\overline{OQ}|}{1} = \frac{|\alpha|}{|\beta|},$$

so $\overline{OQ}$ is of length $|\alpha/\beta|$. ◆

COROLLARY   The set of all constructible real numbers forms a subfield F of the field of real numbers.

PROOF   Proof of this corollary is immediate from Theorem 8.16. ◆

Thus the field F of all constructible real numbers contains $\mathbb{Q}$, the field of rational numbers, since $\mathbb{Q}$ is the smallest subfield of $\mathbb{R}$.

From now on, we proceed analytically. We can construct any rational number. Regarding our given segment

0_____1

of length 1 as the basic unit on an x-axis, we can locate any point (q_1, q_2) in the plane with both coordinates rational. Any further point in the plane that we can locate by using a compass and a straightedge can be found in one of the following three ways:

1. as an intersection of two lines, each of which passes through two known points having rational coordinates,
2. as an intersection of a line that passes through two points having rational coordinates and a circle whose center has rational coordinates and the square of whose radius is rational,
3. as an intersection of two circles whose centers have rational coordinates and the squares of whose radii are rational.

Equations of lines and circles of the type discussed in 1, 2, and 3 are of the form

$$ax + by + c = 0$$

and

$$x^2 + y^2 + dx + ey + f = 0,$$

where $a, b, c, d, e,$ and f are all in $\mathbb{Q}$. Since in Case 3 the intersection of two circles with equations

$$x^2 + y^2 + d_1x + e_1y + f_1 = 0$$

and

$$x^2 + y^2 + d_2x + e_2y + f_2 = 0$$

is the same as the intersection of the first circle having equation

$$x^2 + y^2 + d_1x + e_1y + f_1 = 0,$$

and the line (the common chord) having equation

$$(d_1 - d_2)x + (e_1 - e_2)y + f_1 - f_2 = 0,$$

we see that Case 3 can be reduced to Case 2. For Case 1, a simultaneous solution of two linear equations with rational coefficients can only lead to rational values of x and y, giving us no new points. However, finding a simultaneous solution of a linear equation with rational coefficients and a quadratic equation with rational coefficients, as in Case 2, leads, upon substitution, to a quadratic equation. Such an equation, when solved by the quadratic formula, may have solutions involving square roots of numbers that are not squares in $\mathbb{Q}$.

In the preceding argument, nothing was really used involving $\mathbb{Q}$ except field axioms. If H is the smallest field containing those real numbers constructed so far, the argument shows that the "next new number" constructed lies in a field $H(\sqrt{\alpha})$ for some $\alpha \in H$, where $\alpha > 0$. We have proved half of our next theorem.

THEOREM 8.17 The field F of constructible real numbers consists precisely of all real numbers that we can obtain from $\mathbb{Q}$ by taking square roots of positive numbers a finite number of times and applying a finite number of field operations.

PROOF We have shown that F can contain no numbers except those we obtain from $\mathbb{Q}$ by taking a finite number of square roots of positive numbers and applying a finite number of field operations. However, if $\alpha > 0$ is constructible, then Fig. 8.6 shows that $\sqrt{\alpha}$ is constructible. Let $\overline{OA}$ have length α, and find P on $\overline{OA}$ extended so that $\overline{OP}$ has length 1. Find the midpoint of $\overline{PA}$ and draw a semicircle with $\overline{PA}$ as diameter. Erect a perpendicular to $\overline{PA}$ at O, intersecting the semicircle at Q. Then the

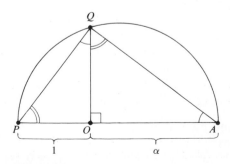

Figure 8.6

triangles OPQ and OQA are similar, so

$$\frac{|\overline{OQ}|}{|\overline{OA}|} = \frac{|\overline{OP}|}{|\overline{OQ}|},$$

and $|\overline{OQ}|^2 = 1\alpha = \alpha$. Thus $\overline{OQ}$ is of length $\sqrt{\alpha}$. Therefore square roots of constructible numbers are constructible.

Theorem 8.16 showed that field operations are possible by construction. ◆

COROLLARY If γ is constructible and $\gamma \notin \mathbb{Q}$, then there is a finite sequence of real numbers $\alpha_1, \ldots, \alpha_n = \gamma$ such that $\mathbb{Q}(\alpha_1, \ldots, \alpha_i)$ is an extension of $\mathbb{Q}(\alpha_1, \ldots, \alpha_{i-1})$ of degree 2. In particular, $[\mathbb{Q}(\gamma):\mathbb{Q}] = 2^r$ for some integer $r \geq 0$.

PROOF The existence of the α_i is immediate from Theorem 8.17. Then

$$2^n = [\mathbb{Q}(\alpha_1, \ldots, \alpha_n):\mathbb{Q}]$$
$$= [\mathbb{Q}(\alpha_1, \ldots, \alpha_n):\mathbb{Q}(\gamma)][\mathbb{Q}(\gamma):\mathbb{Q}],$$

by Theorem 8.10, which completes the proof. ◆

The Impossibility of Certain Constructions

We can now show the impossibility of certain geometric constructions.

THEOREM 8.18 *Doubling the cube is impossible,* that is, given a side of a cube, it is not always possible to construct with a straightedge and a compass the side of a cube that has double the volume of the original cube.

PROOF Let the given cube have a side of length 1, and hence a volume of 1. The cube being sought would have to have a volume of 2, and hence a side of length $\sqrt[3]{2}$. But $\sqrt[3]{2}$ is a zero of irreducible $x^3 - 2$ over $\mathbb{Q}$, so

$$[\mathbb{Q}(\sqrt[3]{2}):\mathbb{Q}] = 3.$$

The corollary of Theorem 8.17 shows that to double this cube of volume 1, we would need to have $3 = 2^r$ for some integer r, but no such r exists. ◆

THEOREM 8.19 *Squaring the circle is impossible*; that is, given a circle, it is not always possible to construct with a straightedge and a compass a square having area equal to the area of the given circle.

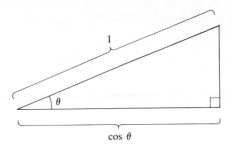

Figure 8.7

PROOF   Let the given circle have a radius of 1, and hence an area of π. We would need to construct a square of side $\sqrt{\pi}$. But π is transcendental over $\mathbb{Q}$, so $\sqrt{\pi}$ is transcendental over $\mathbb{Q}$ also. ♦

THEOREM 8.20   *Trisecting the angle is impossible*; that is, there exists an angle that cannot be trisected with a straightedge and a compass.

PROOF   Figure 8.7 indicates that the angle θ can be constructed if and only if a segment of length $|\cos \theta|$ can be constructed. Now $60°$ is a constructible angle, and we shall show that it cannot be trisected. Note that

$$\cos 3\theta = \cos (2\theta + \theta)$$

$$= \cos 2\theta \cos \theta - \sin 2\theta \sin \theta$$

$$= (2 \cos^2\theta - 1) \cos \theta - 2 \sin \theta \cos \theta \sin \theta$$

$$= (2 \cos^2\theta - 1) \cos \theta - 2 \cos \theta(1 - \cos^2\theta)$$

$$= 4 \cos^3 \theta - 3 \cos \theta.$$

Let $\theta = 20°$, so that $\cos 3\theta = \frac{1}{2}$, and let $\alpha = \cos 20°$. From the identity $4 \cos^3 \theta - 3 \cos \theta = \cos 3\theta$, we see that

$$4\alpha^3 - 3\alpha = \tfrac{1}{2}.$$

Thus α is a zero of $8x^3 - 6x - 1$. This polynomial is irreducible in $\mathbb{Q}[x]$, since, by Theorem 5.20, it is enough to show that it does not factor in $\mathbb{Z}[x]$. But a factorization in $\mathbb{Z}[x]$ would entail a linear factor of the form $(8x \pm 1)$, $(4x \pm 1)$, $(2x \pm 1)$, or $(x \pm 1)$. We can quickly check that none of the numbers $\pm\frac{1}{8}$, $\pm\frac{1}{4}$, $\pm\frac{1}{2}$, and ± 1 is a zero of $8x^3 - 6x - 1$. Thus

$$[\mathbb{Q}(\alpha):\mathbb{Q}] = 3,$$

so by the corollary of Theorem 8.17, α is not constructible. Hence $60°$ cannot be trisected. ♦

Greek mathematicians as far back as the fourth century B.C. had tried without success to find geometric constructions using straightedge and compass to trisect the angle, double the cube, and square the circle. Although they were never able to prove that such constructions were impossible, they did manage to construct the solutions to these problems using other tools, including the conic sections.

It was Carl Gauss in the early nineteenth century who made a detailed study of constructibility in connection with his solution of cyclotomic equations, the equations of the form $x^n - 1 = 0$ with n prime whose roots form the vertices of a regular n-gon. (See Section 9.8.) He showed that although all such equations are solvable using radicals, if $n - 1$ is not a power of 2 then the solutions must involve roots higher than the second. In fact, Gauss asserted that anyone who attempted to find a geometric construction for an n-gon where $n - 1$ is not a power of 2 would "spend his time uselessly." Interestingly, Gauss did not prove the assertion that such constructions were impossible. That was accomplished in 1837 by Pierre Wantzel (1814–1848), who in fact proved the corollary to Theorem 8.17 and also demonstrated Theorems 8.18 and 8.20. The proof of Theorem 8.19, on the other hand, requires a proof that π is transcendental, a result finally achieved in 1882 by Ferdinand Lindemann (1852–1939).

Note that the regular n-gon is constructible for $n \geq 3$ if and only if the angle $2\pi/n$ is constructible, which is the case if and only if a line segment of length $\cos(2\pi/n)$ is constructible. We shall return to the study of the constructibility of certain regular n-gons in Section 9.8.

Exercises 8.4

Concepts

1. Mark each of the following true or false.
   - _____ a. It is impossible to double any cube of constructible edge by compass and straightedge constructions.
   - _____ b. It is impossible to double every cube of constructible edge by compass and straightedge constructions.
   - _____ c. It is impossible to square any circle of constructible radius by straightedge and compass constructions.
   - _____ d. No constructible angle can be trisected by straightedge and compass constructions.

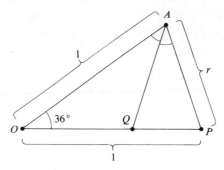

Figure 8.8

_____ e. Every constructible number is of degree 2^r over $\mathbb{Q}$ for some integer $r \geq 0$.

_____ f. We have shown that every real number of degree 2^r over $\mathbb{Q}$ for some integer $r \geq 0$ is constructible.

_____ g. The fact that $\mathbb{Z}$ is a UFD was used strongly at the conclusion of Theorems 8.18 and 8.20.

_____ h. Counting arguments are exceedingly powerful mathematical tools.

_____ i. We can find any constructible number in a finite number of steps by starting with a given segment of unit length and using a straightedge and a compass.

_____ j. We can find the totality of all constructible numbers in a finite number of steps by starting with a given segment of unit length and using a straightedge and a compass.

2. Using Theorem 8.20, show that the regular 9-gon is not constructible.

3. Show _algebraically_ that it is possible to construct an angle of 30°.

4. Referring to Fig. 8.8, where $\overline{AQ}$ bisects angle OAP, show that the regular 10-gon is constructible (and therefore that the regular pentagon is also). [_Hint:_ Triangle OAP is similar to triangle APQ. Show algebraically that r is constructible.]

In Exercises 5 through 8, use the results of Exercise 4 where needed to show that the statement is true.

5. The regular 20-gon is constructible.

6. The regular 30-gon is constructible.

7. The angle 72° can be trisected.

8. The regular 15-gon can be constructed.

8.5

Finite Fields

The purpose of this section is to determine the structure of all finite fields. We shall show that for every prime p and positive integer n, there is exactly one finite field (up to isomorphism) of order p^n. This field $GF(p^n)$ is usually referred to as the **Galois field of order** p^n. We shall be using quite a bit of our material on cyclic groups. The proofs are simple and elegant.

The Structure of a Finite Field

We now show that all finite fields must have prime-power order.

THEOREM 8.21   Let E be a finite extension of degree n over a finite field F. If F has q elements, then E has q^n elements.

PROOF   Let $\{\alpha_1, \ldots, \alpha_n\}$ be a basis for E as a vector space over F. Then every $\beta \in E$ can be *uniquely* written in the form

$$\beta = b_1\alpha_1 + \cdots + b_n\alpha_n$$

for $b_i \in F$. Since each b_i may be any of the q elements of F, the total number of such distinct linear combinations of the α_i is q^n. ◆

COROLLARY   If E is a finite field of characteristic p, then E contains exactly p^n elements for some positive integer n.

PROOF   Every finite field E is a finite extension of a prime field isomorphic to the field $\mathbb{Z}_p$, where p is the characteristic of E. The corollary follows at once from Theorem 8.21. ◆

We now turn to the study of the multiplicative structure of a finite field. The following theorem will show us how any finite field can be formed from the prime subfield.

THEOREM 8.22   Let E be a field of p^n elements contained in an algebraic closure $\bar{\mathbb{Z}}_p$ of $\mathbb{Z}_p$. The elements of E are precisely the zeros in $\bar{\mathbb{Z}}_p$ of the polynomial $x^{p^n} - x$ in $\mathbb{Z}_p[x]$.

PROOF   The set E^* of nonzero elements of E forms a multiplicative group of order $p^n - 1$ under the field multiplication. For $\alpha \in E^*$, the order of α in this group divides the order $p^n - 1$ of the group. Thus for $\alpha \in E^*$, we have $\alpha^{p^n - 1} = 1$, so $\alpha^{p^n} = \alpha$. Therefore, every element in E is a zero of $x^{p^n} - x$. Since $x^{p^n} - x$ can have at most p^n zeros, we see that E contains precisely the zeros of $x^{p^n} - x$ in $\bar{\mathbb{Z}}_p$. ◆

DEFINITION 8.16 (Primitive Root of Unity)   An element α of a field is an **nth root of unity** if $\alpha^n = 1$. It is a **primitive nth root of unity** if $\alpha^n = 1$ and $\alpha^m \neq 1$ for $0 < m < n$.

Thus the nonzero elements of a finite field of p^n elements are all $(p^n - 1)$th roots of unity.

Recall that in Corollary 3 of Theorem 5.18, we showed that the multiplicative group of nonzero elements of a finite field is cyclic. This is a very important fact about finite fields; it has actually been applied to algebraic coding. For the sake of completeness in this section, we now state it here as a theorem, give a corollary, and illustrate with an example.

THEOREM 8.23 The multiplicative group $\langle F^*, \cdot \rangle$ of nonzero elements of a finite field F is cyclic.

PROOF See Corollary 3 of Theorem 5.18. ◆

COROLLARY A finite extension E of a finite field F is a simple extension of F.

PROOF Let α be a generator for the cyclic group E^* of nonzero elements of E. Then $E = F(\alpha)$. ◆

EXAMPLE I Consider the finite field $\mathbb{Z}_{11}$. By Theorem 8.23, $\langle \mathbb{Z}_{11}^*, \cdot \rangle$ is cyclic. Let us try to find a generator of $\mathbb{Z}_{11}^*$ by brute force and ignorance. We

---

◆ **HISTORICAL NOTE** ◆

Although Carl F. Gauss had shown that the set of residues modulo a prime p satisfied the field properties, it was Evariste Galois (1811–1832) who first dealt with what he called "incommensurable solutions" to the congruence $F(x) \equiv 0 \pmod{p}$, where $F(x)$ is an nth degree irreducible polynomial modulo p. He noted in a paper written in 1830 that one should consider the roots of this congruence as "a variety of imaginary symbols" that one can use in calculation just as one uses $\sqrt{-1}$. Galois then showed that if α is any solution of $F(x) \equiv 0 \pmod{p}$, the expression $a_0 + a_1\alpha + a_2\alpha^2 + \cdots + a_{n-1}\alpha^{n-1}$ takes on precisely p^n different values. Finally, he proved results equivalent to Theorem 8.22 and 8.23 of the text.

  Galois's life was brief and tragic. He showed brilliance in mathematics early on, publishing several papers before he was 20 and essentially established the basic ideas of Galois theory (Section 9.6). He was, however, active in French revolutionary politics following the July revolution of 1830. In May of 1831 he was arrested for threatening the life of King Louis-Phillipe. Though he was acquitted, he was rearrested for participating, heavily armed, in a republican demonstration on Bastille Day of that year. Two months after his release from prison the following March, he was killed in a duel, "the victim of an infamous coquette and her two dupes"; the previous night he had written a letter to a friend clarifying some of his work in the theory of equations and requesting that it be studied by other mathematicians. Not until 1846, however, were his major papers published; it is from that date that his work became influential.

start by trying 2. Since $|Z_{11}{}^*| = 10$, 2 must be an element of $Z_{11}{}^*$ of order dividing 10, that is, either 2, 5, or 10. Now

$$2^2 = 4, \qquad 2^4 = 4^2 = 5, \qquad \text{and} \qquad 2^5 = (2)(5) = 10 = -1.$$

Thus neither 2^2 nor 2^5 is 1, but, of course, $2^{10} = 1$, so 2 is a generator of $Z_{11}{}^*$, that is, 2 is a primitive 10th root of unity in Z_{11}. We were lucky.

By the theory of cyclic groups, all the generators of $Z_{11}{}^*$, that is, all the primitive 10th roots of unity in Z_{11}, are then of the form 2^n, where n is relatively prime to 10. These elements are

$$2^1 = 2, \qquad 2^3 = 8,$$
$$2^7 = 7, \qquad 2^9 = 6.$$

The primitive 5th roots of unity in Z_{11} are of the form 2^m, where the gcd of m and 10 is 2, that is,

$$2^2 = 4, \qquad 2^4 = 5,$$
$$2^6 = 9, \qquad 2^8 = 3.$$

The primitive square root of unity in Z_{11} is $2^5 = 10 = -1$. ▲

The Existence of GF(p^n)

We turn now to the question of the existence of a finite field of order p^r for every prime power p^r, $r > 0$. We need the following lemma.

LEMMA 8.2 If F is a finite field of characteristic p with algebraic closure $\bar{F}$, then $x^{p^n} - x$ has p^n distinct zeros in $\bar{F}$.

PROOF We show that $x^{p^n} - x$ has no zeros of multiplicity greater than 1 in $\bar{F}$. Since we have not introduced an algebraic theory of derivatives (see Exercises 13–20 of Section 9.4), this elegant technique is not available to us, so we proceed by long division. Observe that 0 is a zero of $x^{p^n} - x$ of multiplicity 1. Suppose $\alpha \neq 0$ is a zero of $x^{p^n} - x$, and hence is a zero of $f(x) = x^{p^n-1} - 1$. Then $x - \alpha$ is a factor of $f(x)$ in $\bar{F}[x]$, and by long division, we find that

$$\frac{f(x)}{(x - \alpha)} = g(x)$$

$$= x^{p^n-2} + \alpha x^{p^n-3} + \alpha^2 x^{p^n-4} + \cdots + \alpha^{p^n-3}x + \alpha^{p^n-2}.$$

Now $g(x)$ has $p^n - 1$ summands, and in $g(\alpha)$, each summand is

$$\alpha^{p^n - 2} = \frac{\alpha^{p^n - 1}}{\alpha} = \frac{1}{\alpha}.$$

Thus

$$g(\alpha) = [(p^n - 1) \cdot 1]\frac{1}{\alpha} = -\frac{1}{\alpha}.$$

since we are in a field of characteristic p. Therefore, $g(\alpha) \neq 0$, so α is a zero of $f(x)$ of multiplicity 1. ◆

THEOREM 8.24 A finite field $GF(p^n)$ of p^n elements exists for every prime power p^n.

PROOF Let $\bar{\mathbb{Z}}_p$ be an algebraic closure of $\mathbb{Z}_p$, and let K be the subset of $\bar{\mathbb{Z}}_p$ consisting of all zeros of $x^{p^n} - x$ in $\bar{\mathbb{Z}}_p$. Then for $\alpha, \beta \in K$, the equations

$$(\alpha \pm \beta)^{p^n} = \alpha^{p^n} \pm \beta^{p^n} = \alpha \pm \beta$$

and

$$(\alpha\beta)^{p^n} = \alpha^{p^n}\beta^{p^n} = \alpha\beta$$

show that K is closed under addition, subtraction, and multiplication. Now 0 and 1 are zeros of $x^{p^n} - x$. For $\alpha \neq 0$, $\alpha^{p^n} = \alpha$ implies that $(1/\alpha)^{p^n} = 1/\alpha$. Thus K is a subfield of $\bar{\mathbb{Z}}_p$ containing $\mathbb{Z}_p$. Therefore, K is the desired field of p^n elements, since Lemma 8.2 showed that $x^{p^n} - x$ has p^n distinct zeros in $\bar{\mathbb{Z}}_p$. ◆

COROLLARY If F is any finite field, then for every positive integer n, there is an irreducible polynomial in $F[x]$ of degree n.

PROOF Let F have $q = p^r$ elements, where p is the characteristic of F. By Theorem 8.24, there is a field $K \leq \bar{F}$ containing $\mathbb{Z}_p$ (up to isomorphism) and consisting precisely of the zeros of $x^{p^{rn}} - x$. Every element of F is a zero of $x^{p^r} - x$, by Theorem 8.22. Now $p^{rs} = p^r p^{r(s-1)}$. Applying this equation repeatedly to the exponents and using the fact that for $\alpha \in F$ we have $\alpha^{p^r} = \alpha$, we see that for $\alpha \in F$,

$$\alpha^{p^{rn}} = \alpha^{p^{r(n-1)}} = \alpha^{p^{r(n-2)}} = \cdots = \alpha^{p^r} = \alpha.$$

Thus $F \leq K$. Then Theorem 8.21 shows that we must have $[K:F] = n$. We have seen that K is simple over F in the corollary of Theorem 8.23, so $K = F(\beta)$ for some $\beta \in K$. Therefore, irr(β, F) must be of degree n. ◆

Finite fields have been used in algebraic coding. In an article in the *American Mathematical Monthly* 77 (1970): 249–258, Norman Levinson constructs a linear code that can correct up to three errors using a finite field of order 16.

Exercises 8.5

Computations

In Exercises 1 through 3, determine whether there exists a finite field having the given number of elements. (A calculator may be useful.)

1. 4096   **2.** 3127   **3.** 68,921

4. Find the number of primitive 8th roots of unity in GF(9).

5. Find the number of primitive 18th roots of unity in GF(19).

6. Find the number of primitive 15th roots of unity in GF(31).

7. Find the number of primitive 10th roots of unity in GF(23).

Concepts

8. Mark each of the following true or false.

    _____ a. The nonzero elements of every finite field form a cyclic group under multiplication.

    _____ b. The elements of every finite field form a cyclic group under addition.

    _____ c. The zeros in $\mathbb{C}$ of $(x^{28} - 1) \in \mathbb{Q}[x]$ form a cyclic group under multiplication.

    _____ d. There exists a finite field of 60 elements.

    _____ e. There exists a finite field of 125 elements.

    _____ f. There exists a finite field of 36 elements.

    _____ g. The complex number i is a primitive 4th root of unity.

    _____ h. There exists an irreducible polynomial of degree 58 in $\mathbb{Z}_2[x]$.

    _____ i. The nonzero elements of $\mathbb{Q}$ form a cyclic group $\mathbb{Q}^*$ under field multiplication.

    _____ j. If F is a finite field, then every isomorphism mapping F onto a subfield of an algebraic closure $\bar{F}$ of F is an automorphism of F.

Theory

9. Let $\bar{\mathbb{Z}}_2$ be an algebraic closure of $\mathbb{Z}_2$, and let $\alpha, \beta \in \bar{\mathbb{Z}}_2$ be zeros of $x^3 + x^2 + 1$ and of $x^3 + x + 1$, respectively. Using the results of this section, show that $\mathbb{Z}_2(\alpha) = \mathbb{Z}_2(\beta)$.

10. Show that every irreducible polynomial in $\mathbb{Z}_p[x]$ is a divisor of $x^{p^n} - x$ for some n.

11. Let F be a finite field of p^n elements containing the prime subfield $\mathbb{Z}_p$. Show that if $\alpha \in F$ is a generator of the cyclic group $\langle F^*, \cdot \rangle$ of nonzero elements of F, then $\deg(\alpha, \mathbb{Z}_p) = n$.

12. Show that a finite field of p^n elements has exactly one subfield of p^m elements for each divisor m of n.

13. Show that $x^{p^n} - x$ is the product of all monic irreducible polynomials in $\mathbb{Z}_p[x]$ of degree d dividing n.

14. Let p be an odd prime.

    a. Show that for $a \in \mathbb{Z}$, where $a \not\equiv 0 \pmod{p}$, the congruence $x^2 \equiv a \pmod{p}$ has a solution in $\mathbb{Z}$ if and only if $a^{(p-1)/2} \equiv 1 \pmod{p}$. [*Hint:* Formulate an equivalent statement in the finite field $\mathbb{Z}_p$, and use the theory of cyclic groups.]

b. Using part (a), determine whether or not the polynomial $x^2 - 6$ is irreducible in $\mathbb{Z}_{17}[x]$.

15. Show that two finite fields of the same order p^n are isomorphic. [*Hint:* Let $p(x) \in \mathbb{Z}_p[x]$ be irreducible of degree n. Show every field of p^n elements is isomorphic to $\mathbb{Z}_p[x]/\langle p(x)\rangle$.]

8.6

Additional Algebraic Structures†

This expository section will not be used in the remainder of the text. It is designed to give an idea of other important algebraic structures and their relation to the structures we have studied.

Groups with Operators

DEFINITION 8.17 (Group with Operators) A **group with operators** consists of a group G and a set $\mathcal{O}$, the **set of operators**, together with an operation of external multiplication of each element of G by each element of $\mathcal{O}$ on the left such that for all $\alpha, \beta \in G$ and $a \in \mathcal{O}$, the following conditions are satisfied:

1. $(a\alpha) \in G$.
2. $a(\alpha\beta) = (a\alpha)(a\beta)$.

We shall somewhat incorrectly speak of the $\mathcal{O}$-**group** G.

Note how we have followed closely the form of our definition of a vector space from Section 8.2. The operation of external multiplication is really a function $\phi: \mathcal{O} \times G \to G$, where $\phi(a, \alpha)$ is denoted by $a\alpha$ for $\alpha \in G$ and $a \in \mathcal{O}$. Of course, whether you write the operator a on the left of α, as we did here, or on the right is a matter of preference and convenience.

While we did not require any structure on the set $\mathcal{O}$ in our definition, it frequently happens that $\mathcal{O}$ has some natural algebraic structure. Let us give a few examples.

EXAMPLE I   Every abelian group G gives rise to a natural group with operators. Let us write the group operation of G multiplicatively. Letting $\mathcal{O} = \mathbb{Z}$, for $\alpha \in G$ and $n \in \mathbb{Z}$, define $n\alpha = \alpha^n$. Since G is abelian, we have

$$n(\alpha\beta) = (\alpha\beta)^n = \alpha^n\beta^n = (n\alpha)(n\beta).$$

Thus every abelian group G can be regarded as a $\mathbb{Z}$-group. Here $\mathbb{Z}$ has a natural ring structure. ▲

---

† This section is not used in the remainder of the text.

EXAMPLE 2   If V is a vector space over a field F, V can be regarded in a natural way as an F-group. Here F has a field structure. ▲

Consider an $\mathcal{O}$-group G. For a *fixed* $a \in \mathcal{O}$, the mapping $\lambda_a : G \to G$ defined by $\lambda_a(\alpha) = a\alpha$ for $\alpha \in G$ is a homomorphism of G into G, since $a(\alpha\beta) = (a\alpha)(a\beta)$. This suggests our next example.

EXAMPLE 3   Let G be any group, and let $\mathcal{O}$ be any set of homomorphisms of G into itself (such homomorphisms are *endomorphisms* of G). For $\alpha, \beta \in G$ and $\phi \in \mathcal{O}$, the property $\phi(\alpha\beta) = \phi(\alpha)\phi(\beta)$ for the endomorphism ϕ shows that G can be naturally viewed as an $\mathcal{O}$-group. ▲

A substantial chunk of our abelian group theory could have been done for $\mathcal{O}$-groups. Starting back with subgroups, an **admissible subgroup**, or $\mathcal{O}$-**subgroup** of an $\mathcal{O}$-group G, is a subgroup H of $\langle G, \cdot \rangle$ such that $a\alpha \in H$ for all $\alpha \in H$ and $a \in \mathcal{O}$, that is, such that H is closed under external multiplication by elements of $\mathcal{O}$. If we are constantly working with $\mathcal{O}$-groups, we simply drop the term *admissible* and speak of a subgroup of the $\mathcal{O}$-group G, always meaning an $\mathcal{O}$-subgroup. A couple more examples will show how elegantly these ideas relate to our past work.

EXAMPLE 4   Let G be any group, and let $\mathcal{I}$ be the set of all inner automorphisms of G. As in Example 3, G becomes an $\mathcal{I}$-group in a natural way. An (admissible) subgroup H of G then must have the property that $i_g(a) = gag^{-1}$ is in H for all $a \in H$ and all $g \in G$. Thus the $\mathcal{I}$-subgroups H of the $\mathcal{I}$-group G are essentially the normal subgroups of G. ▲

EXAMPLE 5   Let R be a ring. The additive group $\langle R, + \rangle$ of R can be regarded as an R-group, where for a group element a in $\langle R, + \rangle$ and $r \in R$, we define ra by the ring multiplication. The left distributive law in R gives $r(a + b) = (ra) + (rb)$, which is exactly the condition that $\langle R, + \rangle$ be an R-group. An R-subgroup N of the R-group $\langle R, + \rangle$ then must be a subgroup of $\langle R, + \rangle$ satisfying $ra \in N$ for all $a \in N$ and $r \in R$. Thus the R-subgroups are the left ideals of R. If R is a commutative ring, then the R-subgroups are the ideals of R. ▲

We can form the factor group of an $\mathcal{O}$-group G modulo a normal $\mathcal{O}$-subgroup; the factor group also becomes an $\mathcal{O}$-group in a natural way when we define external multiplication on cosets by using representatives. We have the concept of an $\mathcal{O}$-homomorphism from one $\mathcal{O}$-group into another. The exercises ask us to come up with the proper definitions for these ideas.

Finally, we state without proof the Jordan–Hölder theorem for an $\mathcal{O}$-group. The definitions of subnormal series and composition series are analogous to our definitions of Section 3.5; we need only require that every subgroup be an $\mathcal{O}$-subgroup.

THEOREM 8.25 (Jordan–Hölder Theorem) Any two composition series of an $\mathcal{O}$-group G are isomorphic.

Taking $\mathcal{O} = \{\iota\}$, where ι is the identity map of a group G into itself, we recover the Jordan–Hölder theorem of Section 3.5 for composition series. Taking $\mathcal{O} = \mathcal{I}$, the set of inner automorphisms of G, we recover the Jordan–Hölder theorem for a principal series. And finally, but most impressively, taking the Jordan–Hölder theorem for the F-group V, where V is a finite-dimensional vector space over F, we recover the invariance of dimension of V, for a simple F-factor group of the F-group V will be a vector space (an F-group) of dimension 1 over F.

Modules

DEFINITION 8.18 (R-module) Let R be a ring. A **(left)** R-**module** consists of an abelian group M together with an operation of external multiplication of each element of M by each element of R on the left such that for all $\alpha, \beta \in M$ and $r, s \in R$, the following conditions are satisfied:

1. $(r\alpha) \in M$.
2. $r(\alpha + \beta) = r\alpha + r\beta$.
3. $(r + s)\alpha = r\alpha + s\alpha$.
4. $(rs)\alpha = r(s\alpha)$.

We shall somewhat incorrectly speak of the R-**module** M.

An R-module is very much like a vector space except that the scalars need only form a ring. If R is a ring with unity and $1\alpha = \alpha$ for all $\alpha \in M$, then M is a **unitary** R-**module**.

EXAMPLE 6   Every abelian group G can be regarded as a $\mathbb{Z}$-module if we define $n\alpha = \alpha^n$ for $\alpha \in G$ and $n \in \mathbb{Z}$. We have used multiplicative notation for the operation in G. The axioms for a module are readily verified. ▲

EXAMPLE 7   For an ideal n in R, $\langle N, + \rangle$ can be viewed as an R-module, where for $\alpha \in N$ and $r \in R$, $r\alpha$ is the ordinary ring multiplication of r and α, both viewed as elements of the ring R. ▲

We can speak of submodules, quotient modules, and R-homomorphisms from one R-module into another, all by the natural definitions. We can also take a direct sum of R-modules, arriving again at an R-module.

DEFINITION 8.19 (Cyclic Module) An R-module M is **cyclic** if there exists $\alpha \in M$ such that $M = \{r\alpha \mid r \in R\}$.

Thus a cyclic R-module is generated by a single element. The idea of a *set of generators for an R-module* is the natural generalization of the idea of a spanning set of vectors of a vector space. A nice result for us to state without proof and then to illustrate for some special cases, in light of our past work, is the following theorem.

THEOREM 8.26   If R is a PID, then every finitely generated R-module is isomorphic to a direct sum of cyclic R-modules.

Taking $R = Z$ and referring to Example 6, we see from the theorem that every finitely generated abelian group is isomorphic to a direct sum of cyclic groups. This is a large part of the fundamental theorem of finitely generated abelian groups. For $R = F$, where F is a field, applying Theorem 8.26 to a finite-dimensional vector space V over F, we see that V is isomorphic to a direct sum of vector spaces of dimension 1 over F.

Algebras

DEFINITION 8.20 (Algebra)   An **algebra** consists of a vector space V over a field F, together with a binary operation of multiplication on the set V of vectors, such that for all $a \in F$ and $\alpha, \beta, \gamma \in V$, the following conditions are satisfied:

1. $(a\alpha)\beta = a(\alpha\beta) = \alpha(a\beta)$.
2. $(\alpha + \beta)\gamma = \alpha\gamma + \beta\gamma$.
3. $\alpha(\beta + \gamma) = \alpha\beta + \alpha\gamma$.

We shall somewhat incorrectly speak of an **algebra** V **over** F. Also, V is an **associative algebra over** F, if, in addition to the preceding three conditions,

4. $(\alpha\beta)\gamma = \alpha(\beta\gamma)$ for all $\alpha, \beta, \gamma \in V$.

EXAMPLE 8   If E is an extension field of a field F, then $V = E$ can be viewed as an associative algebra over F, where addition and multiplication of elements of V are field addition and multiplication in E, and scalar multiplication by elements of F is again field multiplication E. ▲

EXAMPLE 9   For any group G and field F, the *group algebra* $F(G)$ defined in Section 5.7 is an associative algebra over F. ▲

DEFINITION 8.21 (Division Algebra)   An algebra V over a field F is a **division algebra over** F if V has a unity for multiplication and contains a multiplicative inverse of each nonzero element. (Note that associativity of multiplication is *not* assumed.)

EXAMPLE 10　An extension field E of a field F can be viewed as an associative division algebra over F. Also the quaternions Q of Section 5.7 form an associative division algebra over the real numbers. ▲

We conclude with a statement of some famous results regarding division algebras over the real numbers.

THEOREM 8.27　The real numbers, the complex numbers, and the quaternions are the only (up to isomorphism) associative division algebras over the real numbers (Frobenius, 1878). The only additional division algebra over the real numbers is the Cayley algebra, which is a vector space of dimension 8 over $\mathbb{R}$ (Bott and Milnor, 1957).

Exercises 8.6

Concepts

1. Our definition of an $\mathcal{O}$-group did *not* start

$$\text{An } \mathcal{O}\text{-group}\langle\,,\,,\ldots,\,\rangle \text{ is} \ldots$$

in a fashion similar to our definitions of a group and a ring. If an $\mathcal{O}$-group were defined in this way, what would $\langle\,,\,,\ldots,\,\rangle$ be? Be sure that you get *all* the sets and *all* the operations involved.

2. Repeat Exercise 1 for an R-module.

3. Repeat Exercise 1 for an algebra.

4. Let G be any group, and let $\mathcal{A}$ be the set of *all* automorphisms of G. By Example 3, G can be regarded as an $\mathcal{A}$-group. An $\mathcal{A}$-subgroup of G is a **characteristic subgroup** of G.

   Every subgroup of every abelian group is a normal subgroup, that is, an $\mathcal{I}$-subgroup, but give an example showing that not every subgroup of every abelian group is a characteristic subgroup.

5. Define the concept of an *$\mathcal{O}$-homomorphism of an $\mathcal{O}$-group G into an $\mathcal{O}$-group G'.* Show that the kernel is an $\mathcal{O}$-subgroup of G.

6. Define a *submodule of a (left) R-module M* and a *quotient module of a (left) R-module M modulo a submodule N.*

7. Define an *R-homomorphism of a (left) R-module M into a (left) R-module M'.*

Theory

8. Show that an intersection of admissible subgroups of an $\mathcal{O}$-group G is again an admissible subgroup of G.

9. Show that the factor group of an $\mathcal{O}$-group G modulo an admissible normal subgroup can again be considered an $\mathcal{O}$-group in a natural way. That is, define the external multiplication on the cosets, show it is well defined, and check the axioms for an $\mathcal{O}$-group.

10. Prove that for a left R-module M,

$$0\alpha = 0, \qquad r0 = 0,$$

and

$$(-r)\alpha = -(r\alpha) = r(-\alpha)$$

for every $\alpha \in M$ and $r \in R$.

11. Let M be a left R-module, and let $\alpha \in M$. Show that $L_\alpha = \{a \in R \mid a\alpha = 0\}$ is a left ideal of R.

12. Show that the ring $\langle M_n(F), +, \cdot \rangle$ of $n \times n$ matrices with entries in F becomes an algebra of dimension n^2 over F if one defines scalar multiplication by $b(a_{ij}) = (ba_{ij})$ for $(a_{ij}) \in M_n(F)$ and $b \in F$.

13. Let V be an algebra of finite dimension and with a basis

$$B = \{\beta_i \mid i = 1, \ldots, n\}$$

over a field F. Show that the multiplication of vectors in V is completely determined by the n^2 products $\beta_r\beta_s$ for each ordered pair (β_r, β_s) of basis vectors from B.

14. Let V be an algebra of finite dimension and with a basis

$$B = \{\beta_i \mid i = 1, \ldots, n\}$$

over a field F. Show that V is an associative algebra over F if and only if

$$\beta_r(\beta_s\beta_t) = (\beta_r\beta_s)\beta_t$$

for each of the n^3 ordered triples $(\beta_r, \beta_s, \beta_t)$ of basis vectors from B.

15. Let V be a finite-dimensional vector space over a field F with a basis $B = \{\beta_i \mid i = 1, \ldots, n\}$ over F. Let $\{c_{rst} \mid r, s, t = 1, \ldots, n\}$ be any collection of n^3 scalars in F. Show that there exists exactly one binary operation of multiplication on V such that V is an algebra over F under this multiplication and such that

$$\beta_r\beta_s = \sum_t c_{rst}\beta_t$$

for every ordered pair (β_r, β_s) of basis vectors from B. The scalars c_{rst} are the **structure constants of the algebra.**

◆

CHAPTER NINE

AUTOMORPHISMS AND GALOIS THEORY

◆

In this chapter, we continue our study of extension fields. The principal tools for this study are isomorphic mappings of an extension field E of a field F onto a subfield of $\bar{F}$ that map each element of F into itself. Such mappings exhibit structural similarities of extension fields of F. If we restrict ourselves to such isomorphic mappings of E onto itself, that is, to automorphisms of E leaving each element of F fixed, we can study structural symmetries of the extension E. We will see that these automorphisms form a group. In the final few sections on Galois theory, group theory will come heavily into play, providing a satisfying unification of topics in our text.

9.1

Automorphisms of Fields

The Conjugation Isomorphisms of Algebraic Field Theory

Let F be a field, and let $\bar{F}$ be an algebraic closure of F, that is, an algebraic extension of F that is algebraically closed. Such a field $\bar{F}$ exists, by Theorem 8.14. Our selection of a particular $\bar{F}$ is not critical, since, as we shall show in Section 9.2, any two algebraic closures of F are isomorphic under a map leaving F fixed. *From now on in our work, we shall assume that all algebraic extensions and all elements algebraic over a field F under consideration are contained in one fixed algebraic closure $\bar{F}$ of F.*

Remember that we are engaged in the study of zeros of polynomials. In the terminology of Section 8.3, studying zeros of polynomials in $F[x]$ amounts to studying the structure of algebraic extensions of F and of elements algebraic over F. We shall show that if E is an algebraic extension of F with $\alpha, \beta \in E$, then α and β have the same algebraic properties if and only if $\mathrm{irr}(\alpha, F) = \mathrm{irr}(\beta, F)$. We shall phrase this fact in terms of mappings, as we

445

have been doing all along in field theory. We achieve this by showing the existence of an isomorphism $\psi_{\alpha\beta}$ of $F(\alpha)$ onto $F(\beta)$ that maps each element of F onto itself and maps α onto β, in the case that $\mathrm{irr}(\alpha, F) = \mathrm{irr}(\beta, F)$. The next theorem exhibits this isomorphism $\psi_{\alpha,\beta}$. These isomorphisms will become our fundamental tools for the study of algebraic extensions; they supplant the *evaluation homomorphisms* ϕ_α of Section 5.5, which make their last contribution in defining these isomorphisms. Before stating and proving this theorem, let us introduce some more terminology.

DEFINITION 9.I (Conjugate Elements)   Let E be an algebraic extension of a field F. Two elements $\alpha, \beta \in E$ are **conjugate over** F if $\mathrm{irr}(\alpha, F) = \mathrm{irr}(\beta, F)$, that is, if α and β are zeros of the same irreducible polynomial over F.

EXAMPLE I   The concept of conjugate elements just defined conforms with the classic idea of *conjugate complex numbers* if we understand that by conjugate complex numbers we mean numbers that are *conjugate over* $\mathbb{R}$. If $a, b \in \mathbb{R}$ and $b \neq 0$, the conjugate complex numbers $a + bi$ and $a - bi$ are both zeros of $x^2 - 2ax + a^2 + b^2$, which is irreducible in $\mathbb{R}[x]$. ▲

THEOREM 9.I (The Conjugation Isomorphisms)   Let F be a field, and let α and β be algebraic over F with $\deg(\alpha, F) = n$. The map $\psi_{\alpha,\beta}: F(\alpha) \to F(\beta)$ defined by

$$\psi_{\alpha,\beta}(c_0 + c_1\alpha + \cdots + c_{n-1}\alpha^{n-1}) = c_0 + c_1\beta + \cdots + c_{n-1}\beta^{n-1}$$

for $c_i \in F$ is an isomorphism of $F(\alpha)$ onto $F(\beta)$ if and only if α and β are conjugate over F.

PROOF   Suppose that $\psi_{\alpha,\beta}: F(\alpha) \to F(\beta)$ as defined in the statement of the theorem is an isomorphism. Let $\mathrm{irr}(\alpha, F) = a_0 + a_1x + \cdots + a_nx^n$. Then $a_0 + a_1\alpha + \cdots + a_n\alpha^n = 0$, so

$$\psi_{\alpha,\beta}(a_0 + a_1\alpha + \cdots + a_n\alpha^n) = a_0 + a_1\beta + \cdots + a_n\beta^n = 0.$$

By the last assertion in the statement of Theorem 8.3, this implies that $\mathrm{irr}(\beta, F)$ divides $\mathrm{irr}(\alpha, F)$. A similar argument using the isomorphism $(\psi_{\alpha,\beta})^{-1} = \psi_{\beta,\alpha}$ shows that $\mathrm{irr}(\alpha, F)$ divides $\mathrm{irr}(\beta, F)$. Therefore, since both polynomials are monic, $\mathrm{irr}(\alpha, F) = \mathrm{irr}(\beta, F)$, so α and β are conjugate over F.

Conversely, suppose $\mathrm{irr}(\alpha, F) = \mathrm{irr}(\beta, F) = p(x)$. Then the evaluation homomorphisms $\phi_\alpha: F[x] \to F(\alpha)$ and $\phi_\beta: F[x] \to F(\beta)$ both have the same kernel $\langle p(x) \rangle$. By Theorem 6.7, corresponding to $\phi_\alpha: F[x] \to F(\alpha)$, there is a natural isomorphism ψ_α mapping $F[x]/\langle p(x) \rangle$ onto $\phi_\alpha[F[x]] = F(\alpha)$. Similarly, ϕ_β gives rise to an isomorphism ψ_β mapping

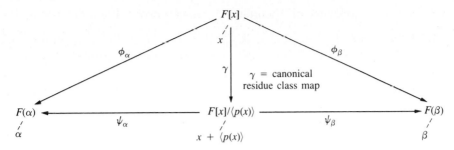

γ = canonical residue class map

Figure 9.1

$F[x]/\langle p(x) \rangle$ onto $F(\beta)$. Let $\psi_{\alpha,\beta} = \psi_\beta(\psi_\alpha)^{-1}$. These mappings are diagrammed in Fig. 9.1 where the dashed lines indicate corresponding elements under the mappings. As the composition of two isomorphisms, $\psi_{\alpha,\beta}$ is again an isomorphism and maps $F(\alpha)$ onto $F(\beta)$. For $(c_0 + c_1\alpha + \cdots + c_{n-1}\alpha^{n-1}) \in F(\alpha)$, we have

$$\psi_{\alpha,\beta}(c_0 + c_1\alpha + \cdots + c_{n-1}\alpha^{n-1})$$
$$= (\psi_\beta\psi_\alpha^{-1})(c_0 + c_1\alpha + \cdots + c_{n-1}\alpha^{n-1})$$
$$= \psi_\beta((c_0 + c_1 x + \cdots + c_{n-1}x^{n-1}) + \langle p(x) \rangle)$$
$$= c_0 + c_1\beta + \cdots + c_{n-1}\beta^{n-1}.$$

Thus $\psi_{\alpha,\beta}$ is the map defined in the statement of the theorem. ◆

The following corollary of Theorem 9.1 is the cornerstone of our proof of the important Isomorphism Extension Theorem of Section 9.2 and of most of the rest of our work.

COROLLARY I Let α be algebraic over a field F. Every isomorphism ψ mapping $F(\alpha)$ onto a subfield of $\bar{F}$ such that $\psi(a) = a$ for $a \in F$ maps α onto a conjugate β of α over F. Conversely, for each conjugate β of α over F, there exists exactly one isomorphism $\psi_{\alpha,\beta}$ of $F(\alpha)$ onto a subfield of $\bar{F}$ mapping α onto β and mapping each $a \in F$ onto itself.

PROOF Let ψ be an isomorphism of $F(\alpha)$ onto a subfield of $\bar{F}$ such that $\psi(a) = a$ for $a \in F$. Let $\mathrm{irr}(\alpha, F) = a_0 + a_1 x + \cdots + a_n x^n$. Then

$$a_0 + a_1\alpha + \cdots + a_n\alpha^n = 0,$$

so

$$0 = \psi(a_0 + a_1\alpha + \cdots + a_n\alpha^n) = a_0 + a_1\psi(\alpha) + \cdots + a_n\psi(\alpha)^n,$$

and $\beta = \psi(\alpha)$ is a conjugate of α.

Conversely, for each conjugate β of α over F, the conjugation isomorphism $\psi_{\alpha,\beta}$ of Theorem 9.1 is an isomorphism with the desired properties. That $\psi_{\alpha,\beta}$ is the only such isomorphism follows from the fact that an isomorphism of $F(\alpha)$ is completely determined by its values on elements of F and its value on α. ◆

As a second corollary of Theorem 9.1, we can prove a familiar result.

COROLLARY 2   Let $f(x) \in \mathbb{R}[x]$. If $f(a + bi) = 0$ for $(a + bi) \in \mathbb{C}$, where $a, b \in \mathbb{R}$, then $f(a - bi) = 0$ also. Loosely, complex zeros of polynomials with real coefficients occur in conjugate pairs.

PROOF   We have seen that $\mathbb{C} = \mathbb{R}(i)$. Now

$$\mathrm{irr}(i, \mathbb{R}) = \mathrm{irr}(-i, \mathbb{R}) = x^2 + 1,$$

so i and $-i$ are conjugate over $\mathbb{R}$. By Theorem 9.1, the conjugation map $\psi_{i,-i} : \mathbb{C} \to \mathbb{C}$ given by $\psi_{i,-i}(a + bi) = a - bi$ is an isomorphism. Thus, if for $a_i \in \mathbb{R}$,

$$f(a + bi) = a_0 + a_1(a + bi) + \cdots + a_n(a + bi)^n = 0,$$

then

$$\begin{aligned} 0 = \psi_{i,-i}(f(a + bi)) &= a_0 + a_1(a - bi) + \cdots + a_n(a - bi)^n \\ &= f(a - bi), \end{aligned}$$

that is, $f(a - bi) = 0$ also. ◆

EXAMPLE 2   Consider $\mathbb{Q}(\sqrt{2})$ over $\mathbb{Q}$. The zeros of $\mathrm{irr}(\sqrt{2}, \mathbb{Q}) = x^2 - 2$ are $\sqrt{2}$ and $-\sqrt{2}$, so $\sqrt{2}$ and $-\sqrt{2}$ are conjugate over $\mathbb{Q}$. According to Theorem 9.1, the map $\psi_{\sqrt{2},-\sqrt{2}} : \mathbb{Q}(\sqrt{2}) \to \mathbb{Q}(\sqrt{2})$ defined by

$$\psi_{\sqrt{2},-\sqrt{2}}(a + b\sqrt{2}) = a - b\sqrt{2}$$

is an isomorphism of $\mathbb{Q}(\sqrt{2})$ onto itself. ▲

Automorphisms and Fixed Fields

As illustrated in the preceding corollary and example, a field may have a nontrivial isomorphism onto itself. *Such maps will be of utmost importance in the work that follows.*

DEFINITION 9.2 (Automorphism)   An isomorphism of a field onto itself is an **automorphism of the field**.

DEFINITION 9.3 (Fixed Field)   If σ is an isomorphism of a field E onto some field, then an element a of E is **left fixed by** σ, if $\sigma(a) = a$. A collection S of isomorphisms of E **leaves a subfield** F of E **fixed** if each $a \in F$ is left fixed by every $\sigma \in S$. If $\{\sigma\}$ leaves F fixed, then σ **leaves** F **fixed.**

EXAMPLE 3   Let $E = Q(\sqrt{2}, \sqrt{3})$. The map $\sigma : E \to E$ defined by

$$\sigma(a + b\sqrt{2} + c\sqrt{3} + d\sqrt{6}) = a + b\sqrt{2} - c\sqrt{3} - d\sqrt{6}$$

for $a, b, c, d \in \mathbb{Q}$ is an automorphism of E; it is the conjugation isomorphism $\psi_{\sqrt{3},-\sqrt{3}}$ of E onto itself if we view E as $(\mathbb{Q}[\sqrt{2}])[\sqrt{3}]$. We see that σ leaves $\mathbb{Q}[\sqrt{2}]$ fixed. ▲

◆ HISTORICAL NOTE ◆

It was Richard Dedekind who first developed the idea of an auto-morphism of a field, what he called a "permutation of the field," in 1894. The earlier application of group theory to the theory of equations had been through groups of permutations of the roots of certain polynomials. Dedekind extended this idea to mappings of the entire field and proved several of the theorems of this section.

Though Heinrich Weber continued Dedekind's approach to groups acting on fields in his algebra text of 1895, this method was not pursued in other texts near the turn of the century. It was not until the 1920s, after Emmy Noether's abstract approach to algebra became influential at Gottingen, that Emil Artin (1898–1962) developed this relationship of groups and fields in great detail. Artin emphasized that the goal of what is now called Galois theory should not be to determine solvability conditions for algebraic equations, but to explore the relationship between field extensions and groups of automorphisms. Artin detailed his approach in a lecture given in 1926; his method was first published in B. L. Van der Waerden's *Modern Algebra* text of 1930 and later by Artin himself in lecture notes in 1938 and 1942. In fact, the remainder of this text is based on Artin's development of Galois theory.

It is our purpose to study the structure of an algebraic extension E of a field F by studying the automorphisms of E that leave fixed each element of F. We shall presently show that these automorphisms form a group in a natural way. We can then apply the results concerning group structure to get information about the structure of our field extension. Thus much of our preceding work is now being brought together. The next three theorems are readily proved, but the ideas contained in them form the foundation for everything that follows. These theorems are therefore of great importance to us. They really amount to observations, rather than theorems; it is the *ideas* contained in them that are important. A big step in mathematics does not always consist of proving a *hard* theorem, but may consist of noticing how certain known mathematics may relate to new situations. Here we are bringing group theory into our study of zeros of polynomials. Be sure to understand the concepts involved. Unlikely as it may seem, they are the key to the solution of our *final goal* in this text.

Final goal (to be more precisely stated later): To show that not all zeros of every quintic (degree 5) polynomial $f(x)$ can be expressed in terms of radicals starting with elements in the field of coefficients of $f(x)$.

If $\{\sigma_i \mid i \in I\}$ is a collection of automorphisms of a field E, the elements of E about which $\{\sigma_i \mid i \in I\}$ gives the least information are those $a \in E$ left fixed by every σ_i for $i \in I$. This first of our three theorems contains almost all that can be said about these fixed elements of E.

THEOREM 9.2   Let $\{\sigma_i \mid i \in I\}$ be a collection of automorphisms of a field E. Then the set $E_{\{\sigma_i\}}$ of all $a \in E$ left fixed by every σ_i for $i \in I$ forms a subfield of E.

PROOF   If $\sigma_i(a) = a$ and $\sigma_i(b) = b$ for all $i \in I$, then

$$\sigma_i(a \pm b) = \sigma_i(a) \pm \sigma_i(b) = a \pm b$$

and

$$\sigma_i(ab) = \sigma_i(a)\sigma_i(b) = ab$$

for all $i \in I$. Also, if $b \neq 0$, then

$$\sigma_i(a/b) = \sigma_i(a)/\sigma_i(b) = a/b$$

for all $i \in I$. Since the σ_i are automorphisms, we have

$$\sigma_i(0) = 0 \qquad \text{and} \qquad \sigma_i(1) = 1$$

for all $i \in I$. Hence $0, 1 \in E_{\{\sigma_i\}}$. Thus $E_{\{\sigma_i\}}$ is a subfield of E. ◆

DEFINITION 9.4 (Fixed Field)   The field $E_{\{\sigma_i\}}$ of Theorem 9.2 is the **fixed field** of $\{\sigma_i \mid i \in I\}$. For a single automorphism σ, we shall refer to $E_{\{\sigma\}}$ as the **fixed field of** σ.

EXAMPLE 4   Consider the automorphism $\psi_{\sqrt{2}, -\sqrt{2}}$ of $\mathbb{Q}(\sqrt{2})$ given in Example 2. For $a, b \in \mathbb{Q}$, we have

$$\psi_{\sqrt{2}, -\sqrt{2}}(a + b\sqrt{2}) = a - b\sqrt{2},$$

and $a - b\sqrt{2} = a + b\sqrt{2}$ if and only if $b = 0$. Thus the fixed field of $\psi_{\sqrt{2}, -\sqrt{2}}$ is $\mathbb{Q}$. ▲

Note that an automorphism of a field E is in particular a one-to-one mapping of E onto E, that is, a *permutation of E*. If σ and τ are automorphisms of E, then the permutation $\sigma\tau$ is again an automorphism of E, since, in general, composition of homomorphisms again yields a homomorphism. This is how group theory makes its entrance.

THEOREM 9.3   The set of all automorphisms of a field E is a group under function composition.

PROOF   Multiplication of automorphisms of E is defined by function composition, and is thus associative (it is *permutation multiplication*). The identity permutation $\iota : E \to E$ given by $\iota(\alpha) = \alpha$ for $\alpha \in E$ is an automorphism of E. If σ is an automorphism, then the permutation σ^{-1} is also an automorphism. Thus all automorphisms of E form a subgroup of S_E, the group of all permutations of E given by Theorem 2.1. ◆

THEOREM 9.4 Let E be a field, and let F be a subfield of E. Then the set $G(E/F)$ of all automorphisms of E leaving F fixed forms a subgroup of the group of all automorphisms of E. Furthermore, $F \le E_{G(E/F)}$.

PROOF For $\sigma, \tau \in G(E/F)$ and $a \in F$, we have

$$(\sigma\tau)(a) = \sigma(\tau(a)) = \sigma(a) = a,$$

so $\sigma\tau \in G(E/F)$. Of course, the identity automorphism ι is in $G(E/F)$. Also, if $\sigma(a) = a$ for $a \in F$, then $a = \sigma^{-1}(a)$, so $\sigma \in G(E/F)$ implies that $\sigma^{-1} \in G(E/F)$. Thus $G(E/F)$ is a subgroup of the group of all automorphisms of E.

Since every element of F is left fixed by every element of $G(E/F)$, it follows immediately that the field $E_{G(E/F)}$ of *all* elements of E left fixed by $G(E/F)$ contains F. ◆

DEFINITION 9.5 (Group of E over F) The group $G(E/F)$ of the preceding theorem is the **group of automorphisms of E leaving F fixed**, or, more briefly, the **group of E over F**.

Do not think of E/F in the notation $G(E/F)$ as denoting a quotient space of some sort, but rather as meaning that E is an extension field of the field F.

The ideas contained in the preceding three theorems are illustrated in the following example. We urge you to study this example carefully.

EXAMPLE 5 Consider the field $\mathbb{Q}(\sqrt{2}, \sqrt{3})$. If we view $\mathbb{Q}(\sqrt{2}, \sqrt{3})$ as $(\mathbb{Q}(\sqrt{3}))(\sqrt{2})$, the conjugation isomorphism $\psi_{\sqrt{2}, -\sqrt{2}}$ of Theorem 9.1 defined by

$$\psi_{\sqrt{2}, -\sqrt{2}}(a + b\sqrt{2}) = a - b\sqrt{2}$$

for $a, b \in \mathbb{Q}(\sqrt{3})$ is an automorphism of $\mathbb{Q}(\sqrt{2}, \sqrt{3})$ having $\mathbb{Q}(\sqrt{3})$ as fixed field. Similarly, we have the automorphism $\psi_{\sqrt{3}, -\sqrt{3}}$ of $\mathbb{Q}(\sqrt{2}, \sqrt{3})$ having $\mathbb{Q}(\sqrt{2})$ as fixed field. Since the product of two automorphisms is an automorphism, we can consider $\psi_{\sqrt{2}, -\sqrt{2}} \psi_{\sqrt{3}, -\sqrt{3}}$, which **moves** both $\sqrt{2}$ and $\sqrt{3}$, that is, leaves neither number fixed. Let

$$\iota = \text{the identity automorphism,}$$

$$\sigma_1 = \psi_{\sqrt{2}, -\sqrt{2}},$$

$$\sigma_2 = \psi_{\sqrt{3}, -\sqrt{3}}, \text{ and}$$

$$\sigma_3 = \psi_{\sqrt{2}, -\sqrt{2}} \psi_{\sqrt{3}, -\sqrt{3}}.$$

The group of all automorphisms of $\mathbb{Q}(\sqrt{2}, \sqrt{3})$ has a fixed field, by

	ι	σ_1	σ_2	σ_3
ι	ι	σ_1	σ_2	σ_3
σ_1	σ_1	ι	σ_3	σ_2
σ_2	σ_2	σ_3	ι	σ_1
σ_3	σ_3	σ_2	σ_1	ι

Table 9.1

Theorem 9.2. This fixed field must contain $\mathbb{Q}$, since every automorphism of a field leaves 1 and hence the prime subfield fixed. A basis for $\mathbb{Q}(\sqrt{2}, \sqrt{3})$ over $\mathbb{Q}$ is $\{1, \sqrt{2}, \sqrt{3}, \sqrt{6}\}$. Since $\sigma_1(\sqrt{2}) = -\sqrt{2}$, $\sigma_1(\sqrt{6}) = -\sqrt{6}$ and $\sigma_2(\sqrt{3}) = -\sqrt{3}$, we see that $\mathbb{Q}$ is exactly the fixed field of $\{\iota, \sigma_1, \sigma_2, \sigma_3\}$. It is readily checked that $G = \{\iota, \sigma_1, \sigma_2, \sigma_3\}$ is a group under automorphism multiplication (function composition). The group table for G is given in Table 9.1. For example,

$$\sigma_1\sigma_3 = \psi_{\sqrt{2}, -\sqrt{2}}\left(\psi_{\sqrt{2}, -\sqrt{2}}\psi_{\sqrt{3}, -\sqrt{3}}\right) = \psi_{\sqrt{3}, -\sqrt{3}} = \sigma_2.$$

The group G is isomorphic to the Klein 4-group. We can show that G is the full group $G(\mathbb{Q}(\sqrt{2}, \sqrt{3})/\mathbb{Q})$, because every automorphism τ of $\mathbb{Q}(\sqrt{2}, \sqrt{3})$ maps $\sqrt{2}$ onto either $\pm\sqrt{2}$, by Corollary 1 of Theorem 9.1. Similarly, τ maps $\sqrt{3}$ onto either $\pm\sqrt{3}$. But since $\{1, \sqrt{2}, \sqrt{3}, \sqrt{2}\sqrt{3}\}$ is a basis for $\mathbb{Q}(\sqrt{2}, \sqrt{3})$ over $\mathbb{Q}$, an automorphism of $\mathbb{Q}(\sqrt{2}, \sqrt{3})$ leaving $\mathbb{Q}$ fixed is determined by its values on $\sqrt{2}$ and $\sqrt{3}$. Now, $\iota, \sigma_1, \sigma_2$, and σ_3 give all possible combinations of values on $\sqrt{2}$ and $\sqrt{3}$, and hence are all possible automorphisms of $\mathbb{Q}(\sqrt{2}, \sqrt{3})$.

Note that $G(\mathbb{Q}(\sqrt{2}, \sqrt{3})/\mathbb{Q})$ has order 4, and $[\mathbb{Q}(\sqrt{2}, \sqrt{3}):\mathbb{Q}] = 4$. *This is no accident*, but rather an instance of a general situation, as we shall see later. ▲

The Frobenius Automorphism

Let F be a finite field. We shall show later that the group of all automorphisms of F is cyclic. Now a cyclic group has by definition a generating element, and it may have several generating elements. For an abstract cyclic group there is no way of distinguishing any one generator as being more important than any other. However, for the cyclic group of all automorphisms of a finite field there is a canonical (natural) generator, the *Frobenius automorphism* (classically, the *Frobenius substitution*). This fact is of considerable importance in some advanced work in algebra. The next theorem exhibits this Frobenius automorphism.

THEOREM 9.5 Let F be a finite field of characteristic p. Then the map $\sigma_p : F \to F$ defined by $\sigma_p(a) = a^p$ for $a \in F$ is an automorphism, the **Frobenius automorphism**, of F. Also, $F_{\{\sigma_p\}} \simeq \mathbb{Z}_p$.

PROOF Let $a, b \in F$. Applying the binomial theorem to $(a + b)^p$, we have

$$(a + b)^p = a^p + (p \cdot 1)a^{p-1}b + \left(\frac{p(p-1)}{2} \cdot 1\right)a^{p-2}b^2$$

$$+ \cdots + (p \cdot 1)ab^{p-1} + b^p$$

$$= a^p + 0a^{p-1}b + 0a^{p-2}b^2 + \cdots + 0ab^{p-1} + b^p$$

$$= a^p + b^p.$$

Thus we have

$$\sigma_p(a + b) = (a + b)^p = a^p + b^p = \sigma_p(a) + \sigma_p(b).$$

Of course,

$$\sigma_p(ab) = (ab)^p = a^p b^p = \sigma_p(a)\sigma_p(b),$$

so σ_p is at least a homomorphism. If $\sigma_p(a) = 0$, then $a^p = 0$, and $a = 0$, so the kernel of σ_p is $\{0\}$, and σ_p is a one-to-one map. Finally, since F is finite, σ_p is onto, by counting. Thus σ_p is an automorphism of F.

The prime field $\mathbb{Z}_p$ must be contained (up to isomorphism) in F, since F is of characteristic p. For $c \in \mathbb{Z}_p$, we have $\sigma_p(c) = c^p = c$, by Fermat's theorem (see the corollary of Theorem 5.8). Thus the polynomial $x^p - x$ has p zeros in F, namely the elements of $\mathbb{Z}_p$. By Corollary 2 of Theorem 5.18, a polynomial of degree n over a field can have at most n zeros in the field. Since the elements fixed under σ_p are precisely the zeros in F of $x^p - x$, we see that

$$\mathbb{Z}_p = F_{\{\sigma_p\}}. \quad \blacklozenge$$

Freshmen in college still sometimes make the error of saying that $(a + b)^n = a^n + b^n$. Here we see that this *freshman exponentiation* $(a + b)^p = a^p + b^p$ with exponent p is actually valid in a field F of characteristic p.

Exercises 9.1

Computations

In Exercises 1 through 8, find all conjugates of the given number over the given field.

1. $\sqrt{2}$ over $\mathbb{Q}$

2. $\sqrt{2}$ over $\mathbb{R}$

3. $3 + \sqrt{2}$ over $\mathbb{Q}$

4. $\sqrt{2} - \sqrt{3}$ over $\mathbb{Q}$

5. $\sqrt{2} + i$ over $\mathbb{Q}$

6. $\sqrt{2} + i$ over $\mathbb{R}$

7. $\sqrt{1 + \sqrt{2}}$ over $\mathbb{Q}$

8. $\sqrt{1 + \sqrt{2}}$ over $\mathbb{Q}(\sqrt{2})$

In Exercises 9 through 14, we consider the field $E = \mathbb{Q}(\sqrt{2}, \sqrt{3}, \sqrt{5})$. In the notation of Theorem 9.1, we have the following conjugation isomorphisms (which are here automorphisms of E):

$$\psi_{\sqrt{2},\,-\sqrt{2}}:(\mathbb{Q}(\sqrt{3},\,\sqrt{5}))(\sqrt{2}) \to (\mathbb{Q}(\sqrt{3},\,\sqrt{5}))(-\sqrt{2}),$$

$$\psi_{\sqrt{3},\,-\sqrt{3}}:(\mathbb{Q}(\sqrt{2},\,\sqrt{5}))(\sqrt{3}) \to (\mathbb{Q}(\sqrt{2},\,\sqrt{5}))(-\sqrt{3}),$$

$$\psi_{\sqrt{5},\,-\sqrt{5}}:(\mathbb{Q}(\sqrt{2},\,\sqrt{3}))(\sqrt{5}) \to (\mathbb{Q}(\sqrt{2},\,\sqrt{3}))(-\sqrt{5}).$$

For shorter notation, let $\tau_2 = \psi_{\sqrt{2},\,-\sqrt{2}}$, $\tau_3 = \psi_{\sqrt{3},\,-\sqrt{3}}$, and $\tau_5 = \psi_{\sqrt{5},\,-\sqrt{5}}$. Compute the indicated element of E.

9. $\tau_2(\sqrt{3})$ **10.** $\tau_2(\sqrt{2} + \sqrt{5})$

11. $(\tau_3 \tau_2)(\sqrt{2} + 3\sqrt{5})$ **12.** $(\tau_5 \tau_3)\left(\dfrac{\sqrt{2} - 3\sqrt{5}}{2\sqrt{3} - \sqrt{2}}\right)$

13. $(\tau_5^2 \tau_3 \tau_2)(\sqrt{2} + \sqrt{45})$ **14.** $\tau_3[\tau_5(\sqrt{2} - \sqrt{3}) + (\tau_2 \tau_5)\sqrt{30})]$

15. Referring to Example 4, find the following fixed fields in $E = \mathbb{Q}(\sqrt{2}, \sqrt{3})$.

    a. $E_{\{\sigma_1, \sigma_3\}}$ *b.* $E_{\{\sigma_3\}}$ *c.* $E_{\{\sigma_2, \sigma_3\}}$

In Exercises 16 through 21, refer to the directions for Exercises 9 through 14 and find the fixed field of the automorphism or set of automorphisms of E.

16. τ_3 **17.** τ_3^2 **18.** $\{\tau_2, \tau_3\}$

19. $\tau_5 \tau_2$ **20.** $\tau_5 \tau_3 \tau_2$ **21.** $\{\tau_2, \tau_3, \tau_5\}$

22. Refer to the directions for Exercises 9 through 14 for this exercise.

    a. Show that each of the automorphisms τ_2, τ_3, and τ_5 is of order 2 in $G(E/\mathbb{Q})$. (Remember what is meant by the *order* of an element of a group.)

    b. Find the subgroup H of $G(E/\mathbb{Q})$ generated by the elements τ_2, τ_3, and τ_5, and give the group table [*Hint:* There are eight elements.]

    c. Just as was done in Example 5, argue that the group H of part (b) is the full group $G(E/\mathbb{Q})$.

Concepts

23. The fields $\mathbb{Q}(\sqrt{2})$ and $\mathbb{Q}(3 + \sqrt{2})$ are the same, of course. Let $\alpha = 3 + \sqrt{2}$.

    a. Find a conjugate $\beta \neq \alpha$ of α over $\mathbb{Q}$.

    b. Referring to part (a), compare the conjugation automorphism $\psi_{\sqrt{2},\,-\sqrt{2}}$ of $\mathbb{Q}(\sqrt{2})$ with the conjugation automorphism $\psi_{\alpha, \beta}$.

24. Describe the value of the Frobenius automorphism σ_2 on each element of the finite field of four elements given in Example 9 of Section 8.1. Find the fixed field of σ_2.

25. Describe the value of the Frobenius automorphism σ_3 on each element of the finite field of nine elements given in Exercise 18 of Section 8.1. Find the fixed field of σ_3.

26. Let F be a field of characteristic $p \neq 0$. Give an example to show that the map

$\sigma_p : F \to F$ given by $\sigma_p(a) = a^p$ for $a \in F$ need not be an automorphism in the case that F is infinite. What may go wrong?

27. Mark each of the following true or false.
_____ a. For all $\alpha, \beta \in E$, there is always an automorphism of E mapping α onto β.
_____ b. For α, β algebraic over a field F, there is always an isomorphism of $F(\alpha)$ onto $F(\beta)$.
_____ c. For α, β algebraic and conjugate over a field F, there is always an isomorphism of $F(\alpha)$ onto $F(\beta)$.
_____ d. Every automorphism of every field E leaves fixed every element of the prime subfield of E.
_____ e. Every automorphism of every field E leaves fixed an infinite number of elements of E.
_____ f. Every automorphism of every field E leaves fixed at least two elements of E.
_____ g. Every automorphism of every field E of characteristic 0 leaves fixed an infinite number of elements of E.
_____ h. All automorphisms of a field E form a group under function composition.
_____ i. The set of all elements of a field E left fixed by a single automorphism of E forms a subfield of E.
_____ j. For fields $F \leq E \leq K$, $G(K/E) \leq G(K/F)$.

Theory

28. Let α be algebraic of degree n over F. Show from Corollary 1 of Theorem 9.1 that there are at most n different isomorphisms of $F(\alpha)$ onto a subfield of $\overline{F}$ and leaving F fixed.

29. Let $F(\alpha_1, \ldots, \alpha_n)$ be an extension field of F. Show that any automorphism σ of $F(\alpha_1, \ldots, \alpha_n)$ leaving F fixed is completely determined by the n values $\sigma(\alpha_i)$.

30. Let E be an algebraic extension of a field F, and let σ be an automorphism of E leaving F fixed. Let $\alpha \in E$. Show that σ induces a permutation of the set of all zeros of $\mathrm{irr}(\alpha, F)$ that are in E.

31. Let E be an algebraic extension of a field F. Let $S = \{\sigma_i \mid i \in I\}$ be a collection of automorphisms of E such that every σ_i leaves each element of F fixed. Show that if S generates the subgroup H of $G(E/F)$, then $E_S = E_H$.

32. We saw in the corollary of Theorem 5.21 that the cyclotomic polynomial

$$\Phi_p(x) = \frac{x^{p-1} - 1}{x - 1} = x^{p-1} + x^{p-2} + \cdots + x + 1$$

is irreducible over $\mathbb{Q}$ for every prime p. Let ζ be a zero of $\Phi_p(x)$, and consider the field $\mathbb{Q}(\zeta)$.

a. Show that $\zeta, \zeta^2, \ldots, \zeta^{p-1}$ are distinct zeros of $\Phi_p(x)$, and conclude that they are all the zeros of $\Phi_p(x)$.

b. Deduce from Corollary 1 of Theorem 9.1 and part (a) of this exercise that $G(\mathbb{Q}(\zeta)/\mathbb{Q})$ is abelian of order $p - 1$.

c. Show that the fixed field of $G(\mathbb{Q}(\zeta)/\mathbb{Q})$ is $\mathbb{Q}$. [*Hint:* Show that

$$\{\zeta, \zeta^2, \ldots, \zeta^{p-1}\}$$

is a basis for $\mathbb{Q}(\zeta)$ over $\mathbb{Q}$, and consider which linear combinations of $\zeta, \zeta^2, \ldots, \zeta^{p-1}$ are left fixed by all elements of $G(\mathbb{Q}(\zeta)/\mathbb{Q})$.]

33. Theorem 9.1 described conjugation isomorphisms for the case where α and β were conjugate algebraic elements over F. Is there a similar isomorphism of $F(\alpha)$ with $F(\beta)$ in the case that α and β are both transcendental over F?

34. Let F be a field, and let x be an indeterminate over F. Determine all automorphisms of $F(x)$ leaving F fixed, by describing their values on x.

35. Prove the following sequence of theorems.

  a. An automorphism of a field E carries elements that are squares of elements in E onto elements that are squares of elements of E.

  b. An automorphism of the field $\mathbb{R}$ of real numbers carries positive numbers onto positive numbers.

  c. If σ is an automorphism of $\mathbb{R}$ and $a < b$, where $a, b \in \mathbb{R}$, then $\sigma(a) < \sigma(b)$.

  d. The only automorphism of $\mathbb{R}$ is the identity automorphism.

9.2

The Isomorphism Extension Theorem

The Extension Theorem

Let us continue studying automorphisms of fields. In this section and the next, we shall be concerned with both the existence and the number of automorphisms of a field E.

Suppose that E is an algebraic extension of F and that we want to find some automorphisms of E. We know from Theorem 9.1 that if $\alpha, \beta \in E$ are conjugate over F, then there is an isomorphism $\psi_{\alpha,\beta}$ of $F(\alpha)$ onto $F(\beta)$. Of course, $\alpha, \beta \in E$ implies both $F(\alpha) \leq E$ and $F(\beta) \leq E$. It is natural to wonder whether the domain of definition of $\psi_{\alpha,\beta}$ can be enlarged from $F(\alpha)$ to a bigger field, perhaps all of E, and whether this might perhaps lead to an automorphism of E. A mapping diagram of this situation is shown in Fig. 9.2.

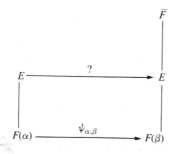

Figure 9.2

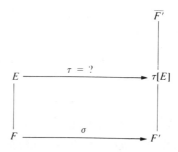

Figure 9.3

Rather than speak of "enlarging the domain of definition of $\psi_{\alpha,\beta}$," it is customary to speak of "**extending the map** $\psi_{\alpha,\beta}$ **to a map** τ," which is a mapping of all of E.

Remember that we are always assuming that all algebraic extensions of F under consideration are contained in a fixed algebraic closure $\bar{F}$ of F. The Isomorphism Extension Theorem shows that the mapping $\psi_{\alpha,\beta}$ can indeed always be extended to an *isomorphism* of E onto a subfield of $\bar{F}$. Whether this extension gives an *automorphism* of E, that, is, maps E into itself, is a question we shall study in Section 9.3. Thus this extension theorem, used in conjunction with our conjugation isomorphisms $\psi_{\alpha,\beta}$, will guarantee the existence of lots of *isomorphism mappings*, at least, for many fields. Extension theorems are very important in mathematics, particularly in algebraic and topological situations.

Let us take a more general look at this situation. Suppose that E is an algebraic extension of a field F and that we have an isomorphism σ of F onto a field F'. Let $\bar{F'}$ be an algebraic closure of $\bar{F'}$. We would like to extend σ to an isomorphism τ of E onto a subfield of F'. This situation is shown in Fig. 9.3. Naively, we pick $\alpha \in E$ but not in F and try to extend σ to $F(\alpha)$. If

$$p(x) = \operatorname{irr}(\alpha, F) = a_0 + a_1 x + \cdots + a_n x^n,$$

let β be a zero in $\bar{F'}$ of

$$q(x) = \sigma(a_0) + \sigma(a_1)x + \cdots + \sigma(a_n)x^n.$$

Here $q(x) \in F'[x]$. Since σ is an isomorphism, we know that $q(x)$ is irreducible in $F'[x]$. It seems reasonable that $F(\alpha)$ can be mapped isomorphically onto $F'(\beta)$ by a map extending σ and mapping α onto β. (This isn't quite Theorem 9.1, but it is close to it; a few elements have been renamed by the isomorphism σ.) If $F(\alpha) = E$, we are done. If $F(\alpha) \neq E$, we have to find another element in E not in $F(\alpha)$ and continue the process. It is a situation very much like that in the construction of an algebraic closure $\bar{F}$ of a field F. Again the trouble is that, in general, where E is not a finite extension, the process may have to be repeated a (possibly large) infinite number of times, so we need Zorn's lemma to handle it. For this reason, we postpone the general proof of Theorem 9.6 to the end of this section.

THEOREM 9.6 (Isomorphism Extension Theorem)   Let E be an algebraic extension of a field F. Let σ be an isomorphism of F onto a field F'. Let $\bar{F'}$ be an algebraic closure of F'. Then σ can be extended to an isomorphism τ of E onto a subfield of $\bar{F'}$ such that $\tau(a) = \sigma(a)$ for all $a \in F$.

We give as a corollary the existence of an extension of one of our conjugation isomorphisms $\psi_{\alpha,\beta}$, as discussed at the start of this section.

COROLLARY I   If $E \leq \bar{F}$ is an algebraic extension of F and $\alpha, \beta \in E$ are conjugate over F, then the conjugation isomorphism $\psi_{\alpha,\beta}: F(\alpha) \to F(\beta)$,

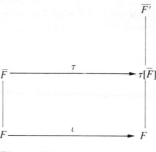

Figure 9.4

given by Theorem 9.1, can be extended to an isomorphism of E onto a subfield of $\bar{F}$.

PROOF Proof of this corollary is immediate from Theorem 9.6 if in the statement of the theorem we replace F by $F(\alpha)$, F' by $F(\beta)$, and $\overline{F'}$ by $\bar{F}$. ◆

As another corollary, we can show, as we promised earlier, that an algebraic closure of F is unique, up to an isomorphism leaving F fixed.

COROLLARY 2 Let $\bar{F}$ and $\overline{F'}$ be two algebraic closures of F. Then $\bar{F}$ is isomorphic to $\overline{F'}$ under an isomorphism leaving each element of F fixed.

PROOF By Theorem 9.6, the identity isomorphism of F onto F can be extended to an isomorphism τ mapping $\bar{F}$ onto a subfield of $\overline{F'}$ that leaves F fixed (see Fig. 9.4). We need only show that τ is onto $\overline{F'}$. But by Theorem 9.6, the map $\tau^{-1}: \tau[\bar{F}] \to \bar{F}$ can be extended to an isomorphism of $\overline{F'}$ onto a subfield of $\bar{F}$. Since τ^{-1} is already onto $\bar{F}$, we must have $\tau[\bar{F}] = \overline{F'}$. ◆

The Index of a Field Extension

Having discussed the question of *existence*, we turn now to the question of *how many*. For a *finite* extension E of a field F, we would like to count how many isomorhisms there are of E onto a subfield of $\bar{F}$ that leave F fixed. We shall show that there are only a finite number of isomorphisms. Since every automorphism in $G(E/F)$ is such an isomorphism, a count of these isomorphisms will include all these automorphisms. Example 5, Section 9.1, showed that $G(\mathbb{Q}(\sqrt{2}, \sqrt{3})/\mathbb{Q})$ has four elements, and that $4 = [\mathbb{Q}(\sqrt{2}, \sqrt{3}):\mathbb{Q}]$. While such an equality is not always true, it is true in a very important case. The next theorem takes the first big step in proving this. We state the theorem in more general terms than we shall need, but it does not make the proof any harder.

THEOREM 9.7 Let E be a finite extension of a field F. Let σ be an isomorphism of F onto a field F', and let $\overline{F'}$ be an algebraic closure of F'.

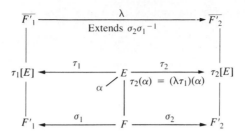

Figure 9.5

Then the number of extensions of σ to an isomorphism τ of E onto a subfield of $\overline{F'}$ is finite, and independent of F', $\overline{F'}$, and σ. That is, the number of extensions is completely determined by the two fields E and F; it is intrinsic to them.

PROOF The diagram in Fig. 9.5 may help us to follow the construction that we are about to make. This diagram is constructed in the following way. Consider two isomorphisms

$$\sigma_1:F \xrightarrow{\text{onto}} F_1', \qquad \sigma_2:F \xrightarrow{\text{onto}} F_2',$$

where $\overline{F_1'}$ and $\overline{F_2'}$ are algebraic closures of F_1' and F_2', respectively. Now $\sigma_2\sigma_1^{-1}$ is an isomorphism of F_1' onto F_2'. Then by Theorem 9.6 and its second corollary, there is an isomorphism

$$\lambda:\overline{F_1'} \xrightarrow{\text{onto}} \overline{F_2'}$$

extending this isomorphism $\sigma_2\sigma_1^{-1}:F' \xrightarrow{\text{onto}} F_2'$. Referring to Fig. 9.5, corresponding to each $\tau_1:E \to \overline{F_1'}$ that extends σ_1, we obtain an iso-morphism $\tau_2:E \to \overline{F_2'}$, by starting at E and going first to the left, then up, and then to the right. Written algebraically,

$$\tau_2(\alpha) = (\lambda\tau_1)(\alpha)$$

for $\alpha \in E$. Clearly τ_2 extends σ_2. The fact that we could have *started* with τ_2 and recovered τ_1 by defining

$$\tau_1(\alpha) = (\lambda^{-1}\tau_2)(\alpha),$$

that is, by chasing the other way around the diagram, shows that the correspondence between $\tau_1:E \to \overline{F_1'}$ and $\tau_2:E \to \overline{F_2'}$ is one to one. In view of this one-to-one correspondence, the number of τ extending σ is independent of F', $\overline{F'}$, and σ.

That the number of mappings extending σ is finite follows from the fact that since E is a finite extension of F, $E = F(\alpha_1, \ldots, \alpha_n)$ for some $\alpha_1, \ldots, \alpha_n$ in E, by Theorem 8.11. There are only a finite number of possible candidates for the images $\tau(\alpha_i)$ in F', for if

$$\mathrm{irr}(\alpha_i, F) = a_{i0} + a_{i1}x + \cdots + a_{im_i}x^{m_i},$$

where $a_{ik} \in F$, then $\tau(\alpha_i)$ must be one of the zeros in $\overline{F'}$ of

$$[\sigma(a_{i0}) + \sigma(a_{i1})x + \cdots + \sigma(a_{im_i})x^{m_i}] \in F'[x]. \quad \blacklozenge$$

DEFINITION 9.6 (Index of E over F) Let E be a finite extension of a field F. The number of isomorphisms of E onto a subfield of $\overline{F}$ leaving F fixed is the **index** $\{E:F\}$ **of** E **over** F.

COROLLARY If $F \le E \le K$, where K is a finite extension field of the field F, then $\{K:F\} = \{K:E\}\{E:F\}$.

PROOF It follows from Theorem 9.7 that each of the $\{E:F\}$ isomorphisms τ_i of E onto a subfield of $\overline{F}$ leaving F fixed has $\{K:E\}$ extensions to an isomorphism of K onto a subfield of $\overline{F}$. $\quad \blacklozenge$

The preceding corollary was really the main thing we were after. Note that it counts something. *Never underestimate a result that counts something*, even if it is only called a "corollary."

We shall show in Section 9.4 that unless F is an infinite field of characteristic $p \ne 0$, we always have $[E:F] = \{E:F\}$ for every finite extension field E of F. For the case $E = F(\alpha)$, the $\{F(\alpha):F\}$ extensions of the identity map $\iota:F \to F$ to maps of $F(\alpha)$ onto a subfield of $\overline{F}$ are given by the conjugation isomorphisms $\psi_{\alpha,\beta}$ for each conjugate β in $\overline{F}$ of α over F. Thus if $\mathrm{irr}(\alpha, F)$ has n *distinct* zeros in $\overline{F}$, we have $\{E:F\} = n$. We shall show later that unless F is infinite and of characteristic $p \ne 0$, the number of distinct zeros of $\mathrm{irr}(\alpha, F)$ is $\deg(\alpha, F) = [F(\alpha):F]$.

EXAMPLE I Consider $E = \mathbb{Q}(\sqrt{2}, \sqrt{3})$ over $\mathbb{Q}$, as in Example 5, Section 9.1. Our work in that example shows that $\{E:\mathbb{Q}\} = [E:\mathbb{Q}] = 4$. Also, $\{E:\mathbb{Q}(\sqrt{2})\} = 2$, and $\{\mathbb{Q}(\sqrt{2}):\mathbb{Q}\} = 2$, so

$$4 = \{E:\mathbb{Q}\} = \{E:\mathbb{Q}(\sqrt{2})\}\{\mathbb{Q}(\sqrt{2}):\mathbb{Q}\} = (2)(2).$$

This illustrates the corollary of Theorem 9.7. $\quad \blacktriangle$

Proof of the Extension Theorem

We restate the extension theorem.

Isomorphism Extension Theorem Let E be an algebraic extension of a field F. Let σ be an isomorphism of F onto a field F'. Let $\overline{F'}$ be an algebraic closure of F'. Then σ can be extended to an isomorphism τ of E onto a subfield of $\overline{F'}$ such that $\tau(a) = \sigma(a)$ for $a \in F$.

PROOF Consider all pairs (L, λ), where L is a field such that $F \leq L \leq E$ and λ is an isomorphism of L onto a subfield of $\overline{F}'$ such that $\lambda(a) = \sigma(a)$ for $a \in F$. The set S of such pairs (L, λ) is nonempty, since (F, σ) is such a pair. Define a partial ordering on S by $(L_1, \lambda_1) \leq (L_2, \lambda_2)$, if $L_1 \leq L_2$ and $\lambda_1(a) = \lambda_2(a)$ for $a \in L_1$. It is readily checked that this relation $\leq$ does give a partial ordering of S.

Let $T = \{(H_i, \lambda_i) \mid i \in I\}$ be a chain of S. We claim that $H = \bigcup_{i \in I} H_i$ is a subfield of E. Let $a, b \in H$, where $a \in H_1$ and $b \in H_2$; then either $H_1 \leq H_2$ or $H_2 \leq H_1$, since T is a chain. If, say, $H_1 \leq H_2$, then $a, b \in H_2$, so $a \pm b$, ab, and a/b for $b \neq 0$ are all in H_2 and hence in H. Since for each $i \in I$, $F \subseteq H_i \subseteq E$, we have $F \subseteq H \subseteq E$. Thus H is a subfield of E.

Define $\lambda : H \to \overline{F}'$ as follows. Let $c \in H$. Then $c \in H_i$ for some $i \in I$, and let

$$\lambda(c) = \lambda_i(c).$$

The map λ is well defined, because if $c \in H_1$ and $c \in H_2$, then either $(H_1, \lambda_1) \leq (H_2, \lambda_2)$ or $(H_2, \lambda_2) \leq (H_1, \lambda_1)$, since T is a chain. In either case, $\lambda_1(c) = \lambda_2(c)$. We claim that λ is an isomorphism of H onto a subfield of $\overline{F}'$. If $a, b \in H$, then there is an H_i such that $a, b \in H_i$, and

$$\lambda(a + b) = \lambda_i(a + b) = \lambda_i(a) + \lambda_i(b) = \lambda(a) + \lambda(b).$$

Similarly,

$$\lambda(ab) = \lambda_i(ab) = \lambda_i(a)\lambda_i(b) = \lambda(a)\lambda(b).$$

If $\lambda(a) = 0$, then $a \in H_i$ for some i implies that $\lambda_i(a) = 0$, so $a = 0$. Therefore, λ is an isomorphism. Thus $(H, \lambda) \in S$, and it is clear from our definitions of H and λ that (H, λ) is an upper bound for T.

We have shown that every chain of S has an upper bound in S, so the hypotheses of Zorn's lemma are satisfied. Hence there exists a maximal element (K, τ) of S. Let $\tau(K) = K'$, where $K' \leq \overline{F}'$. Now if $K \neq E$, let $\alpha \in E$ but $\alpha \notin K$. Now α is algebraic over F, so α is algebraic over K. Also, let $p(x) = \mathrm{irr}(\alpha, K)$. Let ψ_α be the canonical isomorphism

$$\psi_\alpha : K[x]/\langle p(x)\rangle \to K(\alpha),$$

corresponding to the evaluation homomorphism $\phi_\alpha : K[x] \to K(\alpha)$. If

$$p(x) = a_0 + a_1 x + \cdots + a_n x^n,$$

consider

$$q(x) = \tau(a_0) + \tau(a_1)x + \cdots + \tau(a_n)x^n$$

in $K'[x]$. Since τ is an isomorphism, $q(x)$ is irreducible in $K'[x]$. Since $K' \leq \overline{F}'$, there is a zero α' of $q(x)$ in $\overline{F}'$. Let

$$\psi_{\alpha'} : K'[x]/\langle q(x)\rangle \to K'(\alpha')$$

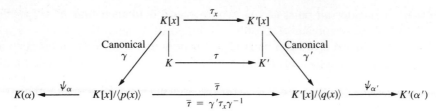

Figure 9.6

be the isomorphism analogous to ψ_α. Finally, let

$$\bar\tau : K[x]/\langle p(x)\rangle \to K'[x]/\langle q(x)\rangle$$

be the isomorphism extending τ on K and mapping $x + \langle p(x)\rangle$ onto $x + \langle q(x)\rangle$. (See Fig. 9.6.) Then the composition of maps

$$\psi_{\alpha'}\bar\tau\psi_\alpha^{-1} : K(\alpha) \to K'(\alpha')$$

is an isomorphism of $K(\alpha)$ onto a subfield of $\overline{F}'$. Clearly, $(K, \tau) < (K(\alpha), \psi_{\alpha'}\bar\tau\psi_\alpha^{-1})$, which contradicts that (K, τ) is maximal. Therefore we must have had $K = E$. ◆

Exercises 9.2

Computations

Let $E = \mathbb{Q}(\sqrt2, \sqrt3, \sqrt5)$. In Exercises 1 through 3, for the given isomorphic mapping of a subfield of E, give all extensions of the mapping to an isomorphic mapping of E onto a subfield of $\overline{\mathbb{Q}}$. Describe the extensions by giving values on the generating set $\{\sqrt2, \sqrt3, \sqrt5\}$ for E over $\mathbb{Q}$.

1. $\iota : \mathbb{Q}(\sqrt2, \sqrt{15}) \to \mathbb{Q}(\sqrt2, \sqrt{15})$, where ι is the identity map
2. $\sigma : \mathbb{Q}(\sqrt2, \sqrt{15}) \to \mathbb{Q}(\sqrt2, \sqrt{15})$ where $\sigma(\sqrt2) = \sqrt2$ and $\sigma(\sqrt{15}) = -\sqrt{15}$
3. $\psi_{\sqrt{30}, -\sqrt{30}} : \mathbb{Q}(\sqrt{30}) \to \mathbb{Q}(\sqrt{30})$

It is a fact, which we can verify by cubing, that the zeros of $x^3 - 2$ in $\mathbb{Q}$ are

$$\alpha_1 = \sqrt[3]2, \qquad \alpha_2 = \sqrt[3]2 \frac{-1 + i\sqrt3}{2}, \qquad \text{and} \qquad \alpha_3 = \sqrt[3]2 \frac{-1 - i\sqrt3}{2},$$

where $\sqrt[3]2$, as usual, is the real cube root of 2. Use this information in Exercises 4 through 6.

4. Describe all extensions of the identity map of $\mathbb{Q}$ to an isomorphism mapping $\mathbb{Q}(\sqrt[3]2)$ onto a subfield of $\overline{\mathbb{Q}}$.
5. Describe all extensions of the identity map of $\mathbb{Q}$ to an isomorphism mapping $\mathbb{Q}(\sqrt[3]2, \sqrt3)$ onto a subfield of $\overline{\mathbb{Q}}$.

6. Describe all extensions of the automorphism $\psi_{\sqrt{3},\,-\sqrt{3}}$ of $\mathbb{Q}(\sqrt{3})$ to an isomorphism mapping $\mathbb{Q}(i, \sqrt{3}, \sqrt[3]{2})$ onto a subfield of $\overline{\mathbb{Q}}$.

7. Let σ be the automorphism of $\mathbb{Q}(\pi)$ that maps π onto $-\pi$.

   a. Describe the fixed field of σ.

   b. Describe all extensions of σ to an isomorphism mapping the field $\mathbb{Q}(\sqrt{\pi})$ onto a subfield of $\overline{\mathbb{Q}(\pi)}$.

Concepts

8. Mark each of the following true or false.
   _____ a. Let $F(\alpha)$ be any simple extension of a field F. Then every isomorphism of F onto a subfield of $\overline{F}$ has an extension to an isomorphism of $F(\alpha)$ onto a subfield of $\overline{F}$.
   _____ b. Let $F(\alpha)$ be any simple algebraic extension of a field F. Then every isomorphism of F onto a subfield of $\overline{F}$ has an extension to an isomorphism of $F(\alpha)$ onto a subfield of $\overline{F}$.
   _____ c. An isomorphism of F onto a subfield of $\overline{F}$ has the same number of extensions to each simple algebraic extension of F.
   _____ d. Algebraic closures of isomorphic fields are always isomorphic.
   _____ e. Algebraic closures of fields that are not isomorphic are never isomorphic.
   _____ f. Any algebraic closure of $\mathbb{Q}(\sqrt{2})$ is isomorphic to any algebraic closure of $\mathbb{Q}(\sqrt{17})$.
   _____ g. The index of a finite extension E over a field F is finite.
   _____ h. The index behaves multiplicatively with respect to finite towers of finite extensions of fields.
   _____ i. Our remarks prior to the first statement of Theorem 9.6 essentially constitute a proof of this theorem for a finite extension E over F.
   _____ j. Corollary 2 of Theorem 9.6 shows that $\mathbb{C}$ is isomorphic to $\overline{\mathbb{Q}}$.

Theory

9. Let K be an algebraic closed field. Show that every isomorphism σ of K onto a subfield of itself such that K is algebraic over $\sigma[K]$ is an automorphism of K, that is, is an onto map. [*Hint*: Apply Theorem 9.6 to σ^{-1}.]

10. Let E be an algebraic extension of a field F. Show that every isomorphism of E onto a subfield of $\overline{F}$ leaving F fixed can be extended to an automorphism of $\overline{F}$.

11. Prove that if E is an algebraic extension of a field F, then two algebraic closures $\overline{F}$ and $\overline{E}$ of F and E, respectively, are isomorphic.

12. Prove that the algebraic closure of $\mathbb{Q}(\sqrt{\pi})$ in $\mathbb{C}$ is isomorphic to any algebraic closure of $\overline{\mathbb{Q}}(x)$, where $\overline{\mathbb{Q}}$ is the field of algebraic numbers and x is an indeterminate.

13. Prove that if E is a finite extension of a field F, then $\{E:F\} \le [E:F]$. [*Hint*: The remarks preceding Example 1 essentially showed this for a simple algebraic extension $F(\alpha)$ of F. Use the fact that a finite extension is a tower of simple extensions, together with the multiplicative properties of the index and degree.]

9.3

Splitting Fields

We are going to be interested chiefly in *automorphisms* of a field E, rather than mere isomorphic mappings of E onto a subfield of $\bar{E}$. It is the *automorphisms* of a field that form a group. We wonder whether for some extension field E of a field F, *every* isomorphic mapping of E onto a subfield of $\bar{F}$ leaving F fixed is actually an automorphism of E.

Suppose E is an algebraic extension of a field F. If $\alpha \in E$ and $\beta \in \bar{F}$ is a conjugate of α over F, then there is a conjugation isomorphism

$$\psi_{\alpha,\beta} : F(\alpha) \to F(\beta)$$

By Corollary 1 of Theorem 9.6, $\psi_{\alpha,\beta}$ can be extended to an isomorphic mapping of E onto a subfield of $\bar{F}$. Now if $\beta \notin E$, such an isomorphic mapping of E can't be an automorphism of E. Thus, *if an algebraic extension E of a field F is such that all its isomorphic mappings onto a subfield of $\bar{F}$ leaving F fixed are actually automorphisms of E, then for every $\alpha \in E$, all conjugates of α over F must be in E also.* This observation seemed to come very easily. We point out that we used a lot of power, namely the existence of the conjugation isomorphisms and the Isomorphism Extension Theorem.

These ideas suggest the formulation of the following definition.

DEFINITION 9.7 (Splitting Field) Let F be a field with algebraic closure $\bar{F}$. Let $\{f_i(x) \mid i \in I\}$ be a collection of polynomials in $F[x]$. A field $E \le \bar{F}$ is the **splitting field of** $\{f_i(x) \mid i \in I\}$ **over** F if E is the smallest subfield of $\bar{F}$ containing F and all the zeros in $\bar{F}$ of each of the $f_i(x)$ for $i \in I$. A field $K \le \bar{F}$ is a **splitting field over** F if it is the splitting field of some set of polynomials in $F[x]$.

EXAMPLE 1 We see that $\mathbb{Q}[\sqrt{2}, \sqrt{3}]$ is a splitting field of $\{x^2 - 2, x^2 - 3\}$, and also of $\{x^4 - 5x^2 + 6\}$. ▲

For one polynomial $f(x) \in F[x]$, we shall often refer to the splitting field of $\{f(x)\}$ over F as the **splitting field of** $f(x)$ **over** F. Note that the splitting field of $\{f_i(x) \mid i \in I\}$ over F in $\bar{F}$ is the intersection of all subfields of $\bar{F}$ containing F and all zeros in $\bar{F}$ of each $f_i(x)$ for $i \in I$. Thus such a splitting field surely does exist.

We now show that splitting fields over F are precisely those fields $E \le \bar{F}$ with the property that all isomorphic mappings of E onto a subfield of $\bar{F}$ leaving F fixed are automorphisms of E. This will be a corollary of the next theorem. *Once more, we are characterizing a concept in terms of mappings.* Remember, we are always assuming that all algebraic extensions of a field F under consideration are in one fixed algebraic closure $\bar{F}$ of F.

THEOREM 9.8 A field E, where $F \leq E \leq \bar{F}$, is a splitting field over F if and only if every automorphism of $\bar{F}$ leaving F fixed maps E onto itself and thus induces an automorphism of E leaving F fixed.

PROOF Let E be a splitting field over F in $\bar{F}$ of $\{f_i(x) \mid i \in I\}$, and let σ be an automorphism of $\bar{F}$ leaving F fixed. Let $\{\alpha_j \mid j \in J\}$ be the collection of all zeros in $\bar{F}$ of all the $f_i(x)$ for $i \in I$. Now our previous work shows that for a fixed α_j, the field $F(\alpha_j)$ has as elements all expressions of the form

$$g(\alpha_j) = a_0 + a_1\alpha_j + \cdots + a_{n_j-1}\alpha_j^{n_j-1}.$$

where n_j is the degree of $\mathrm{irr}(\alpha_j, F)$ and $a_k \in F$. Consider the set S of all *finite* sums of *finite* products of elements of the form $g(\alpha_j)$ for all $j \in J$. The set S is a subset of E closed under addition and multiplication and containing 0, 1, and the additive inverse of each element. Since each element of S is in some $F(\alpha_{j_1}, \ldots, \alpha_{j_r}) \subseteq S$, we see that S also contains the multiplicative inverse of each nonzero element. Thus S is a subfield of E containing all α_j for $j \in J$. By definition of the splitting field E of $\{f_i(x) \mid i \in I\}$, we see that we must have $S = E$. All this work was just to show that $\{\alpha_j \mid j \in J\}$ *generates* E over F, in the sense of taking *finite* sums and *finite* products. Knowing this, we see immediately that the value of σ on any element of E is completely determined by the values $\sigma(\alpha_j)$. But by Corollary 1 of Theorem 9.1, $\sigma(\alpha_j)$ must also be a zero of $\mathrm{irr}(\alpha_j, F)$. By Theorem 8.3, $\mathrm{irr}(\alpha_j, F)$ divides the $f_i(x)$ for which $f_i(\alpha_j) = 0$, so $\sigma(\alpha_j) \in E$ also. Thus σ maps E onto a subfield of E isomorphically. However, the same is true of the automorphism σ^{-1} of $\bar{F}$. Since for $\beta \in E$.

$$\beta = \sigma(\sigma^{-1}(\beta)),$$

we see that σ maps E onto E, and thus induces an automorphism of E.

Suppose, conversely, that every automorphism of $\bar{F}$ leaving F fixed induces an automorphism of E. Let $g(x)$ be an *irreducible* polynomial in $F[x]$ having a zero α in E. If β is any zero of $g(x)$ in $\bar{F}$, then by Theorem 9.1, there is a conjugation isomorphism $\psi_{\alpha,\beta}$ of $F(\alpha)$ onto $F(\beta)$ leaving F fixed. By Theorem 9.6, $\psi_{\alpha,\beta}$ can be extended to an isomorphism τ of $\bar{F}$ onto a subfield of $\bar{F}$. But then

$$\tau^{-1} : \tau[\bar{F}] \to \bar{F}$$

can be extended to an isomorphism mapping $\bar{F}$ onto a subfield of $\bar{F}$. Since the image of τ^{-1} is already all of $\bar{F}$, we see that τ must have been onto $\bar{F}$, so τ is an automorphism of $\bar{F}$ leaving F fixed. Then by assumption, τ induces an automorphism of E, so $\tau(\alpha) = \beta$ is in E. We have shown that if $g(x)$ is an irreducible polynomial in $F[x]$ having one zero in E, then all zeros of $g(x)$ in $\bar{F}$ are in E. Hence if $\{g_k(x)\}$ is the set of *all* irreducible polynomials in $F[x]$ having a zero in E, then E is the splitting field of $\{g_k(x)\}$. ◆

DEFINITION 9.8 (Polynomial Splits in E) Let E be an extension field of a field F. A polynomial $f(x) \in F[x]$ **splits in** E if it factors into a product of linear factors in $E[x]$.

EXAMPLE 2 The polynomial $x^4 - 5x^2 + 6$ in $\mathbb{Q}[x]$ splits in $\mathbb{Q}[\sqrt{2}, \sqrt{3}]$ into $(x - \sqrt{2})(x + \sqrt{2})(x - \sqrt{3})(x + \sqrt{3})$. ▲

COROLLARY 1 If $E \le \bar{F}$ is a splitting field over F, then every irreducible polynomial in $F[x]$ having a zero in E splits in E.

PROOF If E is a splitting field over F in $\bar{F}$, then every automorphism of $\bar{F}$ induces an automorphism of E. The second half of the proof of Theorem 9.8 showed precisely that E is also the splitting field over F of the set $\{g_k(x)\}$ of *all* irreducible polynomials in $F[x]$ having a zero in E. Thus an irreducible polynomial $f(x)$ of $F[x]$ having a zero in E has all its zeros in $\bar{F}$ in E. Therefore, its factorization into linear factors in $\bar{F}[x]$, given by Theorem 8.13, actually takes place in $E[x]$, so $f(x)$ splits in E. ◆

COROLLARY 2 If $E \le \bar{F}$ is a splitting field over F, then every isomorphic mapping of E onto a subfield of $\bar{F}$ and leaving F fixed is actually an automorphism of E. In particular, if E is a splitting field of finite degree over F, then

$$\{E:F\} = |G(E/F)|.$$

PROOF Every isomorphism σ mapping E onto a subfield of $\bar{F}$ leaving F fixed can be extended to an automorphism τ of $\bar{F}$, by Theorem 9.6, together with the *onto* argument of the second half of the proof of Theorem 9.8. If E is a splitting field over F, then by Theorem 9.8, τ restricted to E, that is σ, is an automorphism of E. Thus for a splitting field E over F, every isomorphic mapping of E onto a subfield of $\bar{F}$ leaving F fixed is an automorphism of E.

The equation $\{E:F\} = |G(E/F)|$ then follows immediately for a splitting field E of finite degree over F, since $\{E:F\}$ was defined as the number of different isomorphic mappings of E onto a subfield of $\bar{F}$ leaving F fixed. ◆

EXAMPLE 3 Observe that $\mathbb{Q}(\sqrt{2}, \sqrt{3})$ is the splitting field of

$$\{x^2 - 2, x^2 - 3\}$$

over $\mathbb{Q}$. Example 5, Section 9.1, showed that the mappings $\iota, \sigma_1, \sigma_2$, and σ_3 are all the automorphisms of $\mathbb{Q}(\sqrt{2}, \sqrt{3})$ leaving $\mathbb{Q}$ fixed. (Actually, since every automorphism of a field must leave the prime subfield fixed, we see that these are the only automorphisms of $\mathbb{Q}(\sqrt{2}, \sqrt{3})$.) Then

$$\{\mathbb{Q}(\sqrt{2}, \sqrt{3}):\mathbb{Q}\} = |G(\mathbb{Q}(\sqrt{2}, \sqrt{3})/\mathbb{Q})| = 4.$$

illustrating Corollary 2. ▲

We wish to determine conditions under which

$$|G(E/F)| = \{E{:}F\} = [E{:}F]$$

for finite extensions E of F. This is our next topic. We shall show in the following section that this equation always holds when E is a splitting field over a field F of characteristic 0 or when F is a finite field. This equation need not be true when F is an infinite field of characteristic $p \neq 0$.

EXAMPLE 4   Let $\sqrt[3]{2}$ be the real cube root of 2, as usual. Now $x^3 - 2$ does not split in $\mathbb{Q}(\sqrt[3]{2})$, for $\mathbb{Q}(\sqrt[3]{2}) < \mathbb{R}$, and only one zero of $x^3 - 2$ is real. Thus $x^3 - 2$ factors in $(\mathbb{Q}(\sqrt[3]{2}))[x]$ into a linear factor $x - \sqrt[3]{2}$ and an irreducible quadratic factor. The splitting field E of $x^3 - 2$ over $\mathbb{Q}$ is therefore of degree 2 over $\mathbb{Q}(\sqrt[3]{2})$. Then

$$[E{:}\mathbb{Q}] = [E{:}\mathbb{Q}(\sqrt[3]{2})][\mathbb{Q}(\sqrt[3]{2}){:}\mathbb{Q}] = (2)(3) = 6.$$

We have shown that the splitting field over $\mathbb{Q}$ of $x^3 - 2$ is of degree 6 over $\mathbb{Q}$.
   We can verify by cubing that

$$\sqrt[3]{2}\,\frac{-1 + i\sqrt{3}}{2} \qquad \text{and} \qquad \sqrt[3]{2}\,\frac{-1 - i\sqrt{3}}{2}$$

are the other zeros of $x^3 - 2$ in $\mathbb{C}$. Thus the splitting field E of $x^3 - 2$ over $\mathbb{Q}$ is $\mathbb{Q}(\sqrt[3]{2}, i\sqrt{3})$. (This is *not* the same field as $\mathbb{Q}(\sqrt[3]{2}, i, \sqrt{3})$, which is of degree 12 over $\mathbb{Q}$.) Further study of this interesting example is left to the exercises (see Exercises 7, 8, 9, 14, 19, and 21). ▲

Exercises 9.3

Computations

In Exercises 1 through 6, find the degree over $\mathbb{Q}$ of the splitting field over $\mathbb{Q}$ of the given polynomial in $\mathbb{Q}[x]$.

1. $x^2 + 3$	**2.** $x^4 - 1$	**3.** $(x^2 - 2)(x^2 - 3)$
4. $x^3 - 3$	**5.** $x^3 - 1$	**6.** $(x^2 - 2)(x^3 - 2)$

Refer to Example 4 for Exercises 7 through 9.

7. What is the order of $G(\mathbb{Q}(\sqrt[3]{2})/\mathbb{Q})$?

8. What is the order of $G(\mathbb{Q}(\sqrt[3]{2}, i\sqrt{3})/\mathbb{Q})$?

9. What is the order of $G(\mathbb{Q}(\sqrt[3]{2}, i\sqrt{3})/\mathbb{Q}(\sqrt[3]{2}))$?

10. Let α be a zero of $x^3 + x^2 + 1$ over $\mathbb{Z}_2$. Show that $x^3 + x^2 + 1$ splits in $\mathbb{Z}_2(\alpha)$. [*Hint:* There are eight elements in $\mathbb{Z}_2(\alpha)$. Exhibit two more zeros of $x^3 + x^2 + 1$, in addition to α, among these eight elements. Alternatively, use the results of Section 8.5.]

Concepts

11. Let $f(x)$ be a polynomial in $F[x]$ of degree n. Let $E \leq \bar{F}$ be the splitting field of $f(x)$ over F in $\bar{F}$. What bounds can be put on $[E:F]$?

12. Mark each of the following true or false.
    _____ a. Let $\alpha, \beta \in E$, where $E \leq \bar{F}$ is a splitting field over F. Then there exists an automorphism of E leaving F fixed and mapping α onto β if and only if $\mathrm{irr}(\alpha, F) = \mathrm{irr}(\beta, F)$.
    _____ b. $\mathbb{R}$ is a splitting field over $\mathbb{Q}$.
    _____ c. $\mathbb{R}$ is a splitting field over $\mathbb{R}$.
    _____ d. $\mathbb{C}$ is a splitting field over $\mathbb{R}$.
    _____ e. $\mathbb{Q}(i)$ is a splitting field over $\mathbb{Q}$.
    _____ f. $\mathbb{Q}(\pi)$ is a splitting field over $\mathbb{Q}(\pi^2)$.
    _____ g. For every splitting field E over F, where $E \leq \bar{F}$, every isomorphic mapping of E is an automorphism of E.
    _____ h. For every splitting field E over F, where $E \leq \bar{F}$, every isomorphism mapping E onto a subfield of $\bar{F}$ is an automorphism of E.
    _____ i. For every splitting field E over F, where $E \leq \bar{F}$, every isomorphism mapping E onto a subfield of $\bar{F}$ and leaving F fixed is an automorphism of E.
    _____ j. Every algebraic closure $\bar{F}$ of a field F is a splitting field over F.

13. Show by an example that Corollary 1 of Theorem 9.8 is no longer true if the word *irreducible* is deleted.

14. a. Is $|G(E/F)|$ multiplicative for finite towers of finite extensions, that is, is
$$|G(K/F)| = |G(K/E)||G(E/F)| \qquad \text{for} \quad F \leq E \leq K \leq \bar{F}?$$
    Why? [*Hint:* Use Exercises 7 through 9.]
    b. Is $|G(E/F)|$ multiplicative for finite towers of finite extensions, each of which is a splitting field over the bottom field? Why?

Theory

15. Show that if a finite extension E of a field F is a splitting field over F, then E is a splitting field of one polynomial in $F[x]$.

16. Show that if $[E:F] = 2$, then E is a splitting field over F.

17. Show that for $F \leq E \leq \bar{F}$, E is a splitting field over F if and only if E contains all conjugates over F in $\bar{F}$ for each of its elements.

18. Show that $\mathbb{Q}(\sqrt[3]{2})$ has only the identity automorphism.

19. Referring to Example 4, show that
$$G(\mathbb{Q}(\sqrt[3]{2}, i\sqrt{3})/\mathbb{Q}(i\sqrt{3})) \simeq \langle \mathbb{Z}_3, + \rangle.$$

20. a. Show that an automorphism of a splitting field E over F of a polynomial $f(x) \in F[x]$ permutes the zeros of $f(x)$ in E.
    b. Show that an automorphism of a splitting field E over F of a polynomial

$f(x) \in F[x]$ is completely determined by the permutation of the zeros of $f(x)$ in E given in part (a).

c. Show that if E is a splitting field over F of a polynomial $f(x) \in F[x]$, then $G(E/F)$ can be viewed in a natural way as a certain group of permutations.

21. Let E be the splitting field of $x^3 - 2$ over $\mathbb{Q}$, as in Example 4.

a. What is the order of $G(E/\mathbb{Q})$? [*Hint:* Use Corollary 2 of Theorem 9.8 and the corollary of Theorem 9.6 applied to the tower $\mathbb{Q} \le \mathbb{Q}(i\sqrt{3}) \le E$.]

b. Show that $G(E/\mathbb{Q}) = S_3$, the symmetric group on three letters. [*Hint:* Use Exercise 20, together with part (a).]

22. Show that for a prime p, the splitting field over $\mathbb{Q}$ of $x^p - 1$ is of degree $p - 1$ over $\mathbb{Q}$. [*Hint:* Refer to the corollary of Theorem 5.21.]

23. Let $\bar{F}$ and $\bar{F}'$ be two algebraic closures of a field F, and let $f(x) \in F[x]$. Show that the splitting field E over F of $f(x)$ in $\bar{F}$ is isomorphic to the splitting field E' over F of $f(x)$ in $\bar{F}'$. [*Hint:* Use Corollary 2 of Theorem 9.6.]

9.4

Separable Extensions

Multiplicity of Zeros of a Polynomial

Remember that we are now always assuming that all algebraic extensions of a field F under consideration are contained in one fixed algebraic closure $\bar{F}$ of F.

Our next aim is to determine, for a finite extension E of F, under what conditions $\{E:F\} = [E:F]$. The key to answering this question is to consider the multiplicity of zeros of polynomials.

DEFINITION 9.9 (Multiplicity of a Zero) Let $f(x) \in F[x]$. An element α of $\bar{F}$ such that $f(\alpha) = 0$ is a **zero of $f(x)$ of multiplicity** ν if ν is the greatest integer such that $(x - \alpha)^\nu$ is a factor of $f(x)$ in $\bar{F}[x]$.

The next theorem shows that the multiplicities of the zeros of one given *irreducible* polynomial over a field are all the same. The ease with which we can prove this theorem is a further indication of the power of our conjugation isomorphisms and of our whole approach to the study of zeros of polynomials by means of mappings.

THEOREM 9.9 Let $f(x)$ be irreducible in $F[x]$. Then all zeros of $f(x)$ in $\bar{F}$ have the same multiplicity.

PROOF Let α and β be zeros of $f(x)$ in $\bar{F}$. Then by Theorem 9.1, there is a conjugation isomorphism $\psi_{\alpha,\beta} : F(\alpha) \xrightarrow{\text{onto}} F(\beta)$. By Corollary 1 of Theorem 8.6, $\psi_{\alpha,\beta}$ can be extended to an isomorphism $\tau : \bar{F} \to \bar{F}$. Then τ induces a natural isomorphism $\tau_x : \bar{F}[x] \to \bar{F}[x]$, with $\tau_x(x) = x$. Now τ_x leaves $f(x)$

fixed, since $f(x) \in F[x]$ and $\psi_{\alpha,\beta}$ leaves F fixed. However,

$$\tau_x((x - \alpha)^v) = (x - \beta)^v,$$

which shows that the multiplicity of β in $f(x)$ is greater than or equal to the multiplicity of α. A symmetric argument gives the reverse inequality, so the multiplicity of α equals that of β. ◆

COROLLARY If $f(x)$ is irreducible in $F[x]$, then $f(x)$ has a factorization in $\bar{F}[x]$ of the form

$$a \prod_i (x - \alpha_i)^v,$$

where the α_i are the distinct zeros of $f(x)$ in $\bar{F}$ and $a \in F$.

PROOF The corollary is immediate from Theorem 9.9. ◆

At this point, we should probably show by an example that the phenomenon of a zero of multiplicity greater than 1 of an irreducible polynomial can occur. We shall show later in this section that it can only occur for a polynomial over an infinite field of characteristic $p \neq 0$.

EXAMPLE I Let $E = \mathbb{Z}_p(y)$, where y is an indeterminate. Let $t = y^p$, and let F be the subfield $\mathbb{Z}_p(t)$ of E. (See Fig. 9.7.) Now $E = F(y)$ is algebraic over F, for y is a zero of $(x^p - t) \in F[x]$. By Theorem 8.3, irr(y, F) must divide $x^p - t$ in $F[x]$. [Actually, irr$(y, F) = x^p - t$. We leave a proof of this to the exercises (see Exercise 8).] Since $F(y)$ is not equal to F, we must have the degree of irr$(y, F) \geq 2$. But note that

$$x^p - t = x^p - y^p = (x - y)^p,$$

since E has characteristic p (see Theorem 9.5 and the following comment). Thus y is a zero of irr(y, F) of multiplicity > 1. Actually, $x^p - t = $ irr(y, F), so the multiplicity of y is p. ▲

From here on we rely heavily on Theorem 9.7 and its corollary. Theorem

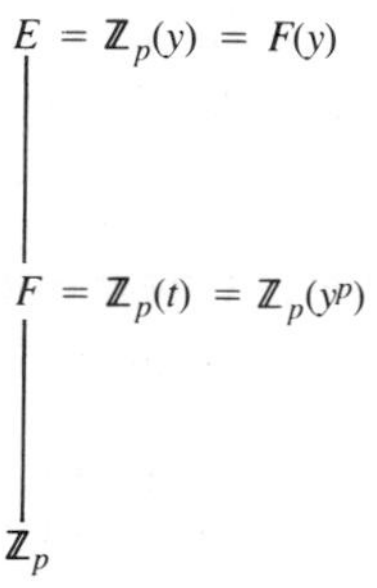

$$E = \mathbb{Z}_p(y) = F(y)$$
$$|$$
$$|$$
$$F = \mathbb{Z}_p(t) = \mathbb{Z}_p(y^p)$$
$$|$$
$$|$$
$$\mathbb{Z}_p$$

Figure 9.7

9.1 and its corollary show that for a simple algebraic extension $F(\alpha)$ of F there is one extension of the identity isomorphism ι mapping F into $\bar{F}$ for every distinct zero of $\mathrm{irr}(\alpha, F)$ and that these are the only extensions of ι. *Thus $\{F(\alpha):F\}$ is the number of distinct zeros of* $\mathrm{irr}(\alpha, F)$.

In view of our work with the theorem of Lagrange and Theorem 8.10, we should recognize the potential of a theorem like this next one.

THEOREM 9.10 If E is a finite extension of F, then $\{E:F\}$ divides $[E:F]$.

PROOF By Theorem 8.11, if E is finite over F, then $E = F(\alpha_1, \ldots, \alpha_n)$, where $\alpha_i \in \bar{F}$. Let $\mathrm{irr}(\alpha_i, F(\alpha_1, \ldots, \alpha_{i-1}))$ have α_i as one of n_i distinct zeros that are all of a common multiplicity v_i, by Theorem 9.9. Then

$$\left[F(\alpha_1, \ldots, \alpha_i):F(\alpha_1, \ldots, \alpha_{i-1})\right] = n_i v_i$$

$$= \{F(\alpha_1, \ldots, \alpha_i):F(\alpha_1, \ldots, \alpha_{i-1})\}v_i.$$

By Theorem 8.10 and the corollary of Theorem 9.7,

$$[E:F] = \prod_i n_i v_i,$$

and

$$\{E:F\} = \prod_i n_i.$$

Therefore, $\{E:F\}$ divides $[E:F]$. ◆

Separable Extensions

DEFINITION 9.10 (Separable Extension) A finite extension E of F is a **separable extension of** F if $\{E:F\} = [E:F]$. An element α of $\bar{F}$ is **separable over** F if $F(\alpha)$ is a separable extension of F. An irreducible polynomial $f(x) \in F[x]$ is **separable over** F if every zero of $f(x)$ in $\bar{F}$ is separable over F.

EXAMPLE 2 The field $E = \mathbb{Q}[\sqrt{2}, \sqrt{3}]$ is separable over $\mathbb{Q}$ since we saw in Example 3 of Section 9.3 that $\{E:\mathbb{Q}\} = 4 = [E:\mathbb{Q}]$. ▲

To make things a little easier, we have restricted our definition of a separable extension of a field F to *finite* extensions E of F. For the corresponding definition for infinite extensions, see Exercise 10.

We know that $\{F(\alpha):F\}$ is the number of distinct zeros of $\mathrm{irr}(\alpha, F)$. Also, the multiplicity of α in $\mathrm{irr}(\alpha, F)$ is the same as the multiplicity of each conjugate of α over F, by Theorem 9.9. *Thus α is separable over F if and only if $\mathrm{irr}(\alpha, F)$ has all zeros of multiplicity 1.* This tells us at once that *an* *irreducible polynomial $f(x) \in F[x]$ is separable over F if and only if $f(x)$ has all zeros of multiplicity 1.*

THEOREM 9.11   If K is a finite extension of E and E is a finite extension of F, that is, $F \leq E \leq K$, then K is separable over F if and only if K is separable over E and E is separable over F.

PROOF   Now

$$[K:F] = [K:E][E:F],$$

and

$$\{K:F\} = \{K:E\}\{E:F\}.$$

Then if K is separable over F, so that $[K:F] = \{K:F\}$, we must have $[K:E] = \{K:E\}$ and $[E:F] = \{E:F\}$, since in each case the index divides the degree, by Theorem 9.10. Thus, if K is separable over F, then K is separable over E and E is separable over F.

For the converse, note that $[K:E] = \{K:E\}$ and $[E:F] = \{E:F\}$ imply that

$$[K:F] = [K:E][E:F] = \{K:E\}\{E:F\} = \{K:F\}. \quad \blacklozenge$$

Theorem 9.11 can be extended in the obvious way, by induction, to any finite tower of finite extensions. The top field is a separable extension of the bottom one if and only if each field is a separable extension of the one immediately under it.

COROLLARY   If E is a finite extension of F, then E is separable over F if and only if each α in E is separable over F.

PROOF   Suppose that E is separable over F, and let $\alpha \in E$. Then

$$F \leq F(\alpha) \leq E,$$

and Theorem 9.11 shows that $F(\alpha)$ is separable over F.

Suppose, conversely, that every $\alpha \in E$ is separable over F. Since E is a finite extension of F, there exist $\alpha_1, \ldots, \alpha_n$ such that

$$F < F(\alpha_1) < F(\alpha_1, \alpha_2) < \cdots < E = F(\alpha_1, \ldots, \alpha_n).$$

Now since α_i is separable over F, α_i is separable over $F(\alpha_1, \ldots, \alpha_{i-1})$, because

$$q(x) = \mathrm{irr}(\alpha_i, F(\alpha_1, \ldots, \alpha_{i-1}))$$

divides $\mathrm{irr}(\alpha_i, F)$, so that α_i is a zero of $q(x)$ of multiplicity 1. Thus

$F(\alpha_1, \ldots, \alpha_i)$ is separable over $F(\alpha_1, \ldots, \alpha_{i-1})$, so E is separable over F, by Theorem 9.11, extended by induction. ◆

Perfect Fields

We now turn to the task of proving that α can fail to be separable over F only if F is an infinite field of characteristic $p \neq 0$. One method is to introduce formal derivatives of polynomials. While this is an elegant technique, and also a useful one, we shall, for the sake of brevity, use the following lemma instead. Formal derivatives are developed in Exercises 13 through 20.

LEMMA 9.1 Let $\bar{F}$ be an algebraic closure of F, and let

$$f(x) = x^n + a_{n-1}x^{n-1} + \cdots + a_1x + a_0$$

be any monic polynomial in $\bar{F}[x]$. If $(f(x))^m \in F[x]$ and $m \cdot 1 \neq 0$ in F, then $f(x) \in F[x]$, that is, all $a_i \in F$.

PROOF We must show that $a_i \in F$, and we proceed, by induction on r, to show that $a_{n-r} \in F$. For $r = 1$,

$$(f(x))^m = x^{mn} + (m \cdot 1)a_{n-1}x^{mn-1} + \cdots + a_0^m.$$

Since $(f(x))^m \in F[x]$, we have, in particular,

$$(m \cdot 1)a_{n-1} \in F.$$

Thus $a_{n-1} \in F$, since $m \cdot 1 \neq 0$ in F.

As induction hypothesis, suppose that $a_{n-r} \in F$ for $r = 1, 2, \ldots, k$. Then the coefficient of $x^{mn-(k+1)}$ in $(f(x))^m$ is of the form

$$(m \cdot 1)a_{n-(k+1)} + g_{k+1}(a_{n-1}, a_{n-2}, \ldots, a_{n-k}),$$

where $g_{k+1}(a_{n-1}, a_{n-2}, \ldots, a_{n-k})$ is a formal polynomial expression in $a_{n-1}, a_{n-2}, \ldots, a_{n-k}$. By the induction hypothesis that we just stated, $g_{k+1}(a_{n-1}, a_{n-2}, \ldots, a_{n-k}) \in F$, so $a_{n-(k+1)} \in F$, since $m \cdot 1 \neq 0$ in F. ◆

We are now in a position to handle fields F of characteristic zero and to show that for a finite extension E of F, we have $\{E:F\} = [E:F]$. By definition, this amounts to proving that every finite extension of a field of characteristic zero is a separable extension. First, we give a definition.

DEFINITION 9.11 (Perfect Field) A field is **perfect** if every finite extension is a separable extension.

THEOREM 9.12 Every field of characteristic zero is perfect.

PROOF Let E be a finite extension of a field F of characteristic zero, and let $\alpha \in E$. Then $f(x) = \mathrm{irr}(\alpha, F)$ factors in $\bar{F}[x]$ into $\prod_i (x - \alpha_i)^\nu$, where the α_i are the distinct zeros of irr(α, F), and, say, $\alpha = \alpha_1$. Thus

$$f(x) = \left(\prod_i (x - \alpha_i) \right)^v,$$

and since $v \cdot 1 \neq 0$ for a field F of characteristic 0, we must have

$$\left(\prod_i (x - \alpha_i) \right) \in F[x],$$

by Lemma 9.1. Since $f(x)$ is irreducible and of minimal degree in $F[x]$ having α as a zero, we then see that $v = 1$. Therefore, α is separable over F for all $\alpha \in E$. By the corollary of Theorem 9.11, this means that E is a separable extension of F. ◆

Lemma 9.1 will also get us through for the case of a finite field, although the proof is a bit harder.

THEOREM 9.13 Every finite field is perfect.

PROOF Let F be a finite field of characteristic p, and let E be a finite extension of F. Let $\alpha \in E$. We need to show that α is separable over F. Now $f(x) = $ irr(α, F) factors in $\bar{F}$ into $\prod_i (x - \alpha_i)^v$, where the α_i are the distinct zeros of $f(x)$, and, say, $\alpha = \alpha_1$. Let $v = p^t e$, where p does not divide e. Then

$$f(x) = \prod_i (x - \alpha_i)^v = \left(\prod_i (x - \alpha_i)^{p^t} \right)^e$$

is in $F[x]$, and by Lemma 9.1, $\prod_i (x - \alpha_i)^{p^t}$ is in $F[x]$ since $e \cdot 1 \neq 0$ in F. Since $f(x) = $ irr(α, F) is of minimal degree over F having α as a zero, we must have $e = 1$.

Theorem 9.5 and the remark following it show then that

$$f(x) = \prod_i (x - \alpha_i)^{p^t} = \prod_i (x^{p^t} - \alpha_i^{p^t}).$$

Thus, if we regard $f(x)$ as $g(x^{p^t})$, we must have $g(x) \in F[x]$. Now $g(x)$ is separable over F with distinct zeros $\alpha_i^{p^t}$. Consider $F(\alpha_1^{p^t}) = F(\alpha^{p^t})$. Then $F(\alpha^{p^t})$ is separable over F. Since $x^{p^t} - \alpha^{p^t} = (x - \alpha)^{p^t}$, we see that α is the only zero of $x^{p^t} - \alpha^{p^t}$ in $\bar{F}$. As a finite-dimensional vector space over a finite field F, $F(\alpha^{p^t})$ must be again a finite field. Hence the map

$$\sigma_p : F(\alpha^{p^t}) \to F(\alpha^{p^t})$$

given by $\sigma_p(a) = a^p$ for $a \in F(\alpha^{p^t})$ is an automorphism of $F(\alpha^{p^t})$ by Theorem 9.5. Consequently, $(\sigma_p)^t$ is also an automorphism of $F(\alpha^{p^t})$, and

$$(\sigma_p)^t(a) = a^{p^t}.$$

Since an automorphism of $F(\alpha^{p^t})$ is an onto map, there is $\beta \in F(\alpha^{p^t})$ such that $(\sigma_p)'(\beta) = \alpha^{p^t}$. But then $\beta^{p^t} = \alpha^{p^t}$, and we saw that α was the only zero of $x^{p^t} - \alpha^{p^t}$, so we must have $\beta = \alpha$. Since $\beta \in F(\alpha^{p^t})$, we have $F(\alpha) = F(\alpha^{p^t})$. Since $F(\alpha^{p^t})$ was separable over F, we now see that $F(\alpha)$ is separable over F. Therefore, α is separable over F and $t = 0$.

We have shown that for $\alpha \in E$, α is separable over F. Then by the corollary of Theorem 9.11, E is a separable extension of F. ◆

We have completed our aim, which was to show that fields of characteristic 0 and finite fields have only separable finite extensions, that is, these fields are perfect. *For finite extensions E of such perfect fields F, we then have* $[E:F] = \{E:F\}$.

The Primitive Element Theorem

The following theorem is a classic of field theory.

THEOREM 9.14 (Primitive Element Theorem) Let E be a finite separable extension of a field F. Then there exists $\alpha \in E$ such that $F = F(\alpha)$. (Such an element α is a **primitive element**.) That is, a finite separable extension of a field is a simple extension.

PROOF If F is a finite field, then E is also finite. Let α be a generator for the cyclic group E^* of nonzero elements of E under multiplication. (See Theorem 8.23.) Clearly $E = F(\alpha)$, so α is a primitive element in this case.

We now assume that F is infinite, and prove of our theorem in the case that $E = F(\beta, \gamma)$. The induction argument from this to the general case is straightforward. Let $\mathrm{irr}(\beta, F)$ have distinct zeros $\beta = \beta_1, \ldots, \beta_n$, and let $\mathrm{irr}(\gamma, F)$ have distinct zeros $\gamma = \gamma_1, \ldots, \gamma_m$ in $\bar{F}$, where all zeros are of multiplicity 1, since E is a separable extension of F. Since F is infinite, we can find $a \in F$ such that

$$a \neq (\beta_i - \beta)/(\gamma - \gamma_j)$$

for all i and j, with $j \neq 1$. That is, $a(\gamma - \gamma_j) \neq \beta_i - \beta$. Letting $\alpha = \beta + a\gamma$, we have $\alpha = \beta + a\gamma \neq \beta_i + a\gamma_j$, so

$$\alpha - a\gamma_j \neq \beta_i$$

for all i and all $j \neq 1$. Let $f(x) - \mathrm{irr}(\beta, F)$, and consider

$$h(x) = f(\alpha - ax) \in (F(\alpha))[x].$$

Now $h(\gamma) = f(\beta) = 0$. However, $h(\gamma_j) \neq 0$ for $j \neq 1$ by construction, since the β_i were the only zeros of $f(x)$. Hence $h(x)$ and $g(x) = \mathrm{irr}(\gamma, F)$ have a common factor in $(F(\alpha))[x]$, namely $\mathrm{irr}(\gamma, F(\alpha))$, which must be linear, since γ is the only common zero of $g(x)$ and $h(x)$. Thus $\gamma \in F(\alpha)$, and therefore $\beta = \alpha - a\gamma$ is in $F(\alpha)$. Hence $F(\beta, \gamma) = F(\alpha)$. ◆

COROLLARY A finite extension of a field of characteristic zero is a simple extension.

PROOF This corollary follows at once from Theorems 9.12 and 9.14. ◆

We see that the only possible "bad case" where a finite extension may not be simple is a finite extension of an infinite field of characteristic $p \neq 0$.

Exercises 9.4

Computations

In Exercises 1 through 4, find α such that the given field is $\mathbb{Q}(\alpha)$. Show that your α is indeed in the given field. Verify by direct computation that the given generators for the extension of $\mathbb{Q}$ can indeed be expressed as formal polynomials in your α with coefficients in $\mathbb{Q}$.

1. $\mathbb{Q}(\sqrt{2}, \sqrt[3]{2})$

2. $\mathbb{Q}(\sqrt[4]{2}, \sqrt[6]{2})$

3. $\mathbb{Q}(\sqrt{2}, \sqrt{3})$

4. $\mathbb{Q}(i, \sqrt[3]{2})$

Concepts

5. Give an example of an $f(x) \in \mathbb{Q}[x]$ that has no zeros in $\mathbb{Q}$ but whose zeros in $\mathbb{C}$ are all of multiplicity 2. Explain how this is consistent with Theorem 9.11, which shows that $\mathbb{Q}$ is perfect.

6. Mark each of the following true or false.

_____ a. Every finite extension of every field F is separable over F.

_____ b. Every finite extension of every finite field F is separable over F.

_____ c. Every field of characteristic 0 is perfect.

_____ d. Every polynomial of degree n over every field F always has n distinct zeros in $\bar{F}$.

_____ e. Every polynomial of degree n over every perfect field F always has n distinct zeros in $\bar{F}$.

_____ f. Every irreducible polynomial of degree n over every perfect field F always has n distinct zeros in $\bar{F}$.

_____ g. Every algebraically closed field is perfect.

_____ h. Every field F has an algebraic extension E that is perfect.

_____ i. If E is a finite separable splitting field extension of F, then $|G(E/F)| = [E:F]$.

_____ j. If E is a finite splitting field extension of F, then $|G(E/F)|$ divides $[E:F]$.

Theory

7. Show that if $\alpha, \beta \in \bar{F}$ are both separable over F, then $\alpha \pm \beta$, $\alpha\beta$, and α/β, if $\beta \neq 0$, are all separable over F. [*Hint:* Use Theorem 9.11 and its corollary.]

8. Show that $\{1, y, \ldots, y^{n-1}\}$ is a basis for $\mathbb{Z}_p(y)$ over $\mathbb{Z}_p(y^p)$, where y is an indeterminate. Referring to Example 1, conclude by a degree argument that $x^p - t$ is irreducible over $\mathbb{Z}_p(t)$, where $t = y^p$.

9. Prove that if E is an algebraic extension of a perfect field F, then E is perfect.

10. A (possibly infinite) algebraic extension E of a field F is a **separable extension of**

F if for every $\alpha \in E$, $F(\alpha)$ is a separable extension of F, in the sense defined in the text. Show that if E is a (possibly infinite) separable extension of F and K is a (possibly infinite) separable extension of E, then K is a separable extension of F.

11. Let E be an algebraic extension of a field F. Show that the set of all elements in E that are separable over F forms a subfield of E, the **separable closure of F in E**. [*Hint:* Use Exercise 7.]

12. Let E be a finite field of order p^n.

   a. Show that the Frobenius automorphism σ_p has order n.

   b. Deduce from part (a) that $G(E/\mathbb{Z}_p)$ is cyclic of order n with generator σ_p. [*Hint:* Remember that

   $$|G(E/F)| = \{E:F\} = [E:F]$$

   for a finite separable splitting field extension E over F.]

Exercises 13 through 20 introduce formal derivatives in $F[x]$.

13. Let F be any field, and let $f(x) = a_0 + a_1x + \cdots + a_ix^i + \cdots + a_nx^n$ be in $F[x]$. The **derivative** $f'(x)$ **of** $f(x)$ is the polynomial

   $$f'(x) = a_1 + \cdots + (i \cdot 1)a_ix^{i-1} + \cdots + (n \cdot 1)a_nx^{n-1},$$

   where $i \cdot 1$ has its usual meaning for $i \in \mathbb{Z}^+$ and $1 \in F$. *These are formal derivatives; no "limits" are involved here.*

   a. Prove that the map $D:F[x] \to F[x]$ given by $D(f(x)) = f'(x)$ is a homomorphism of $\langle F[x], + \rangle$.

   b. Find the kernel of D in the case that F is of characteristic 0.

   c. Find the kernel of D in the case that F is of characteristic $p \neq 0$.

14. Continuing the ideas of Exercise 13, show that:

   a. $D(af(x)) = aD(f(x))$ for all $f(x) \in F[x]$ and $a \in F$.

   b. $D(f(x)g(x)) = f(x)g'(x) + f'(x)g(x)$ for all $f(x), g(x) \in F[x]$. [*Hint:* Use part (a) of this exercise and the preceding exercise and proceed by induction on the degree of $f(x)g(x)$.]

   c. $D((f(x))^m) = (m \cdot 1)f(x)^{m-1}f'(x)$ for all $f(x) \in F[x]$. [*Hint:* Use part (b).]

15. Let $f(x) \in F[x]$, and let $\alpha \in \bar{F}$ be a zero of $f(x)$ of multiplicity v. Show that $v > 1$ if and only if α is also a zero of $f'(x)$. [*Hint:* Apply parts (b) and (c) of Exercise 14 to the factorization $f(x) = (x - \alpha)^v g(x)$ of $f(x)$ in the ring $\bar{F}[x]$.]

16. Show from Exercise 15 that every irreducible polynomial over a field F of characteristic 0 is separable. [*Hint:* Use the fact that irr(α, F) is the *minimal* polynomial for α over F.]

17. Show from Exercise 15 that an irreducible polynomial $q(x)$ over a field F of characteristic $p \neq 0$ is not separable if and only if each exponent of each term of $q(x)$ is divisible by p.

18. Generalize Exercise 15, showing that $f(x) \in F[x]$ has no zero of multiplicity >1 if and only if $f(x)$ and $f'(x)$ have no common factor in $\bar{F}[x]$ of degree >1.

19. Working a bit harder than in Exercise 18, show that $f(x) \in F[x]$ has no zero of multiplicity >1 if and only if $f(x)$ and $f'(x)$ have no common nonconstant factor in $F[x]$. [*Hint:* Use Theorem 7.6 to show that if 1 is a gcd of $f(x)$ and $f'(x)$ in $F[x]$, it is a gcd of these polynomials in $\bar{F}[x]$ also.]

20. Describe a feasible computational procedure for determining whether $f(x) \in F[x]$ has a zero of multiplicity >1, without actually finding the zeros of $f(x)$. [*Hint:* Use Exercise 19.]

9.5

Totally Inseparable Extensions

This section shows that a finite extension E of a field F can be split into two stages: a separable extension K of F, followed by a further extension of K to E that is as far from being separable as one can imagine. Work in this section will not be used in the remainder of the text.

We develop our theory of totally inseparable extensions in a fashion parallel to our development of separable extensions.

DEFINITION 9.12 (Totally Inseparable Extension)   A finite extension E of a field F is a **totally inseparable extension of** F if $\{E:F\} = 1 < [E:F]$. An element α of $\bar{F}$ is **totally inseparable over** F if $F(\alpha)$ is totally inseparable over F.

We know that $\{F(\alpha):F\}$ is the number of distinct zeros of irr(α, F). Thus α is totally inseparable over F if and only if irr(α, F) has only one zero that is of multiplicity >1.

EXAMPLE 1   Referring to Example 1, Section 9.4, we see that $\mathbb{Z}_p(y)$ is totally inseparable over $\mathbb{Z}_p(y^p)$, where y is an indeterminate. ▲

THEOREM 9.15 (Counterpart of Theorem 9.11)   If K is a finite extension of E, E is a finite extension of F, and $F < E < K$, then K is totally inseparable over F if and only if K is totally inseparable over E and E is totally inseparable over F.

PROOF   Since $F < E < K$, we have $[K:E] > 1$ and $[E:F] > 1$. Suppose K is totally inseparable over F. Then $\{K:F\} = 1$, and

$$\{K:F\} = \{K:E\}\{E:F\},$$

so we must have

$$\{K:E\} = 1 < [K:E] \quad \text{and} \quad \{E:F\} = 1 < [E:F].$$

Thus K is totally inseparable over E, and E is totally inseparable over F.

Conversely, if K is totally inseparable over E and E is totally inseparable over F, then

$$\{K:F\} = \{K:E\}\{E:F\} = (1)(1) = 1,$$

and $[K:F] < 1$. Thus K is totally inseparable over F. ◆

Theorem 9.15 can be extended by induction, to any finite proper tower of finite extensions. The top field is a totally inseparable extension of the bottom one if and only if each field is a totally inseparable extension of the one immediately under it.

COROLLARY (Counterpart of the Corollary of Theorem 9.11) If E is a finite extension of F, then E is totally inseparable over F if and only if each α in E, $\alpha \notin F$, is totally inseparable over F.

PROOF Suppose that E is totally inseparable over F, and let $\alpha \in E$, with $\alpha \notin F$. Then

$$F < F(\alpha) \le E.$$

If $F(\alpha) = E$, we are done, by the definition of α totally inseparable over F. If $F < F(\alpha) < E$, then Theorem 9.15 shows that since E is totally inseparable over F, $F(\alpha)$ is totally inseparable over F.

Conversely, suppose that for every $\alpha \in E$, with $\alpha \notin F$, α is totally inseparable over F. Since E is finite over F, there exists $\alpha_1, \ldots, \alpha_n$ such that

$$F < F(\alpha_1) < F(\alpha_1, \alpha_2) < \cdots < E = F(\alpha_1, \ldots, \alpha_n).$$

Now since α_i is totally inseparable over F, α_i is totally inseparable over $F(\alpha_1, \ldots, \alpha_{i-1})$, because $q(x) = \mathrm{irr}(\alpha_i, F(\alpha_1, \ldots, \alpha_{i-1}))$ divides $\mathrm{irr}(\alpha_i, F)$ so that α_i is the only zero of $q(x)$ and is of multiplicity >1. Thus $F(\alpha_1, \ldots, \alpha_i)$ is totally inseparable over $F(\alpha_1, \ldots, \alpha_{i-1})$, and E is totally inseparable over F, by Theorem 9.15, extended by induction. ◆

Thus far we have so closely paralleled our work in Section 9.4 that we could have handled these ideas together.

Separable Closures

We now come to our main reason for including this material.

THEOREM 9.16 Let F have characteristic $p \ne 0$, and let E be a finite extension of F. Then $\alpha \in E$, $\alpha \notin F$, is totally inseparable over F if and only if there is some integer $t \ge 1$ such that $\alpha^{p^t} \in F$. Furthermore, there is a unique extension K of F, with $F \le K \le E$, such that K is separable over F, and either $E = K$ or E is totally inseparable over K.

PROOF Let $\alpha \in E$, $\alpha \notin F$, be totally inseparable over F. Then $\mathrm{irr}(\alpha, F)$ has just one zero α of multiplicity >1, and, as shown in the proof of

Theorem 9.13, $\text{irr}(\alpha, F)$ must be of the form

$$x^{p^t} - \alpha^{p^t}.$$

Hence $\alpha^{p^t} \in F$ for some $t \geq 1$.

Conversely, if $\alpha^{p^t} \in F$ for some $t \geq 1$, where $\alpha \in E$ and $\alpha \notin F$, then

$$x^{p^t} - \alpha^{p^t} = (x - \alpha)^{p^t},$$

and $(x^{p^t} - \alpha^{p^t}) \in F[x]$, showing that $\text{irr}(\alpha, F)$ divides $(x - \alpha)^{p^t}$. Thus $\text{irr}(\alpha, F)$ has α as its only zero and this zero is of multiplicity > 1, so α is totally inseparable over F.

For the second part of the theorem, let $E = F(\alpha_1, \ldots, \alpha_n)$. Then if

$$\text{irr}(\alpha_i, F) = \prod_j \left(x^{p^{t_i}} - \alpha_{ij}^{p^{t_i}}\right),$$

with $\alpha_{i1} = \alpha_i$, let $\beta_{ij} = \alpha_{ij}^{p^{t_i}}$. We have $F(\beta_{11}, \beta_{21}, \ldots, \beta_{n1}) \leq E$, and β_{i1} is a zero of

$$f_i(x) = \prod_j (x - \beta_{ij}),$$

where $f_i(x) \in F[x]$. Now since raising to the power p is an isomorphism σ_p of E onto a subfield of E, raising to the power p^t is the isomorphic mapping $(\sigma_p)^t$ of E onto a subfield of E. Thus since the α_{ij} are all distinct for a fixed i, so are the β_{ij} for a fixed i. Therefore, β_{ij} is separable over F, because it is a zero of a polynomial $f_i(x)$ in $F[x]$ with zeros of multiplicity 1. Then

$$K = F(\beta_{11}, \beta_{21}, \ldots, \beta_{n1})$$

is separable over F, by the proof of the corollary of Theorem 9.11. If all $p^{t_i} = 1$, then $K = E$. If some $p^{t_i} \neq 1$, then $K \neq E$, and $\alpha_i^{p^{t_i}} = \beta_{i1}$ is in K, showing that each $\alpha_i \notin K$ is totally inseparable over K, by the first part of this theorem. Hence $E = K(\alpha_1, \ldots, \alpha_n)$ is totally inseparable over K, by the proof of the corollary of Theorem 9.15.

It follows from the corollaries of Theorems 9.11 and 9.15 that the field K of Theorem 9.16 consists of all elements α in E that are separable over F. Thus K is unique. ◆

DEFINITION 9.13 (Separable Closure) The unique field K of Theorem 9.16 is the **separable closure of F in E**.

The preceding theorem shows the precise structure of totally inseparable extensions of a field of characteristic p. Such an extension can be obtained by repeatedly adjoining pth roots of elements that are not already pth powers.

We remark that Theorem 9.16 is true for infinite algebraic extensions E of F. The proof of the first assertion of the theorem is valid for the case of infinite extensions also. For the second part, since $\alpha \pm \beta$, $\alpha\beta$, and α/β, for $\beta \neq 0$, are all contained in the field $F(\alpha, \beta)$, all elements of E separable over F form a subfield K of E, the **separable closure of F in E**. It follows that an $\alpha \in E$, $\alpha \notin K$, is totally inseparable over K, since α and all coefficients of $\mathrm{irr}(\alpha, K)$ are in a finite extension of F, and then Theorem 9.16 can be applied.

Exercises 9.5

Concepts

1. Let y and z be indeterminates, and let $u = y^{12}$ and $v = z^{18}$. Describe the separable closure of $\mathbb{Z}_3(u, v)$ in $\mathbb{Z}_3(y, z)$.

2. Let y and z be indeterminates, and let $u = y^{12}$ and $v = y^2 z^{18}$. Describe the separable closure of $\mathbb{Z}_3(u, v)$ in $\mathbb{Z}_3(y, z)$.

3. Referring to Exercise 1, describe the totally inseparable closure (see Exercise 6) of $\mathbb{Z}_3(u, v)$ in $\mathbb{Z}_3(y, z)$.

4. Referring to Exercise 2, describe the totally inseparable closure of $\mathbb{Z}_3(u, v)$ in $\mathbb{Z}_3(y, z)$. (See Exercise 6.)

5. Mark each of the following true or false.
___ a. No proper algebraic extension of an infinite field of characteristic $p \neq 0$ is ever a separable extension.
___ b. If $F(\alpha)$ is totally inseparable over F of characteristic $p \neq 0$, then $\alpha^{p^t} \in F$ for some $t > 0$.
___ c. For an indeterminate y, $\mathbb{Z}_5(y)$ is separable over $\mathbb{Z}_5(y^5)$.
___ d. For an indeterminate y, $\mathbb{Z}_5(y)$ is separable over $\mathbb{Z}_5(y^{10})$.
___ e. For an indeterminate y, $\mathbb{Z}_5(y)$ is totally inseparable over $\mathbb{Z}_5(y^{10})$.
___ f. If F is a field and α is algebraic over F, then α is either separable or totally inseparable over F.
___ g. If E is an algebraic extension of a field F, then F has a separable closure in E.
___ h. If E is an algebraic extension of a field F, then E is totally inseparable over the separable closure of F in E.
___ i. If E is an algebraic extension of a field F and E is not a separable extension of F, then E is totally inseparable over the separable closure of F in E.
___ j. If α is totally inseparable over F, then α is the only zero of $\mathrm{irr}(\alpha, F)$.

Theory

6. Show that if E is an algebraic extension of a field F, then the union of F with the set of all elements of E totally inseparable over F forms a subfield of E, the **totally inseparable closure of F in E**.

7. Show that a field F of characteristic $p \neq 0$ is perfect if and only if $F^p = F$, that is, every element of F is a pth power of some element of F.

8. Let E be a finite extension of a field F of characteristic p. In the notation of Exercise 7, show that $E^p = E$ if and only if $F^p = F$. [*Hint:* The map $\sigma_p : E \to E$

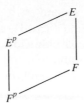

Figure 9.8

defined by $\sigma_p(\alpha) = \alpha^p$ for $\alpha \in E$ is an isomorphism onto a subfield of E. Consider the diagram in Fig. 9.8, and make degree arguments.]

9.6

Galois Theory

Résumé

This section is perhaps the climax in elegance of the subject matter of the entire text. The Galois theory gives a beautiful interplay of group and field theory. Starting with Section 9.1, our work has been aimed at this goal. We shall start by recalling the main results we have developed and should have well in mind.

1. Let $F \le E \le \bar{F}$, $\alpha \in E$, and let β be a conjugate of α over F, that is, irr(α, F) has β as a zero also. Then there is an isomorphism $\psi_{\alpha,\beta}$ mapping $F(\alpha)$ onto $F(\beta)$ that leaves F fixed and maps α onto β.
2. If $F \le E \le \bar{F}$ and $\alpha \in E$, then an automorphism σ of $\bar{F}$ that leaves F fixed *must* map α onto some conjugate of α over F.
3. If $F \le E$, the collection of all automorphisms of E leaving F fixed forms a group $G(E/F)$. For any subset S of $G(E/F)$, the set of all elements of E left fixed by all elements of S is a field E_S. Also, $F \le E_{G(E/F)}$.
4. A field E, $F \le E \le \bar{F}$, is a splitting field over F if and only if every isomorphism of E onto a subfield of $\bar{F}$ leaving F fixed is an automorphism of E. If E is a finite extension and a splitting field over F, then $|G(E/F)| = \{E:F\}$.
5. If E is a finite extension of F, then $\{E:F\}$ divides $[E:F]$. If E is also separable over F, then $\{E:F\} = [E:F]$. Also, E is separable over F if and only if irr(α, F) has all zeros of multiplicity 1 for every $\alpha \in E$.
6. If E is a finite extension of F and is a separable splitting field over F, then $|G(E/F)| = \{E:F\} = [E:F]$.

Normal Extensions

We are going to be interested in finite extensions K of F such that every isomorphism of K onto a subfield of $\bar{F}$ leaving F fixed is an automorphism of

K and such that

$$[K:F] = \{K:F\}.$$

In view of résumé numbers 4 and 5, these are the finite extensions of F that are separable splitting fields over F.

DEFINITION 9.14 (Normal Extension)   A finite extension K of F is a **finite normal extension of** F if K is a separable splitting field over F.

Suppose that K is a finite normal extension of F, where $K \leq \bar{F}$, as usual. Then by result 4, every automorphism of $\bar{F}$ leaving F fixed induces an automorphism of K. As before, we let $G(K/F)$ be the group of all automorphisms of K leaving F fixed. After one more result, we shall be ready to illustrate the main theorem.

THEOREM 9.17   Let K be a finite normal extension of F, and let E be an extension of F, where $F \leq E \leq K \leq \bar{F}$. Then K is a finite normal extension of E, and $G(K/E)$ is precisely the subgroup of $G(K/F)$ consisting of all those automorphisms that leave E fixed. Moreover, two automorphisms σ and τ in $G(K/F)$ induce the same isomorphism of E onto a subfield of $\bar{F}$ if and only if they are in the same left coset of $G(K/E)$ in $G(K/F)$.

PROOF   If K is the splitting field of a set $\{f_i(x) \,|\, i \in I\}$ of polynomials in $F[x]$, then K is the splitting field over E of this same set of polynomials viewed as elements of $E[x]$. Theorem 9.11 shows that K is separable over E, since K is separable over F. Thus K is a normal extension of E. This establishes our first contention.

Now every element of $G(K/E)$ is an automorphism of K leaving F fixed, since it even leaves the possibly larger field E fixed. Thus $G(K/E)$ can be viewed as a subset of $G(K/F)$. Since $G(K/E)$ is a group under function composition also, we see that $G(K/E) \leq G(K/F)$.

Finally, for σ and τ in $G(K/F)$, σ and τ are in the same left coset of $G(K/E)$ if and only if $\tau^{-1}\sigma \in G(K/E)$ or if and only if $\sigma = \tau\mu$ for $\mu \in G(K/E)$. But if $\sigma = \tau\mu$ for $\mu \in G(K/E)$, then for $\alpha \in E$, we have

$$\sigma(\alpha) = (\tau\mu)(\alpha) = \tau(\mu(\alpha)) = \tau(\alpha),$$

since $\mu(\alpha) = \alpha$ for $\alpha \in E$. Conversely, if $\sigma(\alpha) = \tau(\alpha)$ for all $\alpha \in E$, then

$$(\tau^{-1}\sigma)(\alpha) = \alpha$$

for all $\alpha \in E$, so $\tau^{-1}\sigma$ leaves E fixed, and $\mu = \tau^{-1}\sigma$ is thus in $G(K/E)$. ◆

The preceding theorem shows that there is a one-to-one correspondence between left cosets of $G(K/E)$ in $G(K/F)$ and isomorphisms of E onto a subfield of K leaving F fixed. Note that we can't say that these left cosets correspond to *automorphisms* of E over F, since E may not be a splitting field over F. Of course, if E is a *normal* extension of F, then these isomorphisms would be automorphisms of E over F. We might guess that this will happen if and only if $G(K/E)$ is a *normal* subgroup of $G(K/F)$, and this is indeed the case. That is, the two different uses of the word *normal* are really closely related. Thus if E is a normal extension of F, then the left cosets of $G(K/E)$ in $G(K/F)$ can be viewed as elements of the *factor group* $G(K/F)/G(K/E)$, which is then a group of automorphisms acting on E and leaving F fixed. We shall show that this factor group is isomorphic to $G(E/F)$.

The Main Theorem

The Main Theorem of Galois Theory states that for a finite normal extension K of a field F, there is a one-to-one correspondence between the subgroups of $G(K/F)$ and the intermediate fields E, where $F \leq E \leq K$. This correspondence associates with each intermediate field E the subgroup $G(K/E)$. We can also go the other way and start with a subgroup H of $G(K/F)$ and associate with H its fixed field K_H. We shall illustrate this with an example, then state the theorem and discuss its proof.

EXAMPLE I   Let $K = \mathbb{Q}(\sqrt{2}, \sqrt{3})$. Now K is a normal extension of $\mathbb{Q}$, and Example 5, Section 9.1, showed that there are four automorphisms of K leaving $\mathbb{Q}$ fixed. We recall them by giving their values on the basis $\{1, \sqrt{2}, \sqrt{3}, \sqrt{6}\}$ for K over $\mathbb{Q}$.

  ι: This identity map

  σ_1: Maps $\sqrt{2}$ onto $-\sqrt{2}$, $\sqrt{6}$ onto $-\sqrt{6}$, and leaves the others fixed

  σ_2: Maps $\sqrt{3}$ onto $-\sqrt{3}$, $\sqrt{6}$ onto $-\sqrt{6}$, and leaves the others fixed

  σ_3: Maps $\sqrt{2}$ onto $-\sqrt{2}$, $\sqrt{3}$ onto $-\sqrt{3}$, and leaves the others fixed

We saw that $\{\iota, \sigma_1, \sigma_2, \sigma_3\}$ is isomorphic to the Klein 4-group. The complete list of subgroups, with each subgroup paired off with the corresponding intermediate field that it leaves fixed, is as follows:

$$\{\iota, \sigma_1, \sigma_2, \sigma_3\} \leftrightarrow \mathbb{Q},$$
$$\{\iota, \sigma_1\} \leftrightarrow \mathbb{Q}(\sqrt{3}),$$
$$\{\iota, \sigma_2\} \leftrightarrow \mathbb{Q}(\sqrt{2}),$$
$$\{\iota, \sigma_3\} \leftrightarrow \mathbb{Q}(\sqrt{6}),$$
$$\{\iota\} \leftrightarrow \mathbb{Q}(\sqrt{2}, \sqrt{3}).$$

All subgroups of the abelian group $\{\iota, \sigma_1, \sigma_2, \sigma_3\}$ are normal subgroups, and all the intermediate fields are normal extensions of $\mathbb{Q}$. Isn't that elegant?

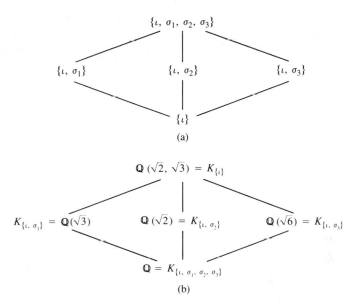

Figure 9.9   (a) Group lattice diagram. (b) Field lattice diagram.

Note that if one subgroup is contained in another, than the larger of the two subgroups corresponds to the smaller of the two corresponding fixed fields. The larger the subgroup, that is, the more automorphisms the smaller the fixed field, that is, the fewer elements left fixed. In Fig. 9.9 we give the corresponding lattice diagrams for the subgroups and intermediate fields. *Note again that the groups near the top correspond to the fields near the bottom.* That is, one lattice looks like the other *inverted* or turned upside down. Since here each lattice actually looks like itself turned upside down, this is not a good example for us to use to illustrate this *lattice inversion principle.* Turn ahead to Fig. 9.13 to see diagrams in which the lattices do not look like their own inversions. ▲

DEFINITION 9.15 (Galois Group)   If K is a finite normal extension of a field F, the $G(K/F)$ is the **Galois group of K over F.**

We shall now state the main theorem, then give another example, and finally, complete the proof of the main theorem.

THEOREM 9.18 (Main Theorem of Galois Theory)   Let K be a finite normal extension of a field F, with Galois group $G(K/F)$. For a field E, where $F \leq E \leq K$, let $\lambda(E)$ be the subgroup of $G(K/F)$ leaving E fixed. Then λ is a one-to-one map of the set of all such intermediate fields E onto the set of all subgroups of $G(K/F)$. The following properties hold for λ:

1. $\lambda(E) = G(K/E)$.

2. $E = K_{G/(K/E)} = K_{\lambda(E)}$.
3. For $H \leq G(K/F)$, $\lambda(E_H) = H$.
4. $[K:E] = |\lambda(E)|$; $[E:F] = \{G(K/F):\lambda(E)\}$, the number of left cosets of $\lambda(E)$ in $G(K/F)$.
5. E is a normal extension of F if and only if $\lambda(E)$ is a normal subgroup of $G(K/F)$. When $\lambda(E)$ is a normal subgroup of $G(K/F)$, then

$$G(E/F) \simeq G(K/F)/G(K/E).$$

6. The lattice of subgroups of $G(K/F)$ is the inverted lattice of intermediate fields of K over F.

Observations on the Proof   We have really already proved a substantial part of this theorem. Let us see just how much we have left to prove.

Property 1 is just the definition of λ found in the statement of the theorem. For Property 2, Theorem 9.4 shows that

$$E \leq K_{G(K/E)}.$$

Let $\alpha \in K$, where $\alpha \notin E$. Since K is a normal extension of E, by using a conjugation isomorphism and the Isomorphism Extension Theorem, we can find an automorphism of K leaving E fixed and mapping α onto a different zero of $\mathrm{irr}(\alpha, F)$. This implies that

$$K_{G(K/E)} \leq E,$$

so $E = K_{G(K/E)}$. This disposes of Property 2 and also tells us that λ is one to one, for if $\lambda(E_1) = \lambda(E_2)$, then by Property 2, we have

$$E_1 = K_{\lambda(E_1)} = K_{\lambda(E_2)} = E_2.$$

Now Property 3 is going to be our main job. This amounts exactly to showing that λ is an onto map. Of course, for $H \leq G(K/F)$, we have $H \leq \lambda(K_H)$, for H surely is included in the set of all automorphisms leaving K_H fixed. Here we will be using strongly our property $[K:E] = \{K:E\}$.

Property 4 follows from $[K:E] = \{K:E\}$, $[E:F] = \{E:F\}$, and the last statement in Theorem 9.17.

We shall have to show that the two senses of the word *normal* correspond for Property 5.

We have already disposed of Property 6 in Example 1. *Thus only Properties 3 and 5 remain to be proved.*

The Main Theorem of Galois Theory is a strong tool in the study of zeros of polynomials. If $f(x) \in F[x]$ is such that every irreducible factor of $f(x)$ is separable over F, then the splitting field K of $f(x)$ over F is a normal extension of F. The Galois group $G(K/F)$ is the **group of the polynomial $f(x)$ over** F. The structure of this group may give considerable information regarding the zeros of $f(x)$. This will be strikingly illustrated in Section 9.9 when we achieve our *final goal*.

Galois Groups over Finite Fields

Let K be a finite extension of a *finite field* F. We have seen that K is a separable extension of F (a finite field is perfect). Suppose that the order of F is p^r and $[K:F] = n$, so the order of K is p^{rn}. Then we have seen that K is the splitting field of $x^{p^{rn}} - x$ over F. Hence K is a normal extension of F.

Now one automorphism of K that leaves F fixed is σ_{p^r}, where for $\alpha \in K$, $\sigma_{p^r}(\alpha) = \alpha^{p^r}$. Note that $(\sigma_{p^r})^i(\alpha) = \alpha^{p^{ri}}$. Since a polynomial of degree p^{ri} can have at most p^{ri} zeros in a field, we see that the smallest power of σ_{p^r} that could possibly leave all p^{rn} elements of K fixed is the nth power. That is, the order of the element σ_{p^r} in $G(K/F)$ is at least n. Therefore, since $|G(K/F)| = [K:F] = n$, it must be that $G(K/F)$ is cyclic and generated by σ_{p^r}. We summarize these arguments in a theorem.

THEOREM 9.19 Let K be a finite extension of degree n of a finite field F of p^r elements. Then $G(K/F)$ is cyclic of order n, and is generated by σ_{p^r}, where for $\alpha \in K$, $\sigma_{p^r}(\alpha) = \alpha^{p^r}$.

We use this theorem to give another illustration of the Main Theorem of Galois Theory.

EXAMPLE 2 Let $F = \mathbb{Z}_p$, and let $K = GF(p^{12})$, so $[K:F] = 12$. Then $G(K/F)$ is isomorphic to the cyclic group $\langle \mathbb{Z}_{12}, + \rangle$. The lattice diagram for the subgroups and for the intermediate fields is given in Fig. 9.10. Again, each lattice is not only the inversion of the other, but unfortunately, also looks like the inversion of itself. Examples where the lattices do not look like their own inversion are given in the next section. We describe the cyclic subgroups of $G(K/F) = \langle \sigma_p \rangle$ by giving generators, e.g.,

$$\langle \sigma_p{}^4 \rangle = \{\iota, \sigma_p{}^4, \sigma_p{}^8\}. \ \blacktriangle$$

Proof of the Main Theorem Completed

We saw that Properties 3 and 5 are all that remain to be proved in the Main Theorem of Galois Theory.

PROOF Turning to Property 3, we must show that for $H \leq G(K/F)$, $\lambda(K_H) = H$. We know that $H \leq \lambda(K_H) \leq G(K/F)$. Thus what we really

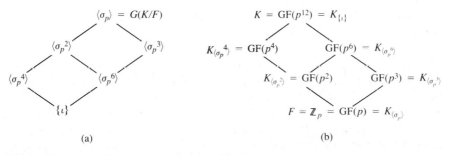

(a)   (b)

Figure 9.10 (a) Group lattice diagram. (b) Field lattice diagram.

must show is that it is impossible to have H a *proper* subgroup of $\lambda(K_H)$. We shall suppose that

$$H < \lambda(K_H).$$

and shall derive a contradiction. As a finite separable extension, $K = K_H(\alpha)$ for some $\alpha \in K$, by Theorem 9.14. Let

$$n = [K:K_H] = \{K:K_H\} = |G(K/K_H)|.$$

Then $H < G(K/K_H)$ implies that $|H| < |G(K/K_H)| = n$. Thus we would have to have $|H| < [K:K_H] = n$. Let the elements of H be $\sigma_1, \ldots, \sigma_{|H|}$, and consider the polynomial

$$f(x) = \prod_{i=1}^{|H|} (x - \sigma_i(\alpha)).$$

Then $f(x)$ is of degree $|H| < n$. Now the coefficients of each power of x in $f(x)$ are *symmetric* expressions in the $\sigma_i(\alpha)$. For example, the coefficient of $x^{|H|-1}$ is $-\sigma_1(\alpha) - \sigma_2(\alpha) - \cdots - \sigma_{|H|}(\alpha)$. Thus these coefficients are invariant under each isomorphism $\sigma_i \in H$, since if $\sigma \in H$, then

$$\sigma\sigma_1, \ldots, \sigma\sigma_{|H|}$$

is again the sequence $\sigma_1, \ldots, \sigma_{|H|}$, except for order, H being a group. Hence $f(x)$ has coefficients in K_H, and since some σ_i is ι, we see that some $\sigma_i(\alpha)$ is α, so $f(\alpha) = 0$. Therefore, we would have

$$\deg(\alpha, K_H) \le |H| < n = [K:K_H] = [K_H(\alpha):K_H].$$

This is impossible. Thus we have proved Property 3.

We turn to Property 5. Every extension E of F, $F \le E \le K$, is separable over F, by Theorem 9.11. Thus E is normal over F if and only if E is a splitting field over F. By the Isomorphism Extension Theorem, every isomorphism of E onto a subfield of $\bar{F}$ leaving F fixed can be extended to an *automorphism* of K, since K is *normal* over F. Thus the automorphisms of $G(K/F)$ induce all possible isomorphisms of E onto a subfield of $\bar{F}$ leaving F fixed. By Theorem 9.8, this shows that E is a splitting field over F, and hence is normal over F, if and only if for all $\sigma \in G(K/F)$ and $\alpha \in E$,

$$\sigma(\alpha) \in E.$$

By Property 2, E is the fixed field of $G(K/E)$, so $\sigma(\alpha) \in E$ if and only if for all $\tau \in G(K/E)$.

$$\tau(\sigma(\alpha)) = \sigma(\alpha).$$

This in turn holds if and only if

$$(\sigma^{-1}\tau\sigma)(\alpha) = \alpha$$

for all $\alpha \in E$, $\sigma \in G(K/F)$, and $\tau \in G(K/E)$. But this means that for all $\sigma \in G(K/F)$ and $\tau \in G(K/E)$, $\sigma^{-1}\tau\sigma$ leaves every element of E fixed, that is,

$$(\sigma^{-1}\tau\sigma) \in G(K/E).$$

This is precisely the condition that $G(K/F)$ be a normal subgroup of $G(K/F)$.

It remains for us to show that when E is a normal extension of F, $G(E/F) \simeq G(K/F)/G(K/E)$. For $\sigma \in G(K/F)$, let σ_E be the *auto-morphism* of E induced by σ (we are assuming that E is a *normal* extension of F). Thus $\sigma_E \in G(E/F)$. The map $\phi: G(K/F) \to G(E/F)$ given by

$$\phi(\sigma) = \sigma_E$$

for $\sigma \in G(K/F)$ is a homomorphism. By the Isomorphism Extension Theorem, every automorphism of E leaving F fixed can be extended to some automorphism of K; that is, it is τ_E for some $\tau \in G(K/F)$. Thus ϕ is onto $G(E/F)$. The kernel of ϕ is $G(K/E)$. Therefore, by the Fundamental Isomorphism Theorem, $G(E/F) \simeq G(K/F)/G(K/E)$. Furthermore, this isomorphism is a natural one. ◆

Exercises 9.6

Computations

The field $K = \mathbb{Q}(\sqrt{2}, \sqrt{3}, \sqrt{5})$ is a finite normal extension of $\mathbb{Q}$. In Exercises 1 through 8, compute the indicated numerical quantity. The notation is that of Theorem 9.18.

1. $[K:\mathbb{Q}]$
2. $|G(K/\mathbb{Q})|$
3. $|\lambda(\mathbb{Q})|$
4. $|\lambda(\mathbb{Q}(\sqrt{2}, \sqrt{3}))|$
5. $|\lambda(\mathbb{Q}(\sqrt{6}))|$
6. $|\lambda(\mathbb{Q}(\sqrt{30}))|$
7. $|\lambda(\mathbb{Q}(\sqrt{2} + \sqrt{6}))|$
8. $|\lambda(K)|$

9. Describe the group of the polynomial $(x^4 - 1) \in \mathbb{Q}[x]$ over $\mathbb{Q}$.

10. Give the order and describe a generator of the group $G(\mathrm{GF}(729)/\mathrm{GF}(9))$.

11. Let K be the splitting field of $x^3 - 2$ over $\mathbb{Q}$. (Refer to Example 4, Section 9.3.)

   a. Describe the six elements of $G(K/\mathbb{Q})$ by giving their values on $\sqrt[3]{2}$ and $i\sqrt{3}$. (By Example 4, Section 9.3, $K = \mathbb{Q}(\sqrt[3]{2}, i\sqrt{3})$.)

   b. To what group we have seen before is $G(K/\mathbb{Q})$ isomorphic?

   c. Using the notation given in the answer to part (a) in the back of the text, give the lattice diagrams for the subfields of K and for the subgroups of $G(K/\mathbb{Q})$, indicating corresponding intermediate fields and subgroups, as we did in Fig. 9.9.

12. Describe the group of the polynomial $(x^4 - 5x^2 + 6) \in \mathbb{Q}[x]$ over $\mathbb{Q}$.

13. Describe the group of the polynomial $(x^3 - 1) \in \mathbb{Q}[x]$ over $\mathbb{Q}$.

Concepts

14. Give an example of two finite normal extensions K_1 and K_2 of the same field F such that K_1 and K_2 are not isomorphic fields but $G(K_1/F) \simeq G(K_2/F)$.

15. Mark each of the following true or false.

_____ a. Two different subgroups of a Galois group may have the same fixed field.

_____ b. In the notation of Theorem 9.18, if $F \leq E < L \leq K$, then $\lambda(E) < \lambda(L)$.

_____ c. If K is a finite normal extension of F, then K is a normal extension of E, where $F \leq E \leq K$.

_____ d. If two finite normal extensions E and L of a field F have isomorphic Galois groups, then $[E:F] = [L:F]$.

_____ e. If E is a finite normal extension of F and H is a normal subgroup of $G(E/F)$, then E_H is a normal extension of F.

_____ f. If E is any finite normal simple extension of a field F, then the Galois group $G(E/F)$ is a simple group.

_____ g. No Galois group is simple.

_____ h. The Galois group of a finite extension of a finite field is abelian.

_____ i. An extension E of degree 2 over a field F is always a normal extension of F.

_____ j. An extension E of degree 2 over a field F is always a normal extension of F if the characteristic of F is not 2.

Theory

16. A finite normal extension K of field F is **abelian over** F if $G(K/F)$ is an abelian group. Show that if K is abelian over F and E is a normal extension of F, where $F \leq E \leq K$, then K is abelian over E and E is abelian over F.

17. Let K be a finite normal extension of a field F. Prove that for every $\alpha \in K$, the **norm of** α **over** F, given by

$$N_{K/F}(\alpha) = \prod_{\sigma \in G(K/F)} \sigma(\alpha),$$

and the **trace of** α **over** F, given by

$$Tr_{K/F}(\alpha) = \sum_{\sigma \in G(K/F)} \sigma(\alpha),$$

are elements of F.

18. Consider $K = \mathbb{Q}(\sqrt{2}, \sqrt{3})$. Referring to Exercise 17, compute each of the following (see Example 1).

a. $N_{K/\mathbb{Q}}(\sqrt{2})$

b. $N_{K/\mathbb{Q}}(\sqrt{2} + \sqrt{3})$

c. $N_{K/\mathbb{Q}}(\sqrt{6})$

d. $N_{K/\mathbb{Q}}(2)$

e. $Tr_{K/\mathbb{Q}}(\sqrt{2})$

f. $Tr_{K/\mathbb{Q}}(\sqrt{2} + \sqrt{3})$

g. $Tr_{K/\mathbb{Q}}(\sqrt{6})$

h. $Tr_{K/\mathbb{Q}}(2)$

19. Let K be a normal extension of F, and let $K = F(\alpha)$. Let

$$\text{irr}(\alpha, F) = x^n + a_{n-1}x^{n-1} + \cdots + a_1x + a_0.$$

Referring to Exercise 17, show that

a. $N_{K/F}(\alpha) = (-1)^n a_0$,   b. $Tr_{K/F}(\alpha) = -a_{n-1}$.

20. Let $f(x) \in F[x]$ be a polynomial of degree n such that each irreducible factor is separable over F. Show that the order of the group of $f(x)$ over F divides $n!$.

21. Let $f(x) \in F[x]$ be a polynomial such that every irreducible factor of $f(x)$ is a separable polynomial over F. Show that the group of $f(x)$ over F can be viewed in a natural way as a group of permutations of the zeros of $f(x)$ in $\bar{F}$.

22. Let F be a field and let ζ be a primitive nth root of unity in $\bar{F}$, where the characteristic of F is either 0 or does not divide n.

   a. Show that $F(\zeta)$ is a normal extension of F.

   b. Show that $G(F(\zeta)/F)$ is abelian. [*Hint:* Every $\sigma \in G(F(\zeta)/F)$ maps ζ onto some ζ^r and is completely determined by this value r.]

23. A finite normal extension K of a field F is **cyclic over** F if $G(K/F)$ is a cyclic group.

   a. Show that if K is cyclic over F and E is a normal extension of F, where $F \le E \le K$, then E is cyclic over F and K is cyclic over E.

   b. Show that if K is cyclic over F, then there exists exactly one field E, $F \le E \le K$, of degree d over F for each divisor d of $[K:F]$.

24. Let K be a finite normal extension of F.

   a. For $\alpha \in K$, show that

$$f(x) = \prod_{\sigma \in G(K/F)} (x - \sigma(\alpha))$$

   is in $F[x]$.

   b. Referring to part (a), show that $f(x)$ is a power of $\mathrm{irr}(\alpha, F)$, and $f(x) = \mathrm{irr}(\alpha, F)$ if and only if $E = F(\alpha)$.

25. The **join** $E \vee L$ of two extension fields of F in $\bar{F}$ is the smallest subfield of $\bar{F}$ containing both E and L. That is, $E \vee L$ is the intersection of all subfields of $\bar{F}$ *containing both* E and L. Let K be a finite normal extension of a field F, and let E and L be extensions of F contained in K, as shown in Fig. 9.11. Describe $G(K/(E \vee L))$ in terms of $G(K/E)$ and $G(K/L)$.

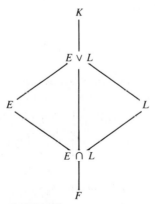

Figure 9.11

26. With reference to the situation in Exercise 25, describe $G\{K/(E \cap L)\}$ in terms of $G(K/E)$ and $G(K/L)$.

9.7

Illustrations of Galois Theory
Symmetric Functions

Let F be a field, and let $y_1, \ldots, y_n$ be n indeterminates. There are some natural automorphisms of $F(y_1, \ldots, y_n)$ leaving f fixed, namely those defined by permutations of $\{y_1, \ldots, y_n\}$. To be more explicit, let σ be a permutation of $\{1, \ldots, n\}$, that is, $\sigma \in S_n$. Then σ gives rise to a natural map $\bar{\sigma} : F(y_1, \ldots, y_n) \to F(y_1, \ldots, y_n)$ given by

$$\bar{\sigma}\left(\frac{f(y_1, \ldots, y_n)}{g(y_1, \ldots, y_n)}\right) = \frac{f(y_{\sigma(1)}, \ldots, y_{\sigma(n)})}{g(y_{\sigma(1)}, \ldots, y_{\sigma(n)})}$$

for $f(y_1, \ldots, y_n), g(y_1, \ldots, y_n) \in F[y_1, \ldots, y_n]$, with $g(y_1, \ldots, y_n) \neq 0$. It is immediate that $\bar{\sigma}$ is an automorphism of $F(y_1, \ldots, y_n)$ leaving F fixed. The elements of $F(y_1, \ldots, y_n)$ left fixed by *all* $\bar{\sigma}$, for all $\sigma \in S_n$, are those rational functions that are *symmetric* in the indeterminates $y_1, \ldots, y_n$.

DEFINITION 9.16 (Symmetric Function) An element of the field $F(y_1, \ldots, y_n)$ is a **symmetric function in** $y_1, \ldots, y_n$ **over** F, if it is left fixed by all permutations of $y_1, \ldots, y_n$, in the sense just explained.

Let $\overline{S}_n$ be the group of all the automorphisms $\bar{\sigma}$ for $\sigma \in S_n$. Observe that $\overline{S}_n$ is naturally isomorphic to S_n. Let K be the subfield of $F(y_1, \ldots, y_n)$, which is the fixed field of $\overline{S}_n$. Consider the polynomial

$$f(x) = \prod_{i=1}^{n} (x - y_i);$$

this polynomial $f(x) \in (F(y_1, \ldots, y_n))[x]$ is a **general polynomial of degree** n. Let $\bar{\sigma}_x$ be the extension of $\bar{\sigma}$, in the natural way, to $(F(y_1, \ldots, y_n))[x]$, where $\bar{\sigma}_x(x) = x$. Now $f(x)$ is left fixed by each map $\bar{\sigma}_x$ for $\sigma \in S_n$; that is,

$$\prod_{i=1}^{n} (x - y_i) = \prod_{i=1}^{n} (x - y_{\sigma(i)}).$$

Thus the coefficients of $f(x)$ are in K; they are symmetric functions in $y_1, \ldots, y_n$. As illustration, note that the constant term of $f(x)$ is

$$(-1)^n y_1 y_2 \cdots y_n,$$

the coefficient of x^{n-1} is $-(y_1 + y_2 + \cdots + y_n)$, and so on. These are symmetric functions in $y_1, \ldots, y_n$.

Thus the first elementary symmetric function in $y_1, \ldots, y_n$ is

$$s_1 = y_1 + y_2 + \cdots + y_n,$$

the second is $s_2 = y_1 y_2 + y_1 y_3 + \cdots + y_{n-1} y_n$, and so on, and the nth is $s_n = y_1 y_2 \cdots y_n$.

Consider the field $E = F(s_1, \ldots, s_n)$. Of course, $E \leq K$, where K is the field of all symmetric functions in $y_1, \ldots, y_n$ over F. But $F(y_1, \ldots, y_n)$ is a finite normal extension of E, namely the splitting field of

$$f(x) = \prod_{i=1}^{n} (x - y_i)$$

over E. Since the degree of $f(x)$ is n, we have at once

$$\left[F(y_1, \ldots, y_n):E\right] \leq n!$$

(see Exercise 11, Section 9.3). However, since K is the fixed field of $\overline{S}_n$ and

$$|\overline{S}_n| = |S_n| = n!,$$

we have also

$$n! \leq \{F(y_1, \ldots, y_n):K\} \leq \left[F(y_1, \ldots, y_n):K\right].$$

Therefore,

$$n! \leq \left[F(y_1, \ldots, y_n):K\right] \leq \left[F(y_1, \ldots, y_n):E\right] \leq n!,$$

so

$$K = E.$$

The full Galois group of $F(y_1, \ldots, y_n)$ over E is therefore $\overline{S}_n$. The fact that $K = E$ shows that every symmetric function can be expressed as a rational function of the elementary symmetric functions $s_1, \ldots, s_n$. We summarize these results in a theorem.

THEOREM 9.20   Let $s_1, \ldots, s_n$ be the elementary symmetric functions in the indeterminates $y_1, \ldots, y_n$. Then every symmetric function of $y_1, \ldots, y_n$ over F is a rational function of the elementary symmetric functions. Also, $F(y_1, \ldots, y_n)$ is a finite normal extension of degree $n!$ of $F(s_1, \ldots, s_n)$, and the Galois group of this extension is naturally isomorphic to S_n.

In view of Cayley's theorem, it can be deduced from Theorem 9.20 that any finite group can occur as a Galois group (up to isomorphism). (See Exercise 13.)

Examples

Let us give our promised example of a finite normal extension having a Galois group whose lattice of subgroups does not look like its own inversion.

EXAMPLE I   Consider the splitting field in $\mathbb{C}$ of $x^4 - 2$ over $\mathbb{Q}$. Now $x^4 - 2$ is irreducible over $\mathbb{Q}$, by Einstein's criterion, with $p = 2$. Let $\alpha = \sqrt[4]{2}$ be the real positive zero of $x^4 - 2$. Then the four zeros of $x^4 - 2$ in $\mathbb{C}$ are α, $-\alpha$, $i\alpha$, and $-i\alpha$, where i is the usual zero of $x^2 + 1$ in $\mathbb{C}$. The splitting field K of $x^4 - 2$ over $\mathbb{Q}$ thus contains $(i\alpha)/\alpha = i$. Since α is a real number,

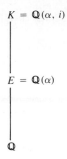

Figure 9.12

$\mathbb{Q}(\alpha) < \mathbb{R}$, so $\mathbb{Q}(\alpha) \neq K$. However, since $\mathbb{Q}(\alpha, i)$ contains all zeros of $x^4 - 2$, we see that $\mathbb{Q}(\alpha, i) = K$. Letting $E = \mathbb{Q}(\alpha)$, we have the diagram in Fig. 9.12.

Now $\{1, \alpha, \alpha^2, \alpha^3\}$ is a basis for E over $\mathbb{Q}$, and $\{1, i\}$ is a basis for K over E. Thus

$$\{1, \alpha, \alpha^2, \alpha^3, i, i\alpha, i\alpha^2, i\alpha^3\}$$

is a basis for K over $\mathbb{Q}$. Since $[K:\mathbb{Q}] = 8$, we must have $|G(K/\mathbb{Q})| = 8$, so we need to find eight automorphisms of K leaving $\mathbb{Q}$ fixed. We know that any such automorphism σ is completely determined by its values on elements of the basis $\{1, \alpha, \alpha^2, \alpha^3, i, i\alpha, i\alpha^2, i\alpha^3\}$, and these values are in turn determined by $\sigma(\alpha)$ and $\sigma(i)$. But $\sigma(\alpha)$ must always be a conjugate of α over $\mathbb{Q}$, that is, one of the four zeros of $\mathrm{irr}(\alpha, \mathbb{Q}) = x^4 - 2$. Likewise, $\sigma(i)$ must be a zero of $\mathrm{irr}(i, \mathbb{Q}) = x^2 + 1$. Thus the four possibilities for $\sigma(\alpha)$, combined with the two possibilities for $\sigma(i)$, must give all eight automorphisms. We describe these in Table 9.2. For example, $\rho_3(\alpha) = -i\alpha$ and $\rho_3(i) = i$, while ρ_0 is the identity automorphism. Now

$$(\mu_1 \rho_1)(\alpha) = \mu_1(\rho_1(\alpha)) = \mu_1(i\alpha) = \mu_1(i)\mu_1(\alpha) = -i\alpha,$$

and, similarly,

$$(\mu_1 \rho_1)(i) = -i$$

so $\mu_1 \rho_1 = \delta_2$. A similar computation shows that

$$(\rho_1 \mu_1)(\alpha) = i\alpha \qquad \text{and} \qquad (\rho_1 \mu_1)(i) = -i.$$

Thus $\rho_1 \mu_1 = \delta_1$, so $\rho_1 \mu_1 \neq \mu_1 \rho_1$ and $G(K/\mathbb{Q})$ is not abelian. Therefore, $G(K/\mathbb{Q})$ must be isomorphic to one of the two nonabelian groups of order 8

	ρ_0	ρ_1	ρ_2	ρ_3	μ_1	δ_1	μ_2	δ_2
$\alpha \rightarrow$	α	$i\alpha$	$-\alpha$	$-i\alpha$	α	$i\alpha$	$-\alpha$	$-i\alpha$
$i \rightarrow$	i	i	i	i	$-i$	$-i$	$-i$	$-i$

Table 9.2

$$G(K/\mathbb{Q})$$

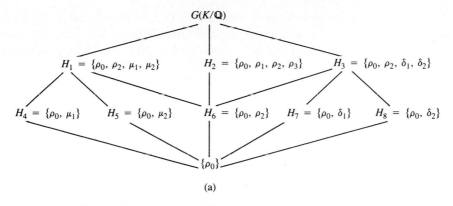

$$H_1 = \{\rho_0, \rho_2, \mu_1, \mu_2\} \qquad H_2 = \{\rho_0, \rho_1, \rho_2, \rho_3\} \qquad H_3 = \{\rho_0, \rho_2, \delta_1, \delta_2\}$$

$$H_4 = \{\rho_0, \mu_1\} \quad H_5 = \{\rho_0, \mu_2\} \quad H_6 = \{\rho_0, \rho_2\} \quad H_7 = \{\rho_0, \delta_1\} \quad H_8 = \{\rho_0, \delta_2\}$$

$$\{\rho_0\}$$

(a)

$$\mathbb{Q}(\sqrt[4]{2}, i) = K = K_{\{\rho_0\}}$$

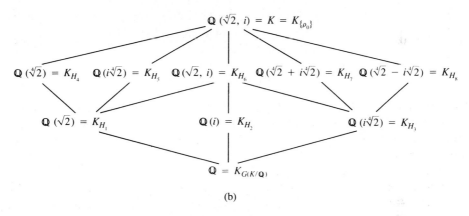

$$\mathbb{Q}(\sqrt[4]{2}) = K_{H_4} \quad \mathbb{Q}(i\sqrt[4]{2}) = K_{H_5} \quad \mathbb{Q}(\sqrt{2}, i) = K_{H_6} \quad \mathbb{Q}(\sqrt[4]{2} + i\sqrt[4]{2}) = K_{H_7} \quad \mathbb{Q}(\sqrt[4]{2} - i\sqrt[4]{2}) = K_{H_8}$$

$$\mathbb{Q}(\sqrt{2}) = K_{H_1} \qquad\qquad \mathbb{Q}(i) = K_{H_2} \qquad\qquad \mathbb{Q}(i\sqrt{2}) = K_{H_3}$$

$$\mathbb{Q} = K_{G(K/\mathbb{Q})}$$

(b)

Figure 9.13 (a) Group lattice diagram. (b) Field lattice diagram.

described in Example 5, Section 4.6. Computing from Table 9.2, we see that ρ_1 is of order 4, μ_1 is of order 2, $\{\rho_1, \mu_1\}$ generates $G(K/\mathbb{Q})$, and $\rho_1\mu_1 = \mu_1\rho_1{}^3 = \delta_1$. Thus $G(K/\mathbb{Q})$ is isomorphic to the group G_1 of Example 5, Section 4.6, the *octic group*. We chose our notation for the elements of $G(K/\mathbb{Q})$ so that its group table would coincide with the table for the octic group in Table 2.2. The lattice of subgroups H_i of $G(K/\mathbb{Q})$ is that given in Fig. 2.7. We repeat it here in Fig. 9.13 and also give the corresponding lattice of intermediate fields between $\mathbb{Q}$ and K. This finally illustrates nicely that one lattice is the inversion of the other.

The determination of the fixed fields K_{H_i} sometimes requires a bit of ingenuity. Let's illustrate. To find K_{H_2}, we merely have to find an extension of $\mathbb{Q}$ of degree 2 left fixed by $\{\rho_0, \rho_1, \rho_2, \rho_3\}$. Since all ρ_j leave i fixed, $\mathbb{Q}(i)$ is the field we are after. To find K_{H_4}, we have to find an extension of $\mathbb{Q}$ of degree 4 left fixed by ρ_0 and μ_1. Since μ_1 leaves α fixed and α is a zero of irr($\alpha, \mathbb{Q}$) = $x^4 - 2$, we see that $\mathbb{Q}(\alpha)$ is of degree 4 over $\mathbb{Q}$ and is left fixed by $\{\rho_0, \mu_1\}$. By *Galois theory*, it is the only such field. Here we are using strongly the one-to-one correspondence given by the Galois theory. If we find one field that fits the bill, it is the one we are after. Finding K_{H_7} requires more ingenuity. Since $H_7 = \{\rho_0, \delta_1\}$ is a group, for any $\beta \in K$ we see that $\rho_0(\beta) + \delta_1(\beta)$ is left fixed by ρ_0 and δ_1. Taking $\beta = \alpha$, we see that $\rho_0(\alpha) + \delta_1(\alpha) = \alpha + i\alpha$ is left fixed by H_7. We can check and see that ρ_0 and δ_1 are the only automorphisms leaving $\alpha + i\alpha$ fixed. Thus by the one-to-one correspondence, we must have

$$\mathbb{Q}(\alpha + i\alpha) = \mathbb{Q}(\sqrt[4]{2} + i\sqrt[4]{2}) = K_{H_7}.$$

Suppose we wish to find $\text{irr}(\alpha + i\alpha, \mathbb{Q})$. If $\gamma = \alpha + i\alpha$, then for every conjugate of γ over $\mathbb{Q}$, there exists an automorphism of K mapping γ into that conjugate. Thus we need only compute the various different values $\sigma(\gamma)$ for $\sigma \in G(K/\mathbb{Q})$ to find the other zeros of $\text{irr}(\gamma, \mathbb{Q})$. By Theorem 9.17, elements σ of $G(K/\mathbb{Q})$ giving these different values can be found by taking a set of representatives of the left cosets of $G(K/\mathbb{Q}(\gamma)) = \{\rho_0, \delta_1\}$ in $G(K/\mathbb{Q})$. A set of representatives for these left cosets is

$$\{\rho_0, \rho_1, \rho_2, \rho_3\}.$$

The conjugates of $\gamma = \alpha + i\alpha$ are thus $\alpha + i\alpha$, $i\alpha - \alpha$, $-\alpha - i\alpha$, and $-i\alpha + \alpha$. Hence

$$
\begin{aligned}
\text{irr}(\gamma, \mathbb{Q}) &= [(x - (\alpha + i\alpha))(x - (i\alpha - \alpha))] \\
&\quad \cdot [(x - (-\alpha - i\alpha))(x - (-i\alpha + \alpha))] \\
&= (x^2 - 2i\alpha x - 2\alpha^2)(x^2 + 2i\alpha x - 2\alpha^2) \\
&= x^4 + 4\alpha^4 = x^4 + 8. \ \blacktriangle
\end{aligned}
$$

We have seen examples in which the splitting field of a quartic (4th degree) polynomial over a field F is an extension of F of degree 8 (Example 1) and of degree 24 (Theorem 9.20, with $n = 4$). The degree of an extension of a field F that is a splitting field of a quartic over F must always divide $4! = 24$. The splitting field of $(x - 2)^4$ over $\mathbb{Q}$ is $\mathbb{Q}$, an extension of degree 1, and the splitting field of $(x^2 - 2)^2$ over $\mathbb{Q}$ is $\mathbb{Q}(\sqrt{2})$, an extension of degree 2. Our last example will give an extension of degree 4 for the splitting field of a quartic.

EXAMPLE 2   Consider the splitting field of $x^4 + 1$ over $\mathbb{Q}$. By Theorem 5.20, we can show that $x^4 + 1$ is irreducible over $\mathbb{Q}$, by arguing that it does not factor in $\mathbb{Z}[x]$. (See Exercise 1.) The work on complex numbers in Section 0.4 shows that the zeros of $x^4 + 1$ are $(1 \pm i)/\sqrt{2}$ and $(-1 \pm i)/\sqrt{2}$. A computation shows that if

$$\alpha = \frac{1 + i}{\sqrt{2}},$$

	σ_1	σ_3	σ_5	σ_7
$\alpha \rightarrow$	α	α^3	α^5	α^7

Table 9.3

then

$$\alpha^3 = \frac{-1 + i}{\sqrt{2}}, \qquad \alpha^5 = \frac{-1 - i}{\sqrt{2}}, \qquad \text{and} \qquad \alpha^7 = \frac{1 - i}{\sqrt{2}}.$$

Thus the splitting field K of $x^4 + 1$ over $\mathbb{Q}$ is $\mathbb{Q}(\alpha)$, and $[K:\mathbb{Q}] = 4$. Let us compute $G(K/\mathbb{Q})$ and give the group and field lattice diagrams. Since there exist automorphisms of K mapping α onto each conjugate of α, and since an automorphism σ of $\mathbb{Q}(\alpha)$ is completely determined by $\sigma(\alpha)$, we see that the four elements of $G(K/\mathbb{Q})$ are defined by Table 9.3. Since

$$(\sigma_j \sigma_k)(\alpha) - \sigma_j(\alpha^k) - (\alpha^j)^k = \alpha^{jk}$$

and $\alpha^8 = 1$, we see that $G(K/\mathbb{Q})$ is isomorphic to the group $\{1, 3, 5, 7\}$ under multiplication modulo 8. This is the group G_8 of Theorem 5.9. Since $\sigma_j^2 = \sigma_1$, the identity, for all j, $G(K/\mathbb{Q})$ must be isomorphic to the Klein 4-group. The lattice diagrams are given in Fig. 9.14.

To find $K_{\{\sigma_1, \sigma_3\}}$, it is only necessary to find an element of K not in $\mathbb{Q}$ left fixed by $\{\sigma_1, \sigma_3\}$, since $[K_{\{\sigma_1, \sigma_3\}}:\mathbb{Q}] = 2$. Clearly $\sigma_1(\alpha) + \sigma_3(\alpha)$ is left fixed by both σ_1 and σ_3, since $\{\sigma_1, \sigma_3\}$ is a group. We have

$$\sigma_1(\alpha) + \sigma_3(\alpha) = \alpha + \alpha^3 = i\sqrt{2}.$$

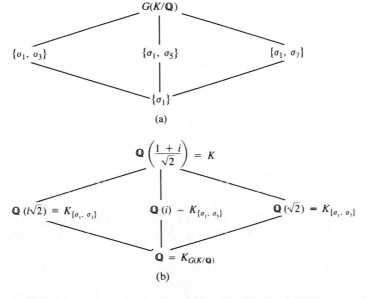

(a)

(b)

Figure 9.14 (a) Group lattice diagram. (b) Field lattice diagram.

Similarly,

$$\sigma_1(\alpha) + \sigma_7(\alpha) = \alpha + \alpha^7 = \sqrt{2}$$

is left fixed by $\{\sigma_1, \sigma_7\}$. This technique is of no use in finding $E_{\{\sigma_1, \sigma_5\}}$, for

$$\sigma_1(\alpha) + \sigma_5(\alpha) = \alpha + \alpha^5 = 0,$$

and $0 \in \mathbb{Q}$. But by a similar argument, $\sigma_1(\alpha)\sigma_5(\alpha)$ is left fixed by both σ_1 and σ_5, and

$$\sigma_1(\alpha)\sigma_5(\alpha) = \alpha\alpha^5 = -i.$$

Thus $\mathbb{Q}(-i) = \mathbb{Q}(i)$ is the field we are after. ▲

Exercises 9.7

Computations (requiring more than the usual amount of theory)

1. Show that $x^4 + 1$ is irreducible in $\mathbb{Q}[x]$, as we asserted in Example 2.

2. Verify that the intermediate fields given in the field lattice diagram in Fig. 9.13 are correct. (Some are verified in the text. Verify the rest.)

3. For each field in the field lattice diagram in Fig. 9.13, find a primitive element generating the field over $\mathbb{Q}$ (see Theorem 9.14) and give its irreducible polynomial over $\mathbb{Q}$.

4. Let ζ be a primitive 5th root of unity in $\mathbb{C}$.

   a. Show that $\mathbb{Q}(\zeta)$ is the splitting field of $x^5 - 1$ over $\mathbb{Q}$.

   b. Show that every automorphism of $K = \mathbb{Q}(\zeta)$ maps ζ onto some power ζ^r of ζ.

   c. Using part (b), describe the elements of $G(K/\mathbb{Q})$.

   d. Give the group and field lattice diagrams for $\mathbb{Q}(\zeta)$ over $\mathbb{Q}$, computing the intermediate fields as we did in Examples 1 and 2.

5. Describe the group of the polynomial $(x^5 - 2) \in (\mathbb{Q}(\zeta))[x]$ over $\mathbb{Q}(\zeta)$, where ζ is a primitive 5th root of unity.

6. Repeat Exercise 4 for ζ a primitive 7th root of unity in $\mathbb{C}$.

7. In the easiest way possible, describe the group of the polynomial

$$(x^8 - 1) \in \mathbb{Q}[x]$$

over $\mathbb{Q}$.

8. Find the splitting field K in $\mathbb{C}$ of the polynomial $(x^4 - 4x^2 - 1) \in \mathbb{Q}[x]$. Compute the group of the polynomial over $\mathbb{Q}$ and exhibit the correspondence between the subgroups of $G(K/\mathbb{Q})$ and the intermediate fields. In other words, do the complete job.

9. Express each of the following symmetric functions in y_1, y_2, y_3 over $\mathbb{Q}$ as a rational function of the elementary symmetric functions s_1, s_2, s_3.

   a. $y_1^2 + y_2^2 + y_3^2$

b. $\dfrac{y_1}{y_2} + \dfrac{y_2}{y_1} + \dfrac{y_1}{y_3} + \dfrac{y_3}{y_1} + \dfrac{y_2}{y_3} + \dfrac{y_3}{y_2}$

c. $(y_1 - y_2)^2(y_1 - y_3)^2(y_2 - y_3)^2$

10. Let α_1, α_2, α_3 be the zeros in $\mathbb{C}$ of the polynomial

$$(x^3 - 4x^2 + 6x - 2) \in \mathbb{Q}[x].$$

Find the polynomial having as zeros precisely the following:

a. $\alpha_1 + \alpha_2 + \alpha_3$

b. $\alpha_1^2, \alpha_2^2, \alpha_3^2$

c. $(\alpha_1 - \alpha_2)^2, (\alpha_1 - \alpha_3)^2, (\alpha_2 - \alpha_3)^2$

Theory

11. Let $f(x) \in F[x]$ be a monic polynomial of degree n having all its irreducible factors separable over F. Let $K \leq \bar{F}$ be the splitting field of $f(x)$ over F, and suppose that $f(x)$ factors in $K[x]$ into

$$\prod_{i=1}^{n} (x - \alpha_i).$$

Let

$$\Delta(f) = \prod_{i<j} (\alpha_i - \alpha_j);$$

the product $(\Delta(f))^2$ is the **discriminant of** $f(x)$.

a. Show that $\Delta(f) = 0$ if and only if $f(x)$ has as a factor the square of some irreducible polynomial in $F[x]$.

b. Show that $(\Delta(f))^2 \in F$.

c. $G(K/F)$ may be viewed as a subgroup of $\bar{S}_n$, where $\bar{S}_n$ is the group of all permutations of $\{\alpha_i \mid i = 1, \ldots, n\}$. Show that $G(K/F)$, when viewed in this fashion, is a subgroup of A_n, the group formed by all even permutations of $\{\alpha_i \mid i = 1, \ldots, n\}$, if and only if $\Delta(f) \in F$.

12. An element of $\mathbb{C}$ is an **algebraic integer** if it is a zero of some *monic* polynomial in $\mathbb{Z}[x]$. Show that the set of all algebraic integers forms a subring of $\mathbb{C}$.

13. Show that every finite group is isomorphic to some Galois group $G(K/F)$ for some finite normal extension K of some field F.

9.8

Cyclotomic Extensions

The Galois Group of a Cyclotomic Extension

This section deals with extension fields of a field F obtained by adjoining to F some roots of unity. The case of a finite field F was covered in Section 8.5, so we shall be primarily concerned with the case where F is infinite.

DEFINITION 9.18 (Cyclotomic Extension) The splitting field of $x^n - 1$ over F is the **nth cyclotomic extension** of F.

Suppose that F is any field, and consider $(x^n - 1) \in F[x]$. By long division, as in the proof of Lemma 8.2, Section 8.5, we see that if α is a zero of $x^n - 1$ and $g(x) = (x^n - 1)/(x - \alpha)$, then $g(\alpha) = (n \cdot 1)(1/\alpha) \neq 0$, provided that the characteristic of F does not divide n. Therefore, under this condition, the splitting field of $x^n - 1$ is a separable and thus a normal extension of F.

Assume from now on that this is the case, and let K be the splitting field of $x^n - 1$ over F. Then $x^n - 1$ has n distinct zeros in K, and by Corollary 3 of Theorem 5.18, these form a cyclic group of order n under the field multiplication. We saw in the corollary of Theorem 1.9 that a cyclic group of order n has $\varphi(n)$ generators, where φ is the Euler phi-function introduced prior to Theorem 5.10. For our situation here, these $\varphi(n)$ generators are exactly the primitive nth roots of unity.

DEFINITION 9.19 (Cyclotomic Polynomial)   The polynomial

$$\Phi_n(x) = \prod_{i=1}^{\varphi(n)} (x - \alpha_i)$$

where the α_i are the primitive nth roots of unity in $\bar{F}$, is the **nth cyclotomic polynomial over** F.

Since an automorphism of the Galois group $G(K/F)$ must permute the primitive nth roots of unity, we see that $\Phi_n(x)$ is left fixed under every element of $G(K/F)$ regarded as extended in the natural way to $K[x]$. Thus $\Phi_n(x) \in F[x]$. In particular, for $F = \mathbb{Q}$, $\Phi_n(x) \in \mathbb{Q}[x]$, and $\Phi_n(x)$ is a divisor of $x^n - 1$. Thus over $\mathbb{Q}$, we must actually have $\Phi_n(x) \in \mathbb{Z}[x]$, by Theorem 5.20. We have seen that $\Phi_p(x)$ is irreducible over $\mathbb{Q}$, in the corollary of Theorem 5.21. While $\Phi_n(x)$ need not be irreducible in the case of the fields $\mathbb{Z}_p$, it can be shown that over $\mathbb{Q}$, $\Phi_n(x)$ is irreducible.

Let us now limit our discussion to characteristic 0, in particular to subfields of the complex numbers. Let i be the usual complex zero of $x^2 + 1$. Our work with complex numbers in Section 0.4 and Example 12, Section 1.2, shows that

$$\left(\cos\frac{2\pi}{n} + i\sin\frac{2\pi}{n}\right)^n = \cos 2\pi + i\sin 2\pi = 1,$$

so $\cos(2\pi/n) + i\sin(2\pi/n)$ is an nth root of unity. The least integer m such that $(\cos(2\pi/n) + i\sin(2\pi/n))^m = 1$ is n. *Thus* $\cos(2\pi/n) + i\sin(2\pi/n)$ *is a primitive nth root of unity, a zero of*

$$\Phi_n(x) \in \mathbb{Q}[x].$$

EXAMPLE 1   A primitive 8th root of unity in $\mathbb{C}$ is

$$\zeta = \cos\frac{2\pi}{8} + i\sin\frac{2\pi}{8}$$

$$= \cos\frac{\pi}{4} + i\sin\frac{\pi}{4}$$

$$= \frac{1}{\sqrt{2}} + i\frac{1}{\sqrt{2}} = \frac{1+i}{\sqrt{2}}.$$

By the theory of cyclic groups, in particular by the corollary of Theorem 1.9, all the primitive 8th roots of unity in $\mathbb{Q}$ are ζ, ζ^3, ζ^5, and ζ^7, so

$$\Phi_8(x) = (x - \zeta)(x - \zeta^3)(x - \zeta^5)(x - \zeta^7).$$

We can compute, directly from this expression, $\Phi_8(x) = x^4 + 1$ (see Exercise 1). Compare this with Example 2, Section 9.7. ▲

Let us still restrict our work to $F = \mathbb{Q}$, and let us assume, without proof, that $\Phi_n(x)$ is irreducible over $\mathbb{Q}$. Let

$$\zeta = \cos\frac{2\pi}{n} + i\sin\frac{2\pi}{n},$$

so that ζ is a primitive nth root of unity. Note that ζ is a generator of the cyclic multiplicative group of order n consisting of *all* nth roots of unity. All the primitive nth roots of unity, that is, all the generators of this group, are of the form ζ^m for $1 \leq m < n$ and m relatively prime to n. The field $\mathbb{Q}(\zeta)$ is the whole splitting field of $x^n - 1$ over $\mathbb{Q}$. Let $K = \mathbb{Q}(\zeta)$. If ζ^m is another primitive nth root of unity, then since ζ and ζ^m are conjugate over $\mathbb{Q}$, there is an automorphism τ_m in $G(K/\mathbb{Q})$ mapping ζ onto ζ^m. Let τ_r be the similar automorphism in $G(K/\mathbb{Q})$ corresponding to the primitive nth root of unity ζ^r. Then

$$(\tau_m\tau_r)(\zeta) = \tau_m(\zeta^r) = (\tau_m(\zeta))^r = (\zeta^m)^r = \zeta^{rm}.$$

This shows that the Galois group $G(K/\mathbb{Q})$ is isomorphic to the group G_n of Theorem 5.9 consisting of elements of $\mathbb{Z}_n$ relatively prime to n under multiplication modulo n. This group has $\varphi(n)$ elements and is abelian.

Special cases of this material have appeared several times in the text and exercises. For example, α of Example 2, Section 9.7, is a primitive 8th root of unity, and we made arguments in that example identical to those given here. We summarize these results in a theorem.

THEOREM 9.21   The Galois group of the nth cyclotomic extension of $\mathbb{Q}$ has $\varphi(n)$ elements and is isomorphic to the group consisting of the positive integers less than n and relatively prime to n under multiplication modulo n.

EXAMPLE 2   Example 2, Section 9.7, illustrates this theorem, for it is easy to see that the splitting field of $x^4 + 1$ is the same as the splitting field of $x^8 - 1$ over $\mathbb{Q}$. This follows from the fact that $\Phi_8(x) = x^4 + 1$ (see Example 1 and Exercise 1). ▲

COROLLARY   The Galois group of the pth cyclotomic extension of $\mathbb{Q}$ for a prime p is cyclic of order $p - 1$.

PROOF   By Theorem 9.21, the Galois group of the pth cyclotomic extension of $\mathbb{Q}$ has $\varphi(p) = p - 1$ elements, and is isomorphic to the group of positive integers less than p and relatively prime to p under multiplication modulo p. This is exactly the multiplicative group $\langle \mathbb{Z}_p{}^*, \cdot \rangle$ of nonzero elements of the field $\mathbb{Z}_p$ under field multiplication. By Theorem 8.23, this group is cyclic. ◆

Constructible Polygons

We conclude with an application determining which regular n-gons are constructible with a compass and a straightedge. We saw in Section 8.4 that the regular n-gon is constructible if and only if $\cos(2\pi/n)$ is a constructible real number. Now let

$$\zeta = \cos\frac{2\pi}{n} + i\sin\frac{2\pi}{n}.$$

Then

$$\frac{1}{\zeta} = \cos\frac{2\pi}{n} - i\sin\frac{2\pi}{n},$$

for

$$\left(\cos\frac{2\pi}{n} + i\sin\frac{2\pi}{n}\right)\left(\cos\frac{2\pi}{n} - i\sin\frac{2\pi}{n}\right) = \cos^2\frac{2\pi}{n} + \sin^2\frac{2\pi}{n} = 1.$$

But then

$$\zeta + \frac{1}{\zeta} = 2\cos\frac{2\pi}{n}.$$

Thus the corollary of Theorem 8.17 shows that the regular n-gon is constructible only if $\zeta + 1/\zeta$ generates an extension of $\mathbb{Q}$ of degree a power of 2.

If K is the splitting field of $x^n - 1$ over $\mathbb{Q}$, then $[K:\mathbb{Q}] = \varphi(n)$, by Theorem 9.21. If $\sigma \in G(K/\mathbb{Q})$ and $\sigma(\zeta) = \zeta^r$, then

$$\sigma\left(\zeta + \frac{1}{\zeta}\right) = \zeta^r + \frac{1}{\zeta^r}$$

$$= \left(\cos\frac{2\pi r}{n} + i\sin\frac{2\pi r}{n}\right) + \left(\cos\frac{2\pi r}{n} - i\sin\frac{2\pi r}{n}\right)$$

$$= 2\cos\frac{2\pi r}{n}.$$

But for $1 < r < n$, we have $2\cos(2\pi r/n) = 2\cos(2\pi/n)$ only in the case that $r = n - 1$. Thus the only elements of $G(K/\mathbb{Q})$ carrying $\zeta + 1/\zeta$ onto itself are the identity automorphism and the automorphism τ, with $\tau(\zeta) = \zeta^{n-1} = 1/\zeta$. This shows that the subgroup of $G(K/\mathbb{Q})$ leaving $\mathbb{Q}(\zeta + 1/\zeta)$ fixed is of order 2, so by Galois theory,

$$\left[\mathbb{Q}\left(\zeta + \frac{1}{\zeta}\right):\mathbb{Q}\right] = \frac{\varphi(n)}{2}.$$

Hence the regular n-gon is constructible only if $\varphi(n)/2$, and therefore also $\varphi(n)$, is a power of 2.

It can be shown by elementary arguments in number theory that if

$$n = 2^v p_1^{s_1} \cdots p_t^{s_t},$$

where the p_i are the distinct odd primes dividing n, then

$$\varphi(n) = 2^{v-1} p_1^{s_1-1} \cdots p_t^{s_t-1}(p_1 - 1) \cdots (p_t - 1). \tag{1}$$

If $\varphi(n)$ is to be a power of 2, then every odd prime dividing n must appear only to the first power and must be one more than a power of 2. Thus we must have each

$$p_i = 2^m + 1$$

for some m. Since -1 is a zero of $x^q + 1$ for q an odd prime, $x + 1$ divides $x^q + 1$ for q an odd prime. Thus, if $m = qu$, where q is an odd prime, then $2^m + 1 = (2^u)^q + 1$ is divisible by $2^u + 1$. Therefore, for $p_i = 2^m + 1$ to be prime, it must be that m is divisible by 2 only, so p_i has to have the form

$$p_i = 2^{(2^k)} + 1,$$

a **Fermat prime**. Fermat conjectured that these numbers $2^{(2^k)} + 1$ were prime for all nonnegative integers k. Euler showed that while $k = 0, 1, 2, 3$, and 4 give the primes 3, 5, 17, 257, and 65537, for $k = 5$, the integer $2^{(2^5)} + 1$ is divisible by 641. It has been shown that for $5 \le k \le 19$, all the numbers $2^{(2^k)} + 1$ are composite. The case $k = 20$ is still unsolved as far as we know. For at least 60 values of k greater than 20, including $k = 9448$, it has been shown that $2^{2^k} + 1$ is composite. It is unknown whether the number of Fermat primes is finite or infinite.

We have thus shown that the only regular n-gons that might be constructible are those where the odd primes dividing n are Fermat primes whose squares do not divide n. In particular, the only regular p-gons that might be constructible for p a prime greater than 2 are those where p is a Fermat prime.

EXAMPLE 3　　The regular 7-gon is not constructible, since 7 is not a Fermat prime. Similarly, the regular 18-gon is not constructible, for while 3 is a Fermat prime, its square divides 18. ▲

It is a fact that we now demonstrate that all these regular n-gons that are candidates for being constructible are indeed actually constructible. Let ζ again be the primitive nth root of unity $\cos(2\pi/n) + i \sin(2\pi/n)$. We saw above that

$$2 \cos \frac{2\pi}{n} = \zeta + \frac{1}{\zeta},$$

and that

$$\left[\mathbb{Q}\left(\zeta + \frac{1}{\zeta} \right) : \mathbb{Q} \right] = \frac{\varphi(n)}{2}.$$

Suppose now that $\varphi(n)$ is a power 2^s of 2. Let E be $\mathbb{Q}(\zeta + 1/\zeta)$. We saw above that $\mathbb{Q}(\zeta + 1/\zeta)$ is the subfield of $K = \mathbb{Q}(\zeta)$ left fixed by $H_1 = \{\iota, \tau\}$, where ι is the identity element of $G(K/\mathbb{Q})$ and $\tau(\zeta) = 1/\zeta$. By Sylow theory, there exist additional subgroups H_j of order 2^j of $G(\mathbb{Q}(\zeta)/\mathbb{Q})$ for $j = 0, 2, 3, \ldots, s$ such that

$$\{\iota\} = H_0 < H_1 < \cdots H_s = G(\mathbb{Q}(\zeta)/\mathbb{Q}).$$

By Galois theory,

$$\mathbb{Q} = K_{H_s} < K_{H_{s-1}} < \cdots < K_{H_1} = \mathbb{Q}\left(\zeta + \frac{1}{\zeta} \right),$$

and $[K_{H_{j-1}} : K_{H_j}] = 2$. Note that $(\zeta + 1/\zeta) \in \mathbb{R}$, so $\mathbb{Q}(\zeta + 1/\zeta) < \mathbb{R}$. If $K_{H_{j-1}} = K_{H_j}(\alpha_j)$, then α_j is a zero of some $(a_j x^2 + b_j x + c_j) \in K_{H_j}[x]$. By the familiar "quadratic formula," we have

$$K_{H_{j-1}} = K_{H_j}\left(\sqrt{b_j^2 - 4a_j c_j} \right).$$

Since we saw in Section 8.4 that construction of square roots of positive constructible numbers can be achieved by a straightedge and a compass, we see that every element in $\mathbb{Q}(\zeta + 1/\zeta)$, in particular $\cos(2\pi/n)$, is constructible. Hence the regular n-gons where $\varphi(n)$ is a power of 2 are constructible.

We summarize our work under this heading in a theorem.

THEOREM 9.22 The regular n-gon is constructible with a compass and a straightedge if and only if all the odd primes dividing n are Fermat primes whose squares do not divide n.

EXAMPLE 4 The regular 60-gon is constructible, since $60 = (2^2)(3)(5)$ and 3 and 5 are both Fermat primes. ▲

Exercises 9.8

Computations

1. Referring to Example 1, complete the indicated computation, showing that $\Phi_8(x) = x^4 + 1$. [*Suggestion:* Compute the product in terms of ζ, and then use the fact that $\zeta^8 = 1$ and $\zeta^4 = -1$ to simplify the coefficients.]

2. Classify the group of the polynomial $(x^{20} - 1) \in \mathbb{Q}[x]$ over $\mathbb{Q}$ according to the fundamental theorem of finitely generated abelian groups. [*Hint:* Use Theorem 9.21.]

3. Using the formula for $\varphi(n)$ in terms of the factorization of n, as given in Eq. (1), compute the formula:

   a. $\varphi(60)$ b. $\varphi(1000)$ c. $\varphi(8100)$

4. Give the first thirty values of $n \geq 3$ of which the regular n-gon is constructible with a straightedge and a compass.

5. Find the smallest angle of integral degree, that is, $1°$, $2°$, $3°$, and so on, constructible with a straightedge and a compass. [*Hint:* Constructing a $1°$ angle amounts to constructing the regular 360-gon, and so on.]

6. Let K be the splitting field of $x^{12} - 1$ over $\mathbb{Q}$.

   a. Find $[K:\mathbb{Q}]$.

   b. Show that for $\sigma \in G(K/\mathbb{Q})$, σ^2 is the identity automorphism. Classify $G(K/\mathbb{Q})$ according to the fundamental theorem of finitely generated abelian groups.

7. Find $\Phi_3(x)$ over $\mathbb{Z}_2$. Find $\Phi_8(x)$ over $\mathbb{Z}_3$.

8. How many elements are there in the splitting field of $x^6 - 1$ over $\mathbb{Z}_3$?

Concepts

9. Mark each of the following true or false.

   _____ a. $\Phi_n(x)$ is irreducible over every field of characteristic 0.
   _____ b. Every zero in $\mathbb{C}$ of $\Phi_n(x)$ is a primitive nth root of unity.
   _____ c. The group of $\Phi_n(x) \in \mathbb{Q}[x]$ over $\mathbb{Q}$ has order n.
   _____ d. The group of $\Phi_n(x) \in \mathbb{Q}[x]$ over $\mathbb{Q}$ is abelian.

_____ e. The Galois group of the splitting field of $\Phi_n(x)$ over $\mathbb{Q}$ has order $\varphi(n)$.
_____ f. The regular 25-gon is constructible with a straightedge and a compass.
_____ g. The regular 17-gon is constructible with a straightedge and a compass.
_____ h. For a prime p, the regular p-gon is constructible if and only if p is a Fermat prime.
_____ i. All integers of the form $2^{(2^k)} + 1$ for nonnegative integers k are Fermat primes.
_____ j. All Fermat primes are numbers of the form $2^{(2^k)} + 1$ for nonnegative integers k.

Theory

10. Show that if F is a field of characteristic not dividing n, then

$$x^n - 1 = \prod_{d\,|\,n} \Phi_d(x)$$

in $F[x]$, where the product is over all divisors d of n.

11. Find the cyclotomic polynomial $\Phi_n(x)$ over $\mathbb{Q}$ for $n = 1, 2, 3, 4, 5$, and 6. [*Hint:* Use Exercise 10.]

12. Find $\Phi_{12}(x)$ in $\mathbb{Q}[x]$. [*Hint:* Use Exercises 10 and 11.]

13. Show that in $\mathbb{Q}[x]$, $\Phi_{2n}(x) = \Phi_n(-x)$ for odd integers $n > 1$. [*Hint:* Use Exercise 10 and the factorization $x^{2n} - 1 = -(x^n - 1)((-x)^n - 1)$. Proceed by induction.]

14. Let $n, m \in \mathbb{Z}^+$ be relatively prime. Show that the splitting field in $\mathbb{C}$ of $x^{nm} - 1$ over $\mathbb{Q}$ is the same as the splitting field in $\mathbb{C}$ of $(x^n - 1)(x^m - 1)$ over $\mathbb{Q}$.

15. Let $n, m \in \mathbb{Z}^+$ be relatively prime. Show that the group of $(x^{nm} - 1) \in \mathbb{Q}[x]$ over $\mathbb{Q}$ is isomorphic to the direct product of the groups of $(x^n - 1) \in \mathbb{Q}[x]$ and of $(x^m - 1) \in \mathbb{Q}[x]$ over $\mathbb{Q}$. [*Hint:* Using Galois theory, show that the groups of $x^m - 1$ and $x^n - 1$ can both be regarded as subgroups of the group of $x^{nm} - 1$. Then use Exercises 46 and 47 of Section 2.4.]

9.9

Insolvability of the Quintic

The Problem

We are familiar with the fact that a quadratic polynomial $f(x) = ax^2 + bx + c$, $a \neq 0$, with real coefficients has $(-b \pm \sqrt{b^2 - 4ac})/2a$ as zeros in $\mathbb{C}$. Actually, this is true for $f(x) \in F[x]$, where F is any field of characteristic $\neq 2$ and the zeros are in $\bar{F}$. Exercise 4 asks us to show this. Thus, for example, $(x^2 + 2x + 3) \in \mathbb{Q}[x]$ has its zeros in $\mathbb{Q}(\sqrt{-2})$. You may wonder whether the zeros of a cubic polynomial over $\mathbb{Q}$ can also always be expressed in terms of radicals. The answer is yes, and, indeed, even the zeros of a polynomial of degree 4 over $\mathbb{Q}$ can be expressed in terms of radicals. After mathematicians had tried for years to find the "radical formula" for zeros of a 5th degree polynomial, it was a triumph when Abel proved that a quintic need not be solvable by radicals. Our first job will be to describe precisely what this means. A large amount of the algebra we have developed is used in the forthcoming discussion.

Extensions by Radicals

DEFINITION 9.20 (Extension by Radicals)   An extension K of a field F is an **extension of F by radicals** if there are elements $\alpha_1, \ldots, \alpha_r \in K$ and positive integers $n_1, \ldots, n_r$ such that $K = F(\alpha_1, \ldots, \alpha_r)$, $\alpha_1^{n_1} \in F$ and $\alpha_i^{n_i} \in F(\alpha_1, \ldots, \alpha_{i-1})$ for $1 < i < r$. A polynomial $f(x) \in F[x]$ is **solvable by radicals over** F if the splitting field E of $f(x)$ over F is contained in an extension of F by radicals.

A polynomial $f(x) \in F(x)$, is thus solvable by radicals over F if we can obtain every zero of $f(x)$ by using a finite sequence of the operations of addition, subtraction, multiplication, division, and taking n_ith roots, starting with elements of F. Now to say that the quintic is not solvable in the classic

---

◆   **H I S T O R I C A L   N O T E**   ◆

The first publication of a formula for solving cubic equations in terms of radicals was in 1545 in the *Ars Magna* of Girolamo Cardano, although the initial discovery of the method is in part also due to Scipione del Ferro and Niccolo Tartaglia. Cardano's student, Lodovico Ferrari, discovered a method for solving quartic equations by radicals, which also appeared in Cardano's work.

After many mathematicians had attempted to solve quintics by similar methods, it was Joseph-Louis Lagrange who in 1770 first attempted a detailed analysis of the general principles underlying the solutions for polynomials of degree 3 and 4, and showed why these methods fail for those of higher degree. His basic insight was that in the former cases there were rational functions of the roots that took on two and three values respectively under all possible permutations of the roots, hence these rational functions could be written as roots of equations of degree less than that of the original. No such functions were evident in equations of higher degree.

The first mathematician to claim to have a proof of the unsolvability of the quintic equation was Paolo Ruffini (1765–1822) in his algebra text of 1799. His proof was along the lines suggested by Lagrange, in that he in effect determined all of the subgroups of S_5 and showed how these subgroups acted on rational functions of the roots of the equation. Unfortunately, there were several gaps in his various published versions of the proof. It was Niels Henrik Abel who in 1824 and 1826 published a complete proof, closing all of Ruffini's gaps and finally settling this centuries-old question.

case, that is, characteristic 0, is not to say that no quintic is solvable, as the following example shows.

EXAMPLE I   The polynomial $x^5 - 1$ is solvable by radicals over $\mathbb{Q}$. The splitting field K of $x^5 - 1$ is generated over $\mathbb{Q}$ by a primitive 5th root ζ of unity. Then $\zeta^5 = 1$, and $K = \mathbb{Q}(\zeta)$. Similarly, $x^5 - 2$ is solvable by radicals over $\mathbb{Q}$, for its splitting field over $\mathbb{Q}$ is generated by $\sqrt[5]{2}$ and ζ, where $\sqrt[5]{2}$ is the real zero of $x^5 - 2$.

To say that the quintic is insolvable in the classic case means that there exists *some* polynomial of degree 5 with real coefficients that is not solvable by radicals. We shall show this. *We assume throughout this chapter that all fields mentioned have characteristic 0.*

The outline of the argument is as follows, and it is worthwhile to try to remember it.

1. *We shall show that a polynomial $f(x) \in F[x]$ is solvable by radicals over F (if and) only if its splitting field E over F has a solvable Galois group.* Recall that a solvable group is one having a composition series with *abelian* quotients. While this theorem goes both ways, we shall not prove the "if" part.

2. *We shall show that there is a subfield F of the real numbers and a polynomial $f(x) \in F[x]$ of degree 5 with a splitting field E over F such that $G(E/F) \simeq S_5$, the symmetric group on 5 letters.* Recall that a composition series for S_5 is $\{\iota\} < A_5 < S_5$. Since A_5 is not abelian, we will be done.

The following lemma does most of our work for step 1.

LEMMA 9.2   Let F be a field of characteristic 0, and let $a \in F$. If K is the splitting field of $x^n - a$ over F, then $G(K/F)$ is a solvable group.

PROOF   Suppose first that F contains all the nth roots of unity. By Corollary 3 of Theorem 5.18, the nth roots of unity form a cyclic subgroup of $\langle F^*, \cdot \rangle$. Let ζ be a generator of the subgroup. (Actually, the generators are exactly the *primitive* nth roots of unity.) Then the nth roots of unity are

$$1, \zeta, \zeta^2, \ldots, \zeta^{n-1}.$$

If $\beta \in \bar{F}$ is a zero of $(x^n - a) \in F[x]$, then all zeros of $x^n - a$ are

$$\beta, \zeta\beta, \zeta^2\beta, \ldots, \zeta^{n-1}\beta.$$

Since $K = F(\beta)$, an automorphism σ in $G(K/F)$ is determined by the value $\sigma(\beta)$ of the automorphism σ on β. Now if $\sigma(\beta) = \zeta^i\beta$ and $\tau(\beta) = \zeta^j\beta$, where $\tau \in G(K/F)$, then

$$(\tau\sigma)(\beta) = \tau(\sigma(\beta)) = \tau(\zeta^i\beta) = \zeta^i\tau(\beta) = \zeta^i\zeta^j\beta,$$

since $\zeta^i \in F$. Similarly,

$$(\sigma\tau)(\beta) = \zeta^i\zeta^i\beta.$$

Thus $\sigma\tau = \tau\sigma$, and $G(K/F)$ is abelian and therefore solvable.

Now suppose that F does not contain a primitive nth root of unity. Let ζ be a generator of the cyclic group of nth roots of unity under multiplication in $\bar{F}$. Let β again be a zero of $x^n - a$. Since β and $\zeta\beta$ are both in the splitting field K of $x^n - a$, $\zeta = (\zeta\beta)/\beta$ is in K. Let $F' = F(\zeta)$, so we have $F < F' \le K$. Now F' is a normal extension of F, since F' is the splitting field of $x^n - 1$. Since $F' = F(\zeta)$, an automorphism η in $G(F'/F)$ is determined by $\eta(\zeta)$, and we must have $\eta(\zeta) = \zeta^i$ for some i, since all zeros of $x^n - 1$ are powers of ζ. If $\mu(\zeta) = \zeta^j$ for $\mu \in G(F'/F)$, then

$$(\mu\eta)(\zeta) = \mu(\eta(\zeta)) = \mu(\zeta^i) = \mu(\zeta)^i = (\zeta^j)^i = \zeta^{ij},$$

and, similarly,

$$(\eta\mu)(\zeta) = \zeta^{ij}.$$

Thus $G(F'/F)$ is abelian. By the Main Theorem of Galois Theory,

$$\{\iota\} \le G(K/F') \le G(K/F)$$

is a normal series and hence a subnormal series of groups. The first part of the proof shows that $G(K/F')$ is abelian, and Galois theory tells us that $G(K/F)/G(K/F')$ is isomorphic to $G(F'/F)$, which is abelian. Exercise 6 shows that if a group has a subnormal series of subgroups with abelian quotient groups, then any refinement of this series also has abelian quotient groups. Thus a composition series of $G(K/F)$ must have abelian quotient groups, so $G(K/F)$ is solvable. ◆

The following theorem will complete Part 1 of our program.

THEOREM 9.23 Let F be a field of characteristic zero, and let $F \le E \le K \le \bar{F}$, where E is a normal extension of F and K is an extension of F by radicals. Then $G(E/F)$ is a solvable group.

PROOF We first show that K is contained in a finite normal extension L of F by radicals and that the group $G(L/F)$ is solvable. Since K is an extension by radicals, $K = F(\alpha_1, \ldots, \alpha_r)$ where $\alpha_i^{n_i} \in F(\alpha_1, \ldots, \alpha_{i-1})$ for $1 < i \le r$ and $\alpha_1^{n_1} \in F$. To form L, we first form the splitting field L_1 of $f_1(x) = x^{n_1} - \alpha_1^{n_1}$ over F. Then L_1 is a normal extension of F, and Lemma 9.2 shows that $G(L_1/F)$ is a solvable group. Now $\alpha_2^{n_2} \in L_1$, and we form the polynomial

$$f_2(x) = \prod_{\sigma \in G(L_1/F)} [(x^{n_2} - \sigma(\alpha_2)^{n_2})].$$

Since this polynomial is invariant under action by any σ in $G(L_1/F)$, we see that $f_2(x) \in F[x]$. We let L_2 be the splitting field of $f_2(x)$ over L_1. Then L_2 is a splitting field over F also and is a normal extension of F by radicals. We can form L_2 from L_1 via repeated steps as in Lemma 9.2, passing to a splitting field of $x^{n_2} - \sigma(\alpha_2)^{n_2}$ at each step. By Lemma 9.2 and Exercise 7, we see that the Galois group over F of each new extension thus formed continues to be solvable. We continue this process of forming splitting fields over F in this manner: at stage i, we form the splitting field of the polynomial

$$f_i(x) = \prod_{\alpha \in G(L_{i-1}/F)} [(x^{n_i} - \sigma(\alpha_i)^{n_i}]$$

over L_{i-1}. We finally obtain a field $L = L_r$ that is a normal extension of F by radicals, and we see that $G(L/F)$ is a solvable group. We see from construction that $K \leq L$.

To conclude we need only note that by Theorem 9.18, we have $G(E/F) \approx G(L/F)/G(L/E)$. Thus $G(E/F)$ is a factor group, and hence a homomorphic image, of $G(L/F)$. Since $G(L/F)$ is solvable, Exercise 16 of Section 4.1 shows that $G(E/F)$ is solvable. ◆

The Insolvability of the Quintic

It remains for us to show that there is a subfield F of the real numbers and a polynomial $f(x) \in F[x]$ of degree 5 such that the splitting field E of $f(x)$ over F has a Galois group isomorphic to S_5.

Let $y_1 \in \mathbb{R}$ be transcendental over $\mathbb{Q}$, $y_2 \in \mathbb{R}$ be transcendental over $\mathbb{Q}(y_1)$, and so on, until we get $y_5 \in \mathbb{R}$ transcendental over $\mathbb{Q}(y_1, \ldots, y_4)$. It can be shown by a counting argument that such transcendental real numbers exist. Transcendentals found in this fashion are **independent transcendental elements over** $\mathbb{Q}$. Let $E = \mathbb{Q}(y_1, \ldots, y_5)$, and let

$$f(x) = \prod_{i=1}^{5} (x - y_i).$$

Thus $f(x) \in E[x]$. Now the coefficients of $f(x)$ are, except possibly for sign, among the *elementary symmetric functions* in the y_i, namely

$$s_1 = y_1 + y_2 + \cdots + y_5,$$

$$s_2 = y_1 y_2 + y_1 y_3 + y_1 y_4 + y_1 y_5 + y_2 y_3$$
$$+ y_2 y_4 + y_2 y_5 + y_3 y_4 + y_3 y_5 + y_4 y_5,$$

$$\vdots$$

$$s_5 = y_1 y_2 y_3 y_4 y_5.$$

The coefficient of x^i in $f(x)$ is $\pm s_{5-i}$. Let $F = \mathbb{Q}(s_1, s_2, \ldots, s_5)$; then

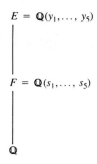

$E = \mathbf{Q}(y_1, \ldots, y_5)$

$F = \mathbf{Q}(s_1, \ldots, s_5)$

$\mathbf{Q}$

Figure 9.15

$f(x) \in F[x]$ (see Fig. 9.15). Then E is the splitting field over F of $f(x)$. Since the y_i behave as indeterminates over $\mathbf{Q}$, for each $\sigma \in S_5$, the symmetric group on five letters, σ induces an automorphism $\bar{\sigma}$ of E defined by $\bar{\sigma}(a) = a$ for $a \in \mathbf{Q}$ and $\bar{\sigma}(y_i) = y_{\sigma(i)}$. Since $\prod_{i=1}^{5} (x - y_i)$ is the same polynomial as $\prod_{i=1}^{5} (x - y_{\sigma(i)})$, we have

$$\bar{\sigma}(s_i) = s_i$$

for each i, so $\bar{\sigma}$ leaves F fixed, and hence $\bar{\sigma} \in G(E/F)$. Now S_5 has order 5!, so

$$|G(E/F)| \geq 5!.$$

Since the splitting field of a polynomial of degree 5 over F has degree at most 5! over F, we see that

$$|G(E/F)| \leq 5!.$$

Thus $|G(E/F)| = 5!$, and the automorphisms $\bar{\sigma}$ make up the full Galois group $G(E/F)$. Therefore, $G(E/F) \simeq S_5$, so $G(E/F)$ is not solvable. This completes our outline, and we summarize in a theorem.

THEOREM 9.24 Let $y_1, \ldots, y_5$ be independent transcendental real numbers over $\mathbf{Q}$. The polynomial

$$f(x) = \prod_{i=1}^{5} (x - y_i)$$

is not solvable by radicals over $F = \mathbf{Q}(s_1, \ldots, s_5)$, where s_i is the ith elementary symmetric function in $y_1, \ldots, y_5$.

It is evident that a generalization of these arguments shows that (*final goal*) a polynomial of degree n need not be solvable by radicals for $n \geq 5$.

In conclusion, we comment that there exist polynomials of degree 5 in $\mathbf{Q}[x]$ that are not solvable by radicals over $\mathbf{Q}$. A demonstration of this is left to the exercises (see Exercise 8).

Exercises 9.9

Concepts

1. Can the splitting field K of $x^2 + x + 1$ over $\mathbb{Z}_2$ be obtained by adjoining a square root to $\mathbb{Z}_2$ of an element in $\mathbb{Z}_2$? Is K an extension of $\mathbb{Z}_2$ by radicals?

2. Is every polynomial in $F[x]$ of the form $ax^8 + bx^6 + cx^4 + dx^2 + e$, where $a \neq 0$, solvable by radicals over F, if F is of characteristic 0? Why?

3. Mark each of the following true or false.

——— a. Let F be a field of characteristic 0. A polynomial in $F[x]$ is solvable by radicals if and only if its splitting field in $\bar{F}$ is contained in an extension of F by radicals.

——— b. Let F be a field of characteristic 0. A polynomial in $F[x]$ is solvable by radicals if and only if its splitting field in $\bar{F}$ has a solvable Galois group over F.

——— c. The splitting field of $x^{17} - 5$ over $\mathbb{Q}$ has a solvable Galois group.

——— d. The numbers π and $\sqrt{\pi}$ are independent transcendental numbers over $\mathbb{Q}$.

——— e. The Galois group of a finite extension of a finite field is solvable.

——— f. No quintic polynomial is solvable by radicals over any field.

——— g. Every 4th degree polynomial over a field of characteristic 0 is solvable by radicals.

——— h. The zeros of a cubic polynomial over a field F of characteristic 0 can always be attained by means of a finite sequence of operations of addition, subtraction, multiplication, division, and taking square roots starting with elements in F.

——— i. The zeros of a cubic polynomial over a field F of characteristic 0 can never be attained by means of a finite sequence of operations of addition, subtraction, multiplication, division, and taking square roots, starting with elements in F.

——— j. The theory of subnormal series of groups plays an important role in applications of Galois theory.

Theory

4. Let F be a field, and let $f(x) = ax^2 + bx + c$ be in $F[x]$, where $a \neq 0$. Show that if the characteristic of F is not 2, the splitting field of $f(x)$ over F is $F(\sqrt{b^2 - 4ac})$. [*Hint:* Complete the square, just as in your high school work, to derive the "quadratic formula."]

5. Show that if F is a field of characteristic different from 2 and

$$f(x) = ax^4 + bx^2 + c,$$

where $a \neq 0$, then $f(x)$ is solvable by radicals over F.

6. Show that for a finite group, every refinement of a subnormal series with abelian quotients also has abelian quotients, thus completing the proof of Lemma 9.2. [*Hint:* Use Theorem 4.3.]

7. Show that for a finite group, a subnormal series with solvable quotient groups can be refined to a composition series with abelian quotients, thus completing the proof of Theorem 9.23. [*Hint:* Use Theorem 4.3.]

8. This exercise exhibits a polynomial of degree 5 in $Q[x]$ that is not solvable by radicals over Q.

a. Show that if a subgroup H of S_5 contains a cycle of length 5 and a transposition τ, then $H = S_5$. [*Hint:* Show that H contains every transposition of S_5, and apply the corollary of Theorem 2.2. See Exercise 51, Section 2.4.]

b. Show that if $f(x)$ is an irreducible polynomial in $Q[x]$ of degree 5 having exactly two complex and three real zeros in C, then the group of $f(x)$ over Q is S_5. [*Hint:* Use Sylow theory to show that the group has an element of order 5. Use the fact that $f(x)$ has exactly two complex zeros to show that the group has an element of order 2. Then apply part (a).]

c. The polynomial $f(x) = 2x^5 - 5x^4 + 5$ is irreducible in $Q[x]$, by the Eisenstein criterion, with $p = 5$. Use the techniques of calculus to find relative maxima and minima and to "graph the polynomial function f" well enough to see that $f(x)$ must have exactly three real zeros in C. Conclude from part (b) and Theorem 9.23 that $f(x)$ is not solvable by radicals over Q.

◆

BIBLIOGRAPHY

◆

Classic Works

1. N. Bourbaki, *Eléments de Mathématique,* Book II of Part I, *Algèbre.* Paris: Hermann, 1942–58.

2. N. Jacobson, *Lectures in Abstract Algebra.* Princeton, N.J.: Van Nostrand, vols. I, 1951, II, 1953, and III, 1964.

3. O. Schreier and E. Sperner, *Introduction to Modern Algebra and Matrix Theory* (English translation), second edition. New York: Chelsea, 1959.

4. B. L. van der Waerden, *Modern Algebra* (English translation). New York: Ungar, vols. I, 1949, and II, 1950.

General Algebra Texts

5. M. Artin, *Algebra.* Englewood Cliffs, N.J.: Prentice-Hall, 1991.

6. A. A. Albert, *Fundamental Concepts of Higher Algebra.* Chicago: University of Chicago Press, 1956.

7. G. Birkhoff and S. MacLane, *A Survey of Modern Algebra,* third edition. New York: Macmillan, 1965.

8. R. A. Dean, *Elements of Abstract Algebra.* New York: Wiley, 1966.

9. J. A. Gallian, *Contemporary Abstract Algebra,* second edition. Lexington, Mass.: D. C. Heath, 1990.

10. I. N. Herstein, *Topics in Algebra.* New York: Blaisdell, 1964.

11. T. W. Hungerford, *Algebra.* New York: Springer, 1974.

12. S. Lang, *Algebra.* Reading, Mass.: Addison-Wesley, 1965.

13. N. H. McCoy, *Introduction to Modern Algebra.* Boston: Allyn and Bacon, 1960.

14. G. D. Mostow, J. H. Sampson, and J. Meyer, *Fundamental Structures of Algebra.* New York: McGraw-Hill, 1963.

15. W. W. Sawyer, *A Concrete Approach to Abstract Algebra.* San Francisco: Freeman, 1959.

16. S. Warner, *Modern Algebra.* Englewood Cliffs, N.J.: Prentice-Hall, vols. I and II, 1965.

Group Theory

17. W. Burnside, *Theory of Groups of Finite Order,* second edition. New York: Dover, 1955.

18. H. S. M. Coxeter and W. O. Moser, *Generators and Relations for Discrete Groups,* second edition. Berlin: Springer, 1965.

19. M. Hall, Jr., *The Theory of Groups.* New York: Macmillan, 1959.

20. A. G. Kurosh, *The Theory of Groups* (English translation). New York: Chelsea, vols. I, 1955, and II, 1956.

21. W. Ledermann, *Introduction to the Theory of Finite Groups,* fourth revised edition. New York: Interscience, 1961.

22. J. G. Thompson and W. Feit, "Solvability of Groups of Odd Order." *Pac. J. Math.,* **13** (1963), 775–1029.

23. M. A. Rabin, "Recursive Unsolvability of Group Theoretic Problems." *Ann. Math.,* **67** (1958), 172–194.

Ring Theory

24. E. Artin, C. J. Nesbitt, and R. M. Thrall, *Rings with Minimum Condition.* Ann Arbor: University of Michigan Press, 1944.

25. N. H. McCoy, *Rings and Ideals* (Carus Monograph No. 8). Buffalo: The Mathematical Association of America; LaSalle, Ill: Open Court, 1948.

26. N. H. McCoy, *The Theory of Rings.* New York: Macmillan, 1964.

Field Theory

27. E. Artin, *Galois Theory* (Notre Dame Mathematical Lecture No. 2), second edition. Notre Dame, Ind.: University of Notre Dame Press, 1944.

28. O. Zariski and P. Samuel, *Commutative Algebra.* Princeton, N.J.: Van Nostrand, vol. I, 1958.

Number Theory

29. G. H. Hardy and E. M. Wright, *An Introduction to the Theory of Numbers,* fourth edition. Oxford: Clarendon Press, 1960.

30. S. Lang, *Algebraic Numbers.* Reading, Mass.: Addison-Wesley, 1964.

31. W. J. LeVeque, *Elementary Theory of Numbers.* Reading, Mass.: Addison-Wesley, 1962.

32. W. J. LeVeque, *Topics in Number Theory.* Reading, Mass.: Addison-Wesley, 2 vols., 1956.

33. T. Nagell, *Introduction to Number Theory.* New York: Wiley, 1951.

34. I. Nivin and H. S. Zuckerman, *An Introduction to the Theory of Numbers.* New York: Wiley, 1960.

35. H. Pollard, *The Theory of Algebraic Numbers* (Carus Monograph No. 9). Buffalo: The Mathematical Association of America: New York: Wiley, 1950.

36. D. Shanks, *Solved and Unsolved Problems in Number Theory.* Washington: Spartan Books, vol. I, 1962.

37. B. M. Stewart, *Theory of Numbers,* second edition. New York: Macmillan, 1964.

38. J. V. Uspensky and M. H. Heaslet, *Elementary Number Theory.* New York: McGraw-Hill, 1939.

39. E. Weiss, *Algebraic Number Theory.* New York: McGraw-Hill, 1963.

Homological Algebra

40. J. P. Jans, *Rings and Homology.* New York: Holt, 1964.

41. S. Mac Lane, *Homology.* Berlin: Springer, 1963.

Other References

42. A. A. Albert (editor), *Studies in Modern Algebra* (MAA Studies in Mathematics, vol. 2). Buffalo: The Mathematical Association of America; Englewood Cliffs, N.J.: Prentice-Hall, 1963.

43. E. Artin, *Geometric Algebra.* New York: Interscience, 1957.

44. R. Courant and R. Robbins, *What Is Mathematics?* Oxford University Press, 1941.

45. H. S. M. Coxeter, *Introduction to Geometry,* second edition. New York: Wiley, 1969.

46. R. H. Crowell and R. H. Fox, *Introduction to Knot Theory.* New York: Ginn, 1963.

NOTATIONS

ANSWERS TO
ODD-NUMBERED
EXERCISES NOT
REQUIRING PROOFS

♦

Chapter 0

Section 0.1

1. proving theorems   **3.** efficiency

7. An integer m is **odd** if there exists an integer n such that $m = 2n + 1$.

9. ABEFGM, CDJ, HKN, I, L, O

11. True

13. False. The equation is satisfied by two integers, 1 and -1.

15. True. $n = 3$ is such an integer.

17. False. $n = 1$ gives a counterexample.

19. True. $(1/2)^2 = 1/4 < 1/2$

21. False. Both 0 and 1 satisfy this equation.

23. True

25. False. Let $n = -2$ and $m = 1$. Then $(-2/1)^2 - 4 > 1$.

27. False. Let $n = -2$ and $m = -1$. Then $(n/m)^3 = 8$ and $(n/m)^2 = 4$, so $(m/n)^3 > (n/m)^2$.

Section 0.2

1. $[-\sqrt{3}, \sqrt{3}]$

3. $\{1, -1, 2, -2, 3, -3, 4, -4, 5, -5, 6, -6, 10, -10, 12, -12, 15, -15, 20, -20, 30, -30, 60, -60\}$

5. Not a set (not well defined)

7. The set $\varnothing$

9. The set $\mathbb{Q}$

11. Not an equivalence relation

13. An equivalence relation; $\bar{0} = \{0\}$, $\bar{a} = \{a, -a\}$ for each nonzero $a \in \mathbb{R}$

15. An equivalence relation;

$\overline{1} = \{1, 2, \ldots, 9\}$,

$\overline{10} = \{10, 11, \ldots, 99\}$,

$\overline{100} = \{100, 101, \ldots, 999\}$, and in general

$\overline{10^n} = \{10^n, 10^n + 1, \ldots, 10^{n+1} - 1\}$

17. An equivalence relation;

$\overline{1} = \{1, 3, 5, 7, \ldots\} = \{2(n - 1) + 1 \mid n \in \mathbb{Z}^+\}$

$\overline{2} = \{2, 4, 6, 8, \ldots\} = \{2n \mid n \in \mathbb{Z}^+\}$

19. $\mathbb{Z}$

21. $\{3n \mid n \in \mathbb{Z}\} = \{\ldots, -9, -6, -3, 0, 3, 6, 9, \ldots\}$

$\{3n + 1 \mid n \in \mathbb{Z}\} = \{\ldots, -8, -5, -2, 1, 4, 7, 10, \ldots\}$

$\{3n + 2 \mid n \in \mathbb{Z}\} = \{\ldots, -7, -4, -1, 2, 5, 8, 11, \ldots\}$

23. $\{8n \mid n \in \mathbb{Z}\}$, $\{8n + 1 \mid n \in \mathbb{Z}\}$, $\{8n + 2 \mid n \in \mathbb{Z}\}$,

$\{8n + 3 \mid n \in \mathbb{Z}\}$, $\{8n + 4 \mid n \in \mathbb{Z}\}$, $\{8n + 5 \mid n \in \mathbb{Z}\}$,

$\{8n + 6 \mid n \in \mathbb{Z}\}$, $\{8n + 7 \mid n \in \mathbb{Z}\}$

25. 1   **27.** 5   **29.** 52

Section 0.3

7. The notion of an "interesting property" has not been made precise; it is not well defined. Also, we work in mathematics with two-valued logic; a statement is either true or false, *but not both*. The assertion that not having an interesting property would be an interesting property seems to contradict this two-valued logic. We would be saying the integer both has and does not have an interesting property.

Section 0.4

1. $6 + 2i$ **3.** $2 + 9i$ **5.** $-i$ **7.** $-i$ **9.** $23 + 7i$

11. $17 - 15i$ **13.** $-4 + 4i$ **15.** $\dfrac{1}{2} + \dfrac{1}{2}i$ **17.** $\dfrac{-7}{10} + \dfrac{1}{10}i$

19. $1 + 0i$ **21.** $2\sqrt{13}$ **23.** $\sqrt{2}\left(-\dfrac{1}{\sqrt{2}} + \dfrac{1}{\sqrt{2}}i\right)$

25. $\sqrt{34}\left(\dfrac{-3}{\sqrt{34}} + \dfrac{5}{\sqrt{34}}i\right)$ **27.** $\pm 1, \pm i$ **29.** $-2, 1 \pm \sqrt{3}\,i$

31. $\dfrac{1}{2}(1 \pm \sqrt{3}\,i), \dfrac{1}{2}(-1 \pm \sqrt{3}\,i), \pm 1$ **33.** $\begin{bmatrix} 2 & 1 \\ 2 & 7 \end{bmatrix}$ **35.** $\begin{bmatrix} -3 + 2i & -1 - 4i \\ 2 & -i \\ 0 & -i \end{bmatrix}$

37. $\begin{bmatrix} 5 & 16 & -3 \\ 0 & -18 & 24 \end{bmatrix}$ **39.** $\begin{bmatrix} 1 & -i \\ 4 - 6i & -2 - 2i \end{bmatrix}$ **41.** $\begin{bmatrix} 8 & -8i \\ 8i & 8 \end{bmatrix}$

43. $\begin{bmatrix} 0 & -1 \\ 1 & 0 \end{bmatrix}$

Chapter 1

Section 1.1

1. e, b, a **3.** $a, c.$ $*$ is not associative.

5. Top row: d; second row; a; fourth row: c, b.

7. Not commutative, not associative

9. Commutative, associative

11. Not commutative, not associative

13. $8, 729, n^{\lfloor n(n+1)/2\rfloor}$ **15.** Yes   **17.** Yes

19. No. Condition 2 is violated: $1 * 1$ is not in $\mathbb{Z}^+$.

21. Let $S = \{?, \Delta\}$. Define $*$ and $*'$ on S by $a * b = ?$ and $a *' b = \Delta$ for all $a, b \in S$. (Other answers are possible.)

23. True

25. True

27. False. Let $f(x) = x^2$, $g(x) = x$, and $h(x) = 2x + 1$. Then
$(f(x) - g(x)) - h(x) = x^2 - 3x - 1$ but
$f(x) - (g(x) - h(x)) = x^2 - (-x - 1) = x^2 + x + 1$.

29. True

31. False. Let $*$ be $+$ and let $*'$ be $\cdot$ on $\mathbb{Z}$.

33. Breaking *both* lists in the two places where the "talk to a friend" occurs, we can write the associative property symbolically as

$$(25 \text{ cents}, 25 \text{ cents})[(25 \text{ cents})(\text{press E, press 9})] =$$

$$[(25 \text{ cents}, 25 \text{ cents})(25 \text{ cents})](\text{press E, press 9}).$$

35.

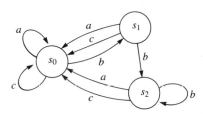

37. Starting in state s_0, this machine finishes in state s_3 if and only if the input string contains exactly three c's.

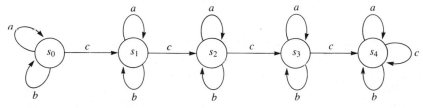

39.

Present state	0	1 (inputs)
s_0	s_0	s_1
s_1	s_0	s_2
s_2	s_0	s_2

Section 1.2

1. No. $\mathcal{G}_3$ fails. **3.** No. $\mathcal{G}_1$ fails. **5.** No. $\mathcal{G}_1$ fails. **9.** No. $\mathcal{G}_3$ fails.
11. Yes **13.** Yes **15.** Yes
17. The three tables are

I.

	e	a	b	c
e	e	a	b	c
a	a	e	c	b
b	b	c	e	a
c	c	b	a	e

II.

	e	a	b	c
e	e	a	b	c
a	a	e	c	b
b	b	c	a	e
c	c	b	e	a

III.

	e	a	b	c
e	e	a	b	c
a	a	b	c	e
b	b	c	e	a
c	c	e	a	b

Tables II and III give the same type of group structure. If we interchange letters a and b in Table II everywhere and then rewrite the resulting table with elements in order e, a, b, c, we obtain Table III.
    a. Yes b. Tables II and III
18. $\mathcal{G}_1\mathcal{G}_3\mathcal{G}_2$, $\mathcal{G}_3\mathcal{G}_1\mathcal{G}_2$, and $\mathcal{G}_3\mathcal{G}_2\mathcal{G}_1$ are not acceptable. The identity e occurs in the statement of $\mathcal{G}_3$, which must not come before e is defined in $\mathcal{G}_2$.
21.

	e	a	b
e	e	a	b
a	a	e	e
b	b	e	e

(Other answers are possible.)

35. a. A semigroup b. A monoid

Section 1.3

1. Yes **3.** Yes **5.** Yes
7. No. If $\det(A) = \det(B) = 2$, then $\det(AB) = 4$.
9. Yes **11.** Yes
13. a. Not a subgroup b. A subgroup
15. a. Not a subgroup b. A subgroup
17. a. Not a subgroup b. Not a subgroup
19. $G_1 \leq G_1$, $G_1 < G_4$
    $G_2 < G_1$, $G_2 \leq G_2$, $G_2 < G_4$, $G_2 < G_7$, $G_2 < G_8$
    $G_3 \leq G_3$, $G_3 < G_2$
    $G_4 \leq G_4$
    $G_5 \leq G_5$
    $G_6 < G_5$, $G_6 \leq G_6$
    $G_7 < G_1$, $G_7 < G_4$, $G_7 \leq G_7$
    $G_8 < G_1$, $G_8 < G_4$, $G_8 < G_7$, $G_8 \leq G_8$
    $G_9 < G_3$, $G_9 < G_5$, $G_9 \leq G_9$
21. $\begin{bmatrix} 0 & -1 \\ -1 & 0 \end{bmatrix}, \begin{bmatrix} 1 & 0 \\ 0 & 1 \end{bmatrix}$ **23.** $\begin{bmatrix} 3^n & 0 \\ 0 & 2^n \end{bmatrix}$ for $n \in \mathbb{Z}$

25. G_1 is cyclic with generators 1 and -1. G_2 is not cyclic. G_3 is not cyclic. G_4 is cyclic with generators 6 and -6. G_5 is cyclic with generators 6 and $\frac{1}{6}$. G_6 is not cyclic.

27. 2   **29.** 5   **31.** 8   **33.** 4

35. a.

	0	1	2	3	4	5
0	0	1	2	3	4	5
1	1	2	3	4	5	0
2	2	3	4	5	0	1
3	3	4	5	0	1	2
4	4	5	0	1	2	3
5	5	0	1	2	3	4

b. $\langle 0 \rangle = \{0\}$
$\langle 2 \rangle = \langle 4 \rangle = \{0, 2, 4\}$
$\langle 3 \rangle = \{0, 3\}$
$\langle 1 \rangle = \langle 5 \rangle = \mathbb{Z}_6$

c. 1 and 5

d.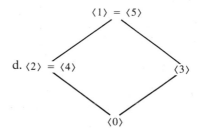

37. The Klein 4-group; V is an example.
39. If $H = \varnothing$, no $a \in H$ exists.
51. True if G is abelian.

Section 1.4

1. $q = 4, r = 6$ **3.** $q = -7, r = 6$ **5.** 8   **7.** 60   **9.** 10
11. 2   **13.** 4   **15.** 16   **17.** 7   **19.** 8
21.

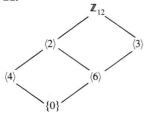

23.

$$\mathbb{Z}_8$$
$$|$$
$$\langle 2 \rangle$$
$$|$$
$$\langle 4 \rangle$$
$$|$$
$$\{0\}$$

25. 1,2,4,8   **27.** 1,2,3,4,5,6,10,12,15,20,30,60
29. 0,1,2,3,4,5,6,7,8,9,10,11
31. 0,2,4,6,8,10,12,14,16
33. $\ldots -24,-18,-12,-6,0,6,12,18,24,\ldots$
35. T F F F T F F F T T   **37.** $\langle \mathbb{Q}, + \rangle$
39. No example exists. Such a group is isomorphic to $\langle \mathbb{Z}, + \rangle$, which has just two generators, 1 and -1.

41. $\pm i$ **43.** $\dfrac{1}{\sqrt{2}}(1 \pm i), \dfrac{1}{\sqrt{2}}(-1 \pm i)$

47. a. Let $G = \{n \in \mathbb{Z} \mid n = rq = st \text{ for some } q, t \in \mathbb{Z}\}$. Then G is a subgroup of $\mathbb{Z}$, and hence cyclic. The **least common multiple** (lcm) of r and s is the positive generator of G.
b. The lcm of r and s is rs if and only if the gcd of r and s is 1.

49. S_3 gives a counterexample.　　　**53.** $p^{r-1}(p-1)$

55. There are exactly d solutions, where d is the gcd of m and n.

59. Not commutative

61. No. It does not contain the identity 0.

63.

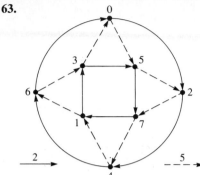

0

65. a. a^3b　　　b. a^2　　　c. a^2

67.

	e	a	b	c	d	f
e	e	a	b	c	d	f
a	a	e	c	b	f	d
b	b	d	e	f	a	c
c	c	f	a	d	e	b
d	d	b	f	e	c	a
f	f	c	d	a	b	e

69.

Chapter 2

Section 2.1

1. $\begin{pmatrix} 1 & 2 & 3 & 4 & 5 & 6 \\ 1 & 2 & 3 & 6 & 5 & 4 \end{pmatrix}$　　　**3.** $\begin{pmatrix} 1 & 2 & 3 & 4 & 5 & 6 \\ 3 & 4 & 1 & 6 & 2 & 5 \end{pmatrix}$

5. $\begin{pmatrix} 1 & 2 & 3 & 4 & 5 & 6 \\ 2 & 6 & 1 & 5 & 4 & 3 \end{pmatrix}$　　　**7.** 2　　　**9.** ι　　　**11.** $\{1, 2, 3, 4\}$

13. ϵ, ρ, and ρ^2 give the three positions of the triangle in Fig. 2.5 obtained by rotations. The permutations ϕ, $\rho\phi$, and $\rho^2\phi$ amount geometrically to turning the triangle over (ϕ) and then rotating it for the other three positions.

15. 6

17. a. $\langle \rho_1 \rangle = \langle \rho_2 \rangle = \{\rho_0, \rho_1, \rho_2\}$
$\langle \mu_1 \rangle = \{\rho_0, \mu_1\}$

　　b.

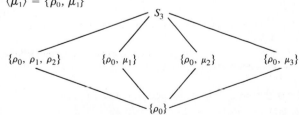

19.

	ρ^0	ρ	ρ^2	ρ^3	ρ^4	ρ^5
ρ^0	ρ^0	ρ	ρ^2	ρ^3	ρ^4	ρ^5
ρ	ρ	ρ^2	ρ^3	ρ^4	ρ^5	ρ^0
ρ^2	ρ^2	ρ^3	ρ^4	ρ^5	ρ^0	ρ
ρ^3	ρ^3	ρ^4	ρ^5	ρ^0	ρ	ρ^2
ρ^4	ρ^4	ρ^5	ρ^0	ρ	ρ^2	ρ^3
ρ^5	ρ^5	ρ^0	ρ	ρ^2	ρ^3	ρ^4

This group is not isomorphic to S_3 since it is an abelian group and S_3 is nonabelian.

21.
$$\begin{bmatrix} 1 & 0 & 0 & 0 \\ 0 & 1 & 0 & 0 \\ 0 & 0 & 1 & 0 \\ 0 & 0 & 0 & 1 \end{bmatrix} \begin{bmatrix} 0 & 0 & 0 & 1 \\ 1 & 0 & 0 & 0 \\ 0 & 1 & 0 & 0 \\ 0 & 0 & 1 & 0 \end{bmatrix} \begin{bmatrix} 0 & 0 & 1 & 0 \\ 0 & 0 & 0 & 1 \\ 1 & 0 & 0 & 0 \\ 0 & 1 & 0 & 0 \end{bmatrix} \begin{bmatrix} 0 & 1 & 0 & 0 \\ 0 & 0 & 1 & 0 \\ 0 & 0 & 0 & 1 \\ 1 & 0 & 0 & 0 \end{bmatrix}$$

$$\begin{bmatrix} 0 & 1 & 0 & 0 \\ 1 & 0 & 0 & 0 \\ 0 & 0 & 0 & 1 \\ 0 & 0 & 1 & 0 \end{bmatrix} \begin{bmatrix} 0 & 0 & 0 & 1 \\ 0 & 0 & 1 & 0 \\ 0 & 1 & 0 & 0 \\ 1 & 0 & 0 & 0 \end{bmatrix} \begin{bmatrix} 0 & 0 & 1 & 0 \\ 0 & 1 & 0 & 0 \\ 1 & 0 & 0 & 0 \\ 0 & 0 & 0 & 1 \end{bmatrix} \begin{bmatrix} 1 & 0 & 0 & 0 \\ 0 & 0 & 0 & 1 \\ 0 & 0 & 1 & 0 \\ 0 & 1 & 0 & 0 \end{bmatrix}$$

23. The Klein 4-group V **25.** $\mathbb{Z}$

27. Not a permutation   **29.** Not a permutation

31. The name *two-to-two function* suggests that such a function f should carry every pair of distinct points into two distinct points. Such a function is one to one in the conventional sense. (If the domain has only one element, a function cannot fail to be two to two, since the only way it can fail to be two to two is to carry two points into one point, and the set does not have two points.) Conversely, every function that is one to one in the conventional sense carries any pair of points into two distinct points. Thus the functions conventionally called one to one are precisely those that carry two points into two points, which is a much more intuitive unidirectional way of regarding them. Also, the standard way of trying to show a function is one to one is precisely to show that it does not carry two points into just one point. Thus, proving a function is one to one becomes more natural in the two-to-two terminology.

33. S_3 is an example.   **35.** Yes   **37.** No (no inverse)   **39.** $|D_n| = 2n$

47. a. All possible products $a *' b$ and all instances of the associative property for $*'$ in G'.

49. a. $T_0(s_0) = s_0,\; T_0(s_1) = s_1$
$T_1(s_0) = s_1,\; T_1(s_1) = s_0$
$T_{11101}(s_0) = s_0,\; T_{11101}(s_1) = s_1$
$T_{1101}(s_0) = s_1,\; T_{1101}(s_1) = s_0$
All transition functions appear. The answer does not change if the empty string is included.

51. $(n + 1)^{n+1}$ **53.**

	T_ϵ	T_1
T_ϵ	T_ϵ	T_1
T_1	T_1	T_ϵ

Yes, we have a group isomorphic to $\mathbb{Z}_2$.

55.

Section 2.2

1. $\{1, 2, 5\}, \{3\}, \{4, 6\}$ **3.** $\{1, 2, 3, 4, 5\}, \{6\}, \{7, 8\}$

5. $\{2n \mid n \in \mathbb{Z}\}, \{2n + 1 \mid n \in \mathbb{Z}\}$

7. $\begin{pmatrix} 1 & 2 & 3 & 4 & 5 & 6 & 7 & 8 \\ 4 & 1 & 3 & 5 & 8 & 6 & 2 & 7 \end{pmatrix}$ **9.** $\begin{pmatrix} 1 & 2 & 3 & 4 & 5 & 6 & 7 & 8 \\ 5 & 4 & 3 & 7 & 8 & 6 & 2 & 1 \end{pmatrix}$

11. $(1, 3, 4)(2, 6)(5, 8, 7) = (1, 4)(1, 3)(2, 6)(5, 7)(5, 8)$

13. F T F F F F T T T F   d. Let $H = \{\iota, (1, 2)(3, 4)(5, 6)\}$

15. a. 4

b. A cycle of length n has order n.

c. σ has order 6; τ has order 4.

d. $(10)6(11)6(12)8$

e. The order of a permutation expressed as a product of disjoint cycles is the least common multiple of the lengths of the cycles.

17. a.

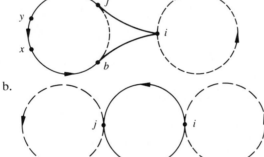

b.

19. n **21.** No. Permutation multiplication is not closed on K.

29.

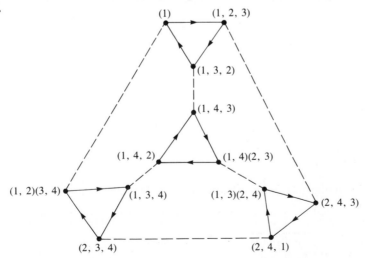

31.

	P	R
P	P	R
R	R	P

33.

35.

37.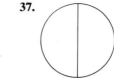

39. Rotations, reflections   **41.** None

45. No. The product of two rotations (about different points) may be a translation, so the set of rotations is not closed under multiplication.

47. Yes. Think of the fixed point as the origin in the plane of complex numbers. Rotations about that point correspond to multiplying by complex numbers z such that $|z| = 1$. The set $\{z \in \mathbb{C} \mid |z| = 1\}$ is a group under multiplication, and the multiplication corresponds to function composition of rotations. The number 1 corresponds to the identity map.

49. No. The product of two glide reflections is orientation preserving, and hence is not a glide reflection.

Section 2.3

1. $4\mathbb{Z} = \{\ldots, -8, -4, 0, 4, 8, \ldots\}$
$1 + 4\mathbb{Z} = \{\ldots, -7, -3, 1, 5, 9, \ldots\}$
$2 + 4\mathbb{Z} = \{\ldots, -6, -2, 2, 6, 10, \ldots\}$
$3 + 4\mathbb{Z} = \{\ldots, -5, -1, 3, 2, 7, 11, \ldots\}$

3. $\langle 2 \rangle = \{0, 2, 4, 6, 8, 10\}$, $1 + \langle 2 \rangle = \{1, 3, 5, 7, 9, 11\}$

5. $\langle 18 \rangle = \{0, 18\}$, $1 + \langle 18 \rangle = \{1, 19\}$, $2 + \langle 18 \rangle = \{2, 20\}, \ldots,$
$17 + \langle 18 \rangle = \{17, 35\}$

7. $\{\rho_0, \mu_2\}, \{\rho_1, \delta_1\}, \{\rho_2, \mu_1\}, \{\rho_3, \delta_2\}$. Not the same.

9. $\{\rho_0, \rho_2\}, \{\rho_1, \rho_3\}, \{\mu_1, \mu_2\}, \{\delta_1, \delta_2\}$

11. Yes, we get a coset group isomorphic to V.

	ρ_0	ρ_2	ρ_1	ρ_3	μ_1	μ_2	δ_1	δ_2
ρ_0	ρ_0	ρ_2	ρ_1	ρ_3	μ_1	μ_2	δ_1	δ_2
ρ_2	ρ_2	ρ_0	ρ_3	ρ_1	μ_2	μ_1	δ_2	δ_1
ρ_1	ρ_1	ρ_3	ρ_2	ρ_0	δ_1	δ_2	μ_2	μ_1
ρ_3	ρ_3	ρ_1	ρ_0	ρ_2	δ_2	δ_1	μ_1	μ_2
μ_1	μ_1	μ_2	δ_2	δ_1	ρ_0	ρ_2	ρ_3	ρ_1
μ_2	μ_2	μ_1	δ_1	δ_2	ρ_2	ρ_0	ρ_1	ρ_3
δ_1	δ_1	δ_2	μ_1	μ_2	ρ_1	ρ_3	ρ_0	ρ_2
δ_2	δ_2	δ_1	μ_2	μ_1	ρ_3	ρ_1	ρ_2	ρ_0

13. 3   **15.** T T T F T F T T F T

17. $G = \mathbb{Z}_2$, subgroup $H = \mathbb{Z}_2$.

19. Impossible. The number of cells must divide the order of the group, and 12 does not divide 6.

25. False. $G = S_3$, $H = \{\rho_0, \mu_1\}$, $a = \rho_1$, $b = \mu_3$ (Table 2.6)

27. True. (You provide the proof.)

39. b. No. $S_{c,d}$ is not closed under permutation multiplication.

    c. Let $\mu \in S_{c,d}$. Then $S_{c,d}$ is the coset $\mu S_{c,c}$ of $S_{c,c}$ in S_A.

Section 2.4

1. Element	**Order**	**Element**	**Order**
$(0,0)$	1	$(0,2)$	2
$(1,0)$	2	$(1,2)$	2
$(0,1)$	4	$(0,3)$	4
$(1,1)$	4	$(1,3)$	4

    The group is not cyclic.

3. 2   **5.** 9   **7.** 60

9. $\{(0,0),(0,1)\}, \ \{(0,0),(1,0)\}, \ \{(0,0),(1,1)\}$

11. $\{(0,0),(0,1),(0,2),(0,3)\}$

    $\{(0,0),(0,2),(1,0),(1,2)\}$

    $\{(0,0),(1,1),(0,2),(1,3)\}$

13. $\mathbb{Z}_{20} \times \mathbb{Z}_3, \mathbb{Z}_{15} \times \mathbb{Z}_4, \mathbb{Z}_{12} \times \mathbb{Z}_5, \mathbb{Z}_5 \times \mathbb{Z}_3 \times \mathbb{Z}_4$

15. $\mathbb{Z}_{12}$. Left cosets: $\mathbb{Z}_{12}$

17. $H = \{0, 2, 4, 6, 8, 10, 12, 14, 16\}$. Left cosets $H, 1 + H$

19. D_4. Left cosets: D_4

21. $\mathbb{Z}_8, \mathbb{Z}_2 \times \mathbb{Z}_4, \mathbb{Z}_2 \times \mathbb{Z}_2 \times \mathbb{Z}_2$

23. $\mathbb{Z}_{32}, \mathbb{Z}_2 \times \mathbb{Z}_{16}, \mathbb{Z}_4 \times \mathbb{Z}_8, \mathbb{Z}_2 \times \mathbb{Z}_2 \times \mathbb{Z}_8, \mathbb{Z}_2 \times \mathbb{Z}_4 \times \mathbb{Z}_4,$

    $\mathbb{Z}_2 \times \mathbb{Z}_2 \times \mathbb{Z}_2 \times \mathbb{Z}_4, \mathbb{Z}_2 \times \mathbb{Z}_2 \times \mathbb{Z}_2 \times \mathbb{Z}_2 \times \mathbb{Z}_2$

25. $\mathbb{Z}_9 \times \mathbb{Z}_{121}, \mathbb{Z}_3 \times \mathbb{Z}_3 \times \mathbb{Z}_{121}, \mathbb{Z}_9 \times \mathbb{Z}_{11} \times \mathbb{Z}_{11}, \mathbb{Z}_3 \times \mathbb{Z}_3 \times \mathbb{Z}_{11} \times \mathbb{Z}_{11}$

29. T T F T F F F F F T

31. a. 1   **b.** Infinitely many

33. T F F T T F T F T T

35. a. Yes; G has exactly one subgroup of order 8. It can be characterized as the subgroup of all elements of G of order dividing 8.

    b. No. For example $\mathbb{Z}_8 \times \mathbb{Z}_9$ has only one subgroup of order 4, while $\mathbb{Z}_2 \times \mathbb{Z}_2 \times \mathbb{Z}_2 \times \mathbb{Z}_9$ has seven subgroups of order 4.

37. 12 for $\mathbb{Z}_4 \times \mathbb{Z} \times \mathbb{Z}_3$; 144 for $\mathbb{Z}_{12} \times \mathbb{Z} \times \mathbb{Z}_{12}$

39. $\left\{ \cos \dfrac{2m\pi}{n} + i \sin \dfrac{2m\pi}{n} \,\middle|\, m, n \in \mathbb{Z}^+, m \le n \right\}$

41. a. 36   **b.** $\mathbb{Z}_2 \times \mathbb{Z}_{12} \times \mathbb{Z}_{60}$

    c. Find an isomorphic group that is a direct product of cyclic groups of prime-power order. For each prime divisor of the order of the group, write the subscripts in the direct product involving that prime in a row in order of increasing magnitude. Keep the right-hand ends of the rows aligned. Then take the product of the numbers down each column of the array. These are the torsion coefficients. Illustrating with the group in (b), we first form $\mathbb{Z}_2 \times \mathbb{Z}_3 \times \mathbb{Z}_3 \times \mathbb{Z}_4 \times \mathbb{Z}_4 \times \mathbb{Z}_5$.

    We now form the array and multiply columns, as in

$$
\begin{array}{rrr}
2 & 4 & 4 \\
 & 3 & 3 \\
 & & 5 \\
\hline
2 & 12 & 60
\end{array}
$$

    obtaining the torsion coefficients 2, 12, 60.

45. S_3

49. The hypothesis of commutativity can be dropped. Just consider the theorem of Lagrange.

53. a. It is commutative if the arrows on both n-gons have the same (clockwise or counterclockwise) direction.
  b. $\mathbb{Z}_2 \times \mathbb{Z}_n$ **c.** It is cyclic if n is odd.
  d. The dihedral group D_n

55. $\sin(2\pi x/\sqrt{3})$ **57.** $\sin(2\pi x/3) + \cos(2\pi y/\sqrt{5})$ **59.** $x^2 + y^2$

63. a. No   **b.** No   **c.** No   **d.** No   **e.** $\mathbb{Z}$

65. a. No   **b.** Yes   **c.** No   **d.** No   **e.** $\mathbb{Z} \times \mathbb{Z}_2$

67. a. Yes   **b.** Yes   **c.** Yes   **d.** No   **e.** $D_\infty \times \mathbb{Z}_2$

69. a. Yes   **b.** No   **c.** Yes   **d.** Yes   **e.** D_∞

71. a. Yes.   180°   **b.** Yes   **c.** No

73. a. No   **b.** Yes   **c.** No

75. a. Yes.   60°, 120°, 180°   **b.** Yes   **c.** No

77. a. No   **b.** No   **c.** Yes   **d.** $(1,0)$ and $(0,1)$

79. a. Yes, 120°   **b.** No   **c.** No   **d.** $(0,1)$ and $(\sqrt{3},1)$

Section 2.5

1. 0001011   0000000   0111001   1111111   1111111   0100101   0000000   0100101
   1111111   0111001

3. $x_4 = x_1 + x_2,\ x_5 = x_1 + x_3,\ x_6 = x_2 + x_3$ **5.** 2

7. $\{111111, 110100, 101010, 100001, 011001, 010010, 001100, 000111\}$

9. $\begin{bmatrix} 1 & 1 & 0 & 1 & 0 & 0 \\ 1 & 0 & 1 & 0 & 1 & 0 \\ 0 & 1 & 1 & 0 & 0 & 1 \end{bmatrix}$

11.

Word	Syndrome	Word	Syndrome
000000	000	000100	100
100000	110	000010	010
010000	101	000001	001
001000	011		

  a. 110   **b.** 001   **c.** 110
  d. Can't tell   **e.** 101

21. $2m + 1$

25. a. 3   **b.** 3   **c.** 4   **d.** 5   **e.** 6   **f.** 7

27. $x_9 = x_1 + x_2 + x_3 + x_6 + x_7,$ $x_{10} = x_5 + x_6 + x_7 + x_8,$
   $x_{11} = x_2 + x_3 + x_4 + x_6 + x_8,$ $x_{12} = x_1 + x_3 + x_4 + x_5$
   (Other answers are possible.)

Chapter 3

Section 3.1

1. Yes   **3.** Yes   **5.** No   **7.** Yes   **9.** Yes   **11.** Yes

13. Yes   **15.** No

17. Using multiplicative notation for G, $\phi(m, n) = h^m k^n$. Using additive notation for G, $\phi(m, n) = mh + nk$, where $mh = h + \cdots + h$ for m summands and nk is similarly defined.

19. Infinitely many   **21.** $g = e$ only

23. TTFTFFTTFF

25. Let $\phi(n)$ be the remainder of n when divided by 4 for $n \in \mathbb{Z}_{12}$.

27. No nontrivial homomorphisms

29. Let $\phi(n)$ be the identity for n even and the transposition $(1, 2)$ for n odd.

31. Let $\phi(2n) = (2n, 2n)$.

33. Let $\phi(\sigma) = \mu$ where $\mu(i) = \sigma(i)$ for $i = 1, 2, 3$ and $\mu(4) = 4$.

39. $\mathrm{Ker}(\mathrm{sgn}_n) = A_n$

41. The image of ϕ is $\langle a \rangle$, and $\mathrm{Ker}(\phi)$ must be some subgroup $n\mathbb{Z}$ of $\mathbb{Z}$.

43. $hk = kh$.

Section 3.2

1. All one-to-one maps ϕ of $\mathbb{Z}_2 \times \mathbb{Z}_2$ onto V satisfying $\phi(0, 0) = e$. There are six of them.

3. $\phi(1, 1)$ must be a generator of $\mathbb{Z}_{10}$, so the possibilities for $\phi(1, 1)$ are $1, 3, 5, 7$. The isomorphisms are determined by $\phi(1, 1)$. For example, if $\phi(1, 1) = 7$, then $\phi(0, 2) = \phi((1, 1) + (1, 1)) = 7 + 7 = 4$, etc.

5. 2   **7.** 2

9. For $\mathbb{Z}_4$, $\lambda_0 = \begin{pmatrix} 0 & 1 & 2 & 3 \\ 0 & 1 & 2 & 3 \end{pmatrix}$, $\lambda_1 = \begin{pmatrix} 0 & 1 & 2 & 3 \\ 1 & 2 & 3 & 0 \end{pmatrix}$, $\lambda_3 = \begin{pmatrix} 0 & 1 & 2 & 3 \\ 2 & 3 & 0 & 1 \end{pmatrix}$,

$\lambda_4 = \begin{pmatrix} 0 & 1 & 2 & 3 \\ 3 & 0 & 1 & 2 \end{pmatrix}$. The table for the left regular representation is the same as

the table for $\mathbb{Z}_4$ with n replaced by λ_n. For S_3, $\rho_{\rho_0} = \begin{pmatrix} \rho_0 & \rho_1 & \rho_2 & \mu_1 & \mu_2 & \mu_3 \\ \rho_0 & \rho_1 & \rho_2 & \mu_1 & \mu_2 & \mu_3 \end{pmatrix}$,

$\rho_{\rho_1} = \begin{pmatrix} \rho_0 & \rho_1 & \rho_2 & \mu_1 & \mu_2 & \mu_3 \\ \rho_1 & \rho_2 & \rho_3 & \mu_2 & \mu_3 & \mu_1 \end{pmatrix}$, etc., where the bottom row in the permutation

ρ_σ consists of the elements of S_3 in the order they appear down the column under σ in Table 2.1. The table for this right regular representation is the same as the table for S_3 with σ replaced by ρ_σ.

11. $\{\mathbb{Z}, 17\mathbb{Z}, 3\mathbb{Z}, \langle \pi \rangle\}$ is a subcollection consisting of isomorphic groups, as are $\{\mathbb{Z}_6, G\}$, $\{\mathbb{Z}_2, S_2\}$, $\{S_6\}$, $\{\mathbb{Q}\}$, $\{\mathbb{R}, \mathbb{R}^+\}$, $\{\mathbb{R}^*\}$, $\{\mathbb{Q}^*\}$, and $\{\mathbb{C}^*\}$.

13. U has an element i of order 4, while $\langle \mathbb{R}, + \rangle$ has no element of order 4.

24. Partial answer for Exercise 25: $\phi(a) = a - 1$

25. a. $a *_1 b = ab + 4a + 4b + 12$

  b. S_2 is the set of all real numbers except $-t$; $a *_2 b = ab + ta + tb + (t^2 - t)$.

  c. S_3 is the set of all real numbers except 1; $a *_3 b = ab - a - b + 2$.

Section 3.3

1. 3   **3.** 4   **5.** 2   **7.** 2   **9.** 4   **11.** 3   **13.** 4

15. 1

17. a. When working with a factor group G/H, you would let a and b be elements of G, not elements of G/H. The student probably does not understand what elements of G/H look like and can write nothing sensible concerning them.

  b. We must show that G/H is abelian. Let aH and bH be two elements of G/H.

19. T T T T T F T F T F

  f. $\mathbb{Z}$ is torsion free but $\mathbb{Z}/n\mathbb{Z} \simeq \mathbb{Z}_n$, a torsion group.

  h. For $n > 2$, S_n is nonabelian, but $S_n/A_n \simeq \mathbb{Z}_2$, and $\mathbb{Z}_2$ is abelian.

  j. $n\mathbb{R} = \mathbb{R}$, so $\mathbb{R}/n\mathbb{R}$ is of order 1.

23. H is normal if and only if $\bar{H} = \{H\}$.

Section 3.4

1. $\mathbb{Z}_2$ **3.** $\mathbb{Z}_4$ **5.** $\mathbb{Z}_4 \times \mathbb{Z}_8$ **7.** $\mathbb{Z}$ **9.** $\mathbb{Z}_3 \times \mathbb{Z} \times \mathbb{Z}_4$
11. $\mathbb{Z}_2 \times \mathbb{Z}$ **13.** $\mathbb{Z}(D_4) = C = \{\rho_0, \rho_2\}$
15. T F F T F T F F T F   **17.** $\{f \in F^* \mid f(0) = 1\}$
19. Yes. Let $f(x) = 1$ for $x \geq 0$ and $f(x) = -1$ for $x < 0$. Then $f(x) \cdot f(x) = 1$ for all
x, so $f^2 \in C^*$ but f is not in C^*. Thus fC^* has order 2 in F^*/C^*.
21. $U/z_0 U$ is isomorphic to $\{e\}$, for $z_0 U = U$.
23. U **25.** $\mathbb{Z}$
27. a. The entire group   b. $\{e\}$

Section 3.5

1. The refinements $\{0\} < 250\mathbb{Z} < 10\mathbb{Z} < \mathbb{Z}$ of $\{0\} < 10\mathbb{Z} < \mathbb{Z}$ and
$\{0\} < 250\mathbb{Z} < 25\mathbb{Z} < \mathbb{Z}$ of $0 < 25\mathbb{Z} < \mathbb{Z}$ are isomorphic.
3. The given series are isomorphic.
5. The refinements
$\{(0, 0)\} < (4800\mathbb{Z}) \times \mathbb{Z} < (240\mathbb{Z}) \times \mathbb{Z} < (60\mathbb{Z}) \times \mathbb{Z} < (10\mathbb{Z}) \times \mathbb{Z} < \mathbb{Z} \times \mathbb{Z}$
of the first series, and
$\{(0, 0)\} < \mathbb{Z} \times (4800\mathbb{Z}) < \mathbb{Z} \times (480\mathbb{Z}) < \mathbb{Z} \times (80\mathbb{Z}) < \mathbb{Z} \times (20\mathbb{Z}) < \mathbb{Z} \times \mathbb{Z}$ of the
second series are isomorphic refinements.
7. $\{0\} < \langle 16 \rangle < \langle 8 \rangle < \langle 4 \rangle < \langle 2 \rangle < \mathbb{Z}_{48}$
$\{0\} < \langle 24 \rangle < \langle 8 \rangle < \langle 4 \rangle < \langle 2 \rangle < \mathbb{Z}_{48}$
$\{0\} < \langle 24 \rangle < \langle 12 \rangle < \langle 4 \rangle < \langle 2 \rangle < \mathbb{Z}_{48}$
$\{0\} < \langle 24 \rangle < \langle 12 \rangle < \langle 6 \rangle < \langle 2 \rangle < \mathbb{Z}_{48}$
$\{0\} < \langle 24 \rangle < \langle 12 \rangle < \langle 6 \rangle < \langle 3 \rangle < \mathbb{Z}_{48}$
9. $\{(\rho_0, 0)\} < A_3 \times \{0\} < S_3 \times (0) < S_3 \times \mathbb{Z}_2$
$\{(\rho_0, 0)\} < \{\rho_0\} \times \mathbb{Z}_2 < A_3 \times \mathbb{Z}_2 < S_3 \times \mathbb{Z}_2$
11. $\{\rho_0\} \times \mathbb{Z}_4$ **13.** $\{\rho_0\} \times \mathbb{Z}_4 \leq \{\rho_0\} \times \mathbb{Z}_4 \leq \{\rho_0\} \times \mathbb{Z}_4 \leq \cdots.$
15. T F T F F T F F T T
   i. The Jordan–Hölder theorem applied to the groups $\mathbb{Z}_n$ implies the Fundamental
   Theorem of Arithmetic.
17. Yes. $\{\rho_0\} < \{\rho_0, \rho_2\} < \{\rho_0, \rho_1, \rho_2, \rho_3\} < D_4$ is a composition (actually a principal)
series and all factor groups are isomorphic to $\mathbb{Z}_2$ and are thus abelian.

Section 3.6

1. $X_{\rho_0} = X$, $X_{\rho_1} = \{C\}$, $X_{\rho_2} = \{m_1, m_2, d_1, d_2, C\}$, $X_{\rho_3} = \{C\}$,
   $X_{\mu_1} = \{s_1, s_3, m_1, m_2, C, P_1, P_3\}$, $X_{\mu_2} = \{s_2, s_4, m_1, m_2, C, P_2, P_4\}$,
   $X_{\delta_1} = \{2, 4, d_1, d_2, C\}$, $X_{\delta_2} = \{1, 3, d_1, d_2, C\}$.
3. $\{1, 2, 3, 4\}, \{s_1, s_2, s_3, s_4\}, \{m_1, m_2\}, \{d_1, d_2\}, \{C\}, \{P_1, P_2, P_3, P_4\}$
5. A transitive G-set has just one orbit.
7. a. $\{s_1, s_2, s_3, s_4\}$ and $\{P_1, P_2, P_3, P_4\}$
11. b. The set of points on the circle with center at the origin and passing through P
   c. The cyclic subgroup $\langle 2\pi \rangle$ of $G = \mathbb{R}$
15. a. $K = g_0 H g_0^{-1}$.
   b. *Conjecture:* H and K should be conjugate subgroups of G.

17.

	X			Y		Z
	a	a	b	a	b	c
0	a	a	b	a	b	c
1	a	b	a	b	c	a
2	a	a	b	c	a	b
3	a	b	a	a	b	c
4	a	a	b	b	c	a
5	a	b	a	c	a	b

There are three of them.

Section 3.7

1. 5   **3.** 2

5. 11,712   **7.** a. 45   b. 231

9. a. 90   b. 6,426

Chapter 4

Section 4.1

1. a. $K = \{0, 3, 6, 9\}$.
  b. $0 + K = \{0, 3, 6, 9\}$, $1 + K = \{1, 4, 7, 10\}$, $2 + K = \{2, 5, 8, 11\}$.
  c. $\psi(0 + K) = 0$, $\psi(1 + K) = 1$, $\psi(2 + K) = 2$.
3. a. $HN = \{0, 2, 4, 6, 8, 10, 12, 14, 16, 18, 20, 22\}$, $H \cap N = \{0, 12\}$.
  b. $0 + N = \{0, 6, 12, 18\}$, $2 + N = \{2, 8, 14, 20\}$, $4 + N = \{4, 10, 16, 22\}$.
  c. $0 + (H \cap N) = \{0, 12\}$, $4 + (H \cap N) = \{4, 16\}$, $8 + (H \cap N) = \{8, 20\}$.
  d. $\psi(0 + N) = 0 + (H \cap N)$, $\psi(2 + N) = 8 + (H \cap N)$,
     $\psi(4 + N) = 4 + (H \cap N)$.
5. a. $0 + H = \{0, 4, 8, 12, 16, 20\}$, $1 + H = \{1, 5, 9, 13, 17, 21\}$,
     $2 + H = \{2, 6, 10, 14, 18, 22\}$, $3 + H = \{3, 7, 11, 15, 19, 23\}$.
  b. $0 + K = \{0, 8, 16\}$, $1 + K = \{1, 9, 17\}$, $2 + K = \{2, 10, 18\}$,
     $3 + K = \{3, 11, 19\}$,
     $4 + K = \{4, 12, 20\}$, $5 + K = \{5, 13, 21\}$, $6 + K = \{6, 14, 22\}$,
     $7 + K = \{7, 15, 23\}$.
  c. $0 + K = \{0, 8, 16\}$, $4 + K = \{4, 12, 20\}$.
  d. $(0 + K) + (H/K) = H/K = \{0 + K, 4 + K\} = \{\{0, 8, 16\}, \{4, 12, 20\}\}$
     $(1 + K) + (H/K) = \{1 + K, 5 + K\} = \{\{1, 9, 17\}, \{5, 13, 21\}\}$
     $(2 + K) + (H/K) = \{2 + K, 6 + K\} = \{\{2, 10, 18\}, \{6, 14, 22\}\}$
     $(3 + K) + (H/K) = \{3 + K, 7 + K\} = \{\{3, 11, 19\}, \{7, 15, 23\}\}$.
  e. $\psi(0 + H) = (0 + K) + (H/K)$, $\psi(1 + H) = (1 + K) + (H/K)$,
     $\psi(2 + H) = (2 + K) + (H/K)$, $\psi(3 + H) = (3 + K) + (H/K)$.

7. *Chain* (3)

$\{0\} \leq \{0\} \leq \langle 12 \rangle$
$\leq \langle 6 \rangle \leq \langle 3 \rangle$
$\leq \langle 3 \rangle \leq \mathbb{Z}_{36}$

Chain (4)

$\{0\} \leq \{0\} < \langle 18 \rangle \leq \langle 18 \rangle$
$\leq \langle 6 \rangle \leq \langle 3 \rangle \leq \mathbb{Z}_{36}$

Isomorphisms

$\{0\}/\{0\} \approx \{0\}/\{0\} \approx \{0\}$
$\langle 12 \rangle/\{0\} \approx \langle 6 \rangle/\langle 18 \rangle \approx \mathbb{Z}_3$
$\langle 6 \rangle/\langle 12 \rangle \approx \langle 18 \rangle/\{0\} \approx \mathbb{Z}_2$
$\langle 3 \rangle/\langle 6 \rangle \approx \langle 3 \rangle/\langle 6 \rangle \approx \mathbb{Z}_2$
$\langle 3 \rangle/\langle 3 \rangle \approx \langle 18 \rangle/\langle 18 \rangle \approx \{0\}$
$\mathbb{Z}_{36}/\langle 3 \rangle \approx \mathbb{Z}_{36}/\langle 3 \rangle \approx \mathbb{Z}_3$

11. b. G/L

Section 4.2

1. 3
3. 1, 3
5. The Sylow 3-subgroups are $\langle (1, 2, 3) \rangle$, $\langle (1, 2, 4) \rangle$, $\langle (1, 3, 4) \rangle$, and $\langle (2, 3, 4) \rangle$. Also $(3, 4)\langle (1, 2, 3) \rangle (3, 4) = \langle (1, 2, 4) \rangle$, etc.
7. T T T F T F T T F F

Section 4.3

1. a. The conjugate classes are $\{\rho_0\}$, $\{\rho_2\}$, $\{\rho_1, \rho_3\}$, $\{\mu_1, \mu_2\}$, $\{\delta_1, \delta_2\}$.
   b. $8 = 2 + 2 + 2 + 2$
3. T T F T T T T T F F
   e. This is somewhat a matter of opinion.
7. e. $p(1) = 1$. $p(2) = 2$. $p(3) = 3$. $p(4) = 5$. $p(5) = 7$. $p(6) = 11$. $p(7) = 15$.
9. $120 = 1 + 10 + 15 + 20 + 20 + 30 + 24$,
   $720 = 1 + 15 + 45 + 40 + 15 + 120 + 90 + 40 + 90 + 144 + 120$.

Section 4.4

1. $\{(1, 1, 1), (1, 2, 1), (1, 1, 2)\}$
3. No. $n(2, 1) + m(4, 1)$ can never yield an odd number for first coordinate.

Section 4.5

1. a. $a^2 b^2 a^3 c^3 b^{-2}$, $b^2 c^{-3} a^{-3} b^{-2} a^{-2}$ **b.** $a^{-1} b^3 a^4 c^6 a^{-1}$, $ac^{-6} a^{-4} b^{-3} a$
3. a. 16   **b.** 36   **c.** 36
5. a. 16   **b.** 36   **c.** 18
9. a. *Partial answer:* $\{1\}$ is a basis for $\mathbb{Z}_4$.   **c.** Yes
11. c. A blop group on S is isomorphic to the *free group* $F[S]$ *on* S.

Section 4.6

1. $(a : a^4 = 1)$; $(a, b : a^4 = 1, b = a^2)$; $(a, b, c : a = 1, b^4 = 1, c = 1)$. (Other answers are possible.)

3. *Octic group:*

	1	a	a^2	a^3	b	ab	a^2b	a^3b
1	1	a	a^2	a^3	b	ab	a^2b	a^3b
a	a	a^2	a^3	1	ab	a^2b	a^3b	b
a^2	a^2	a^3	1	a	a^2b	a^3b	b	ab
a^3	a^3	1	a	a^2	a^3b	b	ab	a^2b
b	b	a^3b	a^2b	ab	1	a^3	a^2	a
ab	ab	b	a^3b	a^2b	a	1	a^3	a^2
a^2b	a^2b	ab	b	a^3b	a^2	a	1	a^3
a^3b	a^3b	a^2b	ab	b	a^3	a^2	a	1

Quaternion group: The same as the table for the octic group except that the sixteen entries in the lower right corner are

a^2	a	1	a^3
a^3	a^2	a	1
1	a^3	a^2	a
a	1	a^3	a^2

5. $\mathbb{Z}_{21}$. $(a, b : a^7 = 1, b^3 = 1, ba = a^2b)$
7. $\mathbb{Z}_{30}$. $D_{15} \simeq (a, b : a^{15} = 1, b^2 = 1, ba = a^{14}b)$
$\mathbb{Z}_3 \times D_5 \simeq (a, b : a^6 = 1, b^5 = 1, ba = a^4b^4)$
$\mathbb{Z}_5 \times S_3 \simeq (a, b : a^{10} = b^3 = 1, ba = ab^2)$
(See Coxeter-Moser [18, p. 134] for a table giving all nonabelian groups of order
<32.)

Chapter 5

Section 5.1

2. 0   **3.** 1   **5.** $(1, 6)$ **7.** Yes   **9.** Yes   **11.** Yes
13. No. $\{ri \mid r \in \mathbb{R}\}$ is not closed under multiplication.
15. $(1, 1), (1, -1), (-1, 1), (-1, -1)$ **17.** All nonzero $q \in \mathbb{Q}$ **19.** 1, 3
21. Let $\mathbb{R} = \mathbb{Z}$ with unity 1 and $R' = \mathbb{Z} \times \mathbb{Z}$ with unity $1' = (1, 1)$. Let $\phi : R \to R'$ be
defined by $\phi(n) = (n, 0)$. Then $\phi(1) = (1, 0) \neq 1'$.
23. $\phi_1 : \mathbb{Z} \to \mathbb{Z}$ where $\phi_1(n) = 0$, $\phi_2 : \mathbb{Z} \to \mathbb{Z}$ where $\phi_2(n) = n$
25. $\phi_1 : \mathbb{Z} \times \mathbb{Z} \to \mathbb{Z}$ where $\phi_1(n, m) = 0$, $\phi_2 : \mathbb{Z} \times \mathbb{Z} \to \mathbb{Z}$ where $\phi_2(n, m) = n$
$\phi_3 : \mathbb{Z} \times \mathbb{Z} \to \mathbb{Z}$ where $\phi_3(n, m) = m$
27. The reasoning is not correct since a product $(X - I_3)(X + I_3)$ of two matrices
may be the zero matrix 0 without having either matrix be 0. Counterexample,
$$\begin{bmatrix} 0 & 0 & 1 \\ 0 & 1 & 0 \\ 1 & 0 & 0 \end{bmatrix}^2 = I_3.$$
29. $a = 2, b = 3$ in $\mathbb{Z}_6$ **31.** T F F F T F T T T T

Section 5.2

1. 0, 3, 5, 8, 9, 11   **3.** No solutions   **5.** 0   **7.** 0   **9.** 12
11. $a^4 + 2a^2b^2 + b^4$ **13.** $a^6 + 2a^3b^3 + b^6$

15. F T F F T T F T F F
17. 1. $\text{Det}(A) = 0$.   2. The column vectors of A are dependent.
3. The row vectors of A are dependent.   4. Zero is an eigenvalue of A.
5. A is not invertible.
27. 9 code words, 81 received words

Section 5.3

1. 3   **3.** 3   **5.** 2
7.

$\varphi(1) = 1$	$\varphi(11) = 10$	$\varphi(21) = 12$
$\varphi(2) = 1$	$\varphi(12) = 4$	$\varphi(22) = 10$
$\varphi(3) = 2$	$\varphi(13) = 12$	$\varphi(23) = 22$
$\varphi(4) = 2$	$\varphi(14) = 6$	$\varphi(24) = 8$
$\varphi(5) = 4$	$\varphi(15) = 8$	$\varphi(25) = 20$
$\varphi(6) = 2$	$\varphi(16) = 8$	$\varphi(26) = 12$
$\varphi(7) = 6$	$\varphi(17) = 16$	$\varphi(27) = 18$
$\varphi(8) = 4$	$\varphi(18) = 6$	$\varphi(28) = 12$
$\varphi(9) = 6$	$\varphi(19) = 18$	$\varphi(29) = 28$
$\varphi(10) = 4$	$\varphi(20) = 8$	$\varphi(30) = 8$

9. $(p - 1)(q - 1)$ **11.** $1 + 4\mathbb{Z}, 3 + 4\mathbb{Z}$ **13.** No solutions
15. No solutions
17. $3 + 65\mathbb{Z}, 16 + 65\mathbb{Z}, 29 + 65\mathbb{Z}, 42 + 65\mathbb{Z}, 55 + 65\mathbb{Z}$
19. 1   **21.** 9   **23.** F T F T T F T F T

Section 5.4

1. $\{q_1 + q_2 i \mid q_1, q_2 \in \mathbb{Q}\}$ **3.** T F T F T T F T T T
13. There are four elements, for 1 and 3 are already units in $\mathbb{Z}_4$.
15. $\left\{ \dfrac{m}{6^n} \,\middle|\, m \in \mathbb{Z}, n \in \mathbb{Z}^+ \right\}$

Section 5.5

1. $f(x) + g(x) = 2x^2 - 3$, $f(x)g(x) = 6x^2 + 4x + 6$
3. $f(x) + g(x) = 5x^2 + 5x + 1$, $f(x)g(x) = x^3 + 5x$
5. 16   **7.** 0   **9.** 2   **11.** 0   **13.** 2, 3   **15.** 0, 2, 4
17. 0, 1, 2, 3
19. $0, x - 5, 2x - 10, x^2 - 25, x^2 - 5x, x^4 - 5x^3$. (Other answers are possible.)
21. T T T T F F T T T F
23. a. They are the units of D.   b. 1, −1   c. 1, 2, 3, 4, 5, 6
25. b. F c. $F[x]$ **29.** a. 4, 27   b. $\mathbb{Z}_2 \times \mathbb{Z}_2, \mathbb{Z}_3 \times \mathbb{Z}_3 \times \mathbb{Z}_3$

Section 5.6

1. $q(x) = x^4 + x^3 + x^2 + x - 2, r(x) = 4x + 3$
3. $q(x) - 6x^4 + 7x^3 + 2x^2$ $x + 2, r(x) = 4$
5. 2, 3   **7.** 3, 10, 5, 11, 14, 7, 12, 6
9. $(x - 1)(x + 1)(x - 2)(x + 2)$
11. $(x - 3)(x + 3)(2x + 3)$

13. Yes. It is of degree 3 with no zeros in $\mathbb{Z}_5$.
$2x^3 + x^2 + 2x + 2$

15. *Partial answer:* $g(x)$ is irreducible over $\mathbb{R}$, but it is not irreducible over $\mathbb{C}$.

19. Yes. $p = 3$ **21.** Yes. $p = 5$ **23.** T T T F T F T T T T

25. $x^2 + x + 1$

27. $x^2 + 1, x^2 + x + 2, x^2 + 2x + 2, 2x^2 + 2, 2x^2 + x + 1, 2x^2 + 2x + 1$

29. $p(p - 1)^2/2$

Section 5.7

1. $1e + 0a + 3b$ **3.** $2e + 2a + 2b$ **5.** j **7.** $(1/50)j - (3/50)k$

9. $\mathbb{R}^*$, that is, $\{a_1 + 0i + 0j + 0k \mid a_1 \in \mathbb{R}, a_1 \neq 0\}$

11. F F F F F F T F T F

c. If $|A| = 1$, then $\mathrm{Hom}(A) = \{0\}$.

e. $0 \in \mathrm{Hom}(A)$ is not in $\mathrm{Iso}(A)$.

Chapter 6

Section 6.1

1. ϕ_1 such that $\phi_1(1) = 1$, ϕ_2 such that $\phi_2(1) = 0$

3. All $n = 2m$ where m is an odd positive integer.

5.

+	$8\mathbb{Z}$	$2 + 8\mathbb{Z}$	$4 + 8\mathbb{Z}$	$6 + 8\mathbb{Z}$
$8\mathbb{Z}$	$8\mathbb{Z}$	$2 + 8\mathbb{Z}$	$4 + 8\mathbb{Z}$	$6 + 8\mathbb{Z}$
$2 + 8\mathbb{Z}$	$2 + 8\mathbb{Z}$	$4 + 8\mathbb{Z}$	$6 + 8\mathbb{Z}$	$8\mathbb{Z}$
$4 + 8\mathbb{Z}$	$4 + 8\mathbb{Z}$	$6 + 8\mathbb{Z}$	$8\mathbb{Z}$	$2 + 8\mathbb{Z}$
$6 + 8\mathbb{Z}$	$6 + 8\mathbb{Z}$	$8\mathbb{Z}$	$2 + 8\mathbb{Z}$	$4 + 8\mathbb{Z}$

	$8\mathbb{Z}$	$2 + 8\mathbb{Z}$	$4 + 8\mathbb{Z}$	$6 + 8\mathbb{Z}$
$8\mathbb{Z}$	$8\mathbb{Z}$	$8\mathbb{Z}$	$8\mathbb{Z}$	$8\mathbb{Z}$
$2 + 8\mathbb{Z}$	$8\mathbb{Z}$	$4 + 8\mathbb{Z}$	$8\mathbb{Z}$	$4 + 8\mathbb{Z}$
$4 + 8\mathbb{Z}$	$8\mathbb{Z}$	$8\mathbb{Z}$	$8\mathbb{Z}$	$8\mathbb{Z}$
$6 + 8\mathbb{Z}$	$8\mathbb{Z}$	$4 + 8\mathbb{Z}$	$8\mathbb{Z}$	$4 + 8\mathbb{Z}$

$2\mathbb{Z}/8\mathbb{Z}$ is not isomorphic to $\mathbb{Z}_4$, for $2\mathbb{Z}/8\mathbb{Z}$ has no unity.

7. Let $\phi:\mathbb{Z} \to \mathbb{Z} \times \mathbb{Z}$ be given by $\phi(n) = (n, 0)$ for $n \in \mathbb{Z}$.

9. R/R and $R/\{0\}$ are not of real interest because R/R is the ring containing only the zero element, and $R/\{0\}$ is isomorphic to R.

11. $\mathbb{Z}$ is an integral domain. $\mathbb{Z}/4\mathbb{Z}$ is isomorphic to $\mathbb{Z}_4$, which has a divisor 2 of 0.

13. $\{(n, n) \mid n \in \mathbb{Z}\}$. (Other answers are possible.)

35. *Partial answer:* By the definition in this exercise, $\sqrt{R} = R$ for every ring R. However, according to the definition in Exercise 31, the radical of R is not always all of R. Thus this terminology is inconsistent if $N = R$.

37. If $\sqrt{N}/N$ is viewed as a subring of R/N, then it is the radical of R/N, in the sense of the definition in Exercise 31.

Section 6.2

1. $\{0, 2, 4\}$ and $\{0, 3\}$ are both prime and maximal.
3. $\{(0, 0), (1, 0)\}$ and $\{(0, 0), (0, 1)\}$ are both prime and maximal.
5. 1   **7.** 2   **9.** 1, 4   **11.** $2\mathbb{Z} \times \mathbb{Z}$ **13.** $4\mathbb{Z} \times \{0\}$
15. Yes. $x^2 - 6x + 6$ is irreducible over $\mathbb{Q}$ by Eisenstein with $p = 2$.
17. Yes. $\mathbb{Z}_2 \times \mathbb{Z}_3$
19. No, Enlarging the domain to a field of quotients, you would have to have a field containing two different prime fields $\mathbb{Z}_p$ and $\mathbb{Z}_q$, which is impossible.

Chapter 7

Section 7.1

1. Yes   **3.** No   **5.** No   **7.** Yes
9. In $\mathbb{Z}[x]$: only $2x - 7, -2x + 7$
     In $\mathbb{Q}[x]$: $4x - 14, x - \frac{7}{2}, 6x - 21, -8x + 28$
     In $\mathbb{Z}_{11}[x]$: $2x - 7, 10x - 2, 6x + 1, 3x - 5, 5x - 1$
11. $(6)(3x^2 - 2x + 8)$ **13.** $(1)(2x^2 - 3x + 6)$
15. T T T F T F F T F T
     i. Either p or one of its associates must appear in every factorization *into irreducibles.*
17. $2x + 4$ is irreducible in $\mathbb{Q}[x]$ but not in $\mathbb{Z}[x]$.
25. *Partial answer:* $x^3 - y^3 = (x - y)(x^2 + xy + y^2)$

Section 7.2

1. Yes   **2.** No. (1) is violated.   **5.** Yes   **7.** 61   **9.** $x^3 + 2x - 1$
11. T F T F T T T F T T
21. *Partial answer:* The equation $ax = b$ has a solution in $\mathbb{Z}_n$ for nonzero $a, b \in \mathbb{Z}_n$ if and only if the positive gcd of a and n in $\mathbb{Z}$ divides b.

Section 7.3

1. $5 = (1 + 2i)(1 + 2i)$ **3.** $4 + 3i = (1 + 2i)(2 - i)$
5. $6 = (2)(3) = (-1 + \sqrt{-5})(-1 - \sqrt{-5})$ **7.** $7 - i$ **15.** $1 + 2i$

Chapter 8

Section 8.1

1. $x^2 - 2x - 1$ **3.** $x^2 - 2x + 2$ **5.** $x^{12} + 3x^8 - 4x^6 + 3x^4 + 12x^2 + 5$
7. Irr$(\alpha, \mathbb{Q}) = x^4 - \frac{2}{3}x^2 - \frac{62}{9}$; deg$(\alpha, \mathbb{Q}) = 4$ **9.** Algebraic, deg$(\alpha, F) = 2$

11. Transcendental   **13.** Algebraic, $\deg(\alpha, F) = 2$
15. Algebraic, $\deg(\alpha, F) = 1$ **17.** $x^2 + x + 1 = (x - \alpha)(x + 1 + \alpha)$
19. T T T T F T F T F T
21. b. $x^3 + x^2 + 1 = (x - \alpha)(x - \alpha^2)[x - (1 + \alpha + \alpha^2)]$
23. It is the monic polynomial in $F[x]$ of *minimal* degree having α as a zero.

Section 8.2

1. $\{(0, 1), (1, 0)\}, \{(1, 1), (-1, 1)\}, \{(2, 1), (1, 2)\}$. (Other answers are possible.)
3. No. $2(-1, 1, 2) - 4(2, -3, 1) + (10, -14, 0) = (0, 0, 0)$
5. $\{1\}$ **7.** $\{1, i\}$ **9.** $\{1, \sqrt[4]{2}, \sqrt{2}, (\sqrt[4]{2})^3\}$
11. T F T T F F F T T T
13. a. The **subspace of V generated by** S is the intersection of all subspaces of V
containing S.
15. *Partial answer:* A basis for F^n is

$$\{(1, 0, \ldots, 0), (0, 1, \ldots, 0), \ldots, (0, 0, \ldots, 1)\}$$

where 1 is the multiplicative identity of F.
21. a. A homomorphism
b. *Partial answer:* The **kernel** (or **nullspace**) of ϕ is $\{\alpha \in V \mid \phi(\alpha) = 0\}$.
c. ϕ is an isomorphism of V with V' if $\mathrm{Ker}(\phi) = \{0\}$ and ϕ maps V onto V'.

Section 8.3

1. $2, \{1, \sqrt{2}\}$ **3.** $8, \{1, \sqrt{3}, \sqrt{5}, \sqrt{15}, \sqrt{2}, \sqrt{6}, \sqrt{10}, \sqrt{30}\}$
5. $6, \{1, \sqrt{2}, \sqrt[3]{2}, \sqrt{2}\,(\sqrt[3]{2}), (\sqrt[3]{2})^2, \sqrt{2}\,(\sqrt[3]{2})^2\}$ **7.** $2, \{1, \sqrt{6}\}$
9. $9, \{1, \sqrt[3]{2}, \sqrt[3]{4}, \sqrt[3]{3}, \sqrt[3]{6}, \sqrt[3]{12}, \sqrt[3]{9}, \sqrt[3]{18}, \sqrt[3]{36}\}$
11. $2, \{1, \sqrt{2}\}$ **13.** $2, \{1, \sqrt{2}\}$ **15.** T F T F F T F F F F
19. *Partial answer:* Extensions of degree 2^n for $n \in \mathbb{Z}^+$ are obtained.

Section 8.4

1. T T T F T F T T T F

Section 8.5

1. Yes   **3.** Yes   **5.** 6   **7.** 0

Section 8.6

1. $\langle G, \mathcal{O}, *, \cdot \rangle$. (Other notations are possible.)
3. $\langle V, F, \oplus, \odot, +, \cdot, \times \rangle$. (Other notations are possible.)
5. *Partial answer:* A map $\phi : G \to G'$ is an $\mathcal{O}$-**homomorphism of the $\mathcal{O}$-group G into
the $\mathcal{O}$-group** G' if for all $\alpha, \beta \in G$ and $a \in \mathcal{O}$, both $\phi(\alpha + \beta) = \phi(\alpha) + \phi(\beta)$
and $\phi(a\alpha) = a\phi(\alpha)$.
7. **A homomorphism of a (left) R-module M into a (left) R-module** M' is a function
$\phi : M \to M'$ such that $\phi(\alpha + \beta) = \phi(\alpha) + \phi(\beta)$ and $\phi(r\alpha) = r\phi(\alpha)$ for all
$\alpha, \beta \in M$ and $r \in R$.

Chapter 9

Section 9.1

1. $\sqrt{2}, -\sqrt{2}$ **3.** $3 + \sqrt{2}, 3 - \sqrt{2}$ **5.** $\sqrt{2} + i, \sqrt{2} - i, -\sqrt{2} + i, -\sqrt{2} - i$
7. $\sqrt{1 + \sqrt{2}}, -\sqrt{1 + \sqrt{2}}, \sqrt{1 - \sqrt{2}}, -\sqrt{1 - \sqrt{2}}$ **9.** $\sqrt{3}$
11. $-\sqrt{2} + \sqrt[3]{5}$ **13.** $-\sqrt{2} + \sqrt[4]{45}$
15. a. $\mathbb{Q}$ b. $\mathbb{Q}(\sqrt{6})$ c. $\mathbb{Q}$
17. $\mathbb{Q}(\sqrt{2}, \sqrt{3}, \sqrt{5})$ **19.** $\mathbb{Q}(\sqrt{3}, \sqrt{10})$ **21.** $\mathbb{Q}$
23. a. $3 - \sqrt{2}$ b. They are the same maps.
25. $\sigma_3(0) = 0, \sigma_3(1) = 1, \sigma_3(2) = 2, \sigma_3(\alpha) = -\alpha, \sigma_3(2\alpha) = -2\alpha,$
$\sigma_3(1 + \alpha) = 1 - \alpha, \sigma_3(1 + 2\alpha) = 1 - 2\alpha, \sigma_3(2 + \alpha) = 2 - \alpha,$
$\sigma_3(2 + 2\alpha) = 2 - 2\alpha; \mathbb{Z}_3(\alpha)_{\{\sigma_3\}} = \mathbb{Z}_3$
27. FFTTFTTTTT   **33.** Yes

Section 9.2

1. The identity map of E onto E;
τ given by $\tau(\sqrt{2}) = \sqrt{2}, \tau(\sqrt{3}) = -\sqrt{3}, \tau(\sqrt{5}) = -\sqrt{5}$
3. τ_1 given by $\tau_1(\sqrt{2}) = \sqrt{2}, \tau_1(\sqrt{3}) = \sqrt{3}, \tau_1(\sqrt{5}) = -\sqrt{5}$;
τ_2 given by $\tau_2(\sqrt{2}) = \sqrt{2}, \tau_2(\sqrt{3}) = -\sqrt{3}, \tau_2(\sqrt{5}) = \sqrt{5}$;
τ_3 given by $\tau_3(\sqrt{2}) = -\sqrt{2}, \tau_3(\sqrt{3}) = \sqrt{3}, \tau_3(\sqrt{5}) = \sqrt{5}$;
τ_4 given by $\tau_4(\sqrt{2}) = -\sqrt{2}, \tau_4(\sqrt{3}) = -\sqrt{3}, \tau_4(\sqrt{5}) = -\sqrt{5}$
5. The identity map of $\mathbb{Q}(\sqrt[3]{2}, \sqrt{3})$ into itself;
τ_1 given by $\tau_1(\alpha_1) = \alpha_1, \tau_1(\sqrt{3}) = -\sqrt{3}$;
τ_2 given by $\tau_2(\alpha_1) = \alpha_2, \tau_2(\sqrt{3}) = \sqrt{3}$;
τ_3 given by $\tau_3(\alpha_1) = \alpha_2, \tau_3(\sqrt{3}) = -\sqrt{3}$;
τ_4 given by $\tau_4(\alpha_1) = \alpha_3, \tau_4(\sqrt{3}) = \sqrt{3}$;
τ_5 given by $\tau_5(\alpha_1) = \alpha_3, \tau_5(\sqrt{3}) = -\sqrt{3}$;
7. a. $\mathbb{Q}(\pi^2)$ b. τ_1 given by $\tau_1(\sqrt{\pi}) = i\sqrt{\pi}, \tau_2$ given by $\tau_2(\sqrt{\pi}) = -i\sqrt{\pi}$

Section 9.3

1. 2   **3.** 4   **5.** 2   **7.** 1   **9.** 2   **11.** $1 \le [E:F] \le n!$
13. Let $F = \mathbb{Q}$ and $E = \mathbb{Q}(\sqrt{2})$. Then

$$f(x) = x^4 - 5x^2 + 6 = (x^2 - 2)(x^2 - 3)$$

has a zero in E, but does not split in E.
21. a. 6

Section 9.4

1. $\alpha = \sqrt[6]{2} = 2/(\sqrt[3]{2}\sqrt{2}). \sqrt{2} = (\sqrt[6]{2})^3, \sqrt[3]{2} = (\sqrt[6]{2})^2.$ (Other answers are possible.)
3. $\alpha = \sqrt{2} + \sqrt{3}. \sqrt{2} = (\frac{1}{2})\alpha^3 - (\frac{9}{2})\alpha, \sqrt{3} = (\frac{11}{2})\alpha - (\frac{1}{2})\alpha^3.$ (Other answers are possible.)
5. $f(x) = x^4 - 4x^2 + 4 = (x^2 - 2)^2.$ Here $f(x)$ is not an irreducible polynomial. Every irreducible factor of $f(x)$ has zeros of multiplicity 1 only.
13. b. The field F c. $F[x^p]$

Section 9.5

1. $\mathbb{Z}_3(y^3, z^9)$ **3.** $\mathbb{Z}_3(y^4, z^2)$ **5.** FTFFFFTFTT

Section 9.6

1. 8   **3.** 8   **5.** 4   **7.** 2

9. The group has two elements, the identity automorphism ι of $\mathbb{Q}(i)$ and σ such that $\sigma(i) = -i$.

11. a. Let $\alpha_1 = \sqrt[3]{2}$, $\alpha_2 = \sqrt[3]{2}\dfrac{-1 + i\sqrt{3}}{2}$, and $\alpha_3 = \sqrt[3]{2}\dfrac{-1 - i\sqrt{3}}{2}$.

The maps are

$$\rho_0, \text{ where } \rho_0 \text{ is the identity map;}$$

$$\rho_1, \text{ where } \rho_1(\alpha_1) = \alpha_2 \text{ and } \rho_1(i\sqrt{3}) = i\sqrt{3};$$

$$\rho_2, \text{ where } \rho_2(\alpha_1) = \alpha_3 \text{ and } \rho_2(i\sqrt{3}) = i\sqrt{3};$$

$$\mu_1, \text{ where } \mu_1(\alpha_1) = \alpha_1 \text{ and } \mu_1(i\sqrt{3}) = -i\sqrt{3};$$

$$\mu_2, \text{ where } \mu_2(\alpha_1) = \alpha_3 \text{ and } \mu_2(i\sqrt{3}) = -i\sqrt{3};$$

$$\mu_3, \text{ where } \mu_3(\alpha_1) = \alpha_2 \text{ and } \mu_3(i\sqrt{3}) = -i\sqrt{3}.$$

b. S_3. The notation in (a) was chosen to coincide with the notation for S_3 in Example 4, Section 2.1.

c.

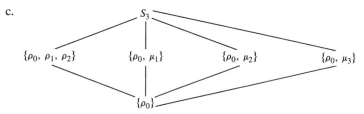

Group lattice diagram

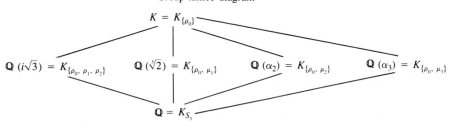

Field lattice diagram

13. The splitting field of $(x^3 - 1) \in \mathbb{Q}[x]$ is $\mathbb{Q}(i\sqrt{3})$, and the group is cyclic of order 2 with elements: ι, where ι is the identity map of $\mathbb{Q}(i\sqrt{3})$, and σ, where $\sigma(i\sqrt{3}) = -i\sqrt{3}$.

15. F F T T T F F T F T

25. *Partial answer:* $G(K/(E \vee L)) = G(K/E) \cap G(K/L)$

Section 9.7

3. $\mathbb{Q}(\sqrt[4]{2}, i)$: $\sqrt[4]{2} + i$, $x^8 + 4x^6 + 2x^4 + 28x^2 + 1$;
  $\mathbb{Q}(\sqrt[4]{2})$: $\sqrt[4]{2}$, $x^4 - 2$;
  $\mathbb{Q}(i\sqrt[4]{2}))$: $i(\sqrt[4]{2})$, $x^4 - 2$;
  $\mathbb{Q}(\sqrt{2}, i)$: $\sqrt{2} + i$, $x^4 - 2x^2 + 9$;
  $\mathbb{Q}(\sqrt[4]{2} + i(\sqrt[4]{2}))$: $\sqrt[4]{2} + i(\sqrt[4]{2})$, $x^4 + 8$;
  $\mathbb{Q}(\sqrt[4]{2} - i(\sqrt[4]{2}))$: $\sqrt[4]{2} - i(\sqrt[4]{2})$, $x^4 + 8$;
  $\mathbb{Q}(\sqrt{2})$: $\sqrt{2}$, $x^2 - 2$;
  $\mathbb{Q}(i)$: i, $x^2 + 1$;
  $\mathbb{Q}(i\sqrt{2})$: $i\sqrt{2}$, $x^2 + 2$;
  $\mathbb{Q}$: 1, $x - 1$

5. The group is cyclic of order 5, and its elements are

	ι	σ_1	σ_2	σ_3	σ_4
$\sqrt[5]{2} \to$	$\sqrt[5]{2}$	$\zeta(\sqrt[5]{2})$	$\zeta^2(\sqrt[5]{2})$	$\zeta^3(\sqrt[5]{2})$	$\zeta^4(\sqrt[5]{2})$

where $\sqrt[5]{2}$ is the real 5th root of 2.

7. The splitting field of $x^8 - 1$ over $\mathbb{Q}$ is the same as the splitting field of $x^4 + 1$ over $\mathbb{Q}$, so a complete description is contained in Example 2. (This is the easiest way to answer the problem.)

9. a. $s_1^2 - 2s_2$ **b.** $\dfrac{s_1 s_2 - 3s_3}{s_3}$ **c.** $s_1^2 s_2^2 - 4s_1^3 s_2 - 4s_2^3 + 18s_1 s_2 s_3 - 27s_3^2$

Section 9.8

3. a. 16 **b.** 400 **c.** 2160

5. 3^0

7. $\Phi_3(x)$ over $\mathbb{Z}_2$ is $x^2 + x + 1$.
  $\Phi_8(x)$ over $\mathbb{Z}_3$ is $x^4 + 1 = (x^2 + x + 2)(x^2 + 2x + 2)$.

9. T T F T T F T T F T

11. $\Phi_1(x) = x - 1$
  $\Phi_2(x) = x + 1$
  $\Phi_3(x) = x^2 + x + 1$
  $\Phi_4(x) = x^2 + 1$
  $\Phi_5(x) = x^4 + x^3 + x^2 + x + 1$
  $\Phi_6(x) = x^2 - x + 1$

Section 9.9

1. No. Yes, K is an extension of $\mathbb{Z}_2$ by radicals.

3. T T T F T F T F F T **i.** $x^3 - 2x$ over $\mathbb{Q}$ gives a counterexample.

INDEX

O

N

P

S